ORNITHOLOGY

VOLUME I

LAND BIRDS

Elibron Classics
www.elibron.com

Elibron Classics series.

ISBN 1-4212-6741-1 (paperback)
ISBN 1-4212-6740-3 (hardcover)

This Elibron Classics Replica Edition is an unabridged facsimile of the edition published in 1870 by the University Press, Cambridge.

ORNITHOLOGY OF CALIFORNIA.

VOL. I.

GEOLOGICAL SURVEY OF CALIFORNIA.

J. D. WHITNEY, State Geologist.

ORNITHOLOGY.

VOLUME I.

LAND BIRDS.

EDITED BY S. F. BAIRD,

FROM THE MANUSCRIPT AND NOTES OF

J. G. COOPER.

PUBLISHED BY AUTHORITY OF THE LEGISLATURE.

1870.

University Press: Welch, Bigelow, & Co.,
Cambridge.

CONTENTS.

INTRODUCTION.

THE volume herewith offered to the public is the first, in the order of publication, of the zoölogical series of the reports of the State Geological Survey of California. This survey, although called simply "geological," was intended to embrace within its scope the natural history and topography of the State as well as its geology. The Act under which the work was organized expressly required of the State Geologist "a full and scientific description of the botanical and zoölogical productions of California." Provision was also made for the collection of specimens in all branches of natural history; these were to be arranged and labelled and held in charge by the Survey, until suitably disposed of by the Legislature. By a later Act the collections of the Survey were ordered to be turned over to the State University, which will be done whenever a building has been prepared for their reception.

In carrying out the intentions of the Legislature with reference to the zoölogical department, an assistant was appointed, whose duty it should be to visit different portions of the State and make collections of animals, studying in the field their distribution and habits, thus providing the materials to be worked up and prepared for publication by specialists in the different branches. The gentleman selected for this position was Dr. J. G. Cooper, who had been employed in the same capacity on the United States Pacific Railroad surveys, and who was already well and favorably known to the scientific world by various publications relating to the botany and zoölogy of the Pacific States.

Dr. Cooper was steadily employed in collecting, from December, 1860, to April, 1862, and during a considerable portion of the year 1863, being assisted for six months of 1862 by Dr. Edward Palmer. The regions examined were chiefly the Colorado Valley near Fort Mojave, the route to the coast from that fort, the vicinity of San Diego and San Pedro, and Santa Barbara and the islands off the coast. In 1864 portions of the Sierra Nevada and the coast from Baulines Bay to Santa Cruz were zoölogically explored. In 1865 Dr. Cooper prepared and presented a series of reports on the higher classes of the animals, in which were embodied all the facts which had been observed by him up to that time. Since the completion of these reports he has, however, been employed, at intervals, at the expense of the Survey, in visiting portions of the State not before sufficiently explored; he has also done a large amount of gratuitous work, in the way of elaborating the materials in various branches of the zoölogy of the Pacific slope, and especially the mollusca, several important papers in reference to their classification and distribution having been published by him in the Proceedings of the California Academy of Sciences and elsewhere.

The report on the birds of California was submitted to Professor Baird for revision and publication, and the first instalment of the work is herewith presented. In preparing it, all the available material of the Smithsonian Institution, in this department, by permission of Professor Henry, has been freely used and much important matter added. The generic and specific technical descriptions are taken almost exclusively from the work of Professor Baird on the Birds of North America; the observations on the habits and distribution of the birds are chiefly from Dr. Cooper. When other authorities have been drawn upon, their names will be found appended to the paragraphs quoted.

As at first proposed, and announced in the Preface to the Geology of California, Vol. I., the ornithological portion of the report was to be comprised in one volume. This has been found impossible, and there will be two; the first being devoted to the LAND BIRDS, while the second will embrace the WATER BIRDS. The scope of these volumes has also been somewhat changed since their preparation was begun. The present one includes all the species of land birds found in North America, north of Mexico, and west of the Rocky Mountains; while the second will contain the water birds of the whole continent north of Mexico. By taking this course, it will be apparent that the volumes will be rendered available to a much larger number of persons, especially to the inhabitants of the entire western half of the United States, than if they were strictly limited to Californian birds, while their size and cost will not be materially augmented. Their circulation will thus be greatly increased, the additional value given to them being out of proportion greater than the additional expense rendered necessary by the course adopted.

Attention is called to the style and execution of the illustrations of the present volume: these are believed to have solved a difficult problem, namely, that of furnishing the means of identifying the species, without making the work very bulky and expensive. The plan here adopted of giving as far as possible life-size figures of the heads of each species, and small full-lengths of each genus, together with generic outline illustrations of the external anatomy, will, we trust, enable even the tyro to refer correctly to genus and species such specimens as may be collected, since the most characteristic parts will be found figured with scrupulous accuracy. Copies colored from nature, both as to the heads alone and the heads and full-lengths together, will be placed at the disposition of the public.

The second volume of the ornithological series will be issued as soon as possible, the illustrations having been all drawn upon the wood, and a part of them already engraved.

All the illustrations introduced are from nature, and, with few exceptions,* prepared expressly for the present work. The full-length generic figures are by Mr. Edwin Sheppard of Philadelphia; the heads by Messrs. J. H. Richard and Henry W. Elliott; the generic outlines by Mr. A. Schönborn; the latter engraved by the peculiar process of Jewett, Chandler, & Co., of Buffalo.

The larger portion of the engraving of this volume has been done by, or under the supervision of, Mr. H. H. Nichols of Washington, and to his artistic skill and fidelity to nature the work is largely indebted for its value and attractiveness. A few excellent cuts have also been made by Mr. Henry Marsh of Cambridge.

J. D. WHITNEY.

CAMBRIDGE, MASS., October 1, 1870.

* A few generic figures, mentioned in the text as they occur, were kindly furnished by the London Society for the Diffusion of Christian Knowledge. Some others also were prepared for an unpublished work by Professor Blasius on the birds of Germany, and obtained from Messrs. Vieweg and Son of Brunswick.

Order OSCINES.

SINGING BIRDS.

Char. Toes, three anterior, one behind; all at the same level, and none versatile, the outer anterior never entirely free to the base. Tail feathers, twelve. Primaries either nine only, or else the first is spurious, or much shorter than the second, making the tenth. Tarsi feathered to the knee; the plates on the anterior face either fused into one, or with distinct divisions; the posterior portion of the sides covered by one continuous plate on either side, meeting in a sharp edge behind, or with only a few divisions inferiorly. Occasionally the hinder side has transverse plates, corresponding in number to the anterior, but there are then usually none on the sides. Larynx provided with a peculiar muscular apparatus for singing, composed of five pairs of muscles.

Most of the species of this division are more or less musical; some, however, have only a harsh voice, though provided with the singing muscles.

Family TURDIDÆ, Thrushes.

Char. Primaries ten; the first very small or not more than half the longest; second, usually shorter than fourth. Wings reaching about to middle of tail. Toes deeply cleft. Tarsi sometimes with the scutellæ united into a continuous plate in front; sometimes distinct. Nostrils oval. Loral and frontal feathers with bristly points; gape provided with bristles. Bill slender; notched at tip.

Food chiefly fruit and insects.

Sub-Family TURDINÆ.

Char. Nostrils oval. Bristles along the base of the bill from gape to nostrils; those of rictus not reaching beyond nostrils. Second quill longer than sixth. Outer lateral toe longer than inner. Tarsi covered anteriorly with a continuous plate. Wings long.

GENUS **TURDUS**, LINNÆUS.

Turdus, LINNÆUS, Systema Naturæ, 1735. (Type, *T. viscivorus*, fide G. R. Gray.)

GEN. CHAR. Bill rather stout; commissure straight to near the tip, which is quite abruptly decurved, and usually distinctly notched; culmen gently convex from base. Bill shorter than the head; both outlines curved. Tarsi longer than the middle toe. Lateral

Turdus ustulatus.

toes nearly equal; outer longer. Wings much longer than the tail, pointed; the first quill spurious and very small, not one fourth the length of longest. Tail short, nearly even, or slightly emarginate.

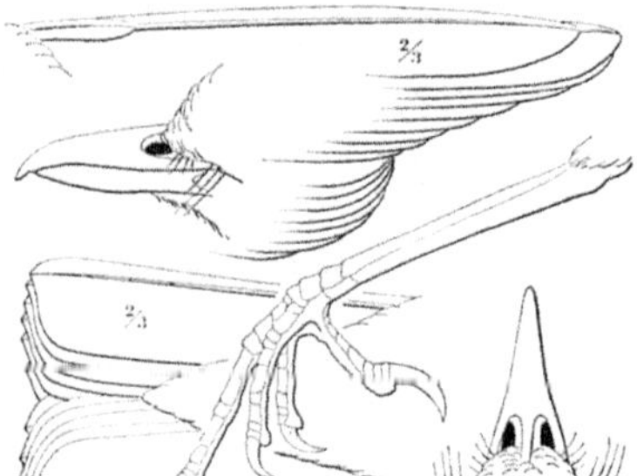

Turdus mustelinus.

The California Thrushes fall into three groups, perhaps not of generic rank, but serving to divide them conveniently into subgenera, as follows: —

HYLOCICHLA. Sexes similar. Bill notched. Bill short, broad at base and much depressed. Tarsi long and slender, longer than middle toe and claw by the length of a claw. Outstretched legs reaching nearly to tip of tail. Color above olivaceous or reddish, beneath whitish, breast spotted; throat without spots. Type, *Turdus mustelinus.*

This group embraces the small Wood Thrushes, all remarkable for their sweet song. All are North American, migrating south in winter, and of wide geographical distribution.

PLANESTICUS. Sexes similar. Bill stouter and higher. Tarsi short, hardly longer than middle toe and claw. Body stout. Under part mostly uniform, the throat white and alone streaked. Type, *T. migratorius.*

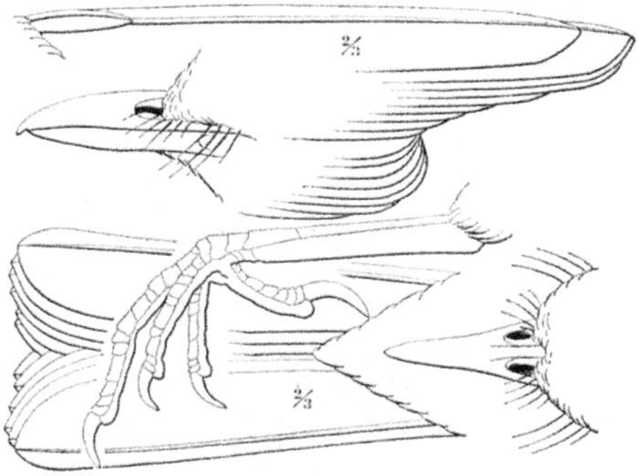

Turdus migratorius.

The common Robin, and its ally from Cape St. Lucas, represent this form in North America; many other species inhabit Middle and South America.

HESPEROCICHLA. Sexes dissimilar. Form of *Planesticus.* Bill not notched. Male, reddish beneath with black collar. Female, paler, with collar indistinct. Type, *T. nævius.*

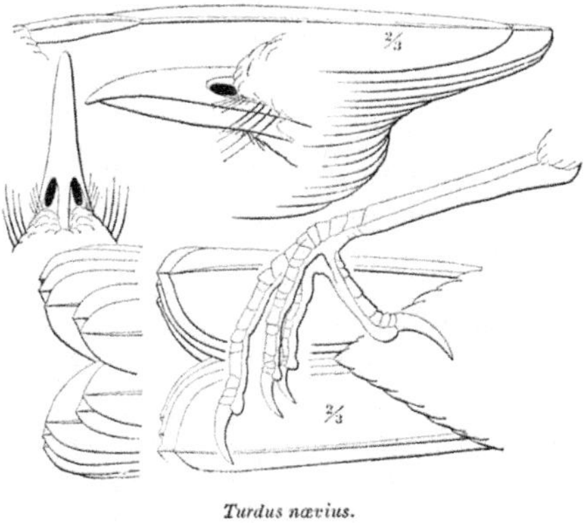

Turdus nævius.

The single species of this section is confined to the Pacific coast.

Turdus nanus, Audubon.

THE DWARF THRUSH.

Turdus nanus, Audubon, Orn. Biogr. V. 1839, 201, pl. 419.—Ib. Birds Amer. III. 1841, 32; pl. 147. (Columbia River.) — Gambel, Pr. A. N. Sc. I. 1843, 262. — Baird, P. R. Rep. IX. Birds, 213. — Heermann, X. vi. 45. — Baird, Rev. N. A. Birds, June, 1864, 16.

Sp. Char. Above light olive-brown, passing into reddish on the upper coverts and tail, slightly also on the wings. Beneath white, with a pale buff tinge on the fore part of breast and sometimes of throat; distinctly defined triangular dusky spots on the sides of throat and across the breast, sides of breast with less distinct and more rounded spots posteriorly. Sides glossed with bluish ash. Tail with a purple tinge. Fourth quill longest. Length, 6.50 inches; extent, 10; wing, 3.30; tail, 2.90; tarsus, 1.10. Iris brown, bill brown, lower mandible yellowish flesh-color at base, feet pale brown.

Hab. Pacific slope of North America, and along valley of Gila to El Paso. North to Fort Crook (about lat. 41°), south to Cape St. Lucas. Rocky Mountains to Fort Bridger.

I saw but few of this species in the Colorado valley, where they seem to remain only for the winter, as I observed none after April 1st. Most of them winter in the same parts of the State in which they spend the summer, chiefly south of San Francisco. They are shy and timid, preferring the dark, shady thickets, and rarely venturing far from them except in the twilight, their large eyes being suited for seeing in dark places. They feed chiefly on the ground, running rapidly, and searching for insects among the leaves and herbage, but not scratching for them. Probably, also, they feed on berries, like others of the family.

About the 25th of April they begin to sing near San Diego, the song consisting of a few low ringing notes, like those of Wilson's Thrush of the East, and *T. ustulatus* of the North, but not so loud. Their usual note of alarm is a single chirp, sometimes loud and ringing, repeated and answered by others for a long distance.

At Santa Cruz, on the 1st of June, I found several of their nests, all built in thickets, under the shade of cottonwood-trees, each about five feet above the ground, and containing eggs in various stages of hatching, from two to four in number, the smaller number probably laid after the destruction of a first set. The nests were built of dry leaves, root fibres, grass, and bark, without mud, lined with decayed leaves; measuring outside 4 inches each way, inside 2.50 wide and 0.20 deep. The eggs measured 0.90 × 0.70, and were pale bluish green, speckled with cinnamon-brown, chiefly at the larger end.

In 1866, at Santa Cruz, I found nests with eggs about May 20th, one on a horizontal branch not more than a foot from the ground, another on an alder-tree fifteen feet up. After raising their young, they all left the vicinity of the town, probably for the moister mountains, where food was more plenty at the end of the dry season.

Turdus ustulatus, NUTTALL.

THE OREGON THRUSH.

Turdus ustulatus, NUTTALL, Man. Orn. I. (2d ed.) 1840, 400. Columbia River (printed *cestulatus*, by a typographical error). — BAIRD, P. R. Rep. IX. Birds of N. Am. 215; pl. 81, f. 2. — IB. Rev. Amer. Birds, 1864, 18. — COOPER and SUCKLEY, P. R. R. XII. iii. Zool. of W. T. 171.

SP. CHAR. Third and fourth quills longest; second intermediate between fourth and fifth. Tail nearly even. Upper parts uniform reddish brown, with a faint olivaceous

tinge. Fore part of the breast tinged with brownish yellow, becoming paler to the chin; the remaining under parts are white. The sides of the throat and the fore part of the breast with small, distinct triangular spots of well-defined brown, much darker than the back; the sides of the breast more obsoletely spotted, and the sides of the body washed with olivaceous yellow-brown. The tibiæ are yellowish brown. Nearly the whole of the lower mandible, except the rami, is brown. Length, 7.25 to 8; extent, 12; wing, 3.75; tail, 3; tarsus, 1.12. Iris and bill, brown; lower mandible yellow at base; legs, pale.

Hab. Coast region of Oregon and Washington Territories. South to San Francisco in winter.

This more northern species is the exact counterpart of *T. nanus* in habits. I found their nests north of the Columbia, about the middle of June, 1854, containing four or five bluish-white eggs, thickly spotted with brown.

Turdus Swainsoni, Cabanis.

THE OLIVE THRUSH.

Turdus Swainsoni, Cabanis, Tschudi, Fauna Peruana, 1844-46, 188.—Baird, Birds N. A. 1858, 216.—Ib. Rev. Amer. Birds, 19.

Turdus olivaceus, Giraud, Birds Long Island, 1843-44, 92.

Sp. Char. Above entirely uniform olivaceous, with a shade of green. Fore part of breast, throat, and chin pale brownish yellow; rest of lower part white; the sides marked with brownish olive. Sides of throat and fore part of breast with rather rounded spots of well-defined brown darker than the back; the rest of the breast, except in the middle,

with rather less distinct spots that are more olivaceous. Tibiæ yellowish brown. Broad ring round the eye, loral region, and a general tinge on the side of the head clear reddish buff. Length, 7.00; wing, 4.15; tail, 3.10; tarsus, 1.00.

Hab. Eastern North America, and westward to Humboldt Valley, the Upper Columbia, and the Yukon Rivers; very rare in California. Southward to Ecuador.

This bird, formerly supposed to be confined to Eastern North America, has of late years been found farther and farther to the west, until its known distribution has become that indicated above. Several specimens were obtained in the vicinity of San Francisco by Mr. Lorquin, and formed part of the collections sent to the Smithsonian Institution by the Russian Telegraph Expedition.

This thrush breeds in the far north, more abundantly about Slave Lake, the Lower Mackenzie and the Upper Yukon Rivers. Like its congeners it is an admirable singer, enlivening the woods with its melody. The nest is placed on a low tree or bush, and the eggs are blue, with numerous reddish spots. Nest and eggs, indeed, are scarcely to be distinguished from those of *T. ustulatus*. The two species, in fact, are closely related, but *T. Swainsoni* will be readily distinguished by the clear olive of the upper parts, instead of reddish brown, and the larger and better defined spots on the breast. (Baird.)

Turdus migratorius, Linnæus.

THE ROBIN.

Turdus migratorius, Linnæus, Syst. Nat. I. 1766, 292. — Wilson, Am. Orn. I. 1808, 35; pl. 2. — Doughty, Cab. Nat. Hist. I. 1830, 133; pl. 12. — Audubon, Orn. Biog. II. 1834, 190; pl. 131. — Ib. Birds Amer. III. 1841, 14; pl. 142. — Newberry, Zool. Cal. and Or. Route, 81; Rep. P. R. R. Surv. VI. 1857. — Heermann, X. vi. 45. — Cooper and Suckley, XII. iii. Zool. of W. T. 172. — *Merula migratoria*, Sw. & Rich., Fauna Bor. Amer. II. 1831, 176. — *T. (Planesticus) migratorius*, Baird, P. R. Rep. IX. Birds, 218. — Ib. Rev. Amer. Birds, 28.

Sp. Char. Third and fourth quills about equal; fifth, a little shorter; second, longer than sixth. Tail slightly rounded. Above olive gray; top and sides of the head black. Chin

and throat white streaked with black. Eyelids, and a spot above the eye anteriorly, white. Under parts and inside of the wings chestnut-brown. The under tail coverts and anal region, with tibiæ, white, showing the plumbeous inner portions of the feathers. Wings dark brown, the feathers edged more or less with pale ash. Tail still darker, the extreme feathers tipped with white. Bill yellow, dusky along the ridge and at the tip. Length, 9.75 to 10.25; extent, 16.50; wing, 5.40; tail, 4.75; tarsus, 1.25.

Hab. Continent of North America to Mexico. Accidental in Europe.

But few of this well-known species came about Fort Mohave in winter, though flocks are said occasionally to wander as far as Fort Yuma. At San Diego they were more numerous in the wet season, and remained until May, when they probably went no farther than the summits of the mountains, which rise 6,000 feet high, east of that place, and are similar in their vegetation and animals, at that height, to far more northern regions.

About the summits of the coast range towards Santa Cruz, elevated about 3,000 feet above the sea, I found Robins quite numerous in May, 1864, and was satisfied that they had nests there. They also abound on the

Turdus migratorius.

higher parts of the Sierra Nevada, but probably do not breed below an elevation of 3,000 feet in lat. 38°, as they have not been seen in the "mining districts" of the west slope in summer. Northward, however, they range lower, and on the Columbia build in all the wooded districts, the young being fledged at Puget Sound as early as June 10th, and two or more broods raised in a season. They prefer the borders of woods, cultivated fields, and orchards, having everywhere a tendency to become familiar, which, with their bright plumage and sweet singing, makes them universal favorites. Their name, associated with that of the European Robin, celebrated for its sociability in winter, and its traditional history, increases the respect felt for our bird. The Robin of Europe, however, is quite different, resembling our red-breasted Blue-birds (*Sialia*), but being olive-colored on the back.

English settlers in the far north have compared our Robin with their

Fieldfare (a species of *Turdus,* which resembles our young Robin with a spotted breast), and have given it that name.

The nest of the Robin is built at various heights on trees, and sometimes, as if relying fully on the nobler traits of humanity, in a shed, stable, or even in a saw-mill, close to the constantly working and noisy machinery. I have found one in the cleft of a split tree which had been broken down by the wind. Nuttall relates instances of their building near a blacksmith's anvil; on the stern timbers of a vessel that was being built at Portsmouth, N. H.; on a harrow, hung up in a cart-shed where three men were at work, and where they took refuge after the destruction of their first nest.

The eggs are four or five, dark bluish green, and unspotted. The nest is composed outside of roots, shavings, etc., then a layer of leaves, moss, and grass, cemented by mud, inside of which is a lining of soft, dry grass.

The Robin's song is loud and sweet, but not much varied. Though usually uttered in spring, it occasionally is heard during other seasons, especially in fine weather, when the musician, having fed heartily in the garden or field, mounts to the top of a tree, and returns thanks for human hospitality in the most pleasing melody.

Their food consists chiefly of insects, especially worms, for which they hop over grassy fields in the spring, watching and listening for the gnawing grub or earthworm, driven to the surface by rain, occasionally seizing one with a quick motion of the head, swallowing it whole, or picking it to pieces. They also feed much on berries, especially during winter, when they eat those of the Madrona (*Arbutus*), and even the bitter fruit of the dogwood (*Cornus*).

Kept in cages they become very familiar, and learn to imitate various tunes and noises. They live sometimes for many years in captivity, and have been made so domestic as to be allowed the free range of the house and surrounding grounds.

Turdus confinis, Baird.

THE CAPE ROBIN.

Turdus confinis, Baird, Rev. Amer. Birds, 1864, 29.

Sp. Char. Above, with sides of head and neck grayish ash, faintly tinged with olivaceous. Chin and throat white streaked with ash brown. Jugulum and breast pale yellowish buff; axillars, inner wing covers, and sides of the breast, more deeply so. Belly and edges of crissal feathers, white; flanks behind, ashy. A whitish stripe from lores over and a quarter of an inch behind the eye. Lower eyelid white. Tail feathers and greater wing coverts narrowly tipped with white. Bill yellowish; upper mandible and tip of lower tinged with dusky. Feet, pale brown. Wing, 5.10 inches; tail, 4.10; tarsus, 1.20; bill, above .60.

Hab. Cape St. Lucas.

A single specimen of this bird was obtained (No. 23,789 of the Smithsonian collection) at Todos Santos, Cape St. Lucas, in 1860, by Mr. John Xantus, who has done so much to extend our knowledge of the zoölogy of both Upper and Lower California. It has a close resemblance to the common Robin, and may indeed be a local variety of it; but, differing appreciably from hundreds of specimens of the latter with which it has been compared,

it merits recognition at least as a strongly marked form. It is much lighter than the Robin, the dark chestnut or cinnamon of the latter being replaced by light buff, and the belly and flanks are much more purely white. The superciliary stripe extends farther behind the eye, and the bill appears to be longer. There is no black or dusky on the head as in the Robin. It was obtained in the summer season, and was probably a resident bird. (BAIRD.)

Turdus nævius, GMELIN.

THE VARIED THRUSH. WESTERN ROBIN.

Turdus nævius, GMELIN, Syst. Nat. I. 1788, 817. — AUDUBON, Orn. Biog. IV. 1838, 489; V. 1839, 284; pl. 369, 433. — IB. Birds Amer. III. 1841, 22; pl. 143. — CABOT, Jour. Bost. Soc. N. H. III. 1848, 17 (spec. shot near Boston). — LAWRENCE, Annals N. Y. Lyc. V. June, 1852, 221 (spec. shot near New York). — NEWBERRY, Zool. Cal. and Or. Route, 81; Rep. P. R. R. Surv. VI. iv. 1857. — HEERMANN, X. vi. 45. — COOPER and SUCKLEY, XII. iii. Zool. of W. T. 172. — *T. (Ixoreus) nævius*, BONAP., Notes Orn. Delattre, in Comptes Rendus, XXVIII. 1854, 269. — BAIRD, P. R. Rep. IX. Birds, 219. — IB. Rev. Amer. Birds, 32. — LORD, Pr. R. Art. Inst. Woolwich, 114.
Orpheus meruloides, RICHARDSON, Fauna Bor. Amer. II. 1831, 187; pl. 38.

SP. CHAR. Fourth quill longest; third and fifth a little shorter; second much longer than sixth. Tail nearly even; the lateral feather shorter. Above rather dark bluish slate; under parts generally, a patch on the upper eyelids continuous with a stripe behind it along the side of the head and neck, the lower eyelids, two bands across the wing coverts and the edges of the quills, in part, rufous orange-brown; middle of belly, white. Sides of the head and neck, continuous with a broad pectoral transverse band, black. Most of tail feathers with a terminal patch of brownish white. Bill, black. Feet, yellow. *Female*

more olivaceous above; the white of the abdomen more extended; the brown beneath, paler; the pectoral band obsolete. Length, 9.75 inches; wing, 5; tail, 3.90; tarsus, 1.25.

Hab. Pacific coast of North America. Accidental on Mackenzie River, on Long Island, and near Boston.

Turdus nævius.

This beautiful bird is a far northern resident, merely visiting the lower country of California in winter, and rarely straggling to the Colorado valley, where one was obtained by Lieutenant Ives's exploring party in 1858. I have not seen them myself south of the Coast Range, near Santa Clara, and there no later than April. It is very probable, however, that some breed in the dark evergreen forests towards the north, as they do near the mouth of the Columbia, though I did not see any about the summits of the Sierra Nevada, in September, at lat. 39°, elevation 7,000 feet.

In October they begin to come down to the valleys, and are quite common in winter near San Francisco. They are then usually timid, but towards spring come more familiarly around houses, and utter their shrill, low notes, which seem much more distant than the bird itself really is. If pursued, they hide, and sit immovable among the foliage. They are said to "warble sweetly" at times, but I have never heard any true song from them. The nest is said to be built much like that of the Robin Thrush, which they much resemble. All these birds left Santa Cruz, about April 1, 1866, and none had returned from the North up to the 1st of November.*

SUB-FAMILY MIMINÆ.

CHAR. Tail long, vaulted at the base; the feathers more or less graduated; size, large; general appearance, thrush-like. Rictus with distinct bristles. Frontal feathers, normal, directed backwards. Anterior half of outer side of tarsi distinctly scutellate.

GENUS **OREOSCOPTES**, BAIRD.

Oroscoptes, BAIRD, P. R. Rep. IX. Birds, 346. (Type, *O. montanus.*)

GEN. CHAR. Culmen only slightly curved towards the tip. Bill longer and slenderer than in *Mimus;* not quite equal to the head. Wings decidedly longer than the tail, rather pointed; the first primary less than half the second, third and fourth longest. Tail rounded; scarcely graduated. Tarsi longer than middle toe and claw.

Oreoscoptes montanus, TOWNSEND.

MOUNTAIN MOCKING-BIRD.

Orpheus montanus, TOWNSEND, J. A. N. Sc. VII. ii. 1837, 192. — AUDUBON, Birds Amer. II. 1841, 194; pl. 139. — *Turdus montanus*, AUDUBON, Orn. Biog. IV. 1838, 437; pl. 369, f. 1. — *Mimus montanus*, BONAPARTE, List, 1838. — NUTTALL, Man. 2d ed. I. 371. — HEERMANN, P. R. Rep. X. vi. 44. — *Oroscoptes montanus*, BAIRD, P. R. Rep. IX. Birds, 347. IB. Review. — KENNERLY, X. iv. 25.

SP. CHAR. First quill rather shorter than the sixth. Tail slightly graduated. Above brownish ash; each feather obsoletely darker in the centre. Beneath, dull white, thickly marked with triangular spots, except on the under tail coverts and around the anus, which

* Nest and eggs of this bird have been received by the Smithsonian Institution from the Yukon (W. H. Dall) and Sitka (Dr. T. T. Minor). The eggs resemble those of the Robin, but differ in being slightly spotted. (BAIRD.)

regions are tinged with yellowish brown. Wing coverts and quills edged with dull white. Tail feathers, brown; the outer edged, and all (except, perhaps, the middle) tipped with white. Length, 8.50 inches; extent, 12.50; wing, 4; tail, 4; tarsus, 1.20. Iris brown; bill brown, white below; feet slaty; soles yellow.

Oreoscoptes montanus.

Hab. Rocky Mountains; south to Mexico. Along valley of Gila and Colorado. Cape St. Lucas.

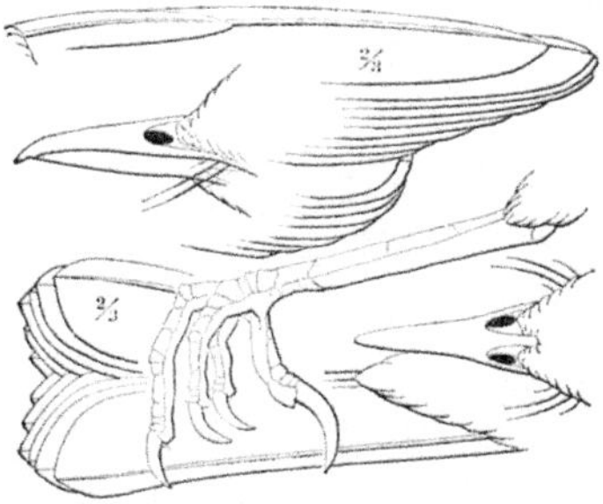

Oreoscoptes montanus.

According to Dr. Heermann, this bird is not rare in some of the cactus districts near San Diego, but I did not succeed in finding any in the southern part of California, although I have found them at Fort Laramie, Nebraska. Their habits are very similar to those of the *Mimus polyglottus*, but they inhabit more barren regions, and, according to Nuttall, are found in the "sage plains" as far north as the Columbia River, at Wallawalla. He describes a nest found in an artemisia-bush, near the sources of the Colorado, as made of small twigs and rough stalks, lined with strips of bark and "bison-wood." The female flew off a short distance, but made no complaint. The eggs were four, almost emerald green, spotted with dark olive of two shades, large and roundish; spots most numerous towards the largest end.

Their song is much like that of the Mocking-bird, but less strong and varied. They feed on berries and insects, and probably migrate but little south of lat. 40°.

GENUS **HARPORHYNCHUS**, CABANIS.

Toxostoma, WAGLER, Isis, 1831, 628. (Type, *T. vetula*, not *Toxostoma*, Raf. 1816.)
Harpes, GAMBEL, Pr. A. N. Sc. II. 1845, 264 (not of *Goldfuss*, 1839).
Harporhynchus, CABANIS, Wiegmann's Archiv, 1848, I. 98. (Type, *Harpes redivivus*.)

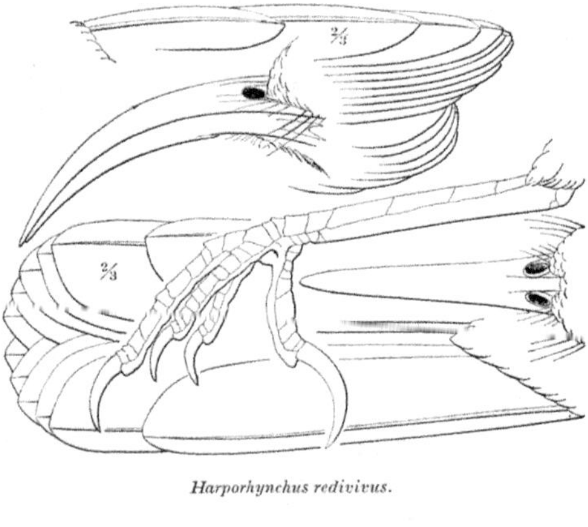

Harporhynchus redivivus.

CHAR. Bill from forehead as long as, or longer than, the head; nearly straight to near the tip, or bow-shaped, without any notch. Tarsus as long as, or longer than, the middle toe, conspicuously scutellate; outer lateral toe a little the longer, not reaching the base of the middle claw. Hind toe longer than lateral; its claw equal to its remaining portion.

Wings short, rounded; the fourth or fifth quill longest; the exposed portion of the first about half that of longest. Tail longer than the wings, broad, more or less graduated.

The type of this genus is *H. redivivus*, but the best known species is the *H. rufus*, or common Brown Thrush of the Eastern United States. Although apparently very different in the structure of the bill, yet there is so

Harporhynchus rufus.

uninterrupted a gradation from the one to the other as to render it impossible to draw a line of distinction. We give here, by way of illustration, a cut of *H. rufus*, and refer to the description of *H. redivivus* for its figure. (BAIRD.)

Harporhynchus redivivus, GAMBEL.

THE SICKLE-BILL THRUSH.

Harpes rediviva, GAMBEL, Pr. A. N. Sc. Phil. II. Aug. 1845, 264. — *Toxostoma rediviva*, GAMBEL, J. A. N. Sc. Phil. 2d ser. I. Dec. 1847, 42. — CASSIN, Illust. I. ix. 1855, 260; pl. 42 (poor figure). — *Harporhynchus redivivus*, CABANIS, Wiegmann's Archiv, 1848, I. 98. — BAIRD, P. R. Rep. IX. Birds, 349. IB. Rev. Amer. Birds, 48. — HEERMANN, X. vi. 45.
"*Promerops de la Californie septentrionale*, LA PEYROUSE, 1797, Atlas Voyage; pl. 37." (GAMBEL.)

SP. CHAR. Wing much rounded; the second quill shorter than the secondaries. Tail much graduated. Bill much decurved, longer than the head. Above, brownish olive, without any shade of green; beneath, pale cinnamon, lightest on the throat, deepening gradually into a brownish rufous on the under tail coverts. The fore part of the breast and sides of the body, brown olive, lighter than the back. An obscure ashy superciliary stripe, and another lighter beneath the eye. Ear coverts and an indistinct maxillary stripe, dark brown; the shafts of the former whitish. Ends and tips of tail feathers obsoletely paler. Length, 12.50 inches; extent, 13; wing, 4.20; tail, 5.75; tarsus, 1.50. Iris and feet brown; bill black.

Hab. California, west of the Sierra Nevada; north to lat. 38°.

This odd-looking and plainly colored bird is common in all the valleys and bushy hillsides south of San Francisco, frequenting chiefly the dense "chapparal," or low thickets, which often cover the gravelly sides of valleys for miles, with an almost impenetrable growth of shrubs from two to six

feet high, and occasionally a small tree rising above them. In such places these birds are constant residents, raising their young in nests built in the low trees or higher shrubs, and finding their subsistence mostly among the dead leaves on the ground, and from the berries which are sometimes numerous about such places.

In Santa Clara valley I found several nests in May, 1864, each containing but three eggs, and built in a low bush about five feet from the ground. They were formed of stout, thorny, or rough twigs, and lined with grass, etc. The eggs are pale bluish green, streaked and specked with faint umber, chiefly at the large end. They measure 1.10 × 0.85 inches. They also build about San Diego, and seem to migrate little if any in all their range, being found in winter at San Francisco. Their short wings are not suited for long flights, and they generally run off on the ground, seeking concealment in the thickets when alarmed. Their bill seems adapted chiefly for scratching among the dead leaves for insects.

Harporhynchus redivivus.

Their song is loud and varied, with frequent imitations of other birds, but they are much less musical than the true Mocking-birds, though often called by that name.

Harporhynchus Lecontii, LAWRENCE.

LECONTE'S THRUSH.

Toxostoma Lecontii, LAWRENCE, Ann. N. Y. Lyc. V. Sept. 1851, 109 (Fort Yuma). — *Harporhynchus Lecontii*, BONAPARTE, Comptes Rendus, XXVIII. 1854, 57. — BAIRD, P. R. Rep. IX. Birds, 350. IB. U. S. and Mex. Bound. Rep. II. iii. 12; pl. 12. IB. Rev. 47.

SP. CHAR. Bill much curved. Second quill about equal to the tenth; exposed portion of the first more than half the longest; outer tail feather an inch shortest. General color above light grayish ash, beneath much paler; the chin and throat above almost white; the sides behind brownish yellow or pale rusty yellow ash, of which color is the crissum and anal region. Tail feathers rather dark brown on the under surface, lighter above, the outer edges and tips of exterior ones obscurely paler. Quills nearly like the back. Length about 10 inches; wing, 3.75; tail, 4.75.

Hab. Fort Yuma, to Mojave River, California.

I found this bird rather common on the deserts along the route between the Colorado Valley and the coast slope of California, wherever there was a thicket of low bushes, generally surrounded by sand-hills They were so

very wild that I could obtain but two. In notes, habits, and general appearance they are like *H. redivivus,* but all I saw had the same pale colors. I found a nest without eggs built in a Yucca, and similar to that of *H. redivivus.*

Harporhynchus crissalis, HENRY.

HENRY'S THRUSH.

Harporhynchus crissalis, HENRY, Pr. A. N. Sc. Phil. May, 1858. — BAIRD, P. R. Rep. IX. Birds, 351, 923. IB. Birds of N. Amer.; pl. 82. IB. Rev. 47.

SP. CHAR. Second quill about as long as the secondaries. Bill much curved, longer than the head. Above olive-brown, with a faint shade of gray; beneath nearly uniform

brownish gray, much paler than the back, passing insensibly into white on the chin; but the under tail coverts dark brownish rufous, and abruptly defined. There is a black maxillary stripe cutting off a white one above it. There do not appear to be any other stripes about the head. There are no bands on the wings, and the tips and outer edges of the tail feathers are very inconspicuously lighter than the remaining portion. Length, 11.50 inches; extent, 11.50; wing, 4.25; tail, 5.60; tarsus, 1.25. Iris and feet brown; bill black.

Hab. Southern Rocky Mountains to Colorado Valley, California.

I found this species rather common at Fort Mojave, but so very shy that I only succeeded in shooting one after much watching for it. Their habits, nest (eggs not seen), and song are closely similar to those of *H. redivivus*, and the colors scarcely differ more than those of birds of other species that vary similarly in specimens from this valley, and those along the coast.

Harporhynchus cinereus, Xantus.

THE ASHY THRUSH.

Harporhynchus cinereus, Xantus, Pr. Phila. Acad. N. S. 1859, 298. — Baird, Rev. Amer. Birds, 1864, 46.

Sp. Char. Above ashy brown, beneath fulvous white, darker on flanks, inside of wing, and crissum. Beneath, except on chin, throat, and from middle of abdomen to crissum, with well-defined V-shaped spots of dark brown at ends of feathers, largest across the

breast. Two narrow whitish bands across tips of greater and middle coverts; quills edged externally with paler. Outer three feathers with a rather obsolete white patch in end of inner web and across tips of outer. Length, 10 inches; wing, 4.10; tail, 4.65; bill from gape, 1.40; tarsus, 1.26.

Hab. Lower California (Cape St. Lucas).

This is one of the new species discovered by Mr. Xantus, in 1859, at Cape St. Lucas.

GENUS **MIMUS**, BOIE.

Mimus, BOIE, Isis, Oct. 1826, 972. (Type, *Turdus polyglottus*.)
Orpheus, SWAINSON, Zool. Jour. III. 1827, 167. (Same type.)

GEN. CHAR. Bill shorter than the head, decurved from the base; distinctly notched at tip. Tarsi longer than the middle toe; lateral toes equal, not reaching the base of the middle claw, and shorter than the hind toe, the claw of which is half the total length.

M. polyglottus.

Tail variable; equal to or longer than the wings, moderately graduated. Wings rounded, the exposed portion of the first nearly or quite half that of the second, which is considerably shorter than the third.

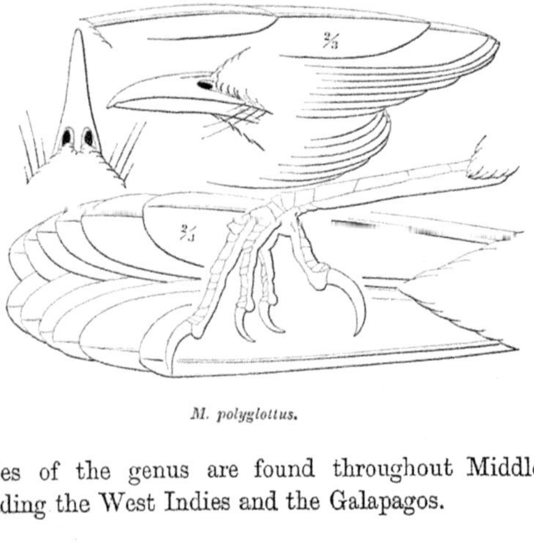

M. polyglottus.

Other species of the genus are found throughout Middle and South America, including the West Indies and the Galapagos.

Mimus polyglottus, Linnæus.

THE MOCKING-BIRD.

Turdus polyglottus, Linnæus, Syst. Nat I. 1766, 293. — Wilson, Am. Orn. II. 1810, 14; pl. x. f. 1. — Bon. Syn. 1828, 76. — Aud. Orn. Biog. I. 1831, 108; V. 1839, 438; pl. 21. — *Mimus polyglottus*, var. *caudatus*, Baird, Rev. Amer. Birds, 48; P. R. Rep. IX. Birds, 344, var. *caudatus*, 245. — Kennerly, X. iv. 25. — Heermann, X. vi. 44.

Sp. Char. Third to sixth quills nearly equal, second shorter than seventh. Tail considerably graduated. Above ashy brown, the feathers very slightly darker centrally, and towards the light plumbeous downy basal portion (scarcely appreciable, except when the feathers are lifted). The under parts are white with a faint brownish tinge, except on the chin, and with a shade of ash across the breast. There is a pale superciliary stripe, but the lores are dusky. The wings and tail are nearly black, except the lesser wing coverts,

which are like the back; the middle and greater tipped with white, forming two bands; the basal portion of the primaries white, most extended on the inner primaries. Tail much graduated, the lateral feathers being 1.20 inch shorter than the middle. The outer tail feather is white; the second is mostly white, except on the outer web and towards the base; the third black with a dull white tip; the rest, except the middle, very slightly tipped with white. The bill and legs are black. Length, 10.25; extent, 14.25; wing, 4.75; tail, 5.75. Iris yellow, bill and feet black.

Hab. West of Rocky Mountains, and north to Monterey, California.

Our Mocking-bird appears to form a peculiar variety, differing from the Eastern in being larger, and having the tail much more graduated, besides some minor differences of color.

In habits I have seen no difference, so far as I have observed them.

At Fort Mojave they were rare in March, and so very wild that I could not get within gunshot of them, flying from the top of one bush to another at a long distance, and very watchful. They winter near the river, probably in some of the bushy cañons chiefly, and seemed to have gone out of the valley in May, when I found them quite common along the Mojave River, and westward to San Diego, but still too wild to shoot. They undoubtedly, also, winter near San Diego, as I have found them common there in February, and they are said to migrate north in small numbers as far as Salinas Valley, near Monterey, which is about as far north as they have been found in the Colorado and San Joaquin Valleys.

They are generally found in the vicinity of the thickets of Cactus (*Opuntia*), both the wild and cultivated species, on the fruit of which they feed much of the time, and which furnishes them with shelter also. I have not been able to discover their nest, but it is probably very similar to that of the Eastern *M. polyglottus*. That bird builds in a thorny bush or thick tree a nest formed of twigs, leaves, and grass, with a thick lining of root-fibres. Their eggs are four or five, pale green, with blotches of brown scattered nearly all over, and they raise two broods annually. (Nuttall.) I have seen similar nests without eggs, in this State.

Their brilliant and endless powers of imitation are too well known to require special description. Most of their song is made up of the notes of other birds, and whatever noises they hear around them. They are frequently brought from San Pedro in cages, and have then all their habits of mimicry unimpaired.

Genus **GALEOSCOPTES**, Cabanis.

Galeoscoptes, Cabanis, Mus. Hein. I. 1850, 52.

Char. Bill shorter than the head, rather broad at base; wings a little shorter than the tail, rounded; secondaries well developed; fourth and fifth quills longest; third and sixth little shorter; first and ninth about equal, and about as long as the secondaries; first quill

G. Carolinensis.

more than half the second, about half the third. Tail graduated. Tarsi longer than middle toe and claw; scutellate anteriorly, more or less distinctly in different specimens. Scutellæ about seven.

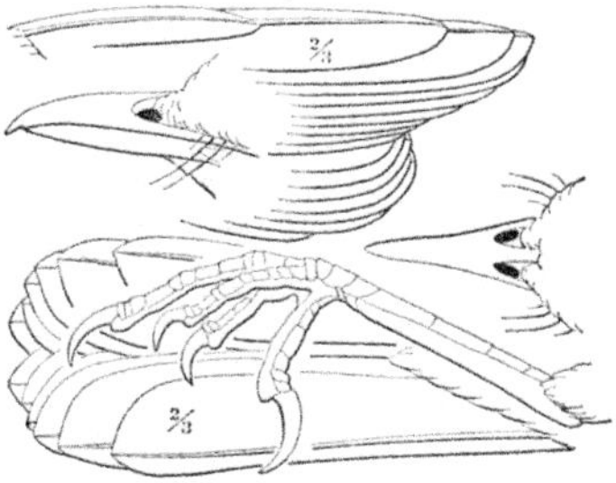

G. Carolinensis.

The single species hitherto assigned to this genus, the common Cat-bird of the United States, closely resembles the Mocking-bird in form, differing mainly in coloration.

Galeoscoptes Carolinensis, LINNÆUS.

THE CAT-BIRD.

Muscicapa Carolinensis, LINNÆUS, Syst. Nat. I. 1776, 328. — ***Orpheus Carolinensis***, AUD., Birds Amer. II. 1841, 195; pl. 140. — *Mimus Carolinensis*, BAIRD, Birds North Amer. 1859, 346. — *Galeoscoptes Carolinensis*, BAIRD, Rev. Amer. Birds, 54.
Turdus felivox, VIEILLOT. *Turdus lividus*, WILSON.

SP. CHAR. Prevailing color dark plumbeous, more ashy beneath; crown of head and nape sooty brown. Under tail coverts brownish chestnut. Length, 8.85; wing, 3.65; tail, 4.00; tarsus, 1.05.

Hab. Head-waters of Columbia River, Central Rocky Mountains, and east to Atlantic; south to Panama. Resident in Bermuda.

The introduction of the Cat-bird, so well known to all the inhabitants of the Eastern States, into the present work results from its occurrence as a common bird on the Columbia River, where it was frequently met with by the naturalists of the Northwestern Boundary Survey. There, as elsewhere, it inhabits low thickets or detached bushes; making its nest of strips of bark, twigs, roots, and such odd scraps as bits of rag, snake-skin, newspaper, etc. The eggs are four or five, and of a deep emerald green, without spots. Their food consists of insects, worms, fruits, and berries. The ordinary call-note resembles the mew of a cat, but it has a very agreeable song, in some respects imitating the notes of other birds. (Baird.)

FAMILY CINCLIDÆ.

CHAR. General characters of *Turdidæ*, ten primaries, of which the first is very short, etc.; the nostrils linear in lower edge of basal membrane. Loral and frontal feathers soft and downy; the mouth entirely without bristles. Body very short and broad. Wings short, rounded and concave.

GENUS **CINCLUS**, BECHSTEIN.

Cinclus, BECHSTEIN, Gemein. Naturg. 1802 (Agassiz).
Hydrobata, VIEILLOT, Analyse, 1816 (Ag.). (Type, *Sturnus cinclus*, LINN., European.)

GEN. CHAR. Bill without any bristles at the base; slender, subulate; the mandible bent slightly upward; the culmen slightly concave to near the tip, which is much curved

C. Mexicanus.

and notched; the commissural edges of the bill finely nicked. Feet large and strong, the toes projecting considerably beyond the tail; the claws large. Lateral toes equal. Tail

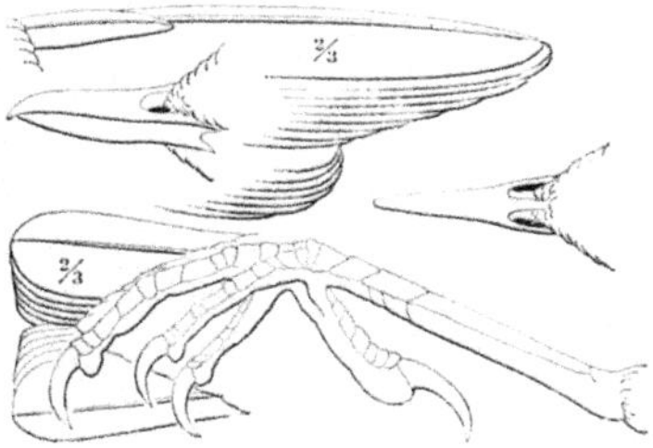

C. Mexicanus.

very short and even; not two thirds the wings, which are concave and somewhat falcate. The first primary is more than one fourth the longest.

Cinclus Mexicanus, SWAINSON.

THE AMERICAN DIPPER; WATER-OUZEL.

Cinclus Pallasii, BONAPARTE, Zool. Jour. II. Jan. 1827, 52. IB. Amer. Orn. II. 1828, 173; pl. xvi. f. 1 (not the Asiatic Pallasii). — *Cinclus Mexicanus*, SWAINSON, Syn. Mex. Birds, in Phil. Mag. 1, May, 1827, 368. — BAIRD, Rev. N. Amer. Birds, 1864, 59. — *Cinclus Americanus*, Sw. & RICH., F. Bor. Am. II. 1831, 173. — NUTTALL, Man. II. 1834, 569. — AUD., Orn. Biog. IV. 1838, 493; V. 1839, 303; pl. 370, 435. IB. Birds Amer. II. 1841, 182; pl. 137. — NEWBERRY, Zool. Cal. and Or. Route, 80; Rep. P. R. R. Surv. VI. iv. 1857. — HEERMANN, X. vi. 44. — *C. Mortoni* and *Townsendii* (AUD.), Townsend Narr. 1839, 337, 346.

Hydrobata Mexicana, BAIRD, P. R. Rep. IX. Birds, 229. — COOPER and SUCKLEY, XII. iii. Zool. of W. T. 175.

SP. CHAR. Above dark plumbeous, beneath paler; head and neck all round a shade of clove, or perhaps a light sooty brown; less conspicuous beneath. A concealed spot of

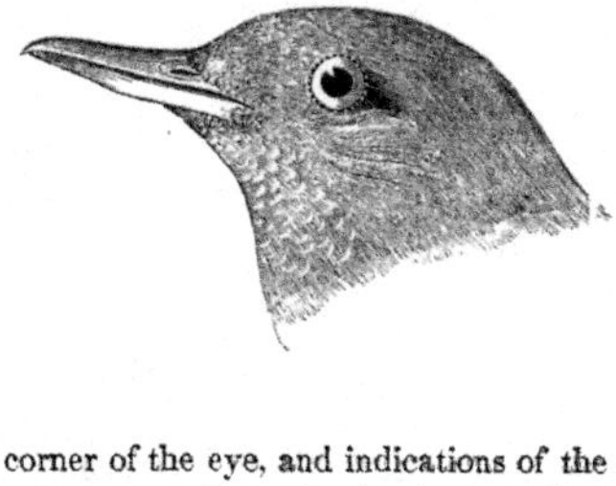

white above the anterior corner of the eye, and indications of the same sometimes on the lower lid. Immature specimens usually with the feathers beneath edged with grayish white; the greater and middle wing coverts and lesser quills tipped with the same. The colors more uniform. Length, 7.50; wing, 4.00; tail, 2.55.

Hab. Rocky Mountains from British America to Mexico, and west to the Pacific.

I first met with this plainly colored, but very interesting bird, at a mountain stream in the Coast Ranges west of Santa Clara, where there was a pair apparently mated, as early as March 10th. About sunset I heard the male singing very melodiously as it sat on one of its favorite rocks in the middle of the foaming rapids, making its delightful melody heard for quite a long distance above the sound of the roaring water. I watched for some time before I could perceive the minstrel, its sombre plumage concealing it in the approaching twilight.

In May, I found the nest of another pair along a stream a few miles farther south, the "Arroyo de los Gatos." It was built near the foot of a mill-dam, and rested on a slight ledge under an overhanging rock, from the top of which water was continually dripping. Its shape was that of an oven, with a small door-way, and it was built externally of green moss (which in that damp spot grew, and prevented its easy discovery), lined with soft grass, and contained young. The eggs have not yet been described, but are probably four to six, and white, like those of the European species. I found a similar nest built in the root of a large tree which lay across a mill-dam, north of the Columbia River, in July, 1854, and was told there that the birds had already raised a brood that season in the same nest, which then contained young.

The strange habits of this bird make it a very remarkable object, and it attracts much attention wherever found. It may be said to combine the form of a sandpiper, the song of a canary, and the aquatic habits of a duck. Its food consists almost wholly of aquatic insects, and these it pursues under water, walking and flying with perfect ease beneath a depth of several feet of water. When they dive below, there is a film of air surrounding them, which looks like silver, and may assist in supporting respiration. They do not, however, swim on the surface, but always dive, and sometimes fly across streams beneath the surface. They prefer clear, noisy mountain streams, but I have seen one on the summit of the Sierra Nevada, catching insects along the shore of a calm mountain lake.

Their flight is rapid and direct, like that of a sandpiper, and when they alight it is always on a rock or log, where they jerk their tails much like that bird. They are found in the Sierra Nevada, at least as far south as Fort Tejon, and northward nearly to the Arctic regions.

FAMILY SAXICOLIDÆ.

CHAR. Wings very long and much pointed, reaching beyond the middle of the short, square, or emarginated tail, and one and a half times (or more) the length of the latter. Spurious primary very short, the second quill longer than the fourth. In the closed wing the outer secondary reaches only about two thirds the length of the longest primary. Nostrils oval. Mouth with bristles.

The genus *Saxicola*, forming the type of this division, is found chiefly in

S. Mexicana.

the Old World, but the *S. œnanthoides*, Vigors, is stated to have come from Northwest America. (See Cassin, Illust. I. 208, 1854, Pl. 34, Nova Scotia specimen.)

GENUS **SIALIA**, SWAINSON.

Sialia, SWAINSON, Zool. Jour. III. Sept. 1827, 173 (*S. sialis*).

GEN. CHAR. Bill, short, stout, broader than high at the base, then compressed; slightly notched at the tip. Rictus with short bristles. Tarsus, not longer than the middle toe.

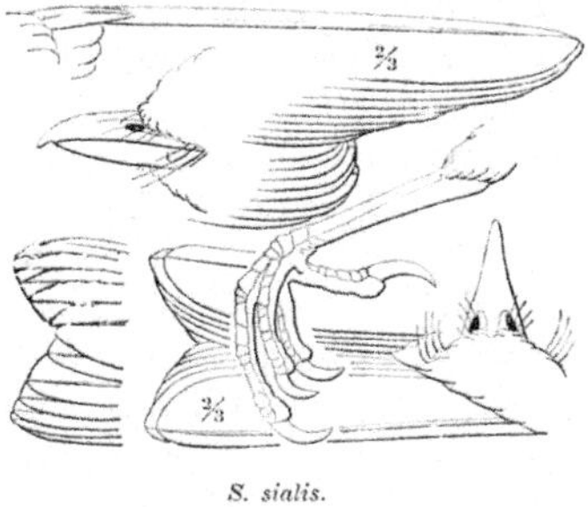

S. sialis.

Claws considerably curved. Wings much longer than the tail; the first primary spurious; not one fourth the longest. Tail moderate; slightly forked.

Sialia Mexicana, SWAINSON.

THE WESTERN BLUEBIRD.

Sialia Mexicana, SWAINSON, F. Bor. Am. II. 1831, 202. — BAIRD, P. R. Rep. IX. Birds, 223. — KENNERLY, X iv. 23. — HEERMANN, X. vi. 43. — COOPER and SUCKLEY, XII. iii. 173. — *Sialia occidentalis*, TOWNSEND, Jour. Ac. Nat. Sc. VII. ii. 1837, 188. — AUD., Birds Amer. II. 1841, 176; pl. 135. — NUTTALL, Man. I. (2d ed.) 1840, 513. — NEWBERRY, Zool. Cal. and Or. Route, 80; Rep. P. R. R. VI. iv. 1857. — *Sylvia occidentalis*, AUD., Orn. Biog. V. 1839, 41; pl. 393. — *Sialia cæruleo-collis*, VIGORS, Zool. Beechey's Voyage, 1839, 18; pl. iii.

SP. CHAR. Bill slender; head and neck all round, and upper parts generally bright azure blue. Interscapular regions, sides and fore part of the breast, and sides of the belly, dark reddish brown. Rest of under parts (with tail coverts) pale bluish, tinged with

gray about the anal region. *Female* duller above; the back brownish; the blue of the throat replaced by ashy brown, with a shade of blue. Length, 6.50; wing, 4.25; tail, 2.90.

Young brownish blue, the breast white, with pale brown spots.

Hab. Pacific Coast, North America, and along the Valley of the Gila to the upper Rio Grande, and south.

This beautiful and interesting bird is abundant in all the wooded districts of California, except high in the mountains, and some probably reside during summer even in the hot climate of the Colorado Valley, where I saw them examining buildings and knot-holes in trees for a nesting place in February.

They are numerous towards the coast, and up to the 49th parallel at least, in summer. Their song being much less sweet and varied than that of the Eastern bluebird (*S. sialis*), they have attracted less attention, and, not being encouraged to build about houses by the little box usually provided for that species, they have not become quite so familiar. I have, however, known of a nest built under the porch of a dwelling-house at Santa Barbara; they only need a little encouragement to become half domesticated.

The nest is usually built in a deserted woodpecker's hole, or other cavity of a tree, and the eggs, very pale blue, are four or five in number. They probably often raise two broods in a season, the first being hatched very early in April.

The song has a peculiar character, sounding as if two different birds were singing together, but it is not loud nor much varied.

Their food consists of insects and berries, and they feed much on grass-hoppers, hovering over the field like the sparrow-hawk, watching for them, and dropping suddenly down on them when seen. In winter they associate in small scattered flocks, flying with a short call-note, and resorting to the Madrona-tree, and others, for berries.

At Santa Cruz this bird is even more confiding than the Eastern species, which rarely frequents such large towns. In the spring, from February 20 to April, many pairs could be seen daily, seeking places for their nests, even in the noisiest streets, inspecting closely every new building that was being erected, as if they supposed it to be for their especial benefit. Some few pairs found nesting-places under porches, in knot-holes, etc., though the inhabitants provided them with no special homes.

One brood came every day during the grape season, and always about noon, to pick up grape-skins thrown out by my door, and were delightfully tame, sitting fearlessly within a few feet of the open window. Their favorite perches during the day were the flag-staffs, weathercocks, and eaves of houses, where they watched for passing insects.

Sialia Arctica, SWAINSON.

THE ROCKY MOUNTAIN BLUEBIRD.

Erythraca (Sialia) Arctica, SWAINSON, F. Bor. Amer. II. 1831, 209 ; pl. xxxix. — *Sialia Arctica*, NUTTALL, Man. II. 1832, 573 ; I. 2d ed. 1840, 514.—AUD., Birds Amer. II. 1841, 178 ; pl. 136. — McCALL, Pr. A. N. Sc. V. June, 1851, 215. — BAIRD, P. R. Rep. IX. Birds, 224. IB. P. R. Rep. X. iii. 13 ; pl. 35. — KENNERLY, X. iv. 24. — HEERMANN, X. vi. 44. — *Sylvia Arctica*, AUDUBON, Orn. Biog. V. 1839, 38 ; pl. 373. — *Sialia macroptera*, BAIRD, Stansbury, Report Exp. Salt Lake, 1852, 314.

SP. CHAR. Azure blue above and below, brightest above ; the belly and undertail coverts white ; the latter tinged with blue at the ends. *Female* showing blue only on the rump, wings, and tail ; a white ring round the eye ; the lores and sometimes a narrow front whitish ; elsewhere replaced by brown. Young like that of *S. Mexicana*, but paler. Length, 7.25 ; extent, 14 ; wing, 4.36 to 5 ; tail, 3.00. Iris brown, bill and feet black.

Hab. From eastern base of Rocky Mountains to Sierra Nevada. North to lat. 64° 30′, and south to Mexico. In winter, along the Pacific coast.

This bird has been found in winter at Fort Yuma, though I saw none at that season east of the sink of the Mojave River.

At San Diego, however, in the severe winter of 1861 – 62, they came down in large numbers to the vicinity of the town, and remained until the end of February, when all suddenly disappeared. They were at that time sitting perched on the low weeds and bushes, about the plains, often quite a flock together, and some constantly hovering like blue butterflies over the grass, at a height often of fifty feet, on the watch for insects. They seem everywhere more gregarious than the other species.

About Lake Tahoe, and the summits of the Sierra Nevada, above six thou-

sand feet elevation, I found this species numerous in September, and they had with them the young lately fledged. They doubtless breed there as they do throughout the Rocky Mountains as far south as Santa Fé, New Mexico, where Colonel McCall found them building in boxes put up for them by the inhabitants. (Heermann.)

They usually build like the other species in holes of trees, sometimes in holes in cliffs. (Nuttall.)

They were remarkably silent at the seasons when I observed them.

Family SYLVIIDÆ, Sylvias.

Char. Bill slender, broad and depressed at the base, distinctly notched and decurved at the tip. Culmen sharp-ridged at base. Frontal feathers reaching to the nostrils, which are oval, with membrane above, and overhung by a few bristles or by a feather. Rictal bristles extending beyond nostrils. Tarsi booted or scutellate. Basal joint of middle toe attached its whole length externally, half-way internally. Primaries ten, spurious primary about half the second, which is shorter than the seventh. Lateral toes equal.

Sub-Family REGULINÆ.

Char. Nostrils oval. The frontal feathers elongated; their bristle shafts with the rictal bristles extending beyond the nostrils, the former scale-like.

Points of loral feathers bristly. Second quill shorter than sixth. Size very small. Wings longer than the emarginated tail. Tarsi booted.

GENUS **REGULUS**, CUVIER.

Regulus, CUVIER, Leçons d'Anat. Comp. 1799 - 1800 (Agassiz). (Type, *Motacilla regulus*, LINN.; *Regulus cristatus*, KOCH.)

GEN. CHAR. Bill slender, much shorter than head, depressed at base, but becoming rapidly compressed; moderately notched at tip. Culmen straight to near the tip, then gently curved. Commissure straight; gonys convex. Rictus well provided

R. satrapa.

with bristles; nostrils covered by a single bristly feather directed forwards. Tarsi elongated, exceeding considerably the middle toe, and without scutellæ. Lateral toes about equal; hind toe with the claw longer than the middle one, and about half

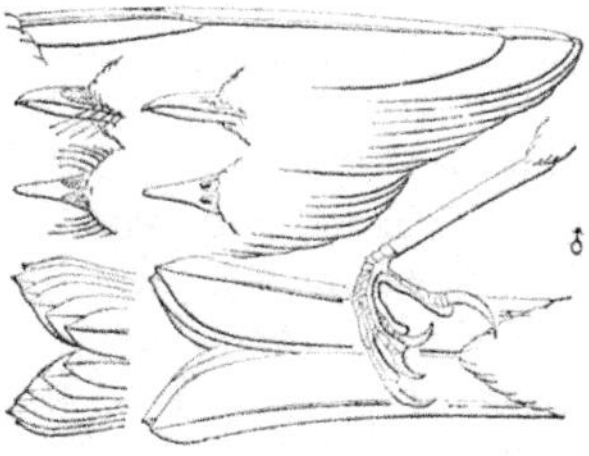

R. satrapa.

the toe. Claws all much curved. First primary about one third as long as the longest; second equal to fifth or sixth. Tail shorter than the wings, moderately forked, the feathers acuminate. Colors olive green above, whitish beneath. Size very small.

Regulus satrapa, Lichtenstein.

THE GOLDEN-CRESTED WREN.

Regulus satrapa, Lichtenstein, Verzeich. Doubl. 1823, No. 410. (Quotes *Parus satrapa*, Illiger, probably a museum name.) — Bonap., List, 1838. Ib. Conspectus, 1850, 291. — Aud., Synopsis, 1839, 82. Ib. Birds Amer. II. 1841, 165; pl. 132. — Baird, P. R. Rep. IX. Birds, 227. — Cooper and Suckley, XII. iii Zool. of W. T. 174.
Sylvia regulus, Wilson, Am. Orn. I. 1808, 126; pl. viii. f. 2. (Not of Latham.)
Regulus tricolor, Nuttall, Man. I. 1832, 420. — Aud., Orn. Biog. II. 1834, 476; pl. 183.

Sp. Char. Above olive green, brightest on the outer edges of the wing; tail feathers tinged with brownish gray toward the head. Forehead, a line over the eye, and a space beneath it, white. Exterior of the crown before and laterally black, embracing a central patch of orange red, encircled by gamboge yellow. A dusky space around the eye.

Eastern variety

Western variety.

Wing coverts with two yellowish white bands, the posterior covering a similar band on the quills, succeeded by a broad dusky one. Under parts dull whitish. Length about 3.80 to 4 inches; extent, 6.25; wing, 2.25; tail, 1.80. Iris brown, bill black, tarsi brown, feet yellow.

Hab. Northern parts of United States from Atlantic to Pacific.

I found a few of this species in September, 1863, at the summits of the Sierra Nevada, above seven thousand feet elevation in lat. 39°, which is the most southern point they have been found at on this coast. In the forests near the Columbia they are abundant, and a few remain in the valleys during summer. Probably they migrate through the whole of the Sierra Nevada in winter, and far towards San Francisco. Their nest and eggs have not yet been described, but a nearly allied European species builds toward the extremity of the branches of fir-trees, the nest being spherical with a small opening at the side; formed of moss and lichen, with a soft lining, and the eggs from six to twelve. While migrating, this species associates with *R. calendula*, and has similar habits. Professor Baird remarks (Rev. Amer. Birds, 65), that, as in several other instances, the Western specimens of this species are much brighter and more olivaceous above, especially on the rump and tail, and names the race or variety, *olivaceus*. There are also some slight differences in the pattern of coloration, represented in the two figures at the head of this article.

Regulus calendula, LINNÆUS.

THE RUBY-CROWNED WREN.

Motacilla calendula, LINNÆUS, Syst. Nat. I. 1766, 337. — *Sylvia calendula*, LATHAM, Ind. Orn. II. 1790, 549. — WILSON, Am. Orn. I. 1808, 83; pl. v. f. 3. — DOUGHTY, Cab. N. H. II. 1832, 61; pl. vi. — *Regulus calendula*, LICHTENSTEIN, Verzeich. 1823, Nos. 408-409. — NUTTALL, Man. I. 1832, 415. — AUDUBON, Orn. Biog. II. 1834, 546; pl. 195. IB. Birds, Amer. II. 1841, 168; pl. 133. — BAIRD, P. R. Rep. IX. Birds, 226. — KENNERLY, X. iii. 24. — HEERMANN, X. vi. 43. — COOPER and SUCKLEY, XII. iii. Zool. of W. T. 174.

SP. CHAR. Above, dark greenish olive, passing into bright olive green on the rump and outer edges of the wings and tail. Crown with large concealed patch of scarlet feathers, which are white at the base. The under parts are grayish white tinged with pale

olive yellow, especially behind. A ring round the eye, two bands on the wing coverts, and the exterior of the inner tertials white. Young without the red on the crown. Length, 4.25 to 4.75; extent, 6.50 to 7.25; wing, 2.40; tail, 1.85. Iris, bill, and feet brown; toes yellow.

Hab. United States from the Atlantic to the Pacific.

These little birds are abundant throughout the State during the colder months, even where there are only bushes, among which they are constantly at work pursuing their insect prey. In May they retire to the mountains, where they breed, but their nest and eggs have never been discovered, as far as I can ascertain, and are probably placed in hollow trees.

They are fearless little creatures, flitting close to a person's head, and eying him with curiosity. In the spring they utter a low, soft warble, often quite extended, and half whispered, as if they feared to disturb the solitudes they inhabit. They, however, often come about houses and gardens, and pursue insects to the tops of the highest trees, catching them by short flights from twig to twig. They feed chiefly on gnats and other Diptera.

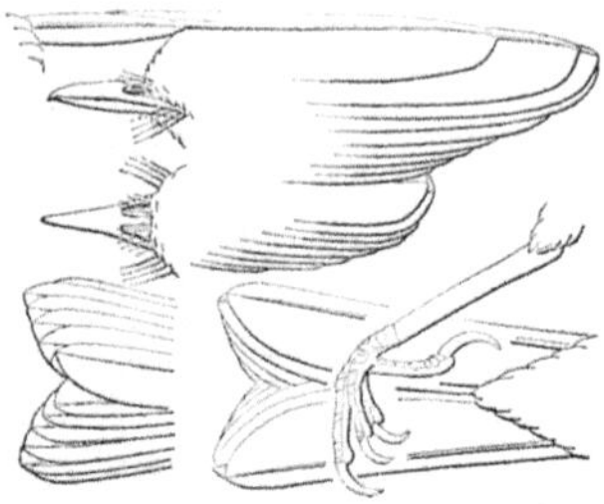

R. calendula.

There is a slight difference between the external anatomy of this species and the preceding, especially in lacking the small feather which covers the nostril. The exact characters are illustrated by the above outline.

SUB-FAMILY POLIOPTILINÆ.

CHAR. Bill slender, elongated, distinctly notched; nostrils not covered by bristly feathers, but exposed; nostrils elongated. Wings about equal to the graduated tail. Tarsi with distinct scutellæ.

The little birds of this sub-family have many points of resemblance to the true wrens and the titmice, but their nearest relationship is probably to the family of *Sylviidæ,* in which Professor Baird has lately placed them.

GENUS **POLIOPTILA**, SCLATER.

Polioptila, SCLATER, Pr. Zool. Soc. 1855, 11. (Type, *Motacilla cærulea.*)
Culicivora, SWAINSON, Class Birds, II. 1837, 243. (Type, *C. atricapilla.*) Not *Culicivora* (type, *stenura*) of Swainson's Zool. Jour. III. 1827, 359.

GEN. CHAR. Bill slender, attenuated, but depressed at the base; nearly as long as the head, distinctly notched at the tip, and provided with moderate rictal bristles. Nostrils rather elongated, not concealed, but anterior to the frontal feathers. Tarsi longer than the middle toe, distinctly scutellate; the toes small, the hinder one scarcely longer than the lateral its claw, scarcely longer than the middle. Outer lateral toe longer than the inner. First primary about one third the longest; second, equal to the seventh. Tail a little longer than the wings, moderately graduated; the feathers rounded.

P. cœrulea.

The species all lead-color above; white beneath, and to a greater or less extent on the exterior of the tail, the rest of which is black. Size very small.

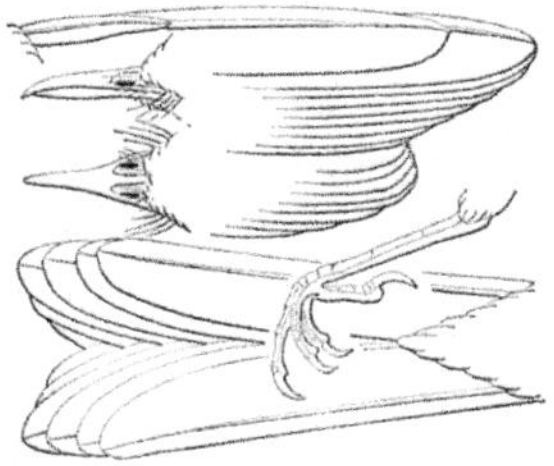

P. cœrulea.

There are nine other closely allied species, besides ours, found in tropical America, described in Professor Baird's "Review of American Birds."

Polioptila cærulea, Linnæus.

THE BLUE-GRAY FLYCATCHER.

Motacilla cœrulea, Linnæus, Syst. Nat. I. 1766, 43. — *Muscicapa cœrulea*, Wilson, Am. Orn. II. 1810, 164; pl. xviii. f. 3. — Aud., Orn. Biog. I. 1831, 431; pl. 84. — Nuttall, I. 1832, 297. — *Culicivora cœrulea*, Bonaparte, List, 1838. — Aud., Birds Amer. I. 1840, 244; pl. 70. — *Sylvania cœrulea*, Nuttall, Man. I. 2d ed. 1840, 337. — *Polioptila cœrulea*, Sclater, Pro. Zool. Soc. Lond. 1853, 11. — Baird, P. R. Rep. IX. Birds, 380. — Heermann, X. vi. 39.

Sp. Char. Above, grayish-blue, gradually becoming bright blue on the crown. A narrow frontal band of black extending backwards over the eye. Under parts and lores bluish-white, tinged with lead-color on the sides. First and second tail feathers white, except at the extreme base, which is black, the color extending obliquely forward on the inner web; third and fourth black, with white tip, very slight on the latter; fifth and sixth entirely black. Upper tail coverts blackish-plumbeous. Quills edged externally with pale

bluish-gray, which is much broader, and nearly white on the tertials. *Female* without any black on the head. Length, 4.50; extent, 6.40; wing, 2.12; tail, 2.25. Iris brown; bill and feet black.

Hab. United States from the Atlantic to the Pacific, south of lat. 42°. South to Guatemala.

This little bird is found in the southern part of California, but seems less common than the peculiarly western *P. melanura*, though they are so nearly similar in habits and appearance that it is nearly impossible to identify without shooting them, except when the black crown of the latter is visible. They all keep about the bushes, hunting insects in pairs or small companies, and seem as much like warblers in habits as titmice. In the East this species also frequents tall trees, and is always rapidly moving about with hanging wings and elevated tail, uttering at times a faint, almost hissing or squeaking note, repeated about three times, like that of the titmice. I found them at Fort Mojave, March 20th, and think some winter there; also at San Diego, in December. In the East they build their nest in May, among twigs, from ten to fifty feet from the ground, forming it of scales of buds, leaves, blossoms, fern-down, and silk of plants, lined with horse-hairs, and covered with lichens. The eggs are four or five, white, with a few reddish dots near the larger end, and they are said to raise two broods in the season. They are not believed to winter north of the Gulf of Mexico. (Nuttall.)

I did not find this, or the other species, at Santa Barbara, but this one was collected as far north as Yreka, California, near lat. 42°, by Mr. Vuille, on May 26th.

Polioptila melanura, Lawrence.

THE BLACK-TAILED FLYCATCHER.

Culicivora atricapilla, Lawrence, Ann. N. Y. Lyceum, V. Sept. 1851, 124. (Not of Swainson.)
Culicivora Mexicana, Cassin, Illust. I. vi. 1854, 164; pl. xxvii. (Not of Bonaparte.)
Polioptila melanura, Lawrence, Ann. N. Y. Lyc. VI. Dec. 1856, 168. — Baird, P. R. Rep. IX. Birds, 382. — Heermann, X. vi. 39.

Sp. Char. Above, ashy-blue; whole crown to bill and eyes, and tail feathers lustrous greenish-black. Beneath, pale bluish-gray, almost white in the middle of the belly; the sides behind, with anal region and under coverts tinged with brown. Edge of eyelids,

and the margin and tip of the outer web of first and second tail feathers, white. *Female* without the black head. Length, 4.50; extent, 6.30; wing, 2.00; tail, 2.10. Iris brown; bill and feet black.

Hab. Valley of the Rio Grande and Gila. West to San Diego. Cape St. Lucas.

This was also a rather common bird during the whole winter at Fort Mojave, as well as at San Diego, and I obtained one in October, on Catalina Island, but did not find it there, or at Santa Barbara, in summer. I saw them about San Diego up to the 20th of May, and young birds, fully fledged, as early as May 2d, but found no nests. Specimens were collected by Mr. F. Gruber as far north as lat. 39°, in the Sierra Nevada. In April I heard them sing a harsh ditty of five parts, somewhat like a wren's song, and another song resembling the notes of a swallow, and exactly like that of *Vireo Belli.* Their scolding note is a faint mew like a cat's.

Polioptila plumbea, Baird.

THE LEAD-COLORED FLYCATCHER.

Culicivora plumbea, Baird, Pr. A. N. Sc. VII. June, 1854, 118. — *Polioptila plumbea*, Baird, P. R. Rep. IX. Birds, 382. — Kennerly, X. iv. 26; pl. 33, f. 1.

Sp. Char. Above, bluish-gray; the forehead uniform with the crown. Eyelids white. A pale grayish-white line over the eye, above which is another of black, much concealed by the feathers, and which does not reach to the bill. Under parts dull white, tinged with bluish on the sides, and with brownish behind. Tail feathers black; the first and second edged and tipped with white; involving the entire outer web of the first, and most of that of the second; the third with only a very faint edging of the same. *Female* without the

black superciliary line. Length, 4.50; extent, 5.50; wing, 1.80; tail, 2.30. Iris brown; bill and feet black.

Hab. Valley of Colorado and Gila.

This species was a winter resident about Fort Mojave, in small numbers, and undistinguishable in habits or general appearance from the two other species found there. I shot one on the 20th of February, which had a cry of alarm much like that of a wren. It seems to represent the *P. melanura,* in the country east of the Colorado, and probably leaves that valley altogether for the mountains in spring. I suspect it is merely the younger plumage of *P. melanura.*

FAMILY CHAMÆADÆ.

CHAR. Size small. Tail very greatly graduated, much longer than the wings. Rictus with long bristles; frontal feathers bristly, directed forward. Whole outer side of tarsi continuous and undivided. Wing much rounded. Primaries ten, sixth longest.

C. fasciata.

GENUS **CHAMÆA**, GAMBEL.

Chamæa, GAMBEL, Pr. A. N. Sc. III. 1847, 154. (Type, *Parus fasciatus*.)

GEN. CHAR. Bill shorter than the head, much compressed. Nostrils linear, covered by an incumbent scale. Tarsus much longer than the toes, without well marked scales.

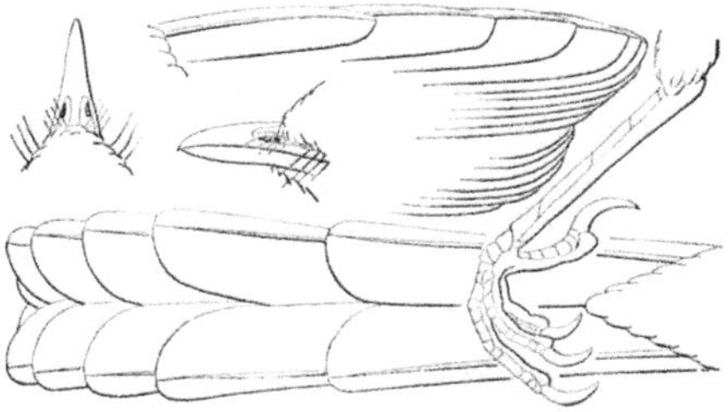

C. fasciata.

Lateral toes equal. Wings short, much rounded; two thirds the length of the tail, which is much graduated; the lateral feathers not two thirds the longest. Feathers narrow, with narrow outer webs. Plumage very soft and lax.

Chamæa fasciata, GAMBEL.

THE GROUND WREN.

Parus fasciatus, GAMBEL, Pr. A. N. Sc. II. Aug. 1845, 265. — *Chamæa fasciata*, GAMBEL, Pr. A. N. Sc. III. Feb. 1847, 154. (Type of genus.) IB. J. A. N. S. 2d Series, I. 1847, 34; pl. viii. f. 3. — CASSIN, Ill. I. ii, 1853, 39; pl. vii. — BAIRD, P. R. Rep. IX. Birds, 370. IB. Rev. Amer. Birds, 76. — HEERMANN, X. vi. 43.

SP. CHAR. Wings scarcely two thirds the length of the tail; both very much graduated. Upper and outer parts generally (including the whole tail) olivaceous-brown, tinged with gray on the head; beneath pale brownish-cinnamon, with obsolete streaks of dusky on the throat and breast. Sides and under tail coverts tinged with olive-brown. Lores

and a spot above the eye obscurely whitish. Tail feathers with obsolete transverse bars. Length, 6.50; extent, 7.25; wing, 2.25; tail, 3.50.

Hab. Coast of California, north to lat. 38°, and foot-hills of the Sierra Nevada.

This curious species is very different in appearance from the North American wrens. The colors are very simple, and the female differs from the above description only in being rather smaller, and with the reddish of the under parts less distinct; the whitish spot over the eye scarcely recognizable.

This interesting link between the wrens and titmice is common everywhere west of the Sierra Nevada, on dry plains and hillsides covered with chapparal and other shrubby undergrowth, but is not found in the forests. It is one of those birds that can live where there is no water, except occasional fogs, for six or eight months together. In these dreary "barrens," its loud trill is heard more or less throughout the year, but especially on spring mornings, when they answer each other from various parts of the thickets. They have a variety of other notes resembling those of the wrens, and correspond with them also in most of their habits, hunting their insect prey in the vicinity of the ground or on low trees, often holding their tails erect, and usually so shy that they can only be seen by patient watching, when curiosity often brings them within a few feet of a person; and as long as he sits quiet, they will fearlessly hop around him, as if fascinated.

In the last week of April, 1862, I found two of their nests near San Diego, built in shrubs about three feet from the ground. They were composed of straws and twigs mixed with feathers, firmly interwoven, the cavity 1.80 inch wide, and 1.70 deep, lined with grass and hair. The eggs were 0.70 × 0.52 inch in size, and pale greenish blue.

Family PARIDÆ.

Char. Bill generally short; conical, not notched nor decurved at tip. Culmen broad and rounded, not sharp-ridged at base. Nostrils rounded, basal, and concealed by dense bristles or bristly feathers. Loral feathers rough and bristly, directed forwards. Tarsi distinctly scutellate; basal joints of anterior toes abbreviated; that of middle toe united about equally for three fourths its length to the lateral; in *Parinæ* forming a kind of palm; outer lateral toe decidedly shorter than inner. Primaries ten, the first much shorter than the second. Tail feathers with soft tips.

This family is now made by most modern ornithologists to include the Nuthatches, notwithstanding the considerable differences in external form.

Sub-Family PARINÆ.

Char. Body compressed. Bill shorter than head, wings rounded, equal to or shorter than the rounded tail. Second quill as short as the tenth. Tarsus longer than middle toe and claw, which are about equal to the hinder; soles of toes widened into a palm. Plumage rather soft and lax.

L. inornatus.

The species of this group, known familiarly as Titmice or Chickadees, belong to the northern portions of the world, and exhibit considerable variety of form and color.

Genus **LOPHOPHANES**, Kaup.

Lophophanes, Kaup, Entw. Gesch. Europ. Thierwelt, 1829 (Agassiz). (Type, *Parus cristatus.*)

Gen. Char. Crown with a conspicuous crest. Bill conical; both upper and lower outlines convex. Wings graduated; first quill very short. Tail moderately long and rounded.

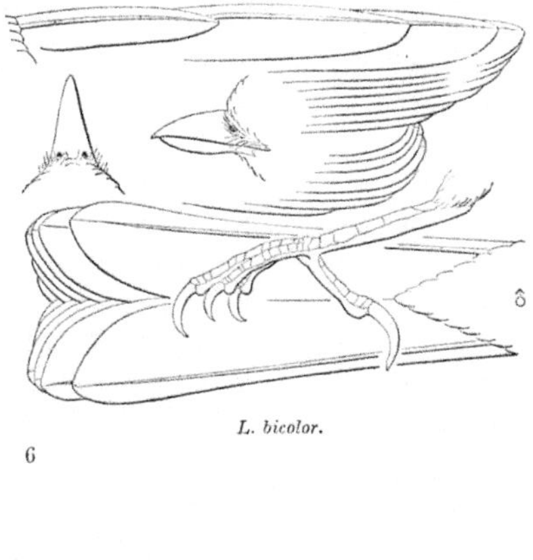

L. bicolor.

Lophophanes inornatus, Gambel.

THE PLAIN-CRESTED TITMOUSE.

Parus inornatus, Gambel, Pr. A. N. Sc. Phila. II. Aug. 1845, 265. (Upper California.) Ib. III. Feb. 1847, 1854. Ib. Jour. Ac. N. Sc. Phila. 2d Series, I. Dec. 1847, 35; pl. viii. — *Lophophanes inornatus*, Cassin, Ill. I. 1853, 19. — Baird, P. R. Rep. IX. Birds, 386. — Heermann, X. vi. 42.

Sp. Char. Crest elongated. Color above olivaceous-ashy, beneath whitish. Sides of body and under tail coverts very faintly tinged with brownish, scarcely appreciable. Sides

of head scarcely different from the crown. Forehead obscurely whitish. Length, 5.60; extent, 9; wing, 2.75. Iris brown; bill black; feet blue.

Hab. Coast of California and southern Rocky Mountains. (Fort Thorn, New Mexico.)

I found a few of this species in February, near San Diego, but none along the Colorado. It prefers the evergreen oak-groves towards the middle of the State, but I have not seen them in the higher Sierra Nevada. They are constant residents near San Francisco, like most of the birds inhabiting the live-oaks, being among the few that brave the fog and winds about San Francisco Bay all the year.

They are generally seen in small parties, scattered about the trees, and frequently calling each other with a variety of rather sweet and loud notes, among which the *chick-a-dee-dee* of its relatives, the *Pari*, is frequent. Some of its notes are almost equal to those of some of our best singers, and, indeed, like the Eastern representative of the genus *L. cristatus*, it probably has some power of imitation, and has the call of *pēto-pēto*, which has given one of its names to that species.

This bird feeds on seeds and acorns as well as insects, and often comes down to the ground, hopping about like a sparrow in search of them. It cracks the acorns with its bill, and hammers at bark and decayed wood after insects, with the industry of a woodpecker.

Their nest is probably in the deserted hole of a woodpecker, or other cavity, but I have never succeeded in finding the eggs. The Eastern species is said to dig a hole for itself at times, and to line it with various warm materials, laying six or eight eggs, white with a few brownish-red specks near the large end. (Nuttall.) That species also kills small birds occasionally, and has been found quite interesting as a cage-bird.

Lophophanes atricristatus, Cassin.

THE BLACK-CRESTED TITMOUSE.

Parus atricristatus, Cassin, Pr A. N. Sc. 1850, 103; pl. ii. — *Lophophanes atricristatus*, Cassin, Ill. Birds Texas, I. 1853; pl. iii. — Baird, Birds N. Amer. 1858, 385. Ib. Rev. Amer. Birds, 78.

Sp. Char. Crest very long (1.25), and much pointed. Above ash-colored. A broad band on the forehead dirty white; rest of head above, and crest black, tinged with ash on

the sides. Color of back shading insensibly into the dull ashy white of the under parts. Sides of body pale brownish-chestnut. *Female* with crest duller black. Iris brown. Length, about 5.25; wing, 3.00.

Hab. Valley of the Rio Grande.

This species has not as yet been detected west of the Rio Grande basin, though it will doubtless be met with in Eastern Arizona. Not much is known of its habits, which, however, probably resemble those of *L. inornatus*. (Baird.)

Lophophanes Wollweberi, Bonaparte.

WOLLWEBER'S TITMOUSE.

Lophophanes Wollweberi, Bonaparte, Comptes Rendus, xxxi. Sept. 1850, 478. — Baird, Birds N. Amer. 1858, 386; pl. liii. f. 1. Ib. Rev. Amer. Birds, 79. — Coues, Pr. A. N. Sc. 1866, 79.

Parus annexus, Cassin, Pr. A. N. Sc. 1850, 103; pl. i.

Lophophanes galeatus, Cabanis, Mus. Hein. 1850, 90.

Sp. Char. Central portion of crest ash, encircled by black, commencing as a frontal band and passing over the eye. Chin, throat, a line from behind the eye and curving round the auriculars to the throat, and some occipital feathers, black. A white line from above the eye margining the crest, with the cheeks below the eye, and under parts gener-

ally, white. A black half collar on the nape. Upper parts of body ashy. Length about 4.50; wing, 2.50.

Hab. Southern Rocky Mountains of New Mexico and Arizona. South into Mexico.

Dr. Coues found this species a permanent and abundant resident at Fort Whipple, Arizona, as well as elsewhere in the Territory. Its habits appear much like those of its allies. (Baird.)

GENUS **PARUS**, LINNÆUS.

Parus, LINNÆUS, Syst. Nat. 1735 (Agassiz).

GEN. CHAR. Head not crested. Body and head stout. Tail moderately long, and slightly rounded. Bill conical, not very stout; the upper and under outlines very gently

P. montanus.

and slightly convex. Tarsus but little longer than middle toe. Crown and throat generally black.

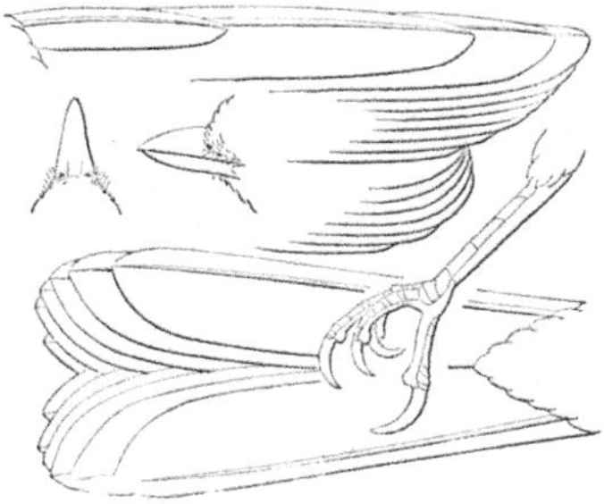

P. atricapillus.

There are five other Eastern species, several very closely allied.

Parus occidentalis, Baird.

THE WESTERN TITMOUSE.

Parus occidentalis, Baird, P. R. Rep. Birds, IX. 391 ; Rev. 81. — Cooper and Suckley, XII. iii. Zool. of W. T. 194.

Sp. Char. Tarsi lengthened. Tail graduated; outer feather about 0.25 of an inch shorter than the middle. Above dark brownish ash; head and neck above and below black, separated on the sides by white; beneath light rusty yellowish brown, scarcely

whiter along the middle of body. Tail and wings not much edged with whitish. Length, about 5 inches; extent, 7.50; wing, 2.50; tail, 2.40. Iris brown. bill black, feet gray.

Hab. North Pacific Coast of the United States.

This species was considered by Nuttall and Newberry to be identical with the Eastern *P. atricapillus*, and indeed it differs very slightly from that bird in plumage, while its habits are almost the same.

I have not met with this species in the southern half of California, but it doubtless visits the northern part towards the coast in winter at least, as it is abundant near the Columbia River. There it is generally found among the deciduous trees along streams and oak-groves, busily seeking food among the leaves and branches, frequently uttering its lively call of *chickadee-dee-dee*, and a variety of other notes, many of them quite tender and musical. It feeds on insects and seeds, and is very fond of fresh meat, fat, and crumbs of bread. They migrate but little, remaining, even when the ground is covered with snow, at the Columbia.

Their nest is made in a hole burrowed near the top of a dead tree in rotten wood, and their eggs, still undescribed, probably resemble those of the Eastern species, which lays from six to twelve, white with brown specks. (Nuttall.)

Parus montanus, Gambel.

THE MOUNTAIN TITMOUSE.

Parus montanus, Gambel, Pr. A. N. Sc. I. April, 1843, 259. (Santa Fe.) Ib. Pr. A. N. Sc. III. Feb. 1847, 155. Ib. Jour. A. N. Sc. 2d Series, I. Dec. 1847, 35; pl. viii. f. 1. — Cassin, Illust. I. 1853, 18. — Baird, P. R. Rep. IX. 394; Rev. 82. — Newberry, VI. iv. 79. — Heermann, X. vi. 42. — Cooper and Suckley, XII. iii. 194.

Sp. Char. Head and neck above, with under part of head and throat, glossy black; forehead, line above the eye and one below it, involving the auriculars, white. These stripes embracing between them a black line through the eye and confluent with the black

of the head. Above, ashy; beneath similar, but paler; the upper part of breast, and middle line of belly, white. Length, about 5.25; extent, 8.00; wing, 2.70; tail, 2.50. Iris brown; bill black; feet lead-color.

Hab. Rocky Mountains, to Sierra Nevada and Cascade Range.

This seems to be the common species of the Sierra Nevada, especially on their east slope, and I have shot one, probably of this species, near San Diego, in February, but the specimen was destroyed by a cat. Its habits, as far as known, are exactly like those of *P. occidentalis*.

Parus rufescens, Townsend.

THE CHESTNUT-BACKED TITMOUSE.

Parus rufescens, Townsend, Jour. A. N. Sc. Phila. VII. ii. 1837, 190. — Audubon, Orn. Biog. IV. 1838, 371; pl. 353. Ib. Birds Amer. II. 1841, 158; pl. 129. — Cassin, Illust. I. 1853, 18. — Baird, P. R. Rep. IX. Birds, 394. — Heermann, X. vi. 42. — Cooper and Suckley, XII. iii. Zool. of W. T. 194.

Sp. Char. Whole head and neck above, and throat from bill to upper part of breast, sooty blackish-brown. Sides of head and neck, upper part of breast, and middle of body,

white; back and sides dark brownish-chestnut. Length, 4.75; extent, 7.50; wing, 2.40; tail, 2.16. Iris brown; bill black; feet lead-color.

Hab. Pacific Coast of the United States, to northern Rocky Mountains.

This is the only species I have found near the coast of this State, and only in the mountains as far south as Santa Cruz, where they are permanent residents, as they are towards the north. They resemble *P. occidentalis* closely in habits, but seem more partial to the evergreen coniferous forests, and their notes are more faint and lisping. They are amusing and familiar little birds, very fond of each other's society. Their eggs I have never found.

Genus **PSALTRIPARUS**, Bonaparte.

Psaltriparus, Bonaparte, Comptes Rendus, XXXI. 1850, 478. (Type, *P. melanotis*.)
Psaltria, Cassin, Illust. N. Amer. Birds, 1853, 19. (Not of Temminck.)

Gen. Char. Size very small and slender. Bill very small, short, compressed, and with its upper outline much curved for the terminal half. Upper mandible much deeper than under. Tail long, slender, much graduated; much longer than the wings; the feathers very narrow. Tarsi considerably longer than the middle toe. No black on the crown or throat.

P. minimus.

The type of this genus is the *P. melanotis*, or the Black-cheeked Titmouse, a species found on the table-lands of Mexico, but although included by some

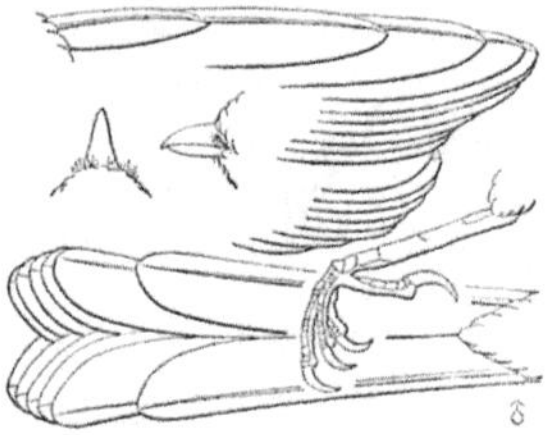

P. melanotis.

in the list of birds of the United States, not yet actually detected within our territory.

Psaltriparus minimus, Townsend.

THE LEAST TITMOUSE.

Parus minimus, Townsend, Jour. A. N. Sc. Phila. VII. ii. 1837, 190. — Audubon, Orn. Biog. IV. 1838, 382; pl. 353, f. 5, 6. Ib. Birds Amer. II. 1841, 160; pl. 130. — *Psaltria minima*, Cassin, Illust. I. 1853, 20. — Heermann, P. R. Rep. X. vi. 38. — *Psaltriparus minimus*, Bonaparte, Comptes Rendus, XXXVIII. 1854 — Baird, P. R. Rep. IX. Birds, 397. — Cooper and Suckley, XII. iii. Zool. of W. T. 195.

Sp. Char. Tail long, feathers graduated. Above, rather dark olivaceous-cinereous; top and sides of head smoky-brown. Beneath, pale whitish-brown, darker on the sides. Length about 4.25 inches; extent, 5.70; wing, 2.00; tail, 2.25. Iris brown; bill and feet black.

Hab. Pacific Coast of the United States, east to the Sierra Nevada.

These little birds frequent the evergreen oaks in small flocks or families throughout the year, as far north at least as San Francisco, and probably much farther, as they appear along the Columbia by the first of April. They are usually observed scattered in the trees near together, keeping up a constant chirping call, and following each other from tree to tree, thoroughly exploring the foliage for their insect food.

As early as March 1st, near San Diego, I found a newly built nest, but without eggs. At that time they are in pairs, but an alarm soon brings others around, and the flock does not probably scatter far during the breeding season, though I have never found nests very near together. The great size of the nest, compared to the bird, would lead us to think that the whole flock united to build one. It is about eight inches long, and three inches in diameter, outside; inside, 5 × 1.50; cylindrical, and suspended by one end

from a low branch. It is in fact a long purse, with an opening near the top, constructed of lichens, gathered from the neighboring branches, mixed with slender stems and heads of weeds, shells of insect larvæ, down of plants, etc. At San Diego, I found two nests on the 9th of May, one containing eggs, deserted when half hatched, and another in the same bush with seven pure white eggs already sat upon, measuring 0.60 × 0.45 inch. The number is from six to eight (Heermann), and Nuttall found them already hatching by the middle of May, in Oregon, so that they probably are usually a month earlier in this State.

When one is killed, the others come round it with great show of anxiety, and call plaintively until they find that it will not follow them. At such times they become so fearless as almost to allow of being taken in the hand.

Psaltriparus plumbeus, Baird.

THE PLUMBEOUS TITMOUSE.

Psaltria plumbea, Baird, Pr. A. N. Sc. Phila. VII. June, 1854, 118. (Little Colorado R. N. Mex.) — *Psaltriparus plumbeus*, Baird, P. R. Rep. IX. Birds, 398. — Rev. Amer. Birds, 79. — Kennerly, X. iv. 25, pl. 33, f. 2. — Coues, Pr. A. N. S. 66, 79.

Sp. Char. Tail long, feathers graduated. Above, rather light olivaceous-cinereous. Top of head rather clearer; forehead, chin, and sides of head, pale smoky-brown. Beneath, brownish-white, scarcely darker on the sides. Length, about 4.20 inches; wing, 2.15; tail, 2.50.

Hab. Southern Rocky Mountains (to Sierra Nevada, Cal. (?) Gruber.)

This species was found by Dr. Kennerly, along the eastern branches of the Colorado, and probably at times comes down to that valley, though I saw none of them at Fort Mojave. According to Dr. Kennerly, they seem to have much the same habits as *P. minimus*, but frequented the tops of the leafless cottonwood-trees, probably where there were few if any others. He notes that specimens from the Little Colorado all had dark eyes, while those from Williams Fork had yellow. This is a difference never noticed before in small birds of the same species, though common among hawks, depending on difference of age.

Dr. Coues found the species abundant at Fort Whipple, Arizona, and noticed the same difference in the color of the eyes.

Genus **AURIPARUS**, Baird.

Auriparus, Baird, Rev. N. Amer. Birds, 85, July, 1854. (Type, *Ægithalus flaviceps*, Sund.)

Gen. Char. Form sylvicoline. Bill conical, nearly straight, and very acute; the commissure very slightly and gently curved. Nostrils concealed by decumbent bristles.

A. flaviceps.

Wings long, little rounded; the first quill half the second; third, fourth, and fifth quills nearly equal and longest. Tail slightly graduated. Lateral toes equal, the anterior united at the extreme base. Hind toe small, about equal to the lateral. Tarsus but little longer than the middle toe.

In the "Review," Professor Baird has formed this genus for one species, on account of its differing from *Paroides* (type, *P. pendulinus*), in a curved

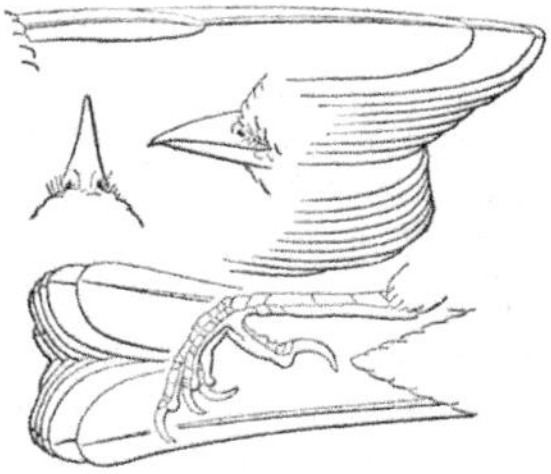

A. flaviceps.

bill, longer tarsi, smaller claws, and rounded tail. It differs from other *Paridæ*, in its longer wings and other characters.

Auriparus flaviceps, Sundevall.

THE YELLOW-HEADED TITMOUSE.

Ægithalus flaviceps, Sundevall, Ofversigt af Vet. Ak. Förhandl. VII. v. 1850, 129. (Sitka or California.) — Heermann, P. R. Rep. X. vi. 43. — *Paroides flaviceps*, Baird, P. R. Rep. IX. Birds, 400. — U. S. and Mex. Bound. Rep.; pl. xv. f. 2. — *Auriparus flaviceps*, Baird, Rev. N. Amer. Birds, 85, July, 1864. — Coues, Pr. A. N. S. 66, 79. *Conirostrum ornatum*, Lawrence, Ann. N. Y. Lyceum, V. May, 1851, 113; pl. v. f. 1. (Texas.)

Sp. Char. Above, cinereous; head, all round yellow; lesser wing coverts chestnut; beneath, brownish-white. Length, 4.50; extent, 6.40; wing, 2.12; tail, 2.35. Iris brown; bill black; feet lead-color.

Hab. Rio Grande to Mojave River, California. Cape St. Lucas.

I found numbers of this beautiful little bird at Fort Mojave during the whole winter, frequenting the thickets of *Algarobia* and other shrubs, and having habits rather intermediate between those of the titmice and warblers, corresponding with their intermediate form. They had something of the same song as the *Parus*, and a loud call, generally uttered as they sat on a high twig, besides a lisping triple note like that of *tsée-tu-tu*.

On the 10th of March, I found a pair building, first forming a wall nearly spherical in outline, out of the thorny twigs of the *Algarobia* (in which tree the nest is usually built), then lining it with softer twigs, leaves, down of plants, and feathers, covering the outside with thorns, until it becomes a mass as large as a man's head, or 9 × 5.50 inches outside, the cavity 4.50 × 2.70, with an opening in one side, just large enough for the bird to enter. On the 27th of March, I found the first nest containing eggs, and afterwards many more. There were in all cases four eggs, pale blue, with numerous small brown spots, chiefly near the large end, though some had very few spots and were much paler; size 0.60 × 0.44 inch. In one nest which I watched, they hatched in about ten days, and in two weeks more the young were ready to leave the nest.

I noticed the nests of this bird in the *Algarobias* that grow in a few places on the mountains west of the Colorado Valley, and along Mojave River as far west as "Point of Rocks," where that tree ceases to grow. They were exceedingly wild in that district, and no doubt leave it for the Colorado Valley in winter, as I saw none of them there in December.

Sub-Family SITTINÆ.

Char. Body depressed. Bill about equal to, or longer than the head. Wings much pointed; much longer than the nearly even tail. Tarsus shorter than the middle toe and claw, which are about equal to the hinder. Plumage more compact than in *Parinæ*.

The *Sittinæ* are represented in America by only a single genus, *Sitta*, with several species, two of which are peculiar to the Western regions, two to the Eastern, and one common to both. Other forms belong to the Old World, one of them, *Sittella*, peculiar to Australia.

The Old World species of *Sitta* amount to about eight, one of these, recently discovered in Northern China, having a very close resemblance to *S. Carolinensis*. There is no species, however, which is found in both America and Europe, or Asia.

GENUS **SITTA**, LINNÆUS.

Sitta, LINNÆUS, Syst. Nat. 1735 (Agassiz).

GEN. CHAR. Bill subulate, acutely pointed, compressed, about as long as the head; culmen and commissure nearly straight; gonys convex and ascending; nostrils covered by a

S. aculeata.

tuft of bristles directed forward. Tarsi stout, scutellate, about equal to the middle toe, much shorter than the hinder, the claw of which is half the total length. Outer lateral toe much longer than inner, and nearly equal to the middle. Tail very short, broad, and

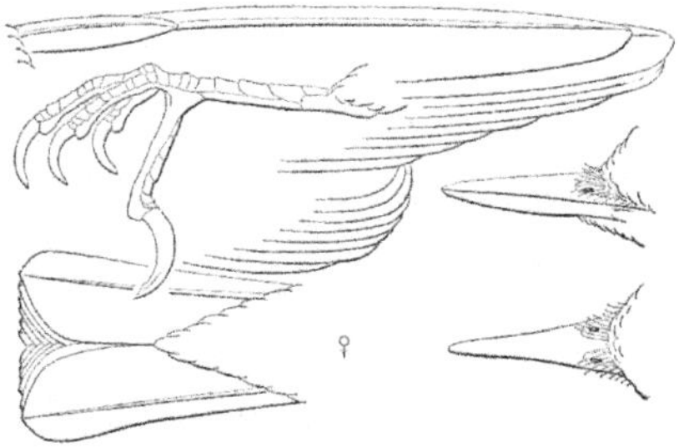

S. Carolinensis.

nearly even; the feathers soft and truncate. Wings reaching nearly to the end of the tail, long and acute, the first primary one third the length of the third, or less.

Sitta aculeata, Cassin.

THE WESTERN NUTHATCH.

Sitta Carolinensis, Gmelin, of West Coast, Nuttall, Man. I. 2d ed. 695.—Newberry, P. R. Rep. VI. iv. 79.

Sitta aculeata, Cassin, Pr. A. N. Sc. Phila. VIII. Oct. 1856, 254.—Baird, P. R. Rep. IX. Birds, 375.—Kennerly, X. iv. 26; pl. 33, f. 4.—Heermann, X. vi. 55.—Cooper and Suckley, XII. iii. Zool. of W. T. 193.

Sp. Char. Above, ashy-blue; top of head and neck black. Under parts and sides of head, to a short distance above the eye, white. Under tail coverts, and tibial feathers brown; concealed primaries white. Bill very slender and long. Length, 6 inches; extent, 10.50; wing, 3.50; tail, 2.12. Iris brown; bill black; feet gray.

Hab. Pacific Coast, and east towards the Rocky Mountains.

This is not a common bird south of San Francisco, and only seen there in the colder months, but I have found them near San Diego in February. I saw none even in the Coast Mountains in summer, near Santa Cruz; but northward they become numerous at that season, frequenting chiefly the groves of deciduous oaks, constantly creeping about their trunks and branches in search of insects, and occasionally seeking for them on the roofs and walls of houses. Their habits are similar to those of *S. Canadensis*, but they are rather slower in movements, and their notes consist of a single harsh call, uttered occasionally, and answered by their comrades. Their nest and eggs are still unknown.

Sitta Canadensis, Linnæus.

THE RED-BELLIED NUTHATCH.

Sitta Canadensis, Linnæus, Syst. Nat. I. 1766, 177.—Nuttall, Man. I. 1832, 583.—Audubon, Orn. Biog. II. 1834, 24; V. 474; pl. 108. Ib. Birds Amer. IV. 179; pl. 248.—Baird, P. R. Rep. IX. Birds, 376.—Cooper and Suckley, XII. iii. Zool. of W. T. 193.

Sitta varia, Wils. Am. Orn. I. 1808, 40; pl. ii.

Sp. Char. Above, ashy-blue. Top of head black; a white line above, and a black one

through the eye. Chin white; rest of under parts rusty-brown. Length about 4.50 inches; extent, 8.25; wing, 2.60.
Hab. North America generally.

I have not myself met with this bird in California, but it has been obtained by Dr. Heermann, in the Sacramento Valley, and a straggler, as far south as Fort Yuma, by Lieutenant Ives's exploring party in 1858–59. It is very common towards the north, in company with *S. aculeata,* and has

similar habits, but is rather quicker in its motions, and with a more varied call of several notes, resembling that of the chicadees (*Parus*). Both occasionally frequent pine forests, especially those that are open. The nest, according to Audubon, is sometimes in the hollow of a low stump or tree (not burrowed out), and the eggs are four, bluish-white, sprinkled with reddish dots.

Sitta pygmæa, Vigors.

THE CALIFORNIAN NUTHATCH.

Sitta pygmæa, Vigors, Zool. Beechey's Voyage, 1839, 25; pl. iv. — Audubon, Orn. Biog. V. 1839, 63; pl. 415. Ib. Birds Amer. IV. 1842, 184; pl. 250. — Newberry, Zool. Cal. Or. Route; P. R. R. Rep. VI. iv. 1857, 79. — Baird, IX. Birds, 378. — Cooper and Suckley, XII. iii. Zool. of W. T. 193.

Sp. Char. Above, ashy-blue; head and upper part of neck greenish ashy-brown, its lower border passing a little below the eye, where it is darker; nape with an obscure whitish spot. Chin and throat whitish; rest of lower parts brownish-white; the sides and behind like the back, but paler. Middle tail feather like the back; its basal half with a long white spot; its outer web edged with black at the base. Length about 4.50 inches; extent, 8; wing, 2.70. Iris brown; bill black; lead-color at base; feet gray.
Hab. Rocky Mountains, lat. 49°, and Pacific Coast south to Monterey.

A specimen from Monterey, — the original locality of the species, — is much smaller than one from the Sierra Nevada, at Lake Tahoe, but may be younger, though both were killed in September. The difference is 4.30 — 7.50 — 2.60 to 4.75 — 8.25 — 2.75, possibly the difference between the size of the sexes.

I have not seen this species south of Monterey, although probably occurring farther south along the mountains. At that place there is an extensive pine forest coming down to the coast, and the summer resort of several northern species, which are rare in other localities of the lower country, but there find a climate rendered quite cool by the sea breeze and frequent fogs. They also frequent the mountains of New Mexico, and one was obtained

near the Colorado River by Lieutenant Ives's exploring party. Northward they avoid the dense forests, preferring those that are open, especially of the true pines, and there wander about in large parties, chirping continually like young chickens, and hunting for insects among the foliage and branches.

Their nest and eggs resemble those of the closely allied *S. pusilla* of the Southern Atlantic States, which builds in hollow trees a few feet from the ground. The eggs, laid in March, are white, with reddish dots, and number from four to six.

Family CERTHIADÆ.

Char. First primary very short, less than half the second; outer lateral toe much the longest; hind toe exceeding both the middle toe and the tarsus, which is scutellate anteriorly, and very short. Bill slender, as long as or longer than the head, without any notch. Entire basal joint of the middle toes united to the lateral.

Sub-Family CERTHIANÆ.

Char. Bill much compressed and greatly decurved; gonys concave. Tail long, cuneate; the feathers stiffened at the tips.

GENUS **CERTHIA**, LINNÆUS.

Certhia, LINNÆUS, Syst. Nat. 1735 (Gray). (Type, *C. familiaris.*)

GEN. CHAR. Bill as long as the head, slender, much compressed and decurved from the base; without notch or rictal bristles. Tarsi distinctly scutellate, very short, not longer

C. Americana.

than the outer lateral toe, which much exceeds the inner, reaching nearly as far as the middle toe. Hind toe longer than the middle one; its claw more than half the total length. Claws all very long and acute. Tail rather longer than the wings, arched or vaulted,

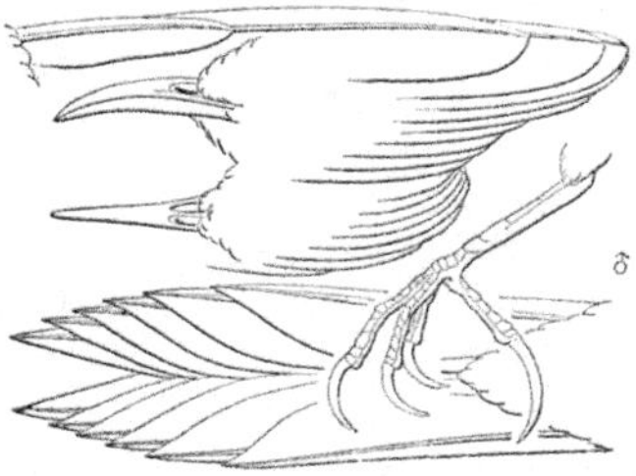

C. Americana.

graduated or cuneate; the feathers very acute at the tips, the shafts stiffened. First primary rather more than one third the fourth or longest one. Color above brown, streaked with white; beneath, white.

Certhia Mexicana, Gloger.

THE WESTERN CREEPER.

"*Certhia Americana*, Bonaparte." Nuttall, Man. 2d ed. I. 701 (referring to Western bird). — Baird, P. R. Rep. IX. Birds, 372 (in part). — Kennerly, X. iv. 26.—Heermann, X. vi. 42. — Cooper and Suckley, XII. iii. Zool. of W. T. 192. — Baird, Birds N. Amer. 372; pl. 83, f. 2. Ib. Rev. N. Amer. Birds, 89, July, 1864. — *Certhia Mexicana*, "Gloger, Handbuch," Reichenbach, Handbuch Spec. Orn. I. 1851, 265; pl. lxii. f. 3841, 3842. — Sclater, Pr. Zool. Soc. 1856, 290. — Baird, P. R. Rep. IX. Birds, 923.

Sp. Char. Bill about the length of the head. Above, dark brown, with a rufous shade, each feather streaked centrally, but not abruptly, with whitish; rump brownish-orange. Beneath, almost silky white; the under tail coverts with a faint rusty tinge. A white streak

over the eye; the ear coverts streaked with whitish. Tail feathers brown centrally, the edges paler yellowish-brown. Wings with a transverse bar of pale reddish-white across both webs. Length, 5.25; extent, 7.50; wing, 2.50; tail, 2.60.

Hab. North America generally.

Certhia Mexicana, Gloger. "A Mexican specimen, and one from Fort Tejon, California, are darker than those from either coast, the rump brownish-orange, the light bars on wings narrower and less distinct. Beneath duller white, the throat similar. Bill considerably longer. Others from the West Coast seem, however, to be intermediate in these characters, though *all* have a darker color and longer bill than Eastern specimens." (Baird.)

The smaller size of Gloger's specimens would correspond with the usual rule as to Southern birds, and the distinction between the species must rest on the darker hues and longer bill of the Western.

This prettily marked but inconspicuous little bird is found in winter throughout the higher mountains and the Coast Ranges, as far south as Santa Cruz, but I have not seen it south of San Francisco in summer, though doubtless living at that time in the higher Sierra Nevada. They frequent chiefly the coniferous trees, creeping up and down their trunks and branches, seeking insects in the crevices of the bark, and so nearly resemble it in general color that they are seen with difficulty when not in motion, and often their shrill wiry notes are heard when the bird itself is scarcely visible without careful search, the cry having the property of seeming quite distant from the bird itself. In March I have heard from them a faint but exceedingly sharp-toned song, somewhat resembling that of a wren.

The nest and eggs of our species have not yet been described; but the Eastern bird, which it so much resembles, is said by Audubon to seek out the deserted hole of a woodpecker, squirrel, or an accidental cavity in a tree, where it makes a nest of grass and lichens, warmly lined with feathers, and lays seven or more eggs of an ashy-white, marked with small reddish-yellow dots.

FAMILY TROGLODYTIDÆ, THE WRENS.

CHAR. No bristles along the gape; the loral feathers with bristly points; the frontal feathers generally not reaching to the nostrils. Nostrils varied; exposed or not covered by feathers, and generally overhung by a scale-like membrane. Bill usually without notch. Wings much rounded, about equal to tail, which is graduated. Primaries ten, the first generally about half the second. Basal joint of middle toe usually united to half the basal joint of inner, and the whole of that or more to the outer. Lateral toes about equal, or the outer a little the longer. Tarsi covered with transverse scales in front.

This family is quite characteristic of America, where it is universally distributed, few being found in the Old World. An unusually large proportion of the species of the United States extend from the Atlantic to the Pacific, although in the extremes of their distribution exhibiting slight permanent variation from each other.

SUB-FAMILY CAMPYLORHYNCHINÆ.

CHAR. Tail plane; nearly even, or slightly rounded, the first and second feathers slightly graduated; the feathers very broad, the longest with the width about one fifth the length. Size medium.

Of this group of wrens but two species are found within the limits of North America. While, however, many belong to Middle and South America, none have yet been detected in the West Indies. The species are much the largest of the *Troglodytidæ*, although having, to a considerable extent, the same habits.

GENUS **CAMPYLORHYNCHUS**, SPIX.

Campylorhynchus, SPIX, Av. Bras. 1824 (Agassiz).

GEN. CHAR. Bill as long as the head; not notched; compressed. Culmen and commissure both greatly decurved; gonys nearly straight. Tarsus longer than middle toe, distinctly scutellate; inner lateral toe a little the longer; hind toe reaching nearly to the middle of the middle claw; shorter than its digit. Wings about as long as tail; exposed

C. brunneicapillus.

portion of first quill about two thirds that of second, and rather more than half the longest, or fourth. Tail feathers very broad, plane; the longest nearly even, with the width about one fifth its length; the two lateral graduated; the outer about five sixths the middle. Plumage soft and loose. Color brown; streaks on the body. Wings and tail transversely barred.

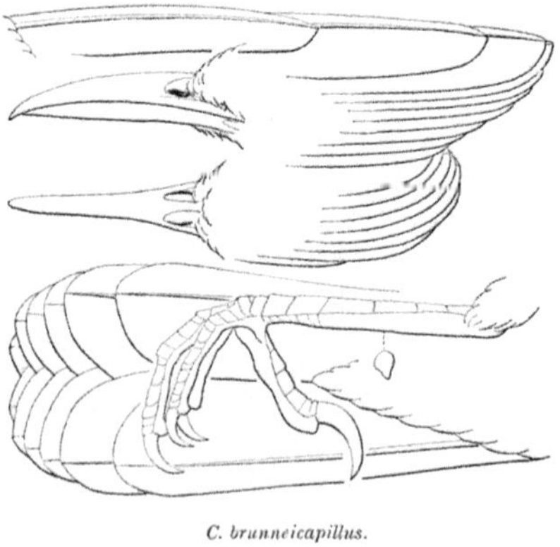

C. brunneicapillus.

Campylorhynchus brunneicapillus, Lafresnaye.

THE CALIFORNIAN CACTUS-WREN.

Picolaptes brunneicapillus, Lafresnaye, Guerin Mag. de Zool. 1835, 61; pl. xlvii. (California.) — Lawrence, Ann. N. Y. Lyc. V. May, 1851, 114. (Texas.) — Heermann, Jour. A. N. Sc. Ph. 2d ser. II. Jan. 1853, 263. — Cassin, Illust. I. 1854, 156; pl. xxv. — *Campylorhynchus brunneicapillus*, Gray, Genera, I. March, 1847, 159. — Baird, P. R. Rep. IX. Birds, 355. — Heermann, X. vi. 41.

Sp. Char. Bill as long as the head. Above, brown; darkest on the head, which is unspotted. Feathers on the back streaked centrally with white. Beneath, whitish, tinged with rusty on the belly; the feathers of the throat and upper parts, and under tail coverts, with large rounded black spots; those of the remaining under parts with smaller, more

linear ones. Chin and line over the eye white. Tail feathers black beneath, barred subterminally (the outer one throughout) with white. Length, 8.50; extent, 11.50; wing, 3.75; tail, 3.25. Iris blood red; bill horn-color, whitish below; feet whitish-brown.

Hab. Valleys of Rio Grande and Gila, to San Diego, California.

I found this interesting bird abundant, and already preparing to build nests, near San Diego, as early as February 26th, but think they had been there during the whole winter. They are, however, easily overlooked, as at most times they keep close in the dense thickets of cactus, where early in the morning the males mount to the top of some low tree, and utter a loud harsh trill, reminding one of the song of the marsh-wren, but much louder and more ringing. Several times after shooting one, I had to cut a path for several yards through the thicket to get the specimen. The males were then very quarrelsome, pursuing each other long distances, with shrill, angry notes of jealousy.

At San Diego, about May 1st, I found several of their nests built in the forks of the most thorny cacti, constructed of grass and fine twigs, with thick matted walls, and in the form of a purse nine and a half inches long and six wide, laid on its side. The eggs, from four to six, are of a pale salmon red or white, mostly very thickly speckled with ashy and red spots. Their size is 1 × 0.68 inch.

On the barren mountains, west of Colorado Valley I found a nest in May, 1861, built so openly that the young, then half-fledged, could be seen through the walls. This was probably adapted for the warmer climate.

In habits it resembles the wren, seeking for insects, and perhaps berries, among the cacti, chiefly on the ground, but I never saw it climbing, as represented in Cassin's plate.

It does not seem to range north of lat. 35° in California.

Campylorhynchus affinis, Xantus.

THE CAPE CACTUS-WREN.

Campylorhynchus affinis, Xantus, Pr. A. N. S. 1859, 298. — Baird, Ib. 1859, 303. Ib. Rev. Amer. Birds, 180.

Sp. Char. This species is very similar to *C. brunneicapillus*, but may be readily distinguished by having the spots in the lower parts more uniformly diffused, instead of being collected on the throat and jugulum. The spots are much larger on the under parts, and the posterior portion of the body lacks the cinnamon tinge. The tail feathers are black, and all barred transversely with white, instead of having them entirely black with single white band near the end (excepting the outer), as in *brunneicapillus*.

Hab. Cape St. Lucas.

This interesting analogue of the Upper California species is extremely abundant at Cape St. Lucas, and will doubtless be detected in time, in Southern California and Arizona. The habits of the two birds are very similar, both building an immense nest of dry grass, and laying a large salmon-colored egg.

GENUS **SALPINCTES**, CABANIS.

Salpinctes, CABANIS, Wiegmann's Archiv, 1847, I. 323.

GEN. CHAR. Bill as long as the head; all the outlines nearly straight to the tip, then decurved; nostrils oval. Feet weak; tarsi decidedly longer than the middle toe; outer lateral toe much longer, reaching to the base of the middle claw, and equal to the hinder.

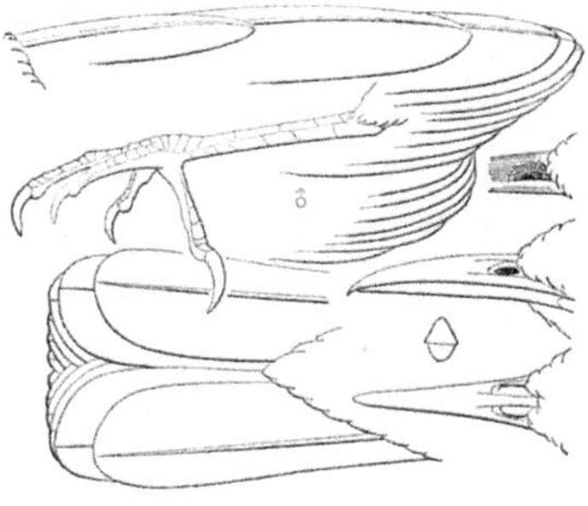

S. obsoletus.

Wings about one fifth longer than the tail; the exposed portion of the first primary about half that of the second, and two fifths of the fourth and fifth. Tail feathers very broad, plane, nearly even or slightly rounded; the lateral moderately graduated.

S. obsoletus.

Salpinctes obsoletus, Say.

THE ROCK WREN.

Troglodytes obsoletus, Say, in Long's Exped. II. 1823, 4. S. Fork of Platte. — Nuttall, Man. I. 1832, 435. — Aud., Orn. Biog. IV. 1838, 443; pl. 360. Ib. Birds Amer. II. 1841, 113; pl. 116. — Newberry, Zool. P. R. Rep. VI. iv. 1857, 80. — Heermann, P. R. Rep. X. vi. 41. — *Salpinctes obsoletus*, Cabanis, Wieg. Arch. I. 323. — Baird, P. R. Rep. IX. Birds, 357.

Sp. Char. Plumage very soft and lax. Bill about as long as the head. Upper parts brownish-gray, each feather with a central line, and (except on the head) transverse bars of dusky, and a small dull brownish-white spot at the end (seen also on the tips of the secondaries). Rump, sides of the body, and posterior part of belly and under tail coverts dull cinnamon, darker above. Rest of under parts dirty white; feathers of throat and breast with dusky central streaks. Lower tail coverts banded broadly with black. Inner

tail feathers like the back; the others with a broad black bar near the end; the tips cinnamon; the outer on each side alternately banded with this color and black. A dull white line above and behind the eye. Length, 6 inches; extent, 9.00; wing, 3.00; tail, 2.40. Iris brown; bill black, yellow, or white below; feet black.

Hab. High central plains through the Rocky Mountains to the Cascades of Columbia River and southward.

This is an abundant species throughout the dry, rocky, and barren districts of the State, especially southward, where it comes to the coast; but towards the north they inhabit farther towards the interior, avoiding the wooded region of the Coast Mountains, and even the warmer valleys, like that of Santa Clara, reappearing towards the Sacramento Valley, and north of this State, again retreating eastward of the Cascade Range. They are numerous in summer throughout all the plains on both sides of the Rocky Mountains, and probably do not migrate much to the south.

Their favorite resorts are the rocks and cañons, among which their loud, shrill chirp of alarm is frequently almost the only sign of life. They are always actively engaged in hunting for insects in the crevices and low herbage, sometimes in autumn in families of five or six.

A nest brought to me from a pile of wood on the bank of the Upper Missouri, in June, 1860, was composed of a loose flooring of sticks, lined with a great quantity of feathers of various birds, and contained nine eggs of a reddish color, thickly spotted with chocolate. I have found nests at San Diego in the cavities under tiled roofs, but always with young hatched, as early as May.

Their song begins to be heard at Fort Mojave in February, and continues through the spring. It is much more like that of the Sickle-bills than that of other wrens, being sweet and varied, but not very loud.

I did not observe them in the Colorado Valley after May 15th, and presume that most of them retired to the cooler mountains. Neither have I seen them towards the summits of the Sierra Nevada, but on their eastern side they doubtless range to a high elevation, having been found at Klamath Lake, Oregon, by Dr. Newberry.

Genus **CATHERPES**, Baird.

Catherpes, Baird, P. R. Rep. IX. Birds, 1858, 356.

Gen. Char. Bill longer than the head, slender, all the outlines nearly straight to the tip, then gently decurved, gonys least so; nostrils linear; tarsus short, about equal to the middle toe, which reaches to the middle of the middle claw. Outer toe considerably

C. Mexicanus.

longer than the inner, reaching beyond the base of the middle claw. Wings a little longer than the tail; the exposed portion of the first primary about half that of the fourth

C. Mexicanus.

and fifth. Tail feathers very broad and perfectly plane, tail nearly even, the two lateral graduated; the outer about eleven twelfths of the middle.

Catherpes Mexicanus, Swainson.

THE WHITE-THROATED WREN.

Thryothorus Mexicanus, Swainson, Zool. Illust. 2d ser. I. 1829; pl. xi. (Real del Monte, Mexico.) — *Troglodytes Mexicanus*, Gray, Genera, I. 1847, 159. — Heermann, J. A. N. Sc. 2d ser. II. 1853, 263. — Cassin, Illust. I. vi. 1854, 173; pl. xxx.
Certhia albifrons, Giraud, 16 Sp. Texan Birds, 1844; pl. viii.
Catherpes Mexicanus, Baird, P. R. Rep. IX. Birds, 356. — Kennerly, X. iv. 26.

Sp. Char. Bill considerably longer than the head; claws large. Head and neck above dark ashy-brown, passing gradually into light rusty-brown on the rump; the sides

of the body, belly, and under tail coverts similar, all these regions marked with small rounded white and dusky spots, the latter in the form of waved bars on the feathers of the back; an obscure white line over the eye. Chin, throat, and upper part of the breast pure white. Tail feathers rusty-red on both sides, with six or eight narrow transverse bars of black. Eye brown; bill and feet slaty. Length, 6.00; extent, 7.65; wing, 2.50; tail, 2.50.

Hab. Valleys of the Rio Grande, Colorado, and Gila, to San Joaquin Valley, California.

I saw but once, at Fort Mojave, in April, a bird which I thought might be of this species, but could not shoot it on account of the dense thicket it frequented. Dr. Kennerly found them along Williams's Fork, among high rocks, darting from one to another, and creeping about their crevices very rapidly, so that he could scarcely kill one, keeping up at the time a strange cry.

Dr. Heermann found them along the Calaveras and Cosumnes Rivers in fall and spring, very active, with a loud, sprightly song, heard at a considerable distance, and often repeated. They were searching for insects among the large boulders along the river. He remarks that their habits and resorts are the same as those of the *Salpinctes obsoletus.*

I have not met with this bird myself in the more northern country, or near the coast, but obtained a specimen collected by Mr. Lorquin, somewhere in the San Joaquin Valley.

Sub-Family TROGLODYTINÆ.

Char. Tail feathers rather narrow; the middle ones less than one sixth as wide as long. Tail more or less vaulted or concave below; usually considerably graduated. Tarsus longer than the middle toe, which exceeds the hinder ones; the lateral toes generally equal, and reaching the base of the middle claw. Hind toe much longer than the lateral. Size diminutive.

T. Bewickii.

Genus THRYOTHORUS, Vieillot.

Thriothorus, Vieillot, Analyse, 1816. (Not of 1819.)
Thryothorus, Vieillot, Nouv. Dict. XXXIV. 1819, 55.
This name was intended by Vieillot to apply to the *Certhia palustris*, Wilson; *Thryothorus arundineus*, Vieillot.

Gen. Char. Bill about as long as the head; nearly straight to near the tip, which is abruptly decurved with an obsolete notch. Gonys nearly straight. Hind toe nearly equal to the middle; the lateral toes equal, reaching to the base of the middle claw. Tarsus

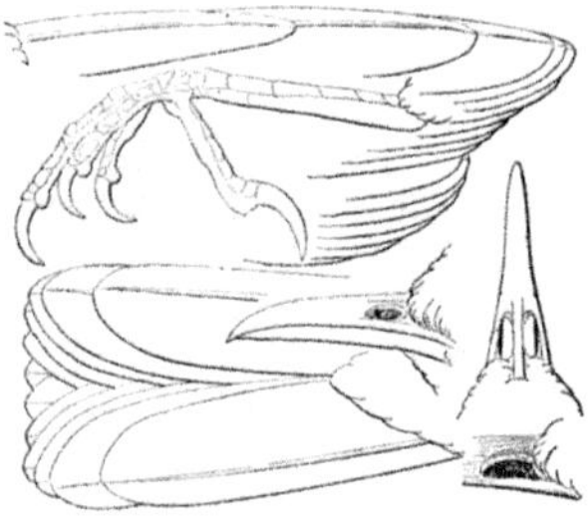

T. Ludovicianus.

longer than the middle toe. Wings about equal to the tail, which is arched, and nearly even; the first or second lateral feathers moderately graduated; the feathers narrow; the width of the longest about one tenth its length.

T. Bewickii.

The type of the genus is *T. Ludovicianus*. The California species belong to a sub-genus *Thryomanes*, the characters of which are expressed in the above diagram. There are several other species of the genus peculiar to Mexico and Central America, and another closely allied form is *Thryophilus*, differing in having the nostril entirely exposed, instead of having a thickened scale-like overhanging process.

Thryothorus spilurus, Vigors.

THE WESTERN MOCKING-WREN.

Troglodytes spilurus, Vigors, Zool. Beechey's Voyage, 18; pl. iv. f. 1, 1839. (California.)
Troglodytes leucogastra, Gould, Pr. Zool. Soc. 1836, 39. (Tamaulipas, Mexico.)
Thryothorus Bewickii, var. *spilurus*, Baird, P. R. Rep. IX. Birds, 363, 1858. — *Troglodytes Bewickii*, Audubon, and other authors in reference to Pacific Coast bird. — Newberry, P. R. Rep. VI. iv. 80. — Heermann, X. vi. 41. — Cooper and Suckley, XII. iii. Zool. of W. T. 189.

Sp. Char. Bill shorter than the head. Tail longer than the wings; much graduated. Upper parts grayish-olive or brown; beneath, grayish-white. A white streak over the eye, the feathers edged above with brown. Exposed surface of the wings and the innermost tail feathers closely barred with dusky; the remaining tail feathers mostly black, barred

or blotched with white at the tips, and on the whole outer web of the exterior feather, and on the under tail coverts. Length, 5.50; extent, 7.00; wing, 2.25; tail, 2.50. Iris brown; bill brown, white, or yellow at base below; feet brown.

Var. *leucogastra*, colors paler, above and below; bill and tail longer.

Hab. Pacific Coast. Var. *leucogastra*, east of the Sierra Nevada, and south into Mexico.

This species abounds throughout the wooded parts of this State and northward, frequenting the densest forests as well as the more open groves. During the winter a few lived in the vicinity of Fort Mojave, but left, probably for the mountains, in April. They winter also, throughout the mild regions towards the coast as far north as Puget's Sound, and do not leave their homes even when there has been considerable snow.

Though resembling closely the so-called mocking wrens of the Atlantic side, I do not think that they really imitate other birds, though having a great variety of songs, some of which resemble those of other birds, and are well calculated to deceive one unaccustomed to them. I have often searched in vain for some new bird, which I thought I heard singing; and after difficult scrambling through thickets in search of the author of the sound, which retreated before me, at last caught a glimpse of the almost invisible performer, to find that it was only this mocker.

Near San Diego, April 21, 1862, I discovered a nest of this species, built in a low bush only three feet from the ground. It was quite open above, formed of twigs, grass, etc., and contained five eggs just ready to hatch; white, with scattered brown specks near the large end. To be sure of the species, I shot and preserved the female.

Genus **TROGLODYTES**, Vieillot.

Troglodytes, Vieillot, Ois. Am. Sept. II. 1807, 52. (Type, *T. ædon.*)

Gen. Char. Wings longer than the tail or nearly equal. Tail rounded; the lateral feathers graduated. Hind claw shorter than the rest of the toe. Back brown, obsoletely waved with dusky.

T. ædon.

Sub-Genus **TROGLODYTES.**

Wings about equal to the tail. Toes reaching to the tip of the tail. Bill nearly as long as the head, compressed, decurved

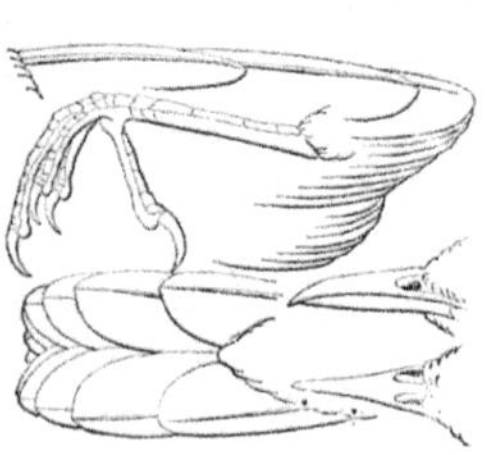

T. ædon.

Troglodytes Parkmanni, Audubon.

PARKMANN'S HOUSE-WREN.

Troglodytes Parkmanni, Audubon, Orn. Biog. V. 1839, 310 (not figured). Ib. Birds Amer. II. 1841, 133; pl. 122. — Nuttall, Man. I. 2d ed. 483. — Baird, P. R. Rep. Birds, IX. 367. — Cooper and Suckley, XII. iii. Zool. of W. T. 191.

Sp. Char. Tail and wings about equal. Bill shorter than the head. Above, dark brown, darker towards the head, brighter on the rump. The feathers everywhere, except on the head and neck, barred with dusky. All the tail feathers barred from the base; the contrast more vivid on the exterior ones. Beneath, grayish-white, tinged with light

brownish across the breast. Under tail coverts whitish, with dusky bars. An indistinct line over the eye, eyelids, and loral region, whitish. Cheeks brown, streaked with whitish. Length, 5.00; extent, 6.50; wing, 2.12; tail, 2.12. Iris brown; bill dark brown, bluish flesh-color at base below; feet whitish or horn-brown.

Hab. Western America, on the plains, and from the Missouri to the Pacific.

This, the exact counterpart in habits of the Eastern house-wren, is common in summer as far north as Puget's Sound; but in winter I have seen but few, and only in the Colorado Valley. There, true to their name, they left the bushes, where they passed the day, to roost at night under the eaves of the garrison buildings. In April they left that valley for the mountains, but some probably also winter towards the coast, as I have heard them at San Francisco as early as the 16th of March, and at Puget's Sound by April 20th.

Throughout the whole coast slope, and probably most of the interior, they build their nests, beginning near San Diego in April. There I found several nests in hollows of trees at various heights, from five to forty feet up, all composed of a floor and barricade of long dry twigs, grass, and bark loosely placed, but interwoven so as to leave only just room for the bird to squeeze in over them. On this is laid a large quantity of feathers, of all kinds of birds, and frequently snake-skins; and the eggs, varying from five to seven (and probably nine), are reddish-white, densely speckled with dark cinnamon dots. Size 0.88 × 0.50 inch. They no doubt raise two broods annually, like the Eastern bird.

I have found a nest built in a horse's skull stuck up on a pole; and they will accept any kind of accommodation provided for them, like the bluebird, being even contented with an old hat with a round hole in the crown, nailed against a wall. The allied Eastern bird (*T. ædon*) will drive the bluebird and martin away from their dwellings, and keep possession, though so much the smaller and weaker. Its pugnacity is so great that no more than one pair can live in the same tree, or about the same house, though several nests may be found, built by the male in leisure hours as an amusement, or to provide against accidents, but rarely if ever to be occupied.

The song of the house-wren is nearly alike in both the Eastern and Western species, but that of the latter is rather less strong and extended. It is, however, lively and pleasing, though not much varied.

They feed on insects, especially spiders, and are attracted about buildings in their search, gathering a wonderful number of them during the day, especially when feeding their young. Like other wrens, they are continually moving about in dark corners, scolding with a harsh chirrup at sight of an enemy, and especially hating cats, which they will follow at a safe distance, with every sign of anger, until far from their nest.

Though few have yet thought of furnishing a dwelling for this lively little songster, it is well worthy of such encouragement, both for its cheerful song and usefulness in destroying insects. Let every one at least take the trouble to stick up some of the skulls so common in the country for the birds to build in.

This species winters in considerable numbers about Santa Cruz, California, where I found them in January, 1866. They have not yet become as familiar about the towns as the *T. ædon* in the East, but will probably be when the houses are surrounded by trees and shrubbery.

Sub-Genus **ANORTHURA.**

"*Anorthura*, Rennie, 1831," in Mont. Ornith. Dict. (Baird.)

Wings much longer than the very short tail. Bill shorter than the head, slender, nearly straight. End of tarsus reaching to the tip of the tail. (Baird.)

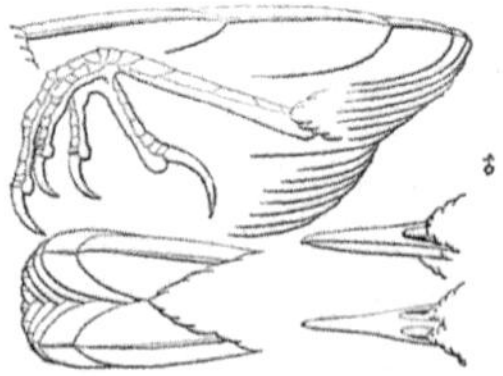

T. hyemalis.

Troglodytes hyemalis, Vieillot.

THE WINTER WREN.

Sylvia troglodytes, Wilson, Am. Orn. I. 1808, 139; pl. viii. f. 6.
Troglodytes hyemalis, Vieillot, Nouv. Dict. XXXIV. 1819, 514. — Swainson, F. B. Am. II. 1831, 318. — Audubon, Orn. Biog. IV. 1838, 430; pl. 360. Ib. Birds Amer. II. 1841, 128; pl. 121. — Nuttall, Man. 2d ed. 481.
Troglodytes Americanus, Heermann, P. R. Rep. X. vi. 41 (not of Audubon ?).
Troglodytes (Anorthura) hyemalis, Baird, P. R. Rep. IX. Birds, 369. — Cooper and Suckley, XII. iii. Zool. of W. T. 191.

Sp. Char. Bill very straight, slender, and conical; shorter than the head. Tail considerably shorter than the wings, which reach to its middle. Upper parts reddish-brown, becoming brighter to the rump and tail; everywhere, except on the head and upper part of the back, with transverse bars of dusky and of lighter. Scapulars and wing coverts with spots of white. Beneath, pale reddish-brown, barred on the posterior half of the

body with dusky and whitish, and spotted with white more anteriorly; outer web of primaries similarly spotted with pale brownish-white. An indistinct pale line over the eye. Length, about, 4.00; extent, 5.75; wing, 1.80; tail, 1.25.

Hab. North America; California, in the mountains, south to Fort Tejon.

This little Northerner spends the summer among the dense evergreen forests of the higher mountains north of lat. 38°, deserting at that season even the lowlands along the Columbia River for more elevated regions.

Audubon states that he found several nests in the mountains of New York and Pennsylvania, which were composed of moss, built into a half-globular form against the lower part of a tree, with a hole in the side, looking so much like the mossy knobs, common in such places, as easily to escape notice. The eggs were six, pale rosy-white, spotted with dark red. The nest was lined with rabbit's fur and feathers of the grouse.

Our bird does not come down to the coast, near San Francisco, even in winter, but may no doubt be found then north of the Bay. Its haunts are the dense woods, piles of logs and brush, fences, etc., where it creeps about like a mouse, rarely flying or mounting the trees. Its song is a rather long and sweet, but low warble, as cheerfully given in gloomy weather as in the sunshine, and in the forest often the only cheering sign of life.*

Genus CISTOTHORUS, Cabanis.

Cistothorus, Cabanis, Mus. Hein. 1850 - 51, 77. (Type, *Troglodytes stellaris.*)
Telmatodytes, Cabanis, Mus. Hein. 1850 - 51, 78. (Type, *Certhia palustris.*)
Thriothorus, Vieillot, Analyse, 1816, according to G. R. Gray. This name would apply better to the habits of *Thryothorus.*

Gen. Char. Bill about as long as the head or much shorter, much compressed, not notched, gently decurved from the middle; the gonys slightly concave or straight. Toes

C. palustris.

reaching to the end of the tail. Tarsus longer than the middle toe. Hind toe longer than the lateral, shorter than the middle. Lateral toes about equal. Hind toe longer than or

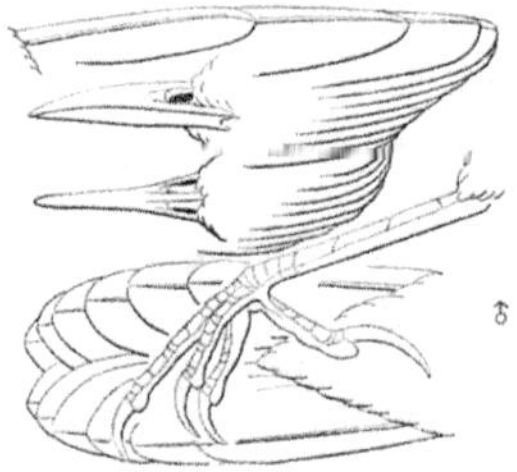

C. palustris.

* *T. Americanus*, Audubon, mentioned by Nuttall and Heermann as an inhabitant of this coast, was probably the present species, since no specimens of the true *Americanus* have been collected. I saw a few of the species about Santa Cruz, lat. 37°, on the coast of California, but not further south.

equal to its digit. Wings rather longer than the tail, all the feathers of which are much graduated; the lateral only two thirds the middle, the feathers narrow. Back black, conspicuously streaked with white.

Cistothorus palustris, Wilson.

THE LONG-BILLED MARSH-WREN.

Certhia palustris, Wilson, Am. Orn. II. 1810, 58; pl. xii. f. 4. — *Troglodytes palustris*, Bonaparte, Obs. Wilson, 1824, No. 66. — Swainson, F. Bor. Am. II. 1832, 319. — Audubon, Orn. Biog. I. 1831, 500; V. 1839, 467; pl. 100. Ib. Birds Amer. II. 1841, 135; pl. 123. — Newberry, Zool. Cal. & Or. Route; P. R. R. Rep. VI. iv. 1857, 80. — Heermann, X. vi. 54. — *Thryothorus palustris*, Nuttall, Man. I. 1832, 439. — Bon. List, 1838. — *Cistothorus* (*Telmatodytes*) *palustris*, Baird, P. R. Rep. IX. Birds, 364.

Sp. Char. Bill about as long as head. Tail and wing nearly equal. Upper parts of a dull reddish-brown, except on the crown, interscapular region, outer surface of tertials, and tail feathers, which are almost black; the first with a median patch like the ground color; the second with short streaks of white, extending round on the sides of the neck;

the third indented with brown; the fourth barred with whitish, decreasing in amount from the outer feather, which is marked from the base, to the fifth, where it is confined to the tips; the two middle feathers above like the back, and barred throughout with dusky. Beneath, rather pure white, the sides and under tail coverts of a lighter shade of brown than the back; a white streak over the eye. Length, 5.25; extent, 6.75; wing, 2.25; tail, 2.00. Iris and feet brown; bill brown, paler below.

Hab. North America from the Atlantic to the Pacific; north to Greenland.

This little bird migrates in winter throughout the State, but I have not observed their nests in the southern portions, and suppose they retire in summer toward the north or the mountain-tops, as I have seen them about Lake Tahoe, over six thousand feet above the sea. They winter near the coast as far north as the Columbia, and are to be found wherever there is a marsh overgrown with the "tulé," (*Scirpus palustris*). Among these rushes they live constantly, running through their dense coverts with great agility, clinging to them sideways where none are prostrate, and rarely flying more than a few yards at a time. Their food consists entirely of insects, which they capture at rest.

In the spring, and less often during the rest of the year, the males may be seen flying up a few feet above the marsh, singing a shrill and quaint, but rather musical ditty, with a sort of gurgling sound, as if coming through the water below them. There is no variety in this song, with time or locality.

The nest is constructed of reeds and sedges, interwoven in an oval form, about six inches long and four wide, having an opening in the side, and a small internal cavity. The rushes are wet when used, and mud is mingled in the walls to give more tenacity. It is lined with fine grass, and sometimes feathers, the whole being suspended among the rushes by strong bands passing round them, or it sometimes rests on those that are bent down, and is also tied to others still standing. The eggs are six or eight, of a dark fawn, almost mahogany color. (Nuttall.)

Few nests are found inhabited near together, and it is supposed that the males spend much of their spare time in building nests that are never used, as does the house-wren. Audubon says that they build a new nest for the second brood annually. Sometimes a single marsh will be seen to contain hundreds of these nests within a short distance of each other.

Family MOTACILLIDÆ.

Char. Bill slender, shorter than the head, notched at the tip; rictus without bristles. Basal joint of middle toe entirely free externally. Tarsi distinctly scutellate, longer than the middle, but nearly equal to the hind toe, which is very long, exceeding all the others; the claw slightly curved. Wing very long, pointed; first quill almost the longest; the tertials considerably longer than the secondaries. Tail emarginate.

This family is represented in North America by three genera: two, *Anthus* and *Neocorys*, belonging to the Western United States. The third genus, *Motacilla*, not unfrequently occurs in the Northeastern States, or near their borders (*Motacilla alba*), and is even recorded as belonging to the country about Behring's Straits, where it is probable that the explorers of the Russian Telegraph Company will find it. It has a very great development in Europe and Asia. South America has several species of *Anthus*, all, however, of a group quite different from that to which *A. Ludovicianus* belongs, and characterized by shorter and more rounded wings.

Genus **ANTHUS**, Bechstein.

Anthus, Bechstein, Gemein. Naturg. Deutschl. 1802 (Agassiz). (Type, *Alauda spinoletta.*)

Gen. Char. Bill slender, much attenuated, and distinctly notched. A few short bristles at the base. Culmen concave at the base. Tarsi quite distinctly scutellate;

A. Ludovicianus.

longer than the middle toe; inner lateral toe the longer. Hind toe rather shorter than the tarsus, but longer than the middle toe, owing to the long, attenuated and moderately curved hind claw, which is considerably more than half the total length of the toe. Tail

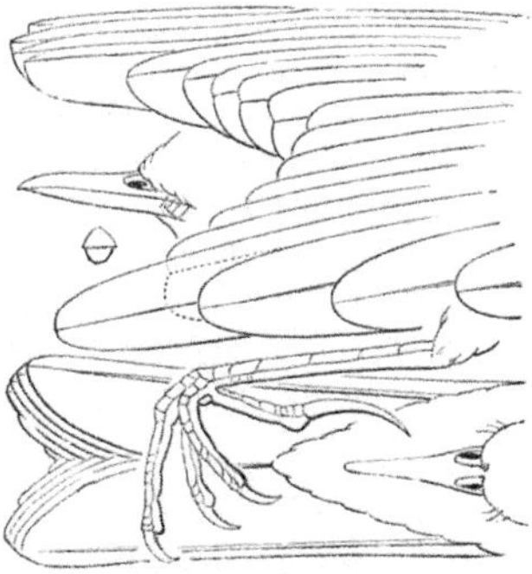

A. Ludovicianus.

rather long, emarginate. Wing very long, considerably longer than the lengthened tail, reaching to its middle. The first primary nearly equal to the longest. The tertials almost as long as the primaries.

Anthus Ludovicianus, Gmelin.

THE TIT-LARK.

Alauda Ludoviciana, Gmelin, Syst. Nat. I. 1788, 793. — *Anthus Ludovicianus*, Lichtenstein, Verz. 1823, 37, No. 421. — Audubon, Birds Amer. III. 1841, 40; pl. 150. — Baird, P. R. Rep. IX. Birds, 232. — Cooper and Suckley, XII. iii. 176.
Alauda rufa, Wilson, Am. Orn. V. 1812, 89; pl. lxxxix.
Anthus spinoletta, Bonaparte, Synopsis, 1828, 90 (not of Linnæus). — Audubon, Orn. Biog. I. 1832, 408; V. 1839, 449; pl. 80. — Nuttall, Man. I. 1832, 450.
Anthus aquaticus, Audubon. Name on pl. x. folio ed. and
Anthus pipiens, Audubon, Orn. Biog. I. 1832, 408; V. 1839, 449; pl. 80. (Young ?)

Sp. Char. (*Female*, in spring.) Above, olive-brown, each feather slightly darker towards the central portion; beneath, pale dull buff, or yellowish-brown, with a maxillary series of dark brown spots and streaks across the breast and along the sides. Ring round the eye, and superciliary stripe yellowish. Central tail feathers like the back,

others dark blackish-brown; the external one white, except at the base within; a white spot at the end of the second. Primaries edged with whitish, other quills with pale brownish. Length, 6.50; wing, 3.45; tail, 2.95.

Hab. North America generally. Greenland (Reinhardt). Accidental in Europe.

This little inconspicuous bird is common throughout the State, particularly in winter, frequenting the barest plains, especially near the water, and often coming into the city, where they perch on the roofs, and occasionally descend into the streets. They seem to live by picking up insects and seeds in places so barren that no other bird will take the trouble to examine them. Their note while here is merely a faint chirp or twitter when disturbed and about to fly; but, according to Audubon, the male sometimes rises on wing to the height of eight or ten yards, uttering a few clear, mellow notes, and then suddenly settles down near its nest, or on some projecting rock. The nest he found in Labrador, made on moss-covered rocks or cliffs near the sea, somewhat sunk in the ground, and formed of fine grass, without any hair lining. The eggs were six, reddish-brown, with numerous dots and lines of a deeper color. It is very likely that they build on the high mountains within this State, or perhaps near the sea-shore, but the nest has never yet been found on this coast. Great numbers often associate together in scattered flocks during winter, and their walk reminds us of the water-thrushes.

GENUS **NEOCORYS**, SCLATER.

Neocorys, SCLATER, P. Z. S. 1857, 5. — BAIRD, Birds N. Amer. 1858, 233.

GEN. CHAR. Bill half as long as the head; the culmen concave at the base, slightly decurved at the tip. Rictus without bristles. Legs stout; tarsi distinctly scutellate, longer than the middle toe; hind toe very long, equal to the tarsus, much longer than the mid-

N. Spraguei.

dle toe, the claw but slightly curved, and about half the total length. Inner lateral toe rather longer than outer. Wings much longer than tail; first quill longest. Tertials considerably longer than the secondaries. Tail rather short, emarginate.

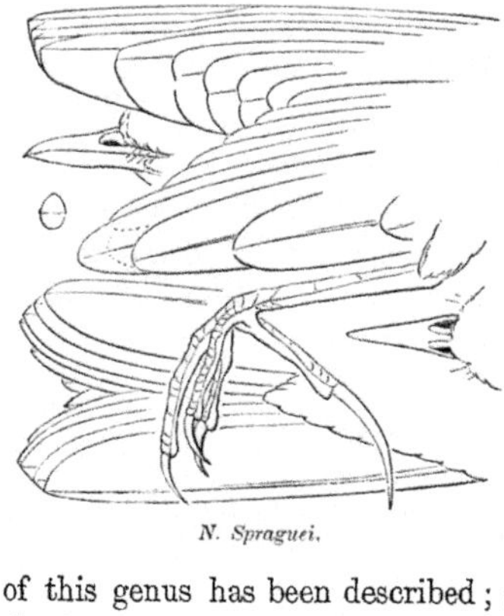

N. Spraguei.

Only one species of this genus has been described; it belongs to the interior plains of North America.

Neocorys Spraguei, Audubon.

SPRAGUE'S LARK.

Alauda Spraguei, Audubon, Orn. Biog. VII. 1843, 335; pl. 486. — *Neocorys Spraguei,* Sclater, P. Z. S. 1857, 5. — Baird, Birds N. Amer. 1858, 234. Ib. Rev. 156.

Sp. Char. Above, brown; all the feathers edged with paler, especially on the neck, where there is a brownish-yellow tinge. Beneath, dull white, with a collar of sharply defined narrow brown streaks across the fore part and along the sides of the breast. Lores and a superciliary line whitish. Tail feathers dark brown, the outer white; the

next white, with the inner margin brown. Outer primary edged with white, and two dull whitish bands across the wings. Bill and feet yellow, the former brown above. Length, 5.75; wing, 3.35; tail, 2.50.

Hab. Missouri and Saskatchewan Plains.

This species has as yet only been found in the plains of the Upper Missouri and the Saskatchewan, though it will probably be met with in the Great Basin. It is said to be abundant in its region, though very few specimens have ever been collected. It is a true skylark, the only one we have, rising high in the air, and singing when out of sight, then descending and concealing itself in the grass. The note is said to be quite as sweet as that of the European skylark. (Baird.)

Family DENDRŒCIDÆ, Warblers.

Sylvicolidæ of Baird and authors.

Char. Primaries nine, the first quill nearly as long as the second or third. Tarsi distinctly scutellate the whole length anteriorly. Bill conical, slender, or depressed, usually half the length of the head; more or less bristled or notched. Nostrils oval or rounded. Lateral toes nearly or quite equal, and shorter than the middle; the basal joint of the middle free nearly to its base, externally; united for about half, internally.

The name *Sylvicola* belonging to a genus of land shells, it is manifestly improper to give the family of birds a name derived from that genus.

This family is pre-eminently American, or New World, and embraces many species. Although a large proportion belong to North America, but few, comparatively, are found in California and the adjacent States.

GENUS **HELMINTHOPHAGA**, CABANIS.

Helminthophaga, CABANIS, Mus. Hein. 1850 - 51, 20. (Type, *Sylvia ruficapilla*.)

GEN. CHAR. Bill elongated, conical, very acute; the outlines very nearly straight, sometimes slightly decurved; no trace of notch at the tip. Wings long and pointed; the

H. celata.

first quill nearly or quite the longest. Tail nearly even or slightly emarginate; short and rather slender. Tarsi longer than the middle toe.

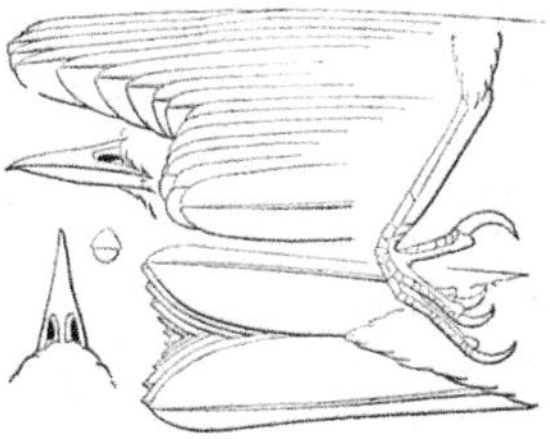

H. ruficapilla.

Helminthophaga ruficapilla, Wilson.

THE NASHVILLE WARBLER.

Sylvia ruficapilla, Wilson, Am. Orn. III. 1811, 120; pl. xxvii. f. 3. — Audubon, Orn. Biog. I. 1832, 450; pl. 89.
Sylvia rubricapilla, Wilson, Am. Orn. VI. 1812, 15. (General Index.) — *Sylvia (Dacnis) rubricapilla*, Nuttall, Man. I. 1832, 412 (*Vermivora*, 2d ed. 472). — *Helinaia rubricapilla*, Audubon, Syn. 1839, 70. Ib. Birds Amer. II. 1841, 103; pl. 113.
Helminthophaga ruficapilla, Baird, P. R. Rep. IX. Birds, 256. Ib. Rev. Amer. Birds, 175.

Sp. Char. Head and neck above and on sides ash-gray, the crown with a patch of concealed dark brownish-orange hidden by ashy tips to the feathers. Upper parts olive-green, brightest on the rump. Under parts generally, with the edge of the wing, deep yellow; the anal region paler; the sides tinged with olive. A broad yellowish white ring

round the eye; the lores yellowish; no superciliary stripe. The inner edges of the tail feathers margined with dull white. *Female* similar, but duller; the under parts paler; but little trace of the red of the crown. Length, 4.65; wing, 2.42; tail, 2.05.

Hab. Eastern North America to the Missouri; Greenland; California.

Though well known as an Eastern species, this was not found in California until 1858, when Mr. J. Xantus found one at Fort Tejon. Afterwards, in the summer of 1863 Mr. F. Gruber collected many specimens in the Sierra Nevada, near Lake Tahoe.

Wilson discovered this species near Nashville, Tennessee, and was attracted by the singular note it made, like the breaking of small dry twigs, or striking two pebbles together six or seven times, loud enough to be heard at forty yards' distance. But little is yet known of their habits, except that, as with others of the genus, the nest is made on or near the ground in the grass, and never on trees, like most of the *Sylvicolidæ*. As a summer bird, it appears to be most abundant in New England and New Brunswick, where its nest and eggs are taken every year by the indefatigable naturalists of that portion of North America. According to Mr. George A. Boardman, of Calais, Maine, the nest is built in the grass, or sunk in a hollow of the ground. The eggs are white, finely sprinkled with reddish, and are the smallest of all laid by our warblers, except, perhaps, those of *Myiodioctes pusillus*.

Helminthophaga celata, Say.

THE ORANGE-CROWNED WARBLER.

Sylvia celata, Say, Long's Exped. R. Mts. I. 1823, 169. — Bonaparte, Am. Orn. I. 1825, 45; pl. v. f. 2. — *Sylvia (Dacnis) celata*, Nuttall, Man. I. 1832, 413, *Vermivora*, 2d ed. 273. — Audubon, Orn. Biog. II. 1834, 449; pl. 178. — *Helinaia celata*, Audubon, Syn. 1839, 69. Ib. Birds Amer. II. 1841, 100; pl. 112. — *Helminthophaga celata*, Baird, P. R. Rep. IX. Birds, 257. — Heermann, X. vi. 40. — Cooper and Suckley, XII. iii. Zool. of W. T. 178.

Sp. Char. Above, olive-green, rather brighter on the rump. Beneath, entirely greenish-yellow, except a little whitish about the anus; the sides tinged with olivaceous. A concealed patch of pale brownish-orange on the crown, hidden by the olivaceous tips to the feathers. Eyelids and an obscure superciliary line yellowish, a dusky obscure streak

through the eye. No white spots on wings or tail of *female*, with little or none of the orange on the crown. Length, 4.75; extent, 7; wing, 2 25; tail, 2.00. Iris, feet, and bill brown; lower mandible yellow. In fall, the bill yellow, and feet and head blue.

Hab. Mississippi River to the Pacific; south to Northern Mexico.

An abundant and constant resident everywhere near the coast south of San Francisco, and extending in summer to the summits of the Sierra Nevada, and north to Puget's Sound.

They frequent the low shrubbery, even where it forms the only vegetation, covering entire mountains, and where there is no water for many miles. There this plain little bird is entirely at home, busily seeking for insects from morning till night, and usually showing its presence only by a harsh chirp when alarmed. About February 1st, at San Diego, and a month later near San Francisco, the males begin to sing their simple trill, which is low, but rather musical, and audible for a long distance in the silent regions they inhabit. They also frequent high trees in open places, but not the evergreen forests; and I have seen one on the barren, waterless island of Santa Barbara.

I have searched frequently for their nests, but never succeeded in finding one. Audubon speaks of finding them in New Brunswick in fir-trees, built of lichens and grass, and lined with fine fibres and feathers. The eggs were four, pale green, with small black spots.

Helminthophaga Luciæ, COOPER.

LUCY'S WARBLER.

Helminthophaga Luciæ, COOPER, Proc. Cal. Acad. Nat. Sc. II. 1862, 120.

SP. CHAR. Above, light ash-gray, with partially concealed spot on vertex, and the upper tail coverts chestnut brown. Quills and tail tinged with brown, edges of primaries and coverts paler; beneath white, tinged with yellowish, this color extending to lores and around eyes, forming a faint line above and behind them. Quills plumbeous beneath, also the tail feathers, the outer of which are edged with white inter-

nally, and with a white patch on the inner web near the end. *Female* differs only in smaller size and duller colors. Iris brown; bill black above, bluish below; feet pale lead-color.

Length, 4.25; extent of wings, 6.50; wing, 2.25; tail, 1.50; tarsus, 0.65; middle toe and claw, 0.95; bill along ridge, 0.35; gape, 0.45; its height at base, 0.15; width, a little more. First quill shorter than the three next, and tail shorter than in other species, proportionally. (Cooper.)

Hab. Colorado Valley, California, to lat. 35°.*

This beautiful little warbler arrived from the south in the vicinity of Fort Mojave on March 25th, when my attention was first struck by its peculiar notes, resembling those of some *Dendrœcas*, but fainter. After considerable watching and scrambling through the dense mesquite thickets, I succeeded in shooting a specimen, and at once saw that it was a new species. Afterwards they became quite numerous, frequenting the tops of the mesquite trees in pursuit of insects, and constantly uttering their short but pleasing song. After the males had been about for ten days, I obtained the first female, for which I had been on the watch daily, and think it probable that they are some time later in their migrations, as is the case with many other small songsters.

Up to the 25th of May, when I left the valley, I watched daily to discover their nests, and therefore shot fewer specimens than I otherwise

* Since obtained by Mr. Holder, about lat. 34°, March 10, 1863, and by Dr. Coues, at Fort Whipple.

would have done, but did not succeed in discovering them building. I obtained five males and one female, all in good plumage.

Helminthophaga Virginiæ, Baird.

VIRGINIA'S WARBLER.

Helminthophaga Virginiæ, Baird, Birds N. Amer. 1860, xi.; pl. 79, f. 1. Ib. Rev. Amer. Birds, 177. — Coues, Pr. Ac. N. Sc. 1866, 70.

Sp. Char. Similar to *H. ruficapilla*. Top and sides of head, back, and wings, light ashy plumbeous, with an almost imperceptible wash of olivaceous-green; quills and tail feathers brown, edged with pure ashy plumbeous, the latter indistinctly and narrowly margined with whitish, internally, and at the end. Rump, with upper and lower tail coverts bright yellow, in vivid contrast with the rest of the body. Crown with a concealed patch

of orange-brown. Rest of under parts brownish-white, with indications of yellow from chin to breast, perhaps entirely yellow there when mature. Inside of wings and axillars whitish. A white ring round the eye. Bill and legs dusky. Length, 5.00; extent, 7.25; wing, 2.50; tail, 2.20; tarsus, 0.67.

Hab. Prescott, Arizona, and Fort Burgwyn, New Mexico.

Of this very rare species but two specimens have been observed, and nothing is known of its habits, which, however, in all probability resemble those of *H. celata, ruficapilla*, etc., in nesting on or near the ground, feeding on minute insects, etc. The plumage described above is not quite mature, and the colors during the breeding season are doubtless considerably brighter and better defined.

The discovery within a few years in Arizona and New Mexico of new warblers, such as *Helminthophaga Virginiæ, Helminthophaga Luciæ*, and *Dendroica Graciæ*, shows that the ornithological fauna of the West is not yet exhausted, and that as its hitherto unexplored regions, especially those towards the southern boundaries of the United States, are investigated, additional varieties will be brought to light. These will in all probability consist of insectivorous oscines and flycatchers, rather than of conirostral birds.

GENUS DENDRŒCA, GRAY.

Sylvicola, GRAY, Genera Birds, 2d ed. 1841, 32. (Not of Humphreys nor Swainson.)
Dendroica, GRAY, Genera Birds, Appendix, 1842, 8.

GEN. CHAR. Bill conical, attenuated, depressed at the base, (where it is, however, scarcely broader than high,) compressed from the middle. Culmen straight for the basal half, then rather rapidly curving, the lower edge of upper mandible also concave. Gonys slightly convex and ascending. A distinct notch near the end of the bill. Bristles, though short, generally quite distinct at the base of the bill. Tarsi long; decidedly longer than

D. Audubonii.

middle toe, which is longer than the hinder one; the claws rather small and much curved; the hind claw nearly as long as its digit. The wings long and pointed; the second quill usually very little longer than the first. The tail slightly rounded and emarginate.

Colors. Tail nearly always with a white spot; its ground color never clear olive-green

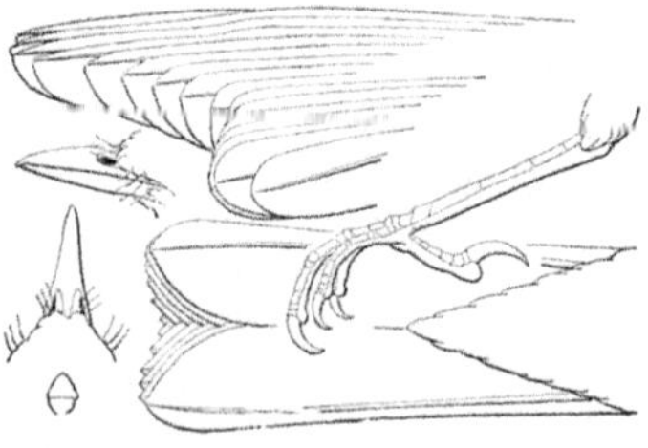

D. coronata.

The most extensive genus of American warblers, there being over twenty species in the United States, of which, however, few are found on this coast.

Dendrœca æstiva, Gmelin.

THE YELLOW WARBLER.

Motacilla æstiva, Gmelin, Syst. Nat. I. 1788, 996. — *Sylvia æstiva*, Latham, Index Orn. II. 1790, 551. — Audubon, Orn. Biog. I. 1831, 476; pl. 95, 35. — Nuttall, Man. I. 1832, 370, 2d ed. 417. — *Sylvicola æstiva*, Swainson, F. Bor. Am. II. 1831, 211. — Audubon, Birds Amer. II. 1841, 50; pl. 88.

Sylvia Childreni, Audubon, Orn. Biog. I. 1831, 180; pl. 35. (Immature.)

Dendroica æstiva, Baird, P. R. Rep. IX. Birds, 282. — Heermann, X. vi. 40. — Cooper and Suckley, XII. iii. Zool. of W. T. 181.

Sp. Char. Head all round, and under parts generally bright yellow; rest of under parts yellow-olivaceous, brightest on the rump. Back with obsolete streaks of dusky reddish-brown. Fore breast and sides of the body streaked with brownish-red. Tail feathers bright yellow; the outer webs and tips, with the whole upper surfaces of the innermost one, brown; extreme outer edges of wing and tail feathers olivaceous, like the

back, the middle and greater coverts and tertials edged with yellow, forming two bands on the wings. *Female* similar, with the crown olivaceous like the back; the streaks wanting on the back, and much restricted on the under parts. Tail with more brown. Length of male, 5.25; wing, 2.66; tail, 2.25. Iris brown; bill horn-color; lower mandible lead-blue; feet yellow.

Hab. The whole of North America from the Atlantic to the Pacific.

This lively and bright little songster is one of the most abundant summer visitors on this coast, as it is in the Atlantic States, and has habits entirely similar. They arrived at Fort Mojave about April 15, 1861, but I saw them at the Straits of Fuca as early as April 8, 1854; so that they must vary their migration very much with the season.

They prefer the warm inland valleys, frequenting chiefly the deciduous trees along rivers, and coming familiarly about gardens and orchards. There they are actively engaged throughout the day seeking their insect food among the leaves, and occasionally warbling their short but pleasing songs, which have considerable variety, and yet a sameness of style and tone which makes the performer easily recognizable.

Their nest is built sometimes in a low bush, at others as high as fifty feet above the ground. It is extremely neat and durable, formed in great part of the down of plants, mixed with fine strips of bark and leaves, sometimes lined with horsehair or feathers. They vary much, however, in the materials chosen, and often select bits of cotton-twine, silk threads and small rags, when building near a house.

The eggs are four or five, of a dull white, thickly sprinkled with pale brown spots near the large end. When the nest is found, the female generally attempts to draw the invader away by feigning lameness, but soon learns to take but little notice, if the eggs are not disturbed.

In autumn they are said to feed much on juicy fruits, but I have not myself observed this.

Dendrœca Audubonii, Townsend.

AUDUBON'S WARBLER.

Sylvia Audubonii, Townsend, J. A. N. Sc. Ph. VII. ii. 1837. — Audubon, Orn. Biog. V. 1839, 52; pl. 395. — *Sylvicola Audubonii*, Bonaparte, List, 1838. — Audubon, Birds Amer. II. 1841, 26; pl. 77. — Nuttall, Man. 2d ed. I. 414. — *Dendroica Audubonii*, Baird, P. R. Rep. IX. Birds, 273. — Kennerly, X. iii. 24. — Heermann, X. vi. 39. — Cooper and Suckley, XII. iii. 181.

Sp. Char. Above bluish-ash, streaked with black, most marked on the middle of the back; on head and neck bluish-ash. Middle of crown, rump, chin, and throat, and a patch on the side of the breast, gamboge-yellow. Space beneath and anterior to the eyes, fore part of breast and sides, black; this color extending behind on the sides in streaks. Middle of belly, under tail coverts, a portion of upper and lower eyelids, and a broad

band on the wings, with a spot on each of the four or five exterior tail feathers, white; rest of tail feathers black. *Female*, brown above; the other markings less conspicuous and less black. Length, 5.50; extent, 9.00; wing, 3.25; tail, 2.25. Iris brown; bill and feet black.

Hab. Pacific Coast of the United States to central Rocky Mountains. South to Mexico.

This is a very abundant species, especially in winter, when large numbers remain in the southern part of California, and are seen flitting about every bush and tree, as well as the tall weeds of the prairie, seeking their insect food and uttering only a sharp chirp occasionally. They are then in the dull plumage of the female, and the males do not obtain their richer hues and black breast until March or April. According to Heermann, they winter as far north as Sacramento, and some, I think, remain near the Columbia River throughout mild winters. About May 1st they all seem to retire towards the north, and I have seen none even in the Coast Mountains, south of San Francisco, after that month, but they begin to appear again in September.

Their song resembles that of the *D. æstiva*, and is heard chiefly in their summer resorts towards the north. I obtained newly fledged young at Lake Tahoe in September, and they probably raise their broods throughout the higher Sierra Nevada.

The nest is believed by Nuttall to be built in the tall coniferous trees, but has not yet been discovered.

This species does not return to the latitude of Santa Cruz (37°, and at the sea level) until the end of September, but some remain there all winter; and about March 20th the males, rapidly changing their plain winter livery for the gayer hues of summer, favor us with a few faint notes before their departure north, the song being much like that of the *D. æstiva* in character. They disappear about April 15th.

Dendrœca coronata, LINNÆUS.

THE YELLOW-CROWNED WARBLER.

Motacilla coronata, LINNÆUS, Syst. Nat. I. 1766, 333. — GMELIN, Syst. Nat. I. 1788, 974 (male). — *Sylvia coronata*, LATHAM, Index Orn. II. 1790, 538. — WILSON, Am. Orn. II. 1810, 138; pl. xvii. f. 4 (summer); II. 356; pl. xlv. f. 3 (winter). — NUTTALL, Man. I. 1832, 361. — AUDUBON, Orn. Biog. II. 1834, 303; pl. cliii. — *Sylvicola coronata*, SWAINSON, F. Bor. Am. II. 1831, 216. — AUDUBON, Birds Amer. II. 1841, 23; pl. lxxvi. — NUTTALL, Man. 2d ed. I. 411. — *Dendroica coronata*, G. R. GRAY, Genera, 2d ed. Suppl. 1842, 8. — BAIRD, P. R. Rep. IX. Birds, 272. — COOPER and SUCKLEY, XII. iii. Zool. of W. T. 180.

SP. CHAR. Above, bluish-ash, streaked with black. Under parts white. The fore part of breast and the sides black, the feathers mostly edged with white. Crown, rump,

and sides of breast yellow. Cheeks and lores black. The eyelids and a superciliary stripe, two bands on the wing, and spots on the outer three tail feathers, white. *Female* of duller plumage and browner above. Length, 5.65; extent, 9.00; wing, 3.00; tail, 2.50.

Hab. Eastern North America to the Missouri Plains, and northward along Yukon Valley to Norton Sound. Stragglers seen on Puget's Sound. California.

This species closely resembles *D. Audubonii,* differing chiefly in having the throat white. I saw a few of them at the Straits of Fuca in April, 1855, and suppose they must migrate through California, though not yet detected there. In the Atlantic States they are numerous, and in habits are the exact counterpart of our *D. Audubonii.* The nest and eggs found in Nova Scotia by Audubon on a fir-tree scarcely differed from those of *D. æstiva.*

Dendrœca nigrescens, TOWNSEND.

THE BLACK-THROATED GRAY WARBLER.

Sylvia nigrescens, TOWNSEND, J. A. N. Sc. Ph. VII. ii. 1837, 191. — AUDUBON, Orn. Biog. V. 1839, 57; pl. 395. — *Vermivora nigrescens,* BONAPARTE, List, 1838. — NUTTALL, Man. I. 2d ed. 1840, 471. — *Sylvicola nigrescens,* AUDUBON, Syn. 1839, 60. In. Birds Amer. II. 1841, 62; pl. 94. — *Dendroica nigrescens,* BAIRD, P. R. Rep. IX. Birds, 270. — HEERMANN, X. vi. 40. — COOPER and SUCKLEY, XII. iii. Zool. of W. T. 180.

SP. CHAR. Head all round, fore part of the breast, and streaks on the side of the body, black; rest of under parts, a stripe on the side of the head, beginning acutely just above the middle of the eye, and another parallel to it, beginning at the base of the under jaw (the stripes of opposite sides confluent on the chin), and running farther back, white. A

yellow spot in front of the eye. Rest of upper parts bluish-gray, the interscapular region and upper tail coverts streaked with black. Wing coverts black, with two narrow white bands; quills and tail feathers brown, the two outer of the latter white, with the shafts and a terminal streak brown; the third brown, with a terminal narrow white streak. Length, 5.00; extent, 7.50; wing, 2.50; tail, 2.10. Iris brown; bill black; feet brown.

Hab. Pacific Coast of the United States; Fort Thorn, New Mexico.

This species appeared near San Diego, April 20th, in small flocks migrating northward, and then uttering only a faint chirp. They frequented low bushes along the coast, but as they go north they take to the deciduous oaks, when the leaves begin to grow early in May, at which time they reach the Columbia River. Their song is rather faint and monotonous, and I have not heard it during their migration north, nor have I seen them in this State after April. According to Townsend, their nest is built in the upper branches of the oak, in Oregon.

Dendrœca Townsendii, NUTTALL.

TOWNSEND'S WARBLER.

Sylvia Townsendii, ("NUTTALL,") TOWNSEND, J. A. N. Sc. Ph. VII. ii. 1837, 191. — AUDUBON, Orn. Biog. V. 1839, 36; pl. 393. — *Sylvicola Townsendii*, BONAPARTE, List, 1838. IB. Consp. 1850, 308. — AUDUBON, Birds Amer. II. 1841, 59; pl. 92. — NUTTALL, Man. I. 2d ed. 1840, 446. — *Dendroica Townsendii*, BAIRD, P. R. Rep. IX. Birds, 269. — COOPER and SUCKLEY, XII. iii. Zool. of W. T. 179.

SP. CHAR. Above, bright olive-green; the feathers all black in the centre, showing more or less as streaks, especially on the crown. Quills, tail, and upper tail covert feathers dark brown, edged with bluish-gray; the wings with two white bands on the coverts; the two outer tail feathers white, with a brown streak near the end; a white

streak only in the end of the third feather. Under parts as far as the middle of the body, with the sides of head and neck, including a superciliary stripe and a spot beneath the eye, yellow; the median portion of the side of the head, the chin and throat, with streaks on the sides of the breast, flanks, and under tail coverts, black; the remainder of the under parts white. Length, 5.00; extent, 8.00; wing, 2.65; tail, 2.25. Iris and feet brown; bill black.

Hab. Pacific Coast, North America; south to Mexico and Guatemala. A straggler taken near Philadelphia.

Small flocks of this species arrivea near San Diego, with *D. nigrescens*, about April 20, 1862, but I have not seen them elsewhere during summer, and they are very scarce at the Columbia River, being supposed to pass by there on their way to more northern regions, or perhaps to the highest parts of the mountains. I shot two in November, 1855, near Santa Clara, and saw what I supposed to be this species at Shoalwater Bay, W. T., as late as December 25, 1854, so that it is very probable they may be more common in this State in winter than summer.

In the north they frequent the higher parts of the lofty firs, and are therefore less easily seen than other species. In this State I found them when migrating among low willows and other bushes.

The spring plumage does not differ materially from that described by Baird from fall specimens.

Dendrœca occidentalis, TOWNSEND.

THE WESTERN WARBLER.

Sylvia occidentalis, TOWNSEND, J. A. N. Sc. VII. ii. 1837, 190. IB. Narrative, 1839, 340. — AUDUBON, Orn. Biog. V. 1839, 55; pl. 55. — *Sylvicola occidentalis*, BONAPARTE, List, 1838. — AUDUBON, Syn. 1839, 60. IB. Birds Amer. II. 1841, 60; pl. 93. — NUTTALL, Man. 2d ed. I. 445. — *Dendroica occidentalis*, BAIRD, P. R. Rep. IX. Birds. 268. — COOPER and SUCKLEY, XII. iii. Zool. of W. T. 178.

SP. CHAR. Crown, with sides of the head and neck, continuous bright yellow, feathers of the former narrowly edged with black; rest of upper parts dark brown, edged with bluish-gray, so much so on the back and rump feathers as to obscure the brown, and with an

olivaceous shade. Chin, throat, and fore part of breast (ending convexly behind in a subcrescentic outline), black; rest of under parts white, faintly streaked on the sides with black. Two white bands on the wing, two outer tail feathers, and the terminal portion of a third, white; the shafts, and an internal streak towards the end, dark brown. Length, 4.75; wing, 2.70; tail, 2.30. Bill jet black; legs brown.

Hab. Pacific Coast, north to Puget's Sound.

This seems to be a very rare species, and I have not seen them in the southern part of California, though they probably migrate into Mexico. One was obtained at Petaluma, by Mr. E. Samuels, on April 1, 1856. They frequent lofty evergreens.

The pair of birds from which the species was described, was obtained by Mr. Townsend near Fort Vancouver, May 28, 1835. They were flitting among the pine trees in the depth of a forest, actively engaged in searching for insects, and hanging frequently from the twigs like titmice. Their note, uttered at distant intervals, resembled very much that of the black-throated blue warbler (*D. cærulescens*).

Dendrœca chrysopareia, Sclater.

THE GOLDEN-CHEEKED WARBLER.

Dendroica chrysopareia, Sclater and Salvin, Pr. Zool. Soc. 1860, 298. Ib. Ibis, 1865, 89. — Baird, Rev. Amer. Birds, 183.

Sp. Char. Similar, in general appearance, to *D. Townsendii*, but the upper parts generally black, with olivaceous edgings on the back, the rump and upper tail coverts pure

black. Sides and lower tail coverts streaked with black. Cheeks yellow; a simple black stripe through the eye; no patch beneath it Bill very short Length, 4.50; wing, 2.50; tail, 2.40; tarsus, 0.75.

Hab. Texas to Guatemala.

This species, originally described from Guatemala, has lately been taken near San Antonio, Texas, and will doubtless be found in New Mexico and Arizona. Nothing is known of its habits, which, however, in all probabilty, resemble those of the Black-throated green and other familiar Eastern species. The specimen described above is the type belonging to Mr. Salvin, of London; those from Texas were obtained by the late Dr. Heermann.

Genus GEOTHLYPIS, Cabanis.

Trichas, Swainson, Zool. Jour. III. July, 1827, 167. (Not of Gloger, March, 1827, equal to *Criniger*, Temm.)
Geothlypis, Cabanis, Wiegmann's Archiv 1847, I. 316, 349.

Gen. Char. Bill sylvicoline, rather depressed, and distinctly notched; rictal bristles very short or wanting. Wings short, rounded, scarcely longer than the tail; the first

G. trichas.

quill shorter than the fourth. Tail long; much rounded or graduated. Legs stout; tarsi elongated, as long as the head. Olive-green above, belly yellow. Tail feathers immaculate. Legs yellow.

G. trichas.

This genus is represented in California by two species, *G. trichas* and *Macgillivrayi*, and a third and a much rarer one, *G. Philadelphia*, occurring in the Eastern States, where *G. trichas* is also abundant. Many others, however, belong to Middle and South America, where they seem to be abundant, and to have much the same habits as do the species of the United States, living in thickets or among the grass, and nesting on or very near the ground.

Geothlypis trichas, LINNÆUS.

THE MARYLAND YELLOW-THROAT.

Turdus trichas, LINNÆUS, Syst. Nat. I. 1766, 293. — *Sylvia trichas*, LATHAM, Ind. Orn. II. 1790. — AUDUBON, Orn. Biog. I. 1832, 120; V. 1838, 463; pls. 23 and 240. — *Geothlypis trichas*, CABANIS, Mus. Hein. 1850, 16. — BAIRD, P. R. Rep. IX. Birds, 170. — COOPER and SUCKLEY, XII. iii. Zool. of W. T. 177.

Sylvia Marilandica, WILSON, Am. Orn. I. 1808, 88; pl. vi. f. 1. — *Trichas Marilandica*, BONAPARTE, List, 1838. — AUDUBON, Syn. 1839, 65. IB. Orn. Biog. II. 1841, 78; pl. 102.

Sylvia Roscoe, AUDUBON, Orn. Biog. I. 1832, 124; pl. 24. (Young male.) — *Trichas Roscoe*, NUTTALL, Man. I. 2d ed. 1840, 457.

Trichas Delafieldii, AUDUBON, Orn. Biog. V. 307. — HEERMANN, P. R. Rep. X. vi. 40.

SP. CHAR. Upper parts olive-green, tinged with brown towards the middle of the crown; chin, throat, and breast as far as the middle of the body, with the under tail coverts, bright yellow. Belly dull whitish-buff. Sides of body strongly tinged with light olive-brown; under coverts glossed with the same. A band of black on the forehead (about 0 20 of an inch wide in the middle), passing backward so as to cover the cheek and ear coverts, and extending a little above the eye; this band bordered behind

by a suffusion of hoary-ash, forming a distinct line above the eye, and widening behind the ear coverts into a larger patch, with a yellow tinge. In winter dress, and in the female, without the black mask, the forehead tinged with brown, the yellow of the throat less extended, the eyelids whitish, and an indistinct superciliary line yellowish. Length of male, 4 80; extent, 6.75; wing, 2.40; tail, 2.20.

Hab. North America from the Atlantic to the Pacific.

A very common little bird during summer, and some possibly winter within the State, though I saw none in the Colorado Valley, in winter. At San Diego I saw the first on the 17th of April, but have before seen them at the Columbia River earlier than that time; and as they are rather scarce at all times in the southern part of the State, I suspect that many winter in the middle portions.

They usually inhabit the densest thickets, and occasionally the reeds about ponds and marshes, being constantly on the move after insects, and rarely showing themselves. In the spring, however, the male often flies upward a few yards, singing a short but lively song, generally in triple bars, and then drops suddenly down again.

Their nest is built in the thicket, often on the ground, very closely concealed; it is made entirely of grass and leaves. The eggs are white, slightly flesh-color, with reddish-brown dots, specks, and lines, mostly near the large end.

They generally keep near brooks and marshes.

Geothlypis Macgillivrayi, Audubon.

MACGILLIVRAY'S WARBLER.

Sylvia Macgillivrayi, Audubon, Orn. Biog. V. 1839, 75; pl. 399. (*Sylvia Philadelphia* on plate.) *Trichas Macgillivrayi*, Audubon, Syn. 1839, 64. Ib. Birds Amer. II. 1841, 74; pl. 100.

Sylvia tolmiæi, Townsend, J. A. N. Sc. VIII. 1839, 149, 159. (Read in 1839, but the volume really not published till 1840.) — *Trichas tolmiæi*, Nuttall, Man. I. 2d ed. 1840, 460. — Heermann, P. R. Rep. X. vi. 40.

Geothlypis Macgillivrayi, Baird, P. R. Rep. IX. Birds of N. Amer. 244; pl. 79, f. 4 (head). — Cooper and Suckley, XII. iii. Zool. of W. T. 177.

Sp. Char. Head and neck all round, throat, and fore part of the breast dark ash-color; a narrow frontlet, loral region, and space round the eye (scarcely complete behind), black. The eyelids above and below the eye (not in a continuous ring), white. The feathers of the chin, throat, and fore breast really black, with ashy-gray tips, more or less concealing the black. Rest of upper parts dark olive-green (sides under the wings paler),

of lower, bright yellow. *Female* with the throat paler and without any black. Young in fall, without the blue hood, and of a dull greenish color generally Length of male. 5.00; extent, 6.75; wing, 2.45; tail, 2.45. Iris brown; bill brown above, yellow below; feet reddish-gray.

Hab. Eastern base of Rocky Mountains to the Pacific, and south to Mexico.

At Fort Mojave I noticed the first of this species April 24th, but they probably arrive earlier, as they reach the Columbia River by May 3d. They differ considerably in habits from *G. trichas*, as they frequent dry localities, hunting for insects, both in low bushes and in the trees, like some of the *Dendrœcas*. Their song is short and rather faint, somewhat resembling that of *G. trichas*. The nest I found at Puget's Sound, in June, was built, without attempt at concealment, about a foot above the ground, and formed wholly of dry grasses, rather loosely put together. According to Nuttall, they also use the bark fibres of the cedar (*Thuja gigantea*), and build among moss. The eggs are white, spotted with reddish.

I met with this species at Lake Tahoe in September, and in the Coast Range during summer, but not near San Diego. It is quite possible that some of them winter in the warmer parts of the State.

They have a very sharp chirp of alarm when they see a strange object, and if watched are very shy, seeking the densest thickets; but if you wait quietly for a short time their curiosity brings them out, and they will approach until within a few feet, keeping up their scolding chirp all the time.

Genus **ICTERIA**, Vieillot.

Icteria, Vieillot, Ois. Amer. Sept. I. 1790, iii. and 85.

Gen. Char. Bill shorter than head; broad at the base, but rapidly becoming compressed or much higher than broad, with the ridge elevated and sharp from the very base of the bill; the upper outline much curved throughout; the commissure less curved

I. virens.

but strongly concave; the gonys nearly straight, the upper edge of the lower jaw as convex as the commissure is concave. No notch in the bill, and the rictal bristles small. Tarsi longer than the toes, without scutellæ, except faint indications on the inner side. Lateral toes about equal; shorter than the hinder. Wings about equal to the tail, rounded; the first quill longer than the secondaries. Tail graduated; above olive, beneath yellow. Abdomen, eyelids, maxillary patch, and line to the bill, white.

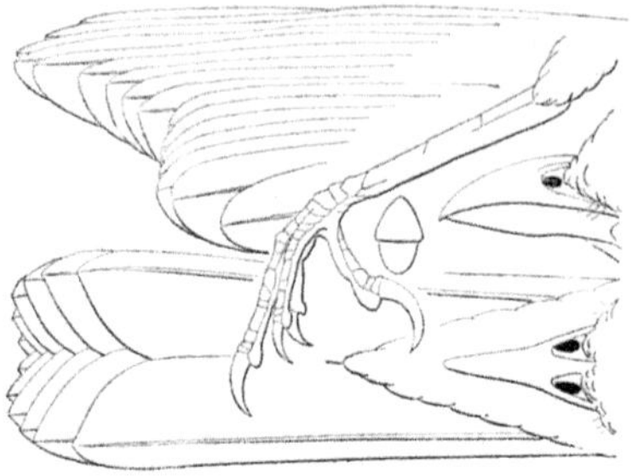

I. virens.

But one well-established species belongs to this genus, although several have been indicated by authors

Icteria longicauda, Lawrence.

THE LONG-TAILED CHAT.

Icteria longicauda, Lawrence, Ann. N. Y. Lyc. VI. iv. Ap. 1853, 4. — Newberry, P. R. Rep. VI. iv. 81; pl. 34, f. 2. — Baird, P. R. Rep. IX. Birds, 249. — Heermann, X. vi. 55.

? *Icteria auricollis* (Licht. Mus. Ber.), Bonaparte, Consp. 1850, 331.

Sp. Char Fourth quill longest; third and fifth shorter; first shorter than the seventh. Above ash-color, tinged with olive on the back and neck; the outer surface of the wings and tail olive. The under parts as far as the middle of the belly bright gamboge yellow, with a tinge of orange; the remaining portions white The superciliary and maxillary

white stripes extend some distance behind the eye. Outer edge of the first primary white. Length, 7.00; extent, 8 50; wing, 3 20; tail, 3 70.

Hab. High central plains of the United States to the Pacific; south to Mexico; north to Walla-walla.

Many of this species arrived at Fort Mojave about April 20, 1861; and at San Diego I saw one on April 26, 1862. Most of them, probably, take an inland route towards the north. They inhabit chiefly the warmer valleys, near streams and marshes; and the only place on the coast where I have seen them was Santa Cruz, where I found a nest, with young, in June. In deference to the opinion of most American naturalists, we have retained a distinct name for the California Chat, as distinguished from *I. virens* of the East. The principal difference, however, is a greater length of tail, which, in view of the variations in this respect observed in many species, appears hardly of specific value.

At Fort Mojave on May 19th I found a nest built in a dense thicket of *Algarobia*, containing three eggs, besides one of the cowbird (*Molothrus*). It was made of slender green twigs and leaves, and lined with grass and hairs. The eggs were white, sprinkled with cinnamon near the large end, somewhat in the form of a ring. Size, 0.75 × 0.64 inch.

The nest is usually closely concealed, but that found at Santa Cruz was in a very open situation, and only about two feet above the ground. I found it by chance, after hunting the dense thickets around thoroughly, and when on the point of giving it up. The old birds are very bold when the nest is approached, keeping up a constant scolding, and almost flying in the intruder's face. At other times they are very shy, and scarcely to be seen without long watching for them.

During the spring both day and night the song of the male is heard, which consists of a great variety of grotesque notes, more odd than musical, and like those of no other bird. They resemble sometimes the noises of ducks, cats, and puppies, but these sounds are not imitated, as they are similar everywhere, and are the same with our species as with the Eastern. Their food consists chiefly of insects and berries. They probably leave the State in September.

A male specimen from Fort Mojave has the tail as short as the Eastern *I. virens* (3.25 inch), but resembles coast specimens in its grayish color. A female like it has since been found there by Mr. Holder. The time of arrival of this bird seems remarkably uniform, as they reached Santa Cruz, in 1866, about April 27th.

A chat from Mexico has been described as *Icteria Velasquezi*, differing principally from the species of the United States in the whitish color of the lower mandible. Young birds, however, of *I. virens* exhibit this same peculiarity, and it is almost certain that this Mexican species has no claim to a distinctive name.

In the general appearance of the *Icteria* there is quite a close relationship to a genus *Teretristis* from Cuba.

GENUS MYIODIOCTES, AUDUBON.

Myiodioctes, AUDUBON, Syn. 1839, 48. (Type, *Motacilla mitrata*.)
Wilsonia, BONAPARTE, List, 1838. (Preoccupied in Botany.)

GEN. CHAR. Bill depressed, flycatcher-like; broader than high at the base; gape with bristles nearly as long as the bill, which is distinctly notched at the tip; both outlines gently convex. Tarsi longer than the head; considerably exceeding the middle toe;

M pusillus

claws all considerably curved. Tail decidedly rounded or slightly graduated; the lateral feathers 0.20 of an inch shorter. Wing very little longer than the tail; the first quill decidedly shorter than the fourth; colors yellow.

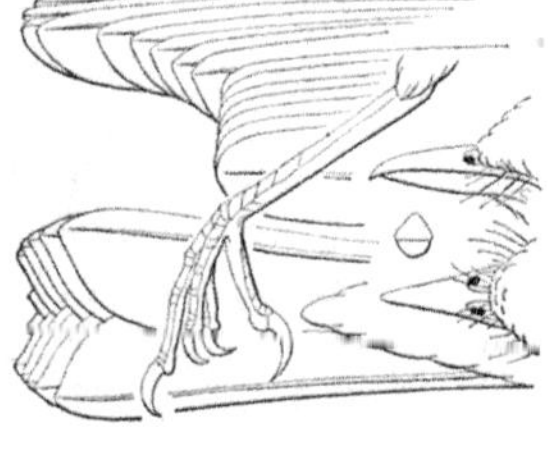

M. mitratus.

Three species besides ours inhabit the eastern side of the continent, all larger and quite differently marked. All have much the habits of fly-catchers, darting after insects and capturing them on the wing. They usually keep in dense thickets, and are generally difficult of approach.

Myiodioctes pusillus, Wilson.

THE GREEN BLACK-CAP WARBLER.

Muscicapa pusilla, Wilson, Am. Orn. III. 1811, 103; pl. xxvi. f. 4. — *Sylvania pusilla*, Nuttall, Man. I. 2d ed. 1840, 335.

Sylvia Wilsonii, Bonaparte, Obs. Wilson, 1826, No. 127. — Nuttall, Man. I. 1832, 408. — *Muscicapa Wilsonii*, Audubon, Orn. Biog. II. 1834, 148; pl. 124. — *Myiodioctes Wilsonii*, Audubon, Syn. 1839, 50. Ib. Birds Amer. II. 1841, 21; pl. 75.

Myiodioctes pusillus, Bonaparte, Conspectus, 1850, 315. — Baird, P. R. Rep. IX. Birds, 293. — Heermann, X. vi. 39. — Cooper and Suckley, XII. iii. Zool. of W. T. 182.

Sp. Char. Forehead, line over and around the eye, and under parts generally bright yellow. Upper parts olive-green; a square patch on the crown lustrous black. Sides of body and cheeks tinged with olive. No white on wings or tail. *Female* similar; the

black of the crown obscured by olive-green. Length, 4.75; wing, 2.25; tail, 2.30. Iris brown; bill brown; feet brownish.

Hab. United States, from the Atlantic to the Pacific; north to Kodiak, and south to Guatemala.

This pretty little bird arrives in California about the first week in May, and passes far to the north, occurring in Sitka and Kodiak. They frequent chiefly the deciduous trees and bushes near water, and have much the same habits and song as the *Dendroica æstiva.*

According to Nuttall, they sometimes reach the Columbia River the first week in May, and he has seen them feeding their young by the 12th of May; so very probably some winter in California. He found a nest on the branch of a service-bush, built chiefly of moss, with a thick lining of grass, and containing four eggs, white with pale brown dots near the larger end, in a sort of circle. They also build in fir-trees, according to Audubon. The nest measures only 3.50 inches wide and 1.50 deep.

The arrival of these little birds at Santa Cruz, in 1866, was about April 20th. I observed them apparently gathering materials for nests at that date, the male singing merrily during the employment. From Nuttall's observations in Oregon, it would seem that they must arrive there quite as early, or else a few remain all winter unnoticed among the shrubbery.

Family HIRUNDINIDÆ, The Swallows.

Char. Bill very broad, short, and much depressed; the culmen less than half the commissure, which opens to beneath the eye. Rictus smooth. Wings very long, greatly exceeding the central tail feathers; the first primary longest. Tarsi shorter than the lateral toes.

Genus **HIRUNDO**, Linnæus.

Hirundo, Linnæus, Syst. Nat. 1735. — Gray, Genera, I. 1845.

Gen. Char. Nostrils basal, small, oblong, and covered partly by a membrane. Tail more or less forked; the outer lateral feather sometimes greatly lengthened. Tarsi

H. horreorum.

naked, shorter than the middle toe, and scutellate. Toes long, slender, the lateral ones unequal. Claws moderate, curved, acute.

H. horreorum.

Hirundo horreorum, BARTON.

THE BARN SWALLOW.

Hirundo horreorum, BARTON, Fragments, N. H. Penn. 1799, 17. — BAIRD, P. R. Rep. IX. Birds, 308. — COOPER and SUCKLEY, XII. iii. Zool. of W. T. 184.

Hirundo rufa, VIEILLOT, Ois. Am. Sept. I. 1807, 60; pl. xxx. (Not of Gmelin.) — CASSIN, Illust. I. 1855, 243. — BREWER, N. Am. Ool. I. 1857, 91; pl. v. f. 63 - 67 (eggs). — NUTTALL, Man. 2d ed. 726. — HEERMANN, P. R. Rep. X. vi. 34.

Hirundo Americana, WILSON, Am. Orn. V. 1812, 34; pl. xxxviii. f. 1, 2. (Not of Gmelin.) — RICH. F. B. A. II. 1831, 329.

Hirundo rustica, AUDUBON, Orn. Biog. II. 1834, 413; pl. 173. IB. Syn. 1839, 35. IB. Birds Amer. I. 1840, 181; pl. 48. (Not of Linnæus.)

SP. CHAR. Tail very deeply forked; outer feathers several inches longer than the inner, very narrow towards the end. Above glossy blue, with concealed white in the middle of the back. Throat chestnut; rest of lower part reddish white, not conspicuously different. A steel-blue collar on the upper part of the breast, interrupted in the middle.

Tail feathers with a white spot near the middle, on the inner web. *Female* with the outer tail feather not quite so long. Length, 6.50; extent, 12.75; wing, 4.75; tail, 4.50. Iris brown; bill black; feet slate-color.

Hab. North America, from the Atlantic to the Pacific.

This well-known and beautiful bird is less abundant along the western than the eastern coast, its place being filled in great part by the *H. lunifrons*, and perhaps also because it does not find so many suitable places for building in. As settlements multiply they seem to be gradually increasing about farms near the coast, building in the barns, and living in perfect harmony with their cousins, which build under the eaves. In wild districts they build in caves, which abound in the bluffs along the sea-shore, from San Diego to the Columbia River.

Their nest is built of mud, plastered up against a rafter, or on some supporting shelf; it is cup-shaped, and lined with fine hay. The saliva of the bird is supposed to render the pellets of mud more adhesive. The eggs are five, white spotted with reddish-brown, and they usually raise two broods in a season, but often leave the last to starve, in their sudden departure south in August or September.

I noticed their arrival at San Diego March 25th, and have found them far more frequent along the sea-coast than inland, probably from the fact that they prefer the vicinity of water, and delight to hunt insects over its surface, sometimes even touching it as they skim along. At Sacramento they were found by Dr. Heermann, and I have seen them near there as late as September 8th. They arrived at Santa Cruz March 21, 1866, showing a remarkable regularity in comparison with 1862, though the latter was a much colder spring. They left Santa Cruz about September 15th.

Hirundo lunifrons, Say.

THE CLIFF SWALLOW.

Hirundo lunifrons, Say, Long's Exped. R. Mts. II. 1823, 47. — Baird, P. R. Rep. IX. Birds, 307. — Heermann, X. vi. 36. — Cooper and Suckley, XII. iii. Zool. of W. T. 184. — Cassin, Illust. I. 1855, 243. — Brewer, N. Am. Ool. I. 1857, 94; pl. 68-73 (Egg).
Hirundo fulva, Bonaparte, Am. Orn. I. 1825, 63; pl. ii. (Not of Vieillot?) — Audubon, Orn. Biog. I. 1831, 353; pl. 58. Ib. Birds Amer. I. 1840, 177; pl. 47. — Nuttall. Man. 2d ed. I. 729.

Sp. Char. Crown and back steel-blue; the upper part of the latter with concealed pale edges to the feathers. Chin, throat, and sides of the head dark chestnut; breast fuscous; belly white. A steel-blue spot on throat. Rump light chestnut; forehead

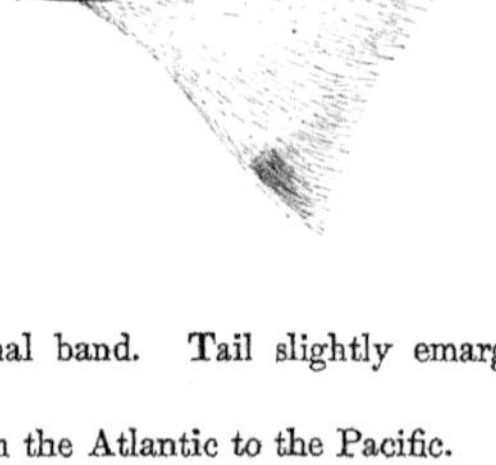

brownish white; a pale nuchal band. Tail slightly emarginate. Length, 5.75; wing, 4.40; tail, 2.20.

Hab. North America, from the Atlantic to the Pacific.

An abundant species throughout California, and as far north as Columbia River, on the coast. I saw the first of them at San Diego March 15, 1862; and at San Francisco they arrive about March 25th, being a week earlier than the barn swallow, and also remaining later in autumn. I have seen them as late as October 5th, and they probably remain longer toward the south. They live almost everywhere during summer, except on the high and wooded mountains, building on the cliffs of the sea-coast, where the cold wind blows, as well as in the hottest valleys, under eaves of houses, and sometimes on the sides of large branches or trunks of trees. Their bottle-shaped nests of mud, lined with straw, are conspicuous objects wherever they are allowed to build them, some even being visible in the noisy city of San Francisco, which only this species visits, sweeping through the crowded streets with entire fearlessness. The eggs are usually four, white, spotted with dusky-brown, and they hatch two broods in the season in most parts of the State. When about the nest, they make a creaking noise very different from the twitter of the barn swallow.

In June I saw a flock of these birds busily catching young grasshoppers on the dry hillside, where these insects were swarming. As I have never heard of other swallows eating grasshoppers, I suppose that this species is specially adapted for such food, other insects being very scarce during the

H. lunifrons.

dry season, and in the dry regions it inhabits so frequently, where other species of swallow are unknown.

This swallow leaves Santa Cruz about September 1st, but probably only goes to the large rivers and lakes of the interior.

To determine the question as to bedbugs being *brought* to houses by these swallows, I allowed about twelve pairs to raise broods under the eaves of the house I lived in at Santa Cruz, in 1866. They built between April 12th and 26th, and the young were fledged July 1st; some also had laid new broods of two and three eggs by the 5th. On tearing down the nests I found bugs (*Cimex*) in every one, whatever part of the roof it occupied, showing that they were *brought by* the birds, none having been observed in the house. But these bugs were evidently a distinct species from the *Cimex lectularius*, being different in form, narrower, and pale yellowish, instead of the characteristic color from which the name "Puce" is derived, through the French name of the insect. Moreover, although many crawled into the cracks of the weather-boards, and could easily have entered the low bedroom windows, none were seen afterwards. So I think we may relieve the swallows of the charge of bringing in these pests, and encourage their building in suitable places, on account of the immense numbers of insects they destroy. As usual, their parasites are peculiar to them, and may be called *Cimex lunifrontis.*

Hirundo bicolor, Vieillot.

THE WHITE-BELLIED SWALLOW.

Hirundo bicolor, Vieillot, Ois. Am. Sept. I. 1807, 61 ; pl. xxxi. — Audubon, Orn. Biog. 1831, 491 ; pl. 98. Ib. Birds Amer. I. 1840, 175 ; pl. 46. — Brewer, N. Am. Ool. I. 1857, 100 ; pl. iv. f. 47 (Egg). — Baird, P. R. Rep. IX. Birds, 310. — Cooper and Suckley, XII. iii. Zool. of W. T. 185. — Heermann, X. vi. 36.

Hirundo viridis, Wilson, Am. Orn. V. 1812, 49 ; pl. xxxviii.

Sp. Char. Glossy metallic-green above ; entirely white beneath. *Female* much duller in color. Length, 6.00 ; extent, 12.50 ; wing, 4.75 ; tail, 2.65. Iris and feet brown ; bill black.

Hab. North America, from the Atlantic to the Pacific.

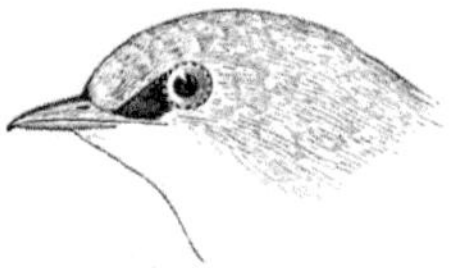

This swallow is to some extent a constant resident in California, a few wintering probably in the extreme southern portion. I first noticed their appearance at Stockton on February 22, 1863, and was informed by Dr. Holden that it does not vary more than three days from that date usually. On February 21, 1862, I had seen the first swallows near San Diego, probably of this species; but they have such powers of flight that they might spend the nights of winter a hundred miles from where they feed at midday. When first arriving they fly so high that it is difficult to distinguish the species.

They are the hardiest of all our swallows, extending their range northward at least to lat. 53°, east of the Rocky Mountains, and preferring the neighborhood of the coast and mountain-tops in this State. On the summit of the Coast Range I found them taking the place of the barn and cliff swallows, and have seen what I supposed this species flying above the summits of the Sierra Nevada as late as September 25th, at an elevation of probably nine thousand feet above the sea, migrating towards the south.

In this State they generally build in the knot-holes of oaks and other trees, but in older settled regions they have been known to prefer the advantages of an old building, or the little dwelling put up for the martins and bluebirds. The nest is made entirely of fine grass, lined with feathers, and the eggs, four or five in number, are pure white. They commonly raise two broods in a season.

They are not very gregarious, few living in the same neighborhood in the breeding season. Audubon found them wintering in Louisiana, roosting in the branches of the wax-myrtle, and feeding on its berries, as a variety to their usual fare of insects. They then associated in large flocks. This indicates more affinity to other insectivorous birds than we find in most swallows.

Five or six pairs of this species wintered at Santa Cruz, and others about sheltered ponds in the vicinity, where I saw them in January, 1866, during the coldest weather of the season. Those at the town roosted in the same knot-holes in the houses where they had raised their young, so that they were probably constant residents there, and not visitors from the north. Like all the swallows, however, they seemed to depart about September 1st, probably on account of the scarcity of insects during the latter part of the dry season. In September, 1865, I saw a few, probably of this species, near Tulare Lake, where they doubtless found food more abundant.

Hirundo thalassina, SWAINSON.

THE VIOLET-GREEN SWALLOW.

Hirundo thalassina, SWAINSON, Taylor's Philos. Mag. I. 1827, 365. — AUDUBON, Orn. Biog. IV. 1838, 597; pl. 385. IB. Birds Amer. I. 1840, 186; pl. 49. — BREWER, N. Am. Oology, I. 1857, 102; pl. v. f. 74 (Egg). — BAIRD, P. R. Rep. IX. Birds, 311. — HEERMANN, X. vi. 36. — COOPER and SUCKLEY, XII. iii. 185.

SP. CHAR. Tail acutely emarginate. Beneath, pure white. Above, soft velvety green, with a very faint shade of purplish-violet concentrated on the nape into a transverse band. Rump rather more vivid green; tail coverts showing a good deal of purple. Colors of female much more obscure. Length, 5.00; extent, 12.00; wing, 4.75; tail, 2.00. Iris brown; bill black; feet brownish.

Hab. Rocky Mountains to the Pacific; south to Mexico.

This little bird, the rival of the humming-birds in brilliancy and delicacy of plumage, arrives in Santa Clara Valley as early as March 15th, and frequents chiefly the groves of oaks along the sides of the valleys, and across the whole Coast Range, excepting the windy and cold neighborhood of the sea. They range at least as far north as the Straits of Fuca, and across the interior to the eastern base of the Rocky Mountains. Their nest is built in the knot-holes of the oaks, and I have never seen them seek the vicinity of

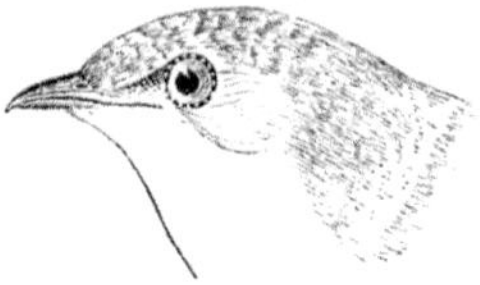

buildings for a dwelling, although not averse to the society of man, but rather more numerous about towns, when their favorite trees grow in the neighborhood. They select such small knot-holes, and generally in such inaccessible situations, that I never succeeded in obtaining their nest and eggs; but according to observers, they lay four, of a pure white color, somewhat like those of the bank swallow. Townsend states that along the sources of the Colorado River they build in the deserted nests of the cliff swallow.

They migrate to the south in September. Their notes consist of a rather faint warbling twitter, uttered as they sit on some low twig, their favorite perch; when flying they seem to be rather silent.

I have not observed this species west of the Coast Range, except when emigrating. They appeared at Santa Cruz, in 1866, on March 19th, and a large flock, with a few *Chætura*, stopped for a few hours on their way southward, October 5th, which dates are probably about the usual ones for their migrations.

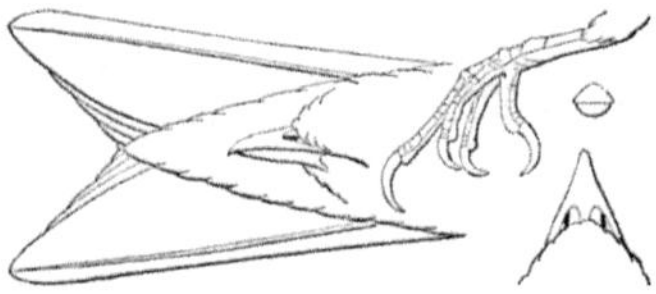

H. thalassina.

Genus COTYLE, Boie.

Cotyle, Boie, Isis, 1822, 550. (Type, *H. riparia.*)

Gen. Char. Bill very flat, extremely broad at the base, and gradually narrowed towards the tip; nostrils prominent and rounded. Tail moderate, nearly straight or some-

C. riparia.

what emarginated. Tarsi rather shorter than the middle toe, slender, and scutellate. Toes very slender, the claws slightly curved. Colors generally dull brown above, without gloss.

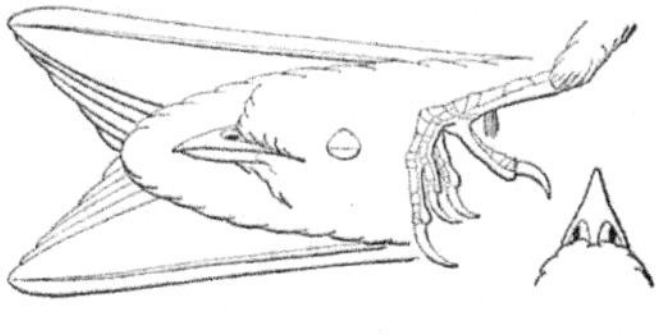

C. riparia.

The birds of this genus are among the plainest colored of the family, lacking entirely the beautiful metallic or velvety lustre characterizing most of the swallows. They belong essentially to the vicinity of the water, always nesting in its banks or near to them, excavating holes, or occupying crevices between rocks and stones.

Cotyle riparia, LINNÆUS.

THE BANK SWALLOW.

Hirundo riparia, LINNÆUS, Syst. Nat. I. 1766, 344. — WILSON, Am. Orn. V. 46; pl. xxxviii. — AUDUBON, Orn. Biog. IV. 1838, 584; pl. 385. IB. Birds Amer. I. 1840, 187; pl. 50. — *Cotyle riparia,* BOIE, Isis, 1822, 550. — BREWER, N. Am. Ool. I. 1857, 105; pl. iv. f. 49 (Egg). — BAIRD, P. R. Rep. IX. Birds, 313.

SP. CHAR. Smallest of American swallows. Tail slightly emarginate. Outer web of first primary soft, without hooks. Lower part of the tarsus with a few scattered feathers. Above, grayish-brown, somewhat fuliginous, with a tendency to paler margins

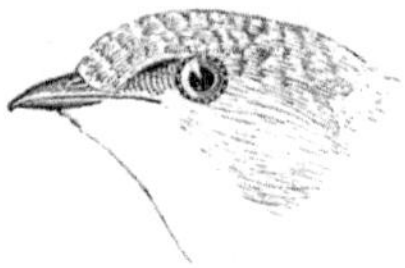

to the feathers. Beneath, pure white, with a band across the breast and sides of the body like the back. Length, 4.75; wing, 4.00; tail, 2.00.

Hab. North America generally.

This species has been obtained at Sacramento by Dr. Heermann, according to Baird, but as it seems rather less common on this coast than the next, and resembles that so closely that they cannot be distinguished without examination of specimens, their habits being exactly alike, I will describe those of both under one heading. It lays white eggs like that species.

The bank swallow is the only species believed to be common to both Europe and America, a careful comparison showing no tangible or permanent difference. The barn swallows of the two hemispheres are closely related, but always distinguishable by the rufous belly and narrow pectoral band of the American variety.

Cotyle serripennis, AUDUBON.

THE ROUGH-WINGED SWALLOW.

Hirundo serripennis, AUDUBON, Orn. Biog. IV. 1838, 593. — IB. Birds Amer. I. 1840, 193; pl. 51. — *Cotyle serripennis,* BONAPARTE, Consp. 1850, 342. — BREWER, N. Am. Ool. I. 1857, 106; pl. iv. f. 50 (Egg). — BAIRD, P. R. Rep. IX. Birds, 313. — KENNERLY, X. iv. 24. — HEERMANN, X. vi. 36. — COOPER and SUCKLEY, XII. iii. Zool. of W. T. 186.

SP. CHAR. Tail slightly emarginate; first primary with the pennulæ of the outer web much stiffened, with their free extremities recurved into a hook, very appreciable to the touch. No feathers on the tarsus and toes. Above, rather light sooty-brown; beneath, whitish-gray, or light brownish-ash, becoming nearly pure white in the middle of the belly

and on the under tail coverts. Length, 5.50; extent, 12.00; wing, 4.50; tail, 2.23. Iris, brown; bill and feet black.

Hab. United States from the Atlantic to the Pacific.

Bank swallows, chiefly of this species, are found in summer through nearly the whole lower portions of the State. I saw them first at Fort Mojave, on the 27th of February, but I have seen them at San Diego on November 9th and January 27th, so that, if they do not winter within the State, they do not go far beyond it.

They frequent chiefly the sandy banks of rivers, and burrow in them to a depth of two or three feet, the holes crowded very near together, and entering a few feet below the upper edge of the bank. At the bottom of the burrow is the nest, composed chiefly of dry grass with a few feathers, containing five white eggs. Sometimes they resort to natural clefts in the bank, or in adobe buildings, and occasionally to knot-holes. In favorable places they congregate in great numbers about one spot, and continue to keep much in flocks during the fall. They have only a faint twittering note when flying. At night they roost in their burrows, and in cold weather have been found almost torpid in them. According to Audubon, the eggs of this species are larger, longer, and more pointed than those of the bank swallows.

The peculiarity of the wings of this species shown in the accompanying figure is shared by several others in Middle and South America, which have lately been grouped under a genus *Stelgidopteryx* (curry-comb wing).

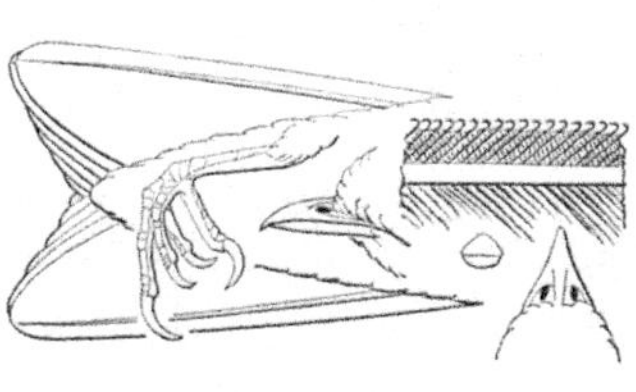

C. serripennis.

Genus PROGNE, Boie.

Progne, Boie, Isis, 1826, 971. (Type, *Hirundo purpurea, L.*)

Char. Bill strong, short, the gape very wide, the sides gradually compressed, the culmen and lateral margins arched to the tip; the latter inflected; the nostrils basal,

P. purpurea

lateral, open, and rounded. Tail considerably forked. Tarsi shorter than the middle toe and claw; about equal to the toe alone. Toes long, strong; lateral ones equal.

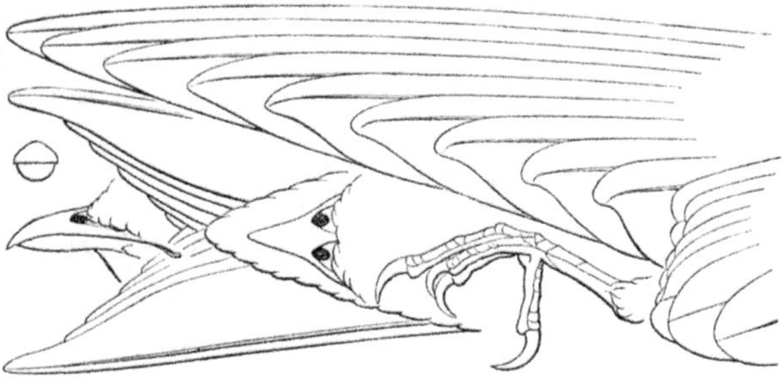

P. purpurea

Many species of this genus occur throughout America, two of them peculiar to the West Indies, and a third to the Galapagos Islands. Some are very local in distribution, though of most powerful flight.

Progne purpurea, Linnæus.

THE PURPLE MARTIN.

Hirundo purpurea, Linnæus, Syst. Nat. I. 1766, 344. — Wilson, Am. Orn. V. 58; pl. 39, f. 1, 2. — Audubon, Orn. Biog. I. 1831, 115; pl. xxiii. Ib. Birds Amer. I. 1840, 170; pl. xli. — *Progne purpurea*, Boie, Isis, 1826, 971. — Bonaparte, List, 1838. — Cassin, Illust. I. 1855, 245. — Brewer, N. Am. Ool. I. 1857, 103; pl. iv. f. 47 (Egg). — Baird. P. R. Rep. IX. Birds, 314. — Heermann, X. vi. 35. — Cooper and Suckley, XII. iii. Zool. of W. T. 186.

Sp. Char. Largest of North American swallows. Closed wings rather longer than the deeply forked tail. Tarsi and toes naked. Color in the old male, everywhere glossy steel-blue, with purple and violet reflections. *Female*, and immature male, less brilliant

above, pale brownish beneath, blotched with darker or with bluish. Length, 8.00; extent, 16.25; wing, 5.80; tail, 3.40. Iris and feet brown; bill black.

Hab. United States generally.

I have not seen the beautiful and sociable martins in the Colorado Valley, nor observed them along the coast earlier than April 29th, when they were migrating through San Francisco, perching for a few hours on lofty flag-staffs during the warm morning, but disappearing when the cold sea-breeze began to blow. They resort chiefly to the warm valleys of the interior, nesting in holes of large trees from near San Diego to Puget's Sound. I also found them nesting on the summits of the Coast Range, in company with the *Hirundo bicolor* and *thalassina*, but preferring the dead tops of the loftiest red woods for their domiciles. They are numerous at Sacramento in summer, and probably through most of the Sierra Nevada, but retire to the south in August.

They have not yet attracted so much attention among our movable and busy people as in the East, where almost every country-house, and even some in large cities, furnish them with a residence, usually a neat little hotel with many apartments, each opening on to a porch, and all mounted together on a high pole. Like pigeons, the martins live in perfect harmony with their neighbors; while our other house-loving pets, the bluebirds and wrens, must have a large range of territory, and drive away intruders too near their homes. Yet the martins have courage enough, as is shown by their occasionally driving away the smaller birds, and even pigeons, in order to appropriate their quarters. They also drive away every hawk or crow that shows itself near their nest, and thus protect poultry.

The loud and sweet twittering song of the martin, though mixed with some harsh notes, is remarkably pleasing, and continues during its whole stay with us, beginning at dawn, and heard at intervals during the day as they pursue their prey through the higher air, generally far above the smaller swallows, though they also sweep occasionally along the ground.

Their nest is made of leaves, straw, hay, and feathers in large quantities, and their eggs, from four to six, are pure white. They probably raise two broods here, as in the East. They prey on the larger flying insects that appear during the day, and none are so swift as to escape them.

According to Audubon, this species arrives in the Middle Atlantic States before the barn swallow.

In 1866 I observed them first on April 26th, at Santa Cruz; but none build in or near the town.

Family VIREONIDÆ, The Greenlets.

Char. Basal joints of anterior toes shortened; that of middle toe shorter than that of inner, and united throughout to the basal joint of both inner and outer toes. Lateral plates of tarsus undivided, except at lower end. Lower edge of lower mandible more than half the length of lower jaw from tip to angle of mouth. Bill conical and compressed, decurved at end. Tarsus longer than middle toe and claw. Lateral toes unequal. Primaries ten; the first short, sometimes wanting.

The family of the vireos, or greenlets, is peculiarly American, and is represented by many species, some or other of them occurring in every part of the continent. But two genera, and these closely allied to each other, belong to the United States.

The relations of the *Vireonidæ* are very close to the *Ampelidæ* and *Laniidæ*, the three forming a special group by themselves.

SUB-FAMILY VIREONINÆ.

CHAR. Bill moderate, cylindrical, somewhat compressed. Wings long, the first primary sometimes wanting. Tail short and nearly even. Sides of the tarsi behind not scutellate.

GENUS **VIREOSYLVIA**, BONAPARTE.

Vireosylvia, BONAPARTE, Gray, Comp. List, 1838. (Type, *Muscicapa olivacea, L.*) — BAIRD, Rev. Amer. Birds, 326.

CHAR. Wings pointed, considerably longer than the nearly even tail. Spurious primary quill either wanting, or very short, not one third the second.

V. olivacea.

The species of this genus are quite numerous, and some of them widely distributed. Several are peculiar to California and Arizona, although others extend over the greater part of the United States.

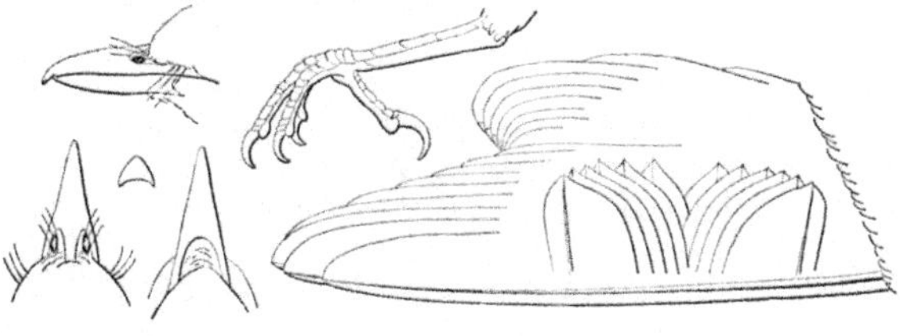

V. olivacea.

Vireosylvia gilva, Var. Swainsoni, Baird.

SWAINSON'S WARBLING GREENLET.

Muscicapa gilva, Vieillot, Ois. I. 1807, 65; pl. xxxiv.
Vireo gilvus, Bonaparte, Obs. Wilson, 1825, No. 123. — Nuttall, I. 1832, 309. — Audubon, Orn. Biog. II. 1834, 114; V. 1839, 433; pl. 118. Ib. Birds Amer. IV. 1842, 149; pl. 241. — Baird, P. R. Rep. IX. Birds, 335. — Heermann, X. vi. 55. — Cooper and Suckley, XII. iii. 188.
Vireo Swainsoni, Baird, Birds N. Amer. 1858, 336. Ib. Rev. Amer. Birds, 343.
Muscicapa melodia, Wilson, Am. Orn. V. 1812, 85; pl. 42, f. 2.

Sp. Char. Third, fourth, and fifth quills nearly equal; second and sixth usually about equal, and about 1.25 of an inch shorter than third; the exposed portion of spurious quill about one fourth the third. Above, greenish-olive; the head and hind neck ashy, the

V. gilvus. *V. Swainsoni.*

back slightly tinged with the same. Lores dusky; a white streak from the base of the upper mandible above and a little behind the eye; beneath the eye whitish. Sides of the head pale yellowish-brown. Beneath, white tinged with very pale yellow on the breast and sides. No light margins whatever on the outer webs of the wings or tail. Spurious

V. Swainsoni.

primary one fourth the length of second. Length about 5.50; extent, 8.60; wings nearly 3.00. Iris brown; bill horn-color and bluish; feet white.

Hab. Atlantic to the Pacific Coast of the United States.

The description just given applies to the *V. gilva*, as found in Eastern North America. The Western bird, however, takes rank as a distinct variety, if not species, under the name of *V. Swainsoni,* and differs in smaller size, more depressed bill, and in the second quill being much shorter than the sixth, instead of equal to it or longer. The colors generally are paler, and the upper mandible is black, not horn-color. The better to illustrate the differences, we give the heads of both, and add the outline of a wing of *Swainsoni,* for comparison with that of *V. gilva.*

This lively and familiar little songster arrives near San Diego about April 10th, and towards the middle of May reaches Puget's Sound, residing through the summer in nearly all the intermediate country, frequenting the deciduous trees along the borders of streams and prairies, coming into gardens and orchards with familiar confidence as soon as cultivation has reclaimed the wilderness.

Their cheerful and varied song may be heard from morning till night, and frequently until late in the autumn, as they industriously seek for their insect prey among the branches, occasionally interrupting the melody by swallowing a victim. They even build their nests in the shade-trees along the streets of busy cities, or in parks; and though their low warbling attracts little attention from the hurrying crowd, they sing on, sure of an audience from some more observant lover of nature's music.

Their nest is built pendant in the forks of a branch high above the ground, sometimes as much as one hundred feet, and is composed of grasses and vegetable fibres neatly interwoven, lined with the silk and down from plants. The eggs are white, with a few blackish spots of various sizes, and straggling lines near the larger end. (Nuttall.)

In the autumn they wander about in families, feeding partly on berries, and uttering a harsh note of complaint or alarm if closely watched, as do the old birds when the nest is approached. They probably leave the State in October.

Vireosylvia solitaria, VIEILLOT.

THE BLUE-HEADED FLYCATCHER.

Muscicapa solitaria, WILSON, Am. Orn. II. 1810, 143; pl. 17, f. 6.
Vireo solitarius, VIEILLOT, Nouv. Dict. 1817. — AUDUBON, Orn. Biog. I. 1831, 147; V. 1839, 432; pl. 23. IB. Syn. 1839. IB. Birds Amer. IV. 1842, 144; pl. 239. — NUTTALL, Man. I. 1832, 305. — BAIRD, P. R. Rep. IX. Birds, 340. — HEERMANN, X. vi. 55. — COOPER and SUCKLEY, XII. iii. Zool. of W. T. 189.
Vireo Cassini, BAIRD, Birds N. Amer. IX. 340; pl. 75, f. 1. — XANTUS, Pr. A. N. S. 1858.

SP. CHAR. Spurious primary very small, not one fourth the second, which is longer than the sixth. Top and sides of the head and upper part of neck dark bluish-ash; rest of upper parts clear olive-green. A white ring round the eye, interrupted in the anterior canthus by a dusky lore, but the white color extending above this spot to the base of the bill. Under parts white; the sides under the wings greenish-yellow. Two bands on the wing coverts, with the edges of the secondaries, greenish-white. Outer tail

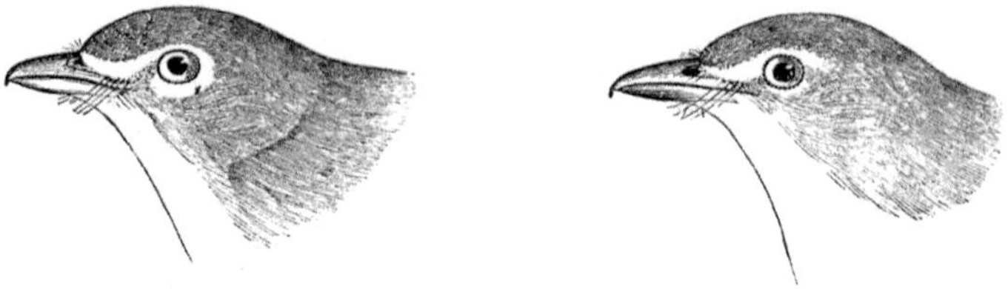

feather with its edge all round, including the whole outer web, whitish. Length about 5.50 inches; extent, 9.00; wing, 3.00. Iris brown; bill black, bluish below; feet lead-color.

Hab. United States, from the Atlantic to the Pacific.

This species reaches Puget's Sound by the 1st of May, but I have not noticed their arrival in the southern part of this State, and have seen none there, except a few at Fort Mojave, Colorado Valley, after May 14th, when they began to make themselves conspicuous by singing, but in a few days seemed to have passed away towards the north. It is quite possible that

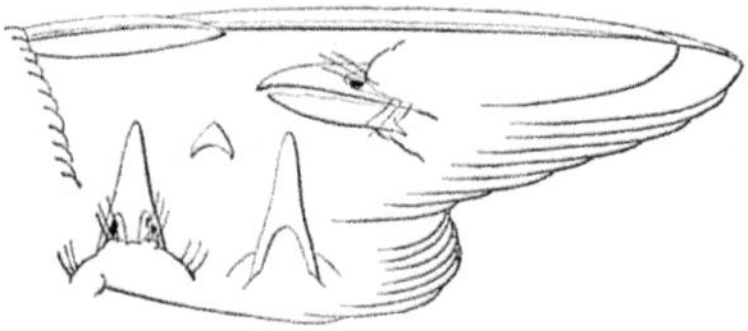

V. solitarius.

they winter in the State. In the Eastern States they are so rare and scattered that their migrations seem to be little understood.

I have seen them, with a nest suspended high in an alder-tree, in May, at the eastern base of the Coast Range, and found them quite common in summer at the Columbia River. According to Nuttall, their nest is sometimes built quite low down, constructed of dead grass, leaves, cobwebs, and externally with shreds of moss (*Hypnum*) glued on, to resemble the branches around. It is lined with fine blades of grass and root-fibres. The eggs, according to Audubon, are four or five, white tinged with flesh-color, and with brownish-red spots near the larger end.

Their favorite resorts are the deciduous oaks, which are most abundant northward.

Their song consists of a variety of notes, slowly delivered in an interrupted manner, and having a rather plaintive but sweet tone, very different from the lively warble of *V. gilva.*

The bird described as *V. Cassini,* by Mr. Xantus, is probably only this species in winter dress. A figure of it is given at the head of the article, side by side with that showing the spring plumage.

Vireosylvia plumbea, Coues.

THE LEAD-COLORED GREENLET.

Vireosylvia plumbea, Coues, Pr. A. N. S. 1866. — Baird, Rev. Amer. Birds, 349.

Sp. Char. Very similar to *V. solitaria*, but larger, with longer wings and tail. The olive-green of *solitaria* is replaced by plumbeous, and the yellowish by white. There is only a faint trace of olive on the lower back and flanks. The olive edgings of the wing

are replaced by ash; of the tail, by whitish. Total length, 6.10; extent of wings, 10.80; wing, 3.25; tarsus, 0.75.

Hab. Rocky Mountains and Arizona to Colima, Mexico.

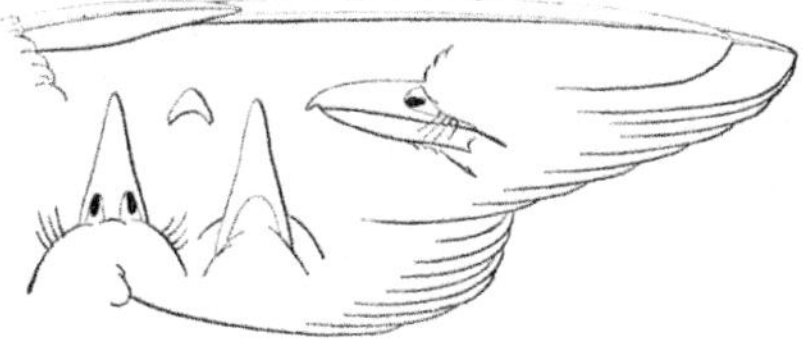

V. plumbea.

This fine species was first described by Dr. Coues, from specimens collected at Prescott, Arizona, where it was very abundant. No notice of its habits has yet been published.

It has been collected at various places through the Rocky Mountains, and appears to be quite abundant at Laramie Peak. The differences from *solitaria* are readily appreciable in comparison. (Baird.)

GENUS **VIREO**, VIEILLOT.

Vireo, VIEILLOT, Orn. Am. Sept. 1, 1807, 83. (Type, *Muscicapa noveboracensis*, BONAP.)

CHAR. Wing rounded, rather longer than the more or less rounded tail; spurious quill lengthened, one third or more as long as the second; second generally, and third almost always, longer than the secondaries; third or fourth quill usually longest. Bill rather compressed.

V. noveboracensis.

This genus is represented in the United States by several species, and by a still larger number in Middle America. All are of plain colors, and, as far

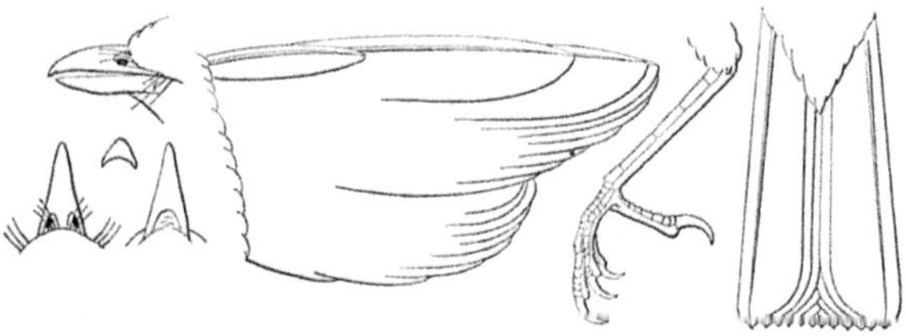

V. noveboracensis.

as we know anything of them, they have the usual gentle and inoffensive characteristics of the type.

Vireo atricapillus, Woodhouse.

THE BLACK-HEADED GREENLET.

Vireo atricapillus, Woodhouse, Pr. Ac. N. Sc. 1852, 60. — Cassin, Illust. 1854, 153; pl. xxiv. — Baird, Birds N. Amer. 1858, 337. Ib. Rev. Birds Amer. 353.

Sp. Char. Top and sides of head and neck black, rest of upper part olive-green. Wing and tail feathers almost black on their upper surface; the quills and rectrices edged

with olive; the wing coverts with two greenish-white bands on a blackish ground. Broad line from bill to and around eye, with under parts white. Bill black. Length, 4.75; extent, 7.25; wing, 2 25; tarsus, 0.75.

Hab Western Texas

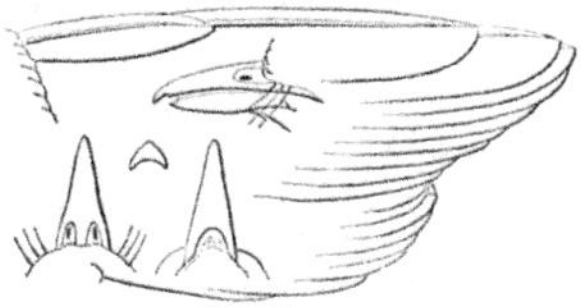

V. atricapillus.

This species, unique among the *Vireonidæ,* is very little known, only two or three specimens having so far been obtained.

Vireo Huttoni, Cassin.

HUTTON'S GREENLET.

Vireo Huttoni, Cassin, Pr. A. N. Sc. Phil. V. Feb. 1851, 150. Ib. VI.; pl. i. f. 1. — Baird, P. R. Rep. IX. Birds N. Amer. 339; pl. 78, f. 2 (eye too small). Ib. Rev. Amer. Birds, 357.

Sp. Char. Fourth, fifth, and sixth quills about equal and longest; third and seventh equal, and 0.10 of an inch shorter; second quill not longer than secondaries; spurious primary large, broad, about half the second. Above, olive-green, becoming considerably darker towards the bill and on sides of head. Beneath, dirty greenish-white, tinged with greenish-yellow posteriorly. A paler ring around the eye. Two broad bands across the

wing coverts and edges of inner tertiaries, with greater portion of outer web of the outer tail feather greenish or olivaceous white. Length, 5.00; extent, 7.75; wing, 2.50. Iris brown; bill black, blue below; feet lead-color.

Hab. California, across by Valley of Gila to Northeastern Mexico.

My attention was first attracted to this bird near San Diego late in February, having before mistaken it for the ruby-crowned wren, which winters there in abundance, and resembles this species closely in appearance and habits. Two of them came down to within a few feet of where I sat, scolding in a harsh tone, when I noticed their larger size and different plumage, besides their remarkably large eyes, and a peculiar slowness and delibera-

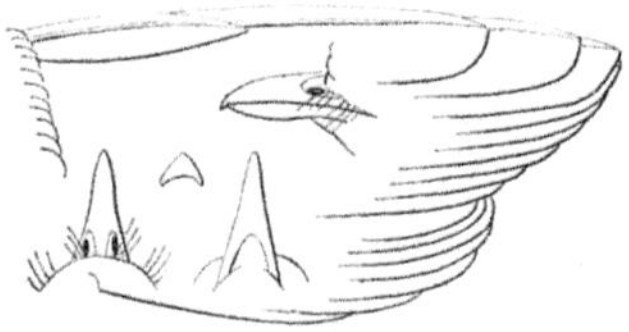

V. Huttoni.

tion in their movements, as they watched and searched the dark foliage for lurking insects. I have since found them wintering plentifully up to latitude 38°; and having observed but few in the Coast Range in May, I think most of them go farther north in summer. At San Diego, however, I shot a female on the 9th of March, containing an egg nearly ready to be laid, but have never been able to find a nest, which is probably built in the dense shade of their favorite evergreen oaks (*Quercus agrifolia*). Their song consists of a few short and quaint notes.

On the table-lands of Mexico, especially about Orizaba, this species is said to be abundant, and resident throughout the year.

Vireo Belli, AUDUBON.

BELL'S VIREO.

Vireo Belli, AUDUBON, Birds Amer. VII. 1844, 333 ; pl. 485 (Missouri). — CASSIN, Pr. A. N. Sc. V. Feb. 1851, 150. — BAIRD, P. R. Rep. IX. Birds, 335.

SP. CHAR. Similar to *V. gilvus*, but smaller. Olive-green above, tinged with ashy on the top and sides of head. A short line from the bill over the eye, and region around lower eyelid, white ; lores dusky. Beneath, yellowish-white ; on the sides of body and

posteriorly, sulphur yellow. Two faint bars of whitish across the wing coverts ; inner tertiaries edged broadly with whitish. Third quill longest, the rest successively shorter, except the second, which is a little shorter than the seventh. Spurious primary about

V. Belli.

two fifths the second, and more than one third of the third. Length, 4.50 ; extent, 7.00 ; wing, 2.25 ; tail, 2.25. Iris brown ; bill horn-color, below pale bluish ; feet lead-color.

Hab. Missouri River and Eastern Texas.

This species appears to belong to the region between the Rocky Mountains and the Missouri River, where it is by no means rare. The nest, as in other species, is built between two forks of a horizontal twig, and with the eggs are much like those of the white-eyed greenlet, *V. noveboracensis.*

Vireo pusillus, Coues.

Vireo pusillus, Coues, Pr. A. N. S. 1866, 76. — Baird, Rev. Amer. Birds, 360.
Vireo Bellii, Cooper, Pr. Cal. Acad. 1861, 122.

Sp. Char. Similar in general appearance to *V. gilvus*, but smaller. Tarsi lengthened; tail graduated. Above, grayish-ash. Beneath white, with a soiled tinge across

the breast. Eyelids and a short line from nostril to eye whitish. Primaries edged faintly with whitish. Length, 4.80; wing, 2.25; tail, 2.25; tarsus, 0.73.

Hab. Cape St. Lucas, San Diego, and Arizona.

This species was first known from specimens collected at Cape St. Lucas by Mr. Xantus. I found it rather common along the upper part of Mojave River in June, 1861, and in the following spring, about April 20th, they began to arrive at San Diego in considerable numbers. In habits, as far as observed, they resemble *V. gilvus*, but differ much in song; those that I

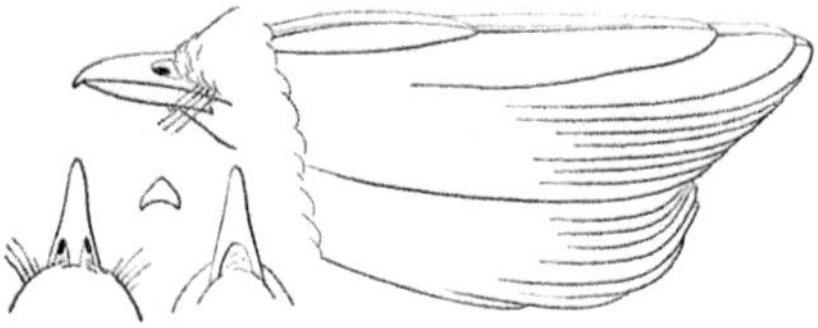

V. pusillus.

heard singing very much like the *Polioptilæ*, which utter a quaint mixture of the notes of the wrens, swallows, and vireos. Like other vireos, they have more or less of an imitative power.

At Sacramento I frequently saw and heard in the willows along the river what, from its peculiar note, I felt pretty certain was this species. As I wished to find its nest, I refrained from shooting specimens, but did not succeed in my object.

Vireo vicinior, Coues.

THE GRAY GREENLET.

Vireo vicinior, Coues, Pr. Ac. N. Sc. 1866, 75. — Baird, Rev. Amer. Birds, 361.

Sp. Char. Above, with sides of head and neck ashy or light plumbeous, faintly olivaceous on the rump. Beneath white, slightly ashy on sides of breast. Flanks and inside of wings showing a faint trace of yellow, only appreciable on raising the wings.

A faint line from bill to eye, and a more distinct ring round the eye, white. No bands on the wings; the quills edged internally with white. Iris brown. Length, 5.10; wing, 2.50; tail, 2.60; tarsus, 0.72.

Hab. Vicinity of Prescott, Arizona.

This strongly marked species, readily distinguishable from any other of its allies, was obtained by Dr. Coues near Prescott, Arizona, and is one of several new species which we owe to his indefatigable industry. Nothing was observed of its habits.

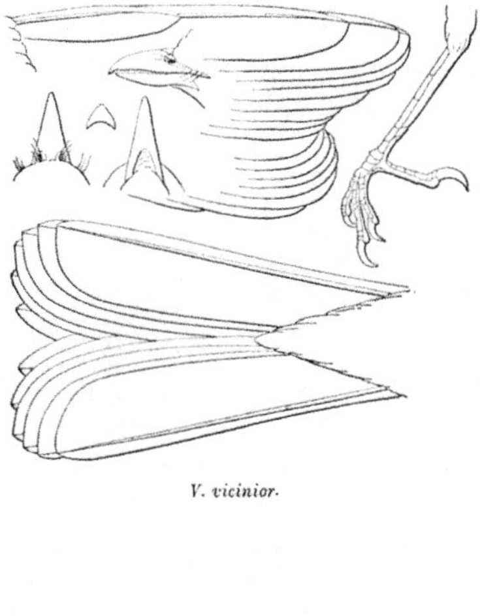

V. vicinior.

Family AMPELIDÆ, The Chatterers.

Char. Basal joint of middle toe about equal to that of inner toe; not united to that of lateral toes, for more than half the length. Lateral plates of tarsi, with tendency to subdivision. Lateral toes nearly equal. Lower edge of lower jaw less than half total length of jaw. Bill short, broad at base, and much depressed; feathers of forehead soft. Nostrils overhung by membrane. Tarsus equal to or less than middle toe and claw.

Sub-Family AMPELINÆ.

Char. Nostrils elongated, linear; the frontal feathers extending close to and concealing them. Wings very long and acute; outer primary very minute and inappreciable. Wings nearly twice the length of the short, nearly even tail. Secondary quills with flat horny appendages at the end of the shafts, like red sealing-wax.

Genus **AMPELIS**, Linnæus.

Ampelis, Linnæus, Syst. Nat. 1735. (Type, *A. garrulus.*)
Bombycilla, Vieillot, Ois. Am. Sept. I. 1807, 88. (Type, *B. cedrorum.*)

Gen. Char. Head with a broad depressed crest. Bill very broad, opening nearly to the eye; a series of short velvety feathers at the base of the bill, with bristles directed

A. garrula.

forwards and covering the nostrils, but none along the rictus. Commissure straight. Culmen and gonys curved, convex; both mandibles notched at tip. Legs stout; tarsi shorter than the middle toe; scutellate anteriorly, and slightly on the lower half of the

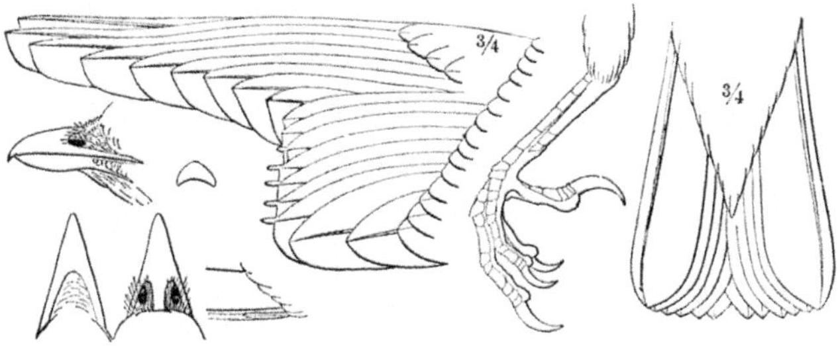

A. garrula.

sides behind; slightly feathered above. Hind toe shorter than the lateral, which are equal. Wings very long, pointed, reaching almost to the tip of the nearly even tail. First primary so short as to be with difficulty discernible; the second quill longest. Tips of secondary quills with horny appendages, like sealing-wax.

Of this genus one species belongs to the northern parts of both the New and Old World, one is peculiar to North America, and a third to Siberia and Japan.

Ampelis garrulus, Linnæus.

THE WAX-WING; THE BOHEMIAN CHATTERER.

Ampelis garrulus, Linnæus, Syst. Nat. 1766, 297. — Bonaparte, 2d List, 1842. Ib. Consp. 1850, 336. — *Bombycilla garrula*, Bonaparte, Zool. Jour. III. 1827, 50. Ib. Am. Orn. III. 1828, pl. xvi. — Rich. F. B. A. II. 1831, 237. — Audubon, Orn. Biog. IV. 1838, 462; pl. 363. Ib. Birds Amer. IV. 1842, 169; pl. 246. — Baird, P. R. Rep. IX. Birds, 317. Ib. Rev. Amer. Birds, 405.

Sp. Char. Much crested. General color brownish-ash, with a faint shade of reddish, especially anteriorly; the forehead, sides of the head, and under tail coverts, brown-

ish-orange; the hinder parts purer ash; the region about the vent white. Primaries and tail feathers plumbeous black, especially towards the tips; the tail with a terminal band of yellow. A narrow frontal line passing backward and involving the eye, and extending above and behind it, chin and upper part of throat black. Tips of the secondary coverts, and a spot on the end of the outer webs of all the quills, white; those on the inner primaries glossed with yellow. Secondaries with red horny tips, like sealing-wax. Side of the lower jaw whitish. Length, 8.00; extent, 13.75; wing, 4.50; tail, 3.00. Iris brown; bill and feet black.

Hab. Northern parts of both continents. Seen in the United States only in severe winters, except along the great lakes. In the Mississippi Valley, south to Fort Riley, Kansas; Colorado Valley to Fort Mojave, lat. 35°.

This very beautiful bird is one of the most boreal, in its preference of a residence, of any of the smaller birds, rarely descending south of lat. 45°, even in the cold of the Eastern winter climate, and ranging entirely around the arctic circle. It is probable, however, that they reside during summer about the summits of some of the loftiest mountains of the interior ranges, if not of the Sierra Nevada, as I have seen them in September at Fort Laramie, and the specimen obtained on the Colorado was a straggler from some neighboring mountains. It appeared on January 10th, after a stormy period which had whitened the tops of the mountains with snow, and was alone, feeding on the berries of the mistletoe, when I shot it. It made no noise, except a sort of hissing cry, and though called "chatterer," it is usually a very silent bird. When in flocks, which are sometimes enormous, they, however, make a loud twittering, and are so irregular in their wanderings as to excite great curiosity when they appear in the more southern parts of Europe. Their food consists chiefly of berries, and of insects during summer.

Ampelis cedrorum, VIEILLOT.

THE CEDAR-BIRD.

Ampelis garrulus, Var. B. LINNÆUS, Syst. Nat. I. 1766, 297.
Bombycilla Carolinensis, BRISSON, Orn. II. 1760, 337. — AUDUBON, Orn. Biog. I. 1831, 227; V. 494; pl. 43. IB. Birds Amer. IV. 1842, 165; pl. 245. — NEWBERRY, P. R. Rep. VI. iv. 81.
Bombycilla cedrorum, VIEILLOT, Ois. Am. Sept. I. 1807, 88; pl. lvii. IB. Galerie Ois. I. 1834, 186; pl. cxviii.
Ampelis Americana, WILSON, Am. Orn. I. 1808, 107; pl. vii.
Ampelis cedrorum, BAIRD, P. R. Rep. IX. Birds, 318. — HEERMANN, X. vi. 56. — COOPER and SUCKLEY, XII. iii. Zool. of W. T. 187.

SP. CHAR. Head crested. General color reddish-olive, passing anteriorly, on the neck, head, and breast, into purplish-cinnamon; posteriorly, on the upper parts, into ash; on the lower, into yellow. Under tail coverts white. Chin dark sooty-black, fading insensibly into the ground color on the throat. Forehead, loral region, space below the eye, and a line above it on the side of the head, intense black. Quills and tail dark plum-

beous, passing behind into dusky; the tail tipped with yellow; the primaries, except the first, margined with hoary. A short maxillary stripe; a narrow crescent on the infero-posterior quarter of the eye white. Secondaries with horny tips, like red sealing-wax. Length, 7.25; extent, 11.75; wing, 4.00; tail, 2.60. Iris brown; bill and feet black.

Hab. North America generally, south to Guatemala, north to lat. 50°.

In the settled districts of this coast the little wax-wing is as yet a rather rare bird, but will probably increase in abundance with cultivation and the abundance of its favorite fruits. The comparative scarcity of wild berries in most parts of the mountains makes it also rather an uncommon bird, but in winter I have seen small flocks as far south as San Diego, feeding on the mistletoe-berries; and they appear then in various parts of the lower country, even at San Francisco. They are at that season very silent, sitting still most of the day after gorging themselves to the throat with berries, rarely uttering their low hissing call.

They are at all seasons silent and gregarious, merely scattering in pairs in early summer, when they build a nest generally in a low tree, composed of fibrous grass, leaves, bark, and down of plants, in which they lay four or five eggs of a pale white color, with a few dark spots at the large

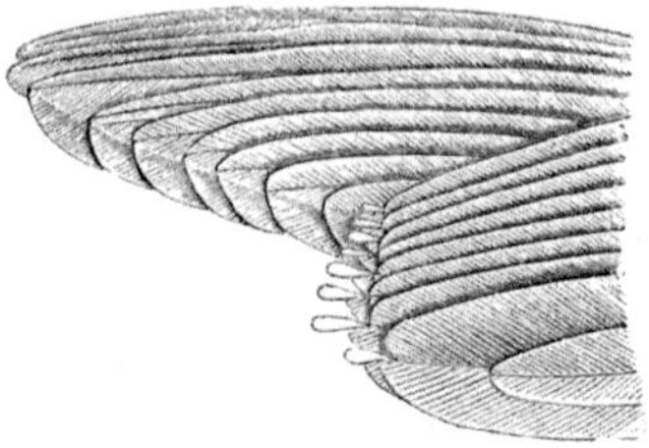

end. (Nuttall.) I have seen a nest at Fort Vancouver, and they are doubtless to be found breeding in this State. They raise two broods, and are easily reared, becoming very tame in a cage.

SUB-FAMILY PTILOGONATINÆ.

CHAR. Legs, moderate. Nostrils oval, naked, not overhung by the frontal feathers. Wings graduated, shorter than the somewhat broad, fan-shaped tail; ten distinct primaries; the first quill nearly half the second.

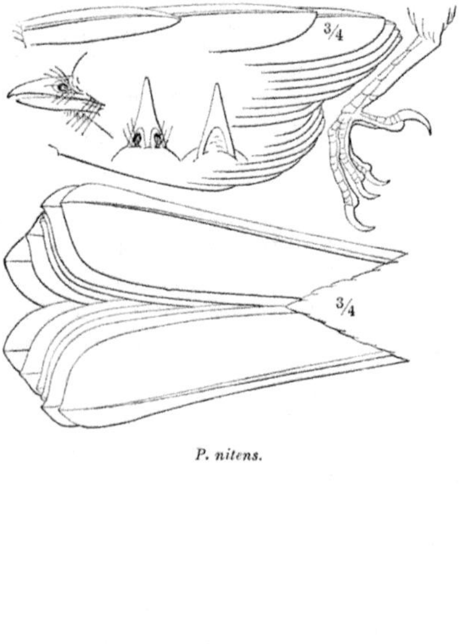

P. nitens.

GENUS **PHAINOPEPLA**, SCLATER.

Phainopepla ("SCLATER"), BAIRD, P. R. Rep. IX. 923, Oct. 1858.

GEN. CHAR. Head with an occipital crest of long narrow feathers. Bill weak, depressed, decidedly narrower than the length of the culmen. Base of bill with short bristles. Tarsi scutellate, bare above; shorter than the middle toe; outer lateral toe

P. nitens.

rather the longer; equal to the hinder. Wings and tail rather long; the former shortest; the first quill half the length of the second, and two fifths the fifth, or longest. Tail feathers broad, widening to the rounded tip; the tail moderately graduated; the middle ones longest.

Phainopepla nitens, SWAINSON.

THE BLACK FLYCATCHER.

Ptilogonys nitens, SWAINSON, Anim. in Menag. 2½ Cent. 1838, 285. — BONAPARTE, Consp. 1850, 335. — HEERMANN, J. A. N. Sc. II. Jan. 1853, 263. IB. P. R. Rep. X. vi. 38. — CASSIN, Illust. I. 1854, 169; pl. xxix. — *Cichlopsis nitens*, BAIRD, P. R. Rep. IX. Birds, 320. — *Phainopepla nitens* ("SCLATER, MSS."), BAIRD, IX. 923. — KENNERLY, X. iv. 22.

SP. CHAR. Head with an elongated occipital crest. Exposed portion of spurious quill about half the length of the second, which equals the secondaries; sixth quill longest. Tail graduated. *Male* throughout of a uniform lustrous black, glossed with green. Inner webs of the primaries white, except at the base, tips, and margins. *Female*, ash-color, paler beneath; the quills, wing, and lower tail coverts and outer tail feathers edged with whitish; rest of tail feathers blackish. Length, 8.00; extent, 11.50; wing, 4.00; tail, 4.30. Iris red; bill and feet black.

Hab. Rio Grande, New Mexico, west to San Diego, California; north to lat. 39°, in California.

This bird, which is in habits and appearance much more like the fly-catchers than the wax-wings, is yet connected with the latter more closely in structure, and has even some sweet notes, indicating a greater affinity to the *Oscines* than to the *Clamatores*.

On the Colorado I found them numerous, especially in winter, and they do not migrate much south of lat. 35°, though found in summer as far north at least as Cosumnes River. (Heermann.) I also found them rather common along the Mojave River in December, and south of Los Angeles to San Diego in summer.

They prefer the vicinity of the trees on which the mistletoe grows, as its berries form much of their food during the whole year, but they also watch for insects from the summit of some low tree, occasionally flying after one, and pursuing it in a zigzag course, very much like the *Sayornis nigricans*. They almost constantly utter a loud cry of alarm or warning, and when pursued are very wild, requiring much artifice in winter to shoot them. If wounded, they conceal themselves so fully in the thick tufts of mistletoe as to be found with much difficulty.

Many left the Colorado Valley early in April, but a few pairs remained, and the males uttered a few short musical notes rather like those of the *Myiarchus*, but sweeter. On the 25th of April I found a nest built on a branch of the mesquite (*Algarobia*), twelve feet from the ground.

When at rest they have the same habit as the pewees of jerking the tail and erecting their crest. When flying, the white spot on the spread wings becomes very conspicuous; and in the deserts along the Mojave River, every thicket of mesquite was frequented by one or more of them, some being constantly on the wing, in their gyratory flight after insects, giving some appearance of life to those otherwise desolate regions in winter.

It was found by the late Captain Feilner, at Fort Crook, in April, 1860, and was probably migratory in that northern and elevated region. It does not seem to occur near the coast in the northern half of California.

GENUS **MYIADESTES**, SWAINSON.

Myiadestes, SWAINSON, Naturalists' Library; Flycatchers, 1838. (Type, *Muscicapa armillata*, VIEILLOT.)

GEN. CHAR. Head not crested. Bill rather narrower than the length of the culmen, much depressed; somewhat attenuated at the end; lateral outline rather concave. Tarsi without feathers above or scutellæ; shorter than the middle toe. Hind toe rather shorter

M. Townsendii.

than the outer lateral toe, which barely reaches the base of the middle claw. Tail and wings very long; the former shorter, quite deeply forked, but the outer lateral feather abruptly graduated, and a little longer than the innermost; the feathers all broad at the

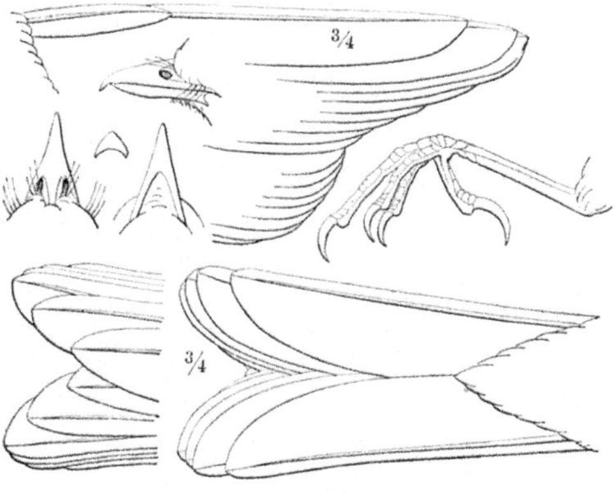

M. Townsendii.

base, and tapering to the tip. Spurious primary nearly one fourth the longest (third), the second a quarter of an inch less than the longest.

There are other species in Mexico.

Myiadestes Townsendii, AUDUBON.

TOWNSEND'S FLYCATCHER.

Ptiliogonys Townsendii, AUDUBON, Orn. Biog. V. May, 1839, 206 ; pl. 419, f. 2. IB. Birds Amer. I. 1840, 243 ; pl. 69. — NUTTALL, Man. I. 2d ed. 1840, 361. — GAMBEL, Pr. A. N. Sc. I. 1843, 261. — NEWBERRY, P R. Rep. VI. iv. 82. — *Myiadestes Townsendii*, CABANIS, Wiegm. Arch. I. 1847, 208. — BAIRD, P. R. Rep. IX. Birds, 321. — KENNERLY, X. iv. 25. — HEERMANN, X. vi. 38. — COOPER and SUCKLEY, XII. iii. Zool. of W. T. 187.

SP. CHAR. Tail rather deeply forked. Exposed portion of spurious quill less than one third that of the second ; fourth quill longest ; second a little longer than the sixth. Head not crested. General color bluish-ash, paler beneath ; under wing coverts white. Quills with a brownish-yellow bar at the base of both webs, mostly concealed, but showing a little below the greater coverts and alulæ ; this succeeded by a bar of dusky, and next to it another of brownish-yellow across the outer webs of the central quills only. Tertials

Adult.

Young.

tipped with white. Tail feathers dark brown ; the middle ones more like the back ; the lateral with the outer web and tip, the second with the tip only, white. A white ring round the eye. Length, 9.00 ; extent, 14.00 ; wing, 4.50 ; tail, 4.00. Iris brown ; bill and feet black.

Hab. United States from Rocky Mountains to the Pacific ; south to the borders of Mexico.

Although plain in plumage, and retiring in habits, this bird is one of the most interesting in the Western country, for, like its not distant relative, the European nightingale, it compensates by its delightful melody for its deficiencies in beauty. Having seen them in the Rocky Mountains, where they seemed merely plain and silent flycatchers, with habits similar to those of *Sayornis sayus*, my astonishment when first I heard one sing in the Sierra Nevada was indeed great; and if I had not shot the bird immediately, I could not have believed that one belonging to the same family as the nearly silent wax-wings and *Phainopepla* could sing with such power, variety, and sweetness. Their song can be compared with nothing uttered by any other bird I have heard in the United States, for it excels that of the mocking-bird in sweetness, besides being entirely original. It has a sort of melancholy slowness, but without the interruptions of that of the Eastern wood-thrush (*Turdus mustelinus*), and agrees better with the descriptions of that of the nightingale of Europe.

I saw only a few of this species among some junipers on the western slope of the mountains, not far from the summit, in September, 1863. The scarcity of the juniper on the western slope, towards the north, seems to be the reason why this bird is not more frequent there; as, according to all accounts, they are found wherever that tree grows in abundance, especially on the mountain ranges of the great interior basin, and their extensions to the north and south.

According to Newberry, thousands of them frequented the east slope of the Cascade Mountains of Oregon, producing even in the fall a perfect chorus of melody from daylight to dark. They feed on the juniper-berries as well as insects, pursuing the latter like flycatchers.

They may be somewhat gregarious, though I have always met with them nearly singly. Dr. T. C. Henry found them at Fort Webster, New Mexico, in large numbers in fall and winter. (Heermann.) They are very scarce northward, west of the Cascade Mountains; and in the Rocky Mountains about lat. 47°, I met with but one. Their home, therefore, seems to be in the vicinity of the great deserts of the central regions, on the cedar-covered mountains that intersect them, where, in a comparatively lifeless and barren wilderness, they make up for the absence of other songsters by a musical talent excelling all.

Their nest and eggs, together with their habits during most of the year, still remain to be described. I have seen a specimen shot at Forest Hill, Nevada County, by Mr. F. Gruber.

FAMILY LANIIDÆ, THE SHRIKES.

CHAR. Feet constructed as in *Ampelidæ*, but much stouter; the tarsus longer than middle toe and claw; sometimes scutellate on both sides. Lower outline of bill, about half the length of lower jaw. Bill very powerful and raptorial, decurved at end, where it is provided with a sharp hook, notch, and tooth. Nostrils concealed by bristly feathers of forehead. Wings much rounded. Tail long and graduated. Primaries ten; the first about half the second.

Genus **COLLURIO**, Vigors.

Collurio, Vigors, Pr. Zool. Soc. 1831, 42. (Type, *Lanius excubitor*, L.) (Gray, Gen. 1844.)
Lanius, of Authors.

Gen. Char. Feathers of forehead stiffened; base of bill, including nostrils, covered by bristly feathers directed forward. Bill shorter than the head, much compressed, and very powerful. Culmen decurved from base, the mandible abruptly bent down in a

C. excubitoroides

powerful hook, with an acute lobe near the tip. Tip of lower mandible bent upwards in a hook; the gonys very convex. Rictus with long bristles. Legs stout; the tarsi rather short, longer than the middle toe; the lateral equal; the claws all very sharp and

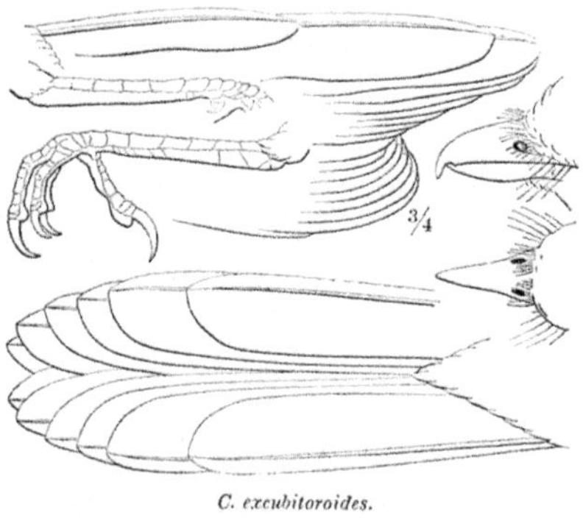

C. excubitoroides.

much curved. Wings rounded; the first primary about half the second, which is equal to the sixth or seventh. Tail longer than the wings, much graduated, the feathers broad.

Collurio borealis, VIEILLOT.

THE NORTHERN SHRIKE; THE BUTCHER-BIRD.

Lanius borealis, VIEILLOT, Ois. Am. Sep. I. 1807, 90; pl. i. — SWAINSON, F. B. Am. II. 1831, 111. — AUDUBON, Syn. 1839, 157. — ID. Birds Amer. IV. 1842, 130; pl. 236.
Collyrio borealis, BAIRD, P. R. Rep. IX. Birds, 324. — COOPER and SUCKLEY, XII. iii. Zool. of W. T. 188.
Lanius excubitor, FORSTER, Phil. Trans. LXII, 1772, 382. — WILSON, I. 1808, 74; pl. v. f. 1. AUDUBON, Orn. Biog. II. 1834, 534; pl. 192.
Lanius septentrionalis, BONAPARTE, Syn. 1828, 72. — NUTTALL, Man. I. 1832, 258. ID. I. 2d ed. 1840, 285. (Not of Gmelin.)

SP. CHAR. Above, light bluish-ash, obscurely soiled with reddish-brown. Forehead, sides of the crown, scapulars, and upper tail coverts, hoary-white. Beneath white, the

Summer.

breast with fine transverse lines. Wings and tail black: the former with a white patch at base of primaries and tips of small quills; the latter with the lateral feathers tipped with white. Bill blackish-brown, considerably lighter at the base. A black stripe from

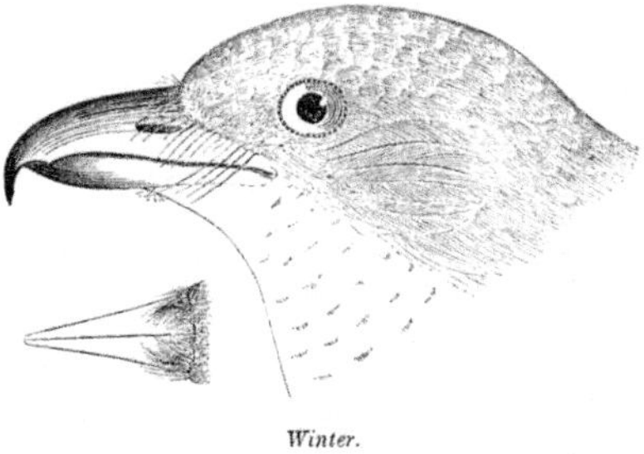

Winter.

the bill through and behind the eye, but beneath the latter interrupted by a whitish crescent. Female and young with the gray soiled with brownish. Length, 10.25; extent, 14.50; wings, 4.50; tail, 4.80; its graduation, 0.90. Iris brown.

Hab. Northern regions from the Atlantic to the Pacific.

This northern species probably visits the northeastern parts of the State in considerable numbers in winter, though I can find no mention of its occurrence in the State, except by Dr. Gambel (Proc. Ac. Nat. Sc. Phil. and Jour. 1846 – 49), who mentions seeing it in the "Californian mountains." They are common at the Columbia River, in October and throughout the winter. It is possible that they reside in the higher mountains in summer.

Their nest and eggs are said to resemble those of *C. excubitoroides*, and their habits while with us are mostly like those of that species. They are, however, more shy and savage in disposition, preying at times on mice and small birds as well as insects, which they impale on thorns, when they are not hungry, for future use. They, however, leave many untouched afterwards. They have been known to pursue and attack cage-birds with all the ferocity of a hawk, and to catch a wounded robin, larger than themselves.

They are said to utter some musical notes, and to imitate other birds. (Nuttall.)

Captain Feilner sent this species to the Smithsonian Institution from Fort Crook, and reported that it was common in the colder months in the northeastern portion of the State. Dr. E. Coues also found it at Fort Whipple, Arizona.

Collurio excubitoroides, SWAINSON.

THE WHITE-RUMPED SHRIKE.

Lanius excubitoroides, SWAINSON, F. Bor. Am. II. 1831, 115. — GAMBEL, Pr. A. N. Sc. III, 1847, 200. — HEERMANN, P. R. Rep. X. vi. 55. — *Collyrio excubitoroides*, BAIRD, P. R. Rep. IX. Birds, 327. IB. Birds N. Amer. pl. 75, f. 2. — KENNERLY, X. iv. 25.

SP. CHAR. Above, rather light pure bluish-ash. Forehead, sides of crown, scapulars, and upper tail coverts, hoary whitish. Beneath plain whitish. Wings and tail black; the former with a white patch at base of primaries and tips of small quills; the latter with the lateral feathers tipped with white, and this extending broadly at the base. Bill, throughout pitch black. A continuous black stripe from the bill through and behind the eye. Length, 9.25; extent, 12.25; wing, 4.00; tail, 4.35. Iris brown; bill and feet black.

Hab. Missouri Plains and fur countries to Pacific Coast. Eastward into Wisconsin, Illinois, and Michigan. (?)

An abundant species in all the plains-region of California and eastward, but it apparently does not extend north to the Columbia River, although found up to lat. 49°, on the east side of the Rocky Mountains. They reside constantly in this State south of lat. 38°, but I have not met with them on any part of the wooded mountains.

I found many about Fort Mojave in winter, which towards spring scattered among the neighboring cañons to build. As early as the 19th of March I found the first nest in a thorny bush, fully two miles from water, containing four eggs. These measured 1 × 0.74 inch, color dusky with specks of pale amber brown thickly scattered, and grouped in a ring round the larger end. (Others from San Diego are paler and a little shorter.)

The old bird was still very shy, keeping out of gunshot, and apparently caring little for its eggs. On April 4th I found another nest similar to the first, and on the 16th found the young hatched, when the female showed more solicitude, making a harsh squeak of complaint, but trying to keep well out of sight.

Near San Diego they were later in building, as I found a nest with fresh eggs, April 20th. I heard there, for the first time, some attempt at a song in December and January, the notes being, however, harsh like those of a jay, but not an imitation, as there were no jays in the vicinity.

I have seen one of this species catch a sparrow, but this is a rare occur-

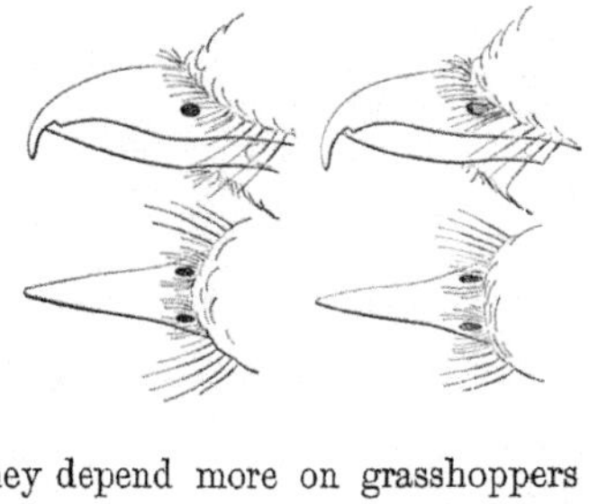

rence, I think. They depend more on grasshoppers and other insects, for which they watch patiently from a fence or other prominent point.

At Santa Cruz I saw none of this species until September, though some probably breed in the warm valleys back from the coast. In fall, however, they were common about the town, perching on the houses and flag-staffs, on the watch for insects, and occasionally uttering their loud harsh cries. One day I saw four of them circling round over the gardens in sportive flight, as if engaged in a game of "tag," screaming and flying more than I ever before noticed.

Specimens exhibit considerable differences in the size and shape of the bill, as shown by the preceding figures. One of the figures represents the amount of white on the secondary quills.

Collurio elegans, Swainson.

THE WHITE-WINGED SHRIKE.

Lanius elegans, Swainson, Fauna Bor. Am. II. 1831, 122. — Nuttall, Man. I. 2d ed. 1840, 287. — Gambel, Pr. A. N. Sc. I. 1843, 261. — Baird, P. R. Rep. IX. Birds, 328 Ib. Birds N. Amer. pl. lxxv. f. 1.

Sp. Char. "Clear bluish-gray, beneath unspotted white, frontlet the same color with the head; a broad white band across the wing; a slender and very cruciform tail, entirely bordered with white; the second quill longer than the sixth, the fourth longest;

tarsi exceeding the length of bill, measured from the angle of mouth. Length, 9.75; tail, 4.42; bill from angle of mouth, 0.92; above, 0.67; tarsus, 1.21." (Swainson.)

Hab. Of original specimen, uncertain, but somewhere in Western North America.

In his "Birds of North America" Professor Baird figures a specimen obtained by Dr. Gambel, somewhere in "California," as probably this species, and refers to the same specimen, on page 327, as differing in some respects from both *C. excubitoroides* and "*C. elegans*." It is darker above than the former, head like back, rump but little paler, black extending farther down cheeks, the under parts more bluish on the sides, and the secondaries with a white patch on base of inner web visible from above. The bill is one fifth longer (0.75 to 0.60), the nostril a fourth farther from the tip (0.60 to 0.46). This makes the bill longer even than that of Swainson's specimen (0.75 to 0.67). Such specimens being very rare would seem, perhaps, nothing more than a variety, if not a far northern species. All that I shot at Fort Mojave and San Diego were typically like *C. excubitoroides*.

The figure on the opposite page represents the extent of white on the secondary quill.

Family TANAGRIDÆ, The Tanagers.

Char. The precise position of the tanagers is a matter oi much uncertainty, the relationship to the American *Fringillidæ* being very close. Both have nine primaries and scutellate tarsi, and the bill in some genera resembles that of unquestionable finches; it is, however, usually longer, and though stout at the base, is not strictly conical, and lacks the great strength necessary for a hard vegetable instead of soft animal diet, or one of berries and fruits. The bill is also notched behind the tip; its margins are not acute or inflexed, as in the *Fringillidæ*.

P. Ludoviciana.

Genus PYRANGA, Vieillot.

Pyranga, Vieillot, Ois. Am. Sept. I. 1807, iv.
Phænisoma, Swainson, Class Birds, ii. 1837, 284.

Gen. Char. Bill somewhat straight; sub-conical, cylindrical, notched at tip; culmen moderately curved; commissure with a median acute lobe. Wings elongated; the four first primaries about equal. Tail moderate, slightly forked. Colors of the male chiefly scarlet, of the female yellowish.

Many species related to these inhabit the tropics of America, nearly all of which have very brilliant plumage.

Pyranga Cooperi, Ridgway.

COOPER'S TANAGER.

Pyranga æstiva, Cooper, Pr. Cal. Acad. Sc. 1861, 122.
Pyranga hepatica, Coues, Pr. Acad. Nat. Sc. Phil. 1866, 71.
Pyranga Cooperi, Ridgway, " " " 1869, 130.

Sp. Char. *Male*. Length, 8.60; extent, 13.50; wing, 4.24; tail, 3.68; bill (along culmen), 0.84; tarsus, 0.80. Generally rich pure vermilion, similar to that of *P. æstiva*, but brighter than in any Eastern examples of this, and less rosaceous than in Central American specimens. Upper surface less marked, purplish-dusky; the head and neck above being

Male.

abruptly lighter than the back, in some specimens scarcely darker than the throat. Exposed tips of primaries pure slaty-amber, each faintly margined terminally with whitish.

Female. Above, deep orange-olivaceous; beneath, olivaceous orange-yellow, purest medially; crissum pure deep Indian-yellow, the inner webs of tail feathers margined with the same. Distinct line of orange-yellow over lores. (Ridgway.)

Hab. Southern portion of Middle Province of United States, and south to Colima and Mazatlan.

This species, dedicated by Mr. Robert Ridgway to Dr. J. G. Cooper, in an analytical monograph of the red *Pyrangas*, may be readily distinguished from *P. æstiva* upon slight comparison. The size is much larger (length, 8.60, instead of 7.25; wing, 4.24, instead of 3.81), the bill especially so, being 0.84 along the culmen, instead of 0.70, as in *P. æstiva*. The colors

Female.

also differ very materially from those of *æstiva;* the red of the head above being scarcely different from that of the throat, instead of being purplish-dusky like the back, as in *æstiva*. The wing also is much more pointed than in *æstiva*, the primaries extending 1.16 beyond the tertiaries, instead of 0.84. The third quill is generally longest, but in some specimens the second and third are equal. In *P. æstiva* the second is generally longest. (Baird.)

This beautiful bird I found quite common, after April 25, 1861, at Fort Mojave, Colorado Valley, lat. 35°. They then frequented chiefly the tall cotton-woods, feeding on insects, and occasionally flying down to the *Larrea* bushes after a kind of bee found on them. Dr. Coues obtained the same species at Los Pinos in New Mexico, and it has also been found at Mazatlan and Colima.

They have a call-note, sounding like " Ke-dik," which singularly enough means, in the language of the Mojave natives, " Come here." They also sing in a loud, clear tone, and a style much like that of the robin, but with the faculty of making the sound appear very far distant, which is a protection to birds of such bright plumage.

Pyranga hepatica, Swainson.

THE LIVER-COLORED TANAGER.

Pyranga hepatica, Swainson, Phil. Mag. I. 1827, 438. — Baird, Birds N. Amer. 1858, 302. — Coues, Pr. A. N. S. 1866, 71.

Pyranga azaræ, Woodhouse, Sitgraves Expl. 1853, 82.

Sp. Char. Length, 8 00; wing, 4.12; tail, 3.36; bill (along culmen), 0.68; tarsus, 0.84. Bill shorter than in *P. æstiva*, as well as higher and broader at the base, becoming compressed toward the tip. A distinct prominent tooth about the middle of the commissure; color plumbeous-black, paler on lower mandible. Head above brownish scarlet, purer on forehead, the tips of the feathers more brownish. Rest of upper surface, with sides, brownish-ashy, tinged with dull hepatic-red; the wings, upper tail coverts, and tail inclin-

Male.

ing to dull hepatic-scarlet. Beneath, fine light scarlet, purest on the throat, paler posteriorly, the lateral parts much obscured by grayish.

Female. Above, greenish ashy-olivaceous, brightest on forehead; edges of the primaries, upper tail coverts, and tail more ashy on interscapular region. Beneath, nearly uniform greenish olivaceous-yellow, purest medially. Lores ashy, superciliary stripe olivaceous yellow.

Female.

Young Male. Much like the female, but forehead and crown olivaceous-orange, deepest anteriorly; superciliary stripe bright orange, purest over lore. Whole throat, maxillæ, and abdomen medially, rich orange, most intense, and tinged with orange chrome, on throat.

The red of this species is entirely different from that of any other, having a peculiar red-lead tinge; it is also more restricted in extent than in any other species, except the *P. testacea.* (Ridgway, MSS.)

Hab. Southern Rocky Mountains of United States, and mountainous regions of Mexico.

For the first introduction of this species to the fauna of the United States we are indebted to Dr. Woodhouse, who obtained a specimen in the San Francisco mountains of Arizona, and published it as *P. azaræ*.

Pyranga Ludoviciana, Wilson.

THE LOUISIANA TANAGER.

Tanagra Ludoviciana, Wilson, Am. Orn. III. 1811, 27; pl. xx. f. 1. — Audubon, Orn. Biog. IV. 1838, 385; V. 1839, 90; pl. 354, 400. — *Tanagra (Pyranga) Ludoviciana*, Nuttall, Man. I. 1832, 471, 2d ed. 543. — *Pyranga Ludoviciana*, Audubon, Syn. 1839, 137. Ib. Birds Amer. III. 1841, 211; pl. 210. — Sclater, Pr. Zool. Soc. 1856, 125. — Baird, P. R. Rep. IX. Birds, 303. — Heermann, X. vi. 52. — Cooper and Suckley, XII. iii. Zool. of W. T. 182.

Sp. Char. Bill shorter than the head. Tail slightly forked; first three quills nearly equal. *Male*, yellow; the middle of the back, the wings, and the tail black. Head and

Male.

neck all round strongly tinged with red; least so on the sides. A band of yellow across the middle coverts, and of yellowish-white across the greater ones; the tertials more or less edged with whitish.

Female.

Female, olive-green above, yellowish beneath; the feathers of the interscapular region dusky, margined with olive. The wings and tail rather dark brown, the former with the same marks as the male. Length, 7.50; extent, 11.50; wing, 3.75; tail, 2.85. Iris brown; bill horn-color, greenish below; feet lead-color.

Hab. From the Upper Missouri to the Pacific, south to Mexico.

I noticed the arrival of this beautiful bird, near San Diego, in small parties on the 24th of April, and they reach Puget's Sound about May 15th. The males come some time in advance, clothed in their full summer livery, and are more bold and conspicuous than the females, which are rarely seen without close watching. They frequent trees, feeding on insects and berries, and singing much in the same manner as the other species, and more like the robin and grosbeaks than any other birds.

I saw none of them in the Coast Range towards Santa Cruz, or at Santa Barbara, in summer, and suppose they must seek the higher and more northern regions at that season. Their nest and eggs have not been described. I found this species in September, 1860, in the higher Rocky Mountains, near the sources of the Columbia, in lat. 47°, and they probably remain until October within this State. In the fall the young and old, all in the same dull greenish plumage, associate in families, and feed on elderberries and other kinds, without that timidity which they have in spring. Although found as far east as the Rocky Mountains, I saw none along the Colorado Valley, probably because they migrate more in the line of mountain ranges.

FAMILY FRINGILLIDÆ, THE FINCHES.

CHAR. Primaries nine. Bill very short, abruptly conical and robust. Commissure strongly angulated at base of bill. Tarsi scutellate anteriorly, but the sides with two undivided plates meeting behind along the median line, as a sharp posterior ridge.

SUB-FAMILY FRINGILLINÆ.

CHAR. Wings very long and much pointed; generally one third longer than the more or less forked tail; first quill usually nearly as long as, or longer than the second. Tertiaries but little longer, or equal to the secondaries, and always much exceeded by the primaries. Bill very variable in shape and size, the upper mandible, however, as broad as the lower; nostrils rather more lateral than usual; and always more or less concealed by a series of small bristly feathers applied along the base of the upper mandible; no bristles at the base of the bill. Feet short and rather weak. Hind claw usually considerably longer than the middle anterior one; sometimes nearly the same size.

C. Americana.

This sub-family is composed of the brightest colored and largest of American sparrow-like birds, mostly inhabiting trees, and feeding on their seeds. They have usually large, stout bills, and sing better than those of most of the other divisions.

The sexes usually differ very much in colors, and the young males, as well as many of those in winter plumage, have the plain garb of the females.

Genus CURVIROSTRA, Scopoli.

Loxia, Linnæus, Syst. Nat. 1758. (Type, *Loxia curvirostra*, L.)
Curvirostra, "Scopoli, 1777." (Type, *Loxia curvirostra*.)

Gen. Char. Mandibles much elongated, compressed and attenuated; greatly curved or falcate, the points crossing or overlapping to a greater or less degree. Tarsi very

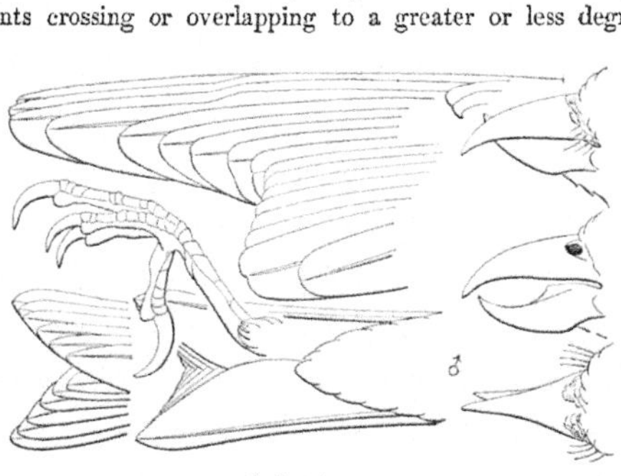

C. Americana.

short; claws all very long, the lateral extending beyond the middle of the central; hind claw longer than its digit. Wings very long and much pointed, reaching beyond the middle of the narrow, forked tail. Colors reddish in the male.

Curvirostra Americana, Wilson.

THE RED CROSSBILL.

Curvirostra Americana, Wilson, Am. Orn. IV. 1841, 44; pl. xxxi. f. 1, 2. — Baird, Birds N. Amer. 427. — *Loxia Americana*, Bonaparte, List, 1838. — Newberry, Zool. Cal. and Or. Route, P. R. R. Rep. VI. iv. 1857 87.

Loxia curvirostra, Forster Phil. Trans. LXII. 1772, No. 23. — Audubon, Biog. II. 1834, 559; V. 511; pl. 197. Ib. Birds Amer. III. 1841, 186; pl. 200.

Loxia Mexicana, Strickland, Baird, P. R. Rep. IX. Birds, 427, 924.

Sp. Char. *Male*, dull red; darkest across the back; wings and tail dark blackish-brown. *Female*, dull greenish-olive above, each feather with a dusky centre; rump and crown bright greenish-yellow. Beneath grayish; tinged, especially on the sides of the

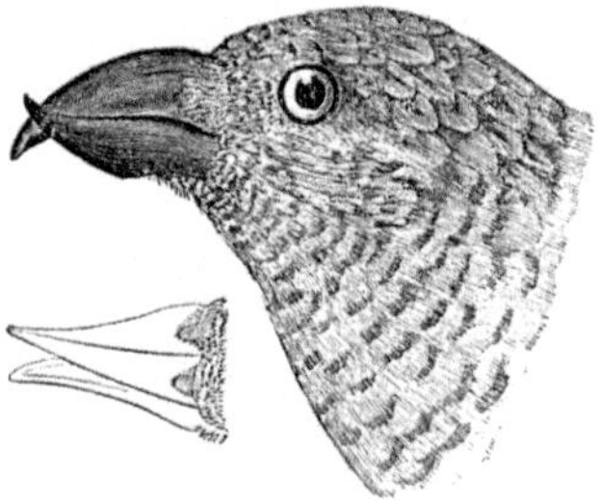

Male. California.

Female. Puget Sound.

body, with greenish-yellow. Young entirely brown; paler beneath. (Bill of the Western about 0.20 inch longer than that of the Eastern bird.)

Male, about 6 inches; extent, 10.25; wing, 3.30; tail, 2.25. Iris, bill, and feet, brown.

Hab. North America generally, coming southward in winter. Resident in the mountains of Pennsylvania and the Sierra Nevada, California.

I found this curious bird in considerable numbers about the summits of the Sierra Nevada, lat. 39°, in September, 1863, and have heard of their being shot near San Francisco in winter, when they also probably wander as far south as Monterey. Their migrations are guided chiefly by the prevalence of the coniferous trees, whose seeds constitute their principal food, especially those with small seeds and cones easily broken, such as the spruces, cypresses, and redwoods. I have also seen flocks of them descend to the ground in the Rocky Mountains in quest of some seeds of small plants, and they often come down to drink, especially about sunset.

When flying they have a loud, clear, and sharp call-note, but are so silent when feeding on the tops of the trees as to be found with difficulty. In snowy regions they become very tame, coming about houses in search of food, and seem at all times very unsuspicious of danger. In orchards they have been known to destroy apples merely for their seeds, and sometimes to eat the buds of trees.

In the spring the males have something of a song, but not very long nor connected. Their gay plumage at that season looks so much like the young red cones of the spruces, among which they sit, that they are not easily distinguished.

I never found their nests myself, but they are said to be built in the forks of fir-trees, as early as January or February, and the eggs, four or five, are greenish-gray, with a circle of reddish-brown spots, points, and lines, chiefly at the larger end, often also extending over the whole surface. (Nuttall.)

Male. Pennsylvania.

Female.

Their feet are unusually strong, and they use them as well as their bills in climbing like parrots, hanging to the cones, head downwards, while twisting off the scales by a dexterous turn of the bill, which, though looking deformed, is admirably adapted for this purpose.

Specimens from the mountains of California all appear to come very near to the variety with large bill, named *C. Mexicana* by Strickland, and resident in the mountains of Mexico. Others, however, from Washington Territory, have the bill more like the Eastern form.

Curvirostra leucoptera, GMELIN.

THE WHITE-WINGED CROSSBILL.

Loxia leucoptera, GMELIN, Syst. Nat. I. 1788, 540. — AUDUBON, Birds Amer. III. 190; pl. 201. — *Curvirostra leucoptera*, WILSON, Am. Orn. IV.; pl. xxxi. — BAIRD, Birds N. Amer. 427.

SP. CHAR. *Male*, carmine-red, tinged with dusky across the back. Sides of body under wings streaked with brown; from the middle of the belly to tail coverts whitish, the latter streaked with brown. Scapulars, wings, and tail black; two broad white bands on the wings across the ends of greater and middle coverts.

Female, brownish, tinged with olive-green in places; feathers of back and crown with

Male.

Female.

dusky centres; rump bright brownish-yellow. Two white bands on wings. Length, 6.25; wing, 3.50; tail, 2.60.

Hab. Northern part of America.

This species, though not as yet actually detected within the limits of the State of California, is yet so abundant in the regions north of it as to render its occurrence there highly probable. Its habits are much like those of the common crossbill. (Baird.)

Genus PINICOLA, Vieillot.

Pinicola, Vieillot, Ois. Am. Sept. I. 1807, p. iv.; pl. i. f. 13.
"*Corythus*," Cuvier, R. An. 1817.

Gen. Char. Bill short, nearly as high as long; upper outline much curved from the base; the margins of the mandibles rounded; the commissure gently concave, and abruptly

P. Canadensis.

deflexed at the tip; base of the upper mandible much concealed by the bristly feathers covering the basal third. Tarsus rather shorter than the middle toe; lateral toe short, but their long claws reach the base of the middle one, which is longer than the hind claw.

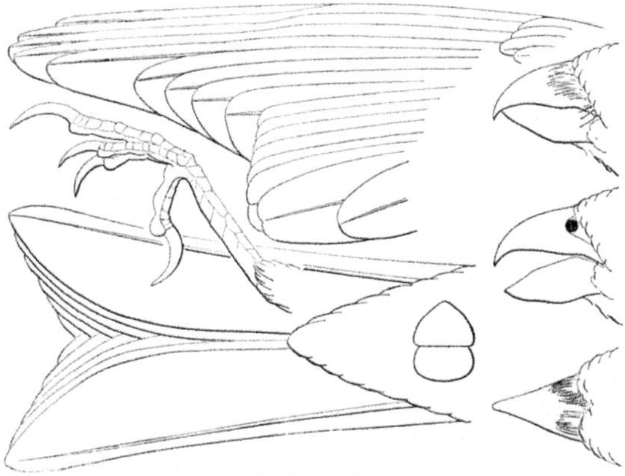

P. Canadensis.

Wings moderate; the first quill rather shorter than the second, third, and fourth. Tail rather shorter than the wings; nearly even.

Pinicola Canadensis, Brisson.

THE PINE GROSBEAK.

Loxia enucleator, Forster, Phil. Trans. LXII. 1772, 383. — Wilson, Am. Orn. I. 1808, 80; pl. v. — *Pyrrhula enucleator*, Audubon, Orn. Biog. IV. 1838, 414; pl. 358. — *Corythus enucleator*, Audubon, Birds Amer. III. 1841, 179; pl. 199.
Pinicola Canadensis, Cabanis, Mus. Hein. 1851, 167. — Baird, P. R. Rep. IX. Birds, 410.

Sp. Char. General color carmine-red, not continuous above however, except on the head; the feathers showing brownish centres on the back, where, too, the red is darker.

Male. New York.

Loral region, base of lower jaw all round, sides, and posterior part of body, with under tail coverts, ashy, whitest behind. Wing with two white bands across the tips of the greater and middle coverts; the outer edges of the quills also white, broadest on the tertiaries.

Female, ashy, brownish above, tinged with greenish-yellow beneath; top of head, rump,

Female.

and upper tail coverts, brownish gamboge-yellow. Wings as in the male. Length about 8 inches; extent, 12.75; wing, 4.62; tail, 3.50. Iris brown; bill brown; legs black.

Hab. Arctic America. South to United States, in severe winters. Resident in Sierra Nevada, California.

This beautiful bird is not uncommon near the summits of the Sierra Nevada, lat. 39°, in September, and doubtless breeds there, as I obtained two fine specimens in the young plumage. These seem rather smaller than the largest Eastern one mentioned by Baird, but their wing is longer than the average (4.50), while the tail is half an inch shorter.

They were feeding on spruce seeds when I first saw them, and still lingered about, after two had been shot, as if waiting for their comrades, soon descending to some shrubby alders to eat their seeds, then to the

ground, where they hopped about for some time, uttering a low chirping note, and allowing me to go within a few feet of them.

About Hudson's Bay they are said to be quite musical in spring, and to build nests in trees, not very high up, of twigs externally, lined with feathers. The eggs are said to be white. Besides seeds of trees, they feed on their buds, also on berries, and probably on insects during the breeding season, like other seed-eaters.

In Labrador, Audubon found them very unsuspicious, and even stupid, as when he shot one, the others, instead of flying away, came down within a few feet of him to examine him closely. They have been kept in cages, and are very musical.

Although the American pine grosbeak has received a different name from the European, it is very difficult to distinguish them, and it is probable that they are quite identical. Western specimens have larger bills than Eastern, as shown by the lower figure on the opposite page, which is from a California bird. Those from Kodiak have still larger bills.

GENUS **CARPODACUS**, KAUP.

Carpodacus, KAUP, "Entw. Europ. Thierw. 1829."
Erythrospiza, BONAPARTE, Saggio di una dist. met. 1831.

GEN. CHAR. Bill short, stout, and vaulted; the culmen decurved towards the end; the commissure nearly straight to the slightly decurved end. A slight development of bristly feathers along the sides of the bill, concealing the nostrils. Tarsus shorter than the middle toe; lateral claws reaching to the base of the middle one. Claw of hind toe

C. frontalis.

much curved, smaller than the middle one, and rather less than the digital portion Wings long and pointed, reaching to the middle of the tail, which is considerably shorter than the wing, and moderately forked. Colors red, or red and brown.

These birds all have more or less purple colors, which are supposed by

C. frontalis.

some to characterize the females as well as the males at times, but they do not obtain it the first year, and are said to lose it as they grow old.

Carpodacus Californicus, Baird.

THE WESTERN PURPLE FINCH.

Carpodacus purpureus, of West Coast, Newberry, P. R. Rep. VI. iv. 88. — Heermann, X. vi. 50.

Carpodacus Californicus, Baird, P. R. Rep. IX. Birds, 413. Ib. Birds of N. Amer.; pl. 72. — Cooper and Suckley, XII. iii. Zool. of W. T. 196.

Sp. Char. Third quill longest; first shorter than fourth. Body crimson, palest on the breast, darkest across the middle of back and wing coverts, where the feathers have dusky centres. The red extends below continuously to the lower part of the breast, and in spots to the tibiæ. The belly and under tail coverts white, streaked faintly with brown, except

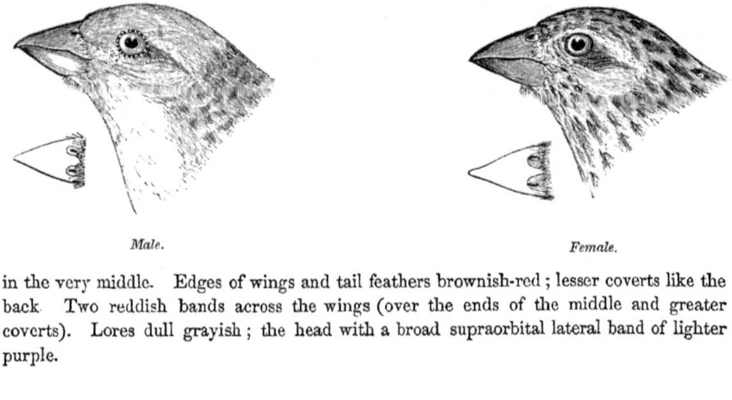

Male. *Female.*

in the very middle. Edges of wings and tail feathers brownish-red; lesser coverts like the back. Two reddish bands across the wings (over the ends of the middle and greater coverts). Lores dull grayish; the head with a broad supraorbital lateral band of lighter purple.

Female, olivaceous-brown; brighter on the rump. Beneath white. All the feathers everywhere streaked with brown, except on the middle of the belly and under coverts; a superciliary light stripe. No light edging to wing coverts. Length, 6.25; extent, 10.25; wing, 3.25; tail, 2.50. Iris brown; bill and feet paler brown.

Hab. Pacific Coast of the United States, chiefly northward.

This brightly plumaged little bird is rather a northern species, being common at the Columbia River and northward, while in this State it has not yet been found south of Monterey on the coast, and Fort Tejon in the Sierra Nevada. In summer they frequent the mountain forests, especially those composed in part of *Coniferæ*, while in winter they descend to the valleys, and associate with their more southern cousins, the *C. frontalis*. I found them in May on the summits of the Coast Range towards Santa Cruz, but not very numerous. They then had nests, but I did not succeed in finding any. That of the closely allied *C. purpureus* of the Eastern States is built in a low tree, composed of coarse grass lined with root fibres, and the eggs, five in number, are of a rather pale green, with scattered dots and streaks of dark brown or dull purple.

The song of this bird is quite loud and varied, often resembling that of different birds, such as *Vireos* and *Dendroicas*, for which I have mistaken it. This would doubtless succeed as well in a cage as the other species; but I have not seen any in captivity, though the Eastern species is often sold in cages by the name of linnets. Their food consists of all such seeds and berries as they can obtain, besides buds of trees in times of scarcity.

Carpodacus Cassinii, BAIRD.

CASSIN'S PURPLE FINCH.

Carpodacus Cassinii, BAIRD, Pr. Ac. Nat. Sc. Phila. VII. June, 1854, 119. IB. P. R. Rep. IX. Birds, 414. — KENNERLY, X. iv. 27; pl. xxvii. f. 1.

SP. CHAR. Larger than *C. Californicus*. Bill 0.55 of an inch above. Second and third quills longest; first longer than fourth. Above pale grayish-brown, the feathers

Male.

Female.

streaked with darker brown, and with only an occasional gloss of reddish, except on the crown, which is uniform deep crimson, and on the rump. Sides of the head and neck, throat, and upper part of breast, with rump, pale rose-color; rest of under parts white, very faintly and sparsely streaked with brown.

Female without any red, and streaked on the head and upper parts with brown. Length, 6.50; extent, 10.75; wing, 3.60; tail, 2.60. Iris, bill, and feet brown.

Hab. Rocky Mountains and valley of the Colorado to the Sierra Nevada, lat. 39°, California. Resident in the mountains of Mexico.

I found this bird in large numbers about Lake Tahoe; but all were in the brown plumage, and seemed so much like the *C. Californicus* in habits that I mistook them for that species. I noticed, however, a call-note when they flew which reminded me of that of the *Pyrangæ*, and seemed different from the other *Carpodaci*. Their song, as I have since heard it, was also louder and finer than that of the *C. Californicus*, and more original in its style.

Nothing has yet been published of their other habits. They may frequent the Colorado Valley at times, but I did not find them there.

Carpodacus frontalis, SAY.

THE BURION; THE HOUSE FINCH.

Fringilla frontalis, SAY, Long's Exped. II. 1824, 40. — AUDUBON, Orn. Biog. V. 1839, 230; pl. 424. — *Erythrospiza frontalis*, AUDUBON, Birds Amer. III. 1841, 175; pl. 197. — GAMBEL, Jour. A. N. S. 2d Series, I. 1847, 53. — *Carpodacus frontalis*, GRAY, Genera, 1844-49. — M'CALL, Pr. A. N. Sc. V. 1851, 219 — BAIRD, P. R. Rep. IX. Birds, 415. — NEWBERRY, VI. iv. 88. — KENNERLY, X. iv. 28.

Carpodacus familiaris, M'CALL, Pr. A. N. Sc. VII. April, 1852, 61, Sante Fé. — CASSIN, Ill. I. 73; pl. xiii. 1854. — HEERMANN, X. vi. 50.

SP. CHAR. *Male.* Bill short, much curved. Forehead, for nearly the length of the bill, a broad superciliary stripe extending to the nape, side of lower jaw, chin, throat, and upper part of the breast, crimson-red; rump, paler. Rest of upper parts and

Male.

Female.

sides of neck grayish-brown, with an occasional gloss of red externally on the crown, and with scarcely appreciable darker brown towards the centres of the feathers. Belly, under tail coverts, and sides whitish, conspicuously streaked with light brown; sometimes red to the middle of the former. Length, 6.00; extent, 9.50; wing, 3.25; tail, 2.80. Iris, bill, and feet brown.

Hab. Rocky Mountains to the Pacific. Not north of California.

This lively and musical little bird abounds in nearly all the southern portions of California, and, according to Newberry, throughout the valleys northward up to Oregon. It is everywhere the species most peculiar to the valleys, while the other two frequent the forest-clad mountains.

I have found this species on the barren rocky hills near the Colorado, and in plains near the coast, where there is no plant higher than the wild mustard, on the seeds of which it feeds. It frequents groves also, and open forests on the summit of the Coast Ranges, in small numbers, in company with *C. Californicus*, and at times feeds on buds of trees, and seeds of the cottonwood and other plants.

It is principally abundant about ranches and gardens, where it does much mischief by destroying seeds and young plants, fruit, etc., for which depredations even its cheerful and constant song does not compensate, and the angry gardener wages unrelenting war against the race.

At San Diego they build as early as the 15th of March, or perhaps even sooner. The situation and materials of their nest are exceedingly variable. I have found them in trees, on logs and rocks, the top rail of a picket-fence, inside a window-shutter, in the holes of walls, under tile or thatch roofs, in haystacks and barns, in the interstices between the sticks of a hawk's nest, and in an old nest of the oriole. About houses they always seek the protection of man, as if quite unconscious of having made him their

C. hæmorrhous, Wagler. *Male.*

enemy. Heermann mentions also, as locations of nests, the thorny cactus and deserted woodpeckers' holes. The materials are usually coarse grass or weeds, with a lining of hair or fine roots. The eggs, from four to six, are bluish-white, with spots and lines of black, chiefly towards the larger end. They measure 0.78 × 0.56 inch.

The songs of this species differ very much from those of the others. They are very lively and varied, though short, and are heard throughout the year. Cage-birds, usually called "California linnets," are easily kept and frequently to be seen, but generally their purple changes to yellow after long confinement.

They raise two, if not three, broods annually. These assemble in large flocks in autumn, but migrate very little if any to the south.

The house-finch of California is represented in Mexico by a closely allied species, if not a mere variety, the *C. hæmorrhous* of Wagler, in which the colors are more sharply defined, and the red more restricted, as shown in the preceding figure.

Genus ÆGIOTHUS, Cabanis.

Ægiothus, Cabanis, Mus. Hein. 1851, 161. (Type, *Fringilla linaria*, L.)

Gen. Char. Bill very short, conical, and acutely pointed; the outlines even concave; the commissure straight; the base of the upper mandible, and the nostrils concealed by stiff, appressed, bristly feathers; middle of the mandible having several ridges parallel

A. linaria.

with the culmen. Inner lateral toe the longer, its claw reaching to the middle of the central claw; the hind toe rather longer than the digital portion. Wings very long, reaching the middle of the tail; second quill a little longer than the first and third. Tail deeply forked.

A. linaria.

Species of this genus belong to the northern parts of both hemispheres. Though spending the summer in the highest latitudes, during the winter they generally migrate in large troops to a somewhat milder climate.

Ægiothus linaria, LINNÆUS.

THE LESSER REDPOLL.

Fringilla linaria, LINNÆUS, Syst. Nat. I. 1766, 322. — *Ægiothus linaria*, CABANIS, Mus. Hein. 1851, 161. — BAIRD, Birds N. Amer. 428.
Linaria minor, AUDUBON, Birds Amer. III. 122 ; pl. 179.

SP. CHAR. Above, light yellowish, each feather streaked with dark brown. Crown dark crimson. Upper part of breast and sides of the body tinged with a lighter tint of the same; the rump and under tail coverts are also similar, but less vivid, and with dusky streaks. Rest of under parts white, streaked on the sides with brown. Loral region and chin dusky; cheeks (brightest over the eye) and a narrow front, whitish. Wing feathers

edged externally, and tail feathers all round, with white. Two yellowish-white bands across the wing coverts; secondaries and tertiaries edged broadly with the same. Bill yellowish, tinged with brown on the culmen and gonys; the basal bristles brown, reaching over half the bill. Length, 5.50; wing, 3.10; tail, 2.70.

Hab. Throughout Eastern North America, coming south in winter. Washington Territory. Northern Europe and Asia.

The specimen described above is a male in winter dress. The spring plumage has much more of the red. The winter specimens of the female lack the rose of the under parts and rump; the breast is streaked across with dusky.

During the breeding season this interesting bird is very abundant in the northernmost regions of America. On the Western coast it comes southward as far as Washington Territory, and still lower on the Atlantic coast. They are very familiar, entering cities and towns, and feeding unsuspiciously in yards and gardens. They subsist partially on the seeds of pine, birch, linden, alder, and other trees, and very much resemble the chickadees in their way of feeding.

Genus LEUCOSTICTE, Swainson.

Leucosticte, Swainson, Fauna Bor. Amer. II. 1831, 265. (Type, *Linaria tephrocotis*, Sw.)

Gen. Char. Bill conical, rounded, rather blunt at the tip; the culmen slightly convex; the commissure slightly concave; the nostrils concealed by depressed bristly feathers; a

L. tephrocotis.

depressed ridge extending about parallel with the culmen above the middle of the bill. Another more conspicuously angulated ridge, extending forward from the lower posterior angle of the side of the lower mandible, nearly parallel with the gonys. Tarsus about

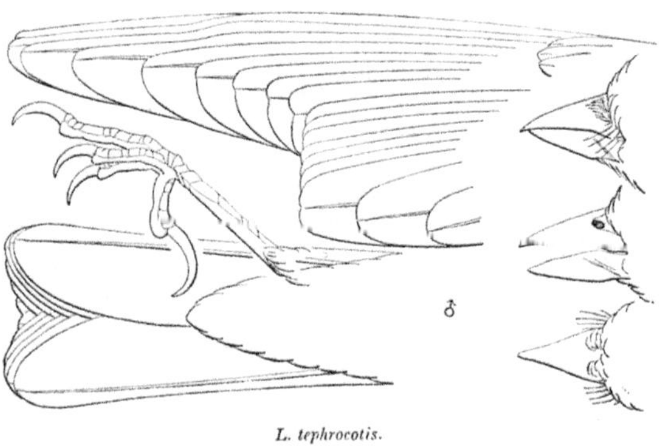

L. tephrocotis.

equal to the middle toe. Inner and outer toes nearly equal, their claws not reaching beyond the base of the middle one. Hind toe rather longer, its claw longer than the digital portion. Wings very long; first quill longest. Tail forked.

Leucosticte griseinucha, Brandt.

THE GRAY-EARED FINCH.

Passer arctous, var. Pallas, Zoog. Rosso-Asiat. II. 1831, 23.
Fringilla (Linaria) griseinucha, Brandt, Bull. Acad. St. Pet. Nov. 1844, 36. — *Montifringilla (Leucosticte) griseinucha*, Bon. and Schl. Mon. Loxiens, 1850, 35; pl. 41. — *Leucosticte griseinucha*, Baird, Birds N. Amer. 430. — Kittlitz, Denkwürdig. I. 1858, 291 (Nesting). — *Leucosticte griseigenys*, Gould, Pr. Zool. Soc. 1843, July 25, 104. Ib. Voy. Sulphur, 42; pl. xxiii.

Sp. Char. General color dark brownish-chocolate anteriorly, the feathers of the back rather darker in the centre, and with paler edges. Forehead and crown black; rest of head, including the cheeks and ears, of a rather silvery gray; throat blackish, shading off insensibly into the chocolate of breast. Feathers of abdomen (and under part of breast to a less degree), flanks, and crissum, with the rump and upper tail coverts and lesser middle wing coverts tipped with dark pomegranate or rose-red, allowing more or less

of their dusky bases to be visible, especially above, where there is an appearance of bars. Wing and tail feathers brown, nearly all, including the greater wing coverts, edged with pale yellowish-gray, with only a faint tinge of rose. Bill dusky, darkest at tip. Legs black. Total length, 7.50; wing, 4.80; tail, 3.50; exposed portion of first primary, 3.50. Bill: length from forehead, 0.69; from nostril, 0.42. Legs: tarsus, 0.95; middle toe and claw, 0.92; claw alone, 0.35; hind toe and claw, 0.69; claw alone, 0.38.

Hab. Aleutian Islands. (St. George's and Aonalaska.)

This is considerably the largest of the American species of *Leucosticte*, and has the longest bill. It also has the chocolate and rose colors darker, and the rose extending farther forward on the breast than in other species. It could only be confounded with *C. littoralis* as to color, both having the head above and on the sides ashy, covering the whole ear coverts; but the dusky patch in the crown is more extended, the gray ash of chin more restricted, and the throat darker. The rose of the abdomen extends over on to the breast, and the tints are different.

A specimen, apparently young, perhaps a female, differs in having a duller tint, and a tinge of olivaceous-yellow on the middle of the abdomen and crissum. The lining of the wings is without any rose-color. Bonaparte and Schlegel describe the young of this species also without any tinge of rose.

Specimens of this bird were obtained by Dr. T. T. Miner at Aonalaska, and by Mr. Dall at St. George's Island, Behring's Sea, with the eggs, which are white. (Baird.)

Leucosticte littoralis, Baird.

HEPBURN'S FINCH.

Leucosticte griseinucha, Elliot, Illust. Birds Amer. I. pl. 11, 1868.
Leucosticte littoralis, Baird, MSS.

Sp. Char. Body chocolate-brown; the feathers narrowly margined with paler, those of the back with rather darker, centres. Abdomen, flanks, crissum, rump, upper tail coverts, wing coverts, and quills, edged with rose-red, more or less continuous (least so in the rump), the outer edges of secondaries and tail feathers pale fulvous, the latter with a rosy shade. Head silvery-gray; the forehead and patch on crown black; the chin ashy, like

and continuous with the cheek; the throat dusky-brown, shading into the chocolate of breast. Bill reddish, with extreme tip dusky. Total length, 7.00; wing, 4.30; tail, 3.10; exposed portion of first primary, 3.40. Bill: length from forehead, 0.60; from nostril, 0.35. Legs: tarsus, 0.76.

Hab. Sitka (Bischoff), Fort Simpson, British Columbia (Hepburn).

This species, believed to be new, bears much resemblance to *L. griseinucha*, but is considerably smaller, the colors brighter and lighter, more like those of *tephrocotis*, the bill shorter and more conical; the dark patch on the head more restricted; the whole chin ashy. From *tephrocotis* it is distinguished by the extension of the ash of the head below the eye, and from *campestris* by having the ear coverts ashy, instead of the anterior portion of cheeks only; there is also apparently a greater extent of gray on the chin.

The only specimens I have seen are one from Sitka, collected by Bischoff, and one obtained a year or two previously at Fort Simpson, by Mr. Hepburn, an eminent English naturalist. long time resident at San Francisco and Victoria. (Baird.)

Leucosticte campestris, BAIRD.

THE GRAY-CHEEKED FINCH.

Leucosticte campestris, BAIRD, MSS.

SP. CHAR. Body light chocolate-brown; the feathers edged with paler, those of the back with rather darker, centres. Feathers of anal region, flanks behind, crissum, rump, and upper tail coverts, wing coverts, and primary quills, edged with rose-red; secondary quills and tail feathers with pale fulvous; little or no trace of rose on under wings. Fore-

head and patch on crown blackish; the hind head to nape, cheeks, immediately under the eye (but not including the auriculars, except perhaps the most anterior), and base of lower mandible all round, ashy-gray. Throat dusky. Bill yellowish, with dusky tip. Legs dusky. Specimen figured above from near Denver City, Colorado, January, 1862. Dr. C. Wernigk. (Baird.)

Total length, 7.00; wing, 4.00; tail, 3.00; exposed portion of first primary, 3.10. Bill: length from forehead, 0.60; from nostril, 0.40. Legs: tarsus, 0.75; middle toe and claw, 0.80; claw alone, 0.24; hind toe and claw, 0.80; claw alone, 0.37.

Hab. Colorado Territory. (Dr. Wernigk.)

This species bears a close resemblance to *L. tephrocotis*, and may indeed be a variety of it, but as it differs in the characters that appear generally to be those most constant in *Leucosticte*, and as in over fifty skins of the *tephrocotis* I have seen nothing like it, I am inclined to consider them distinct. The size and general appearance are much the same, the difference being that in *tephrocotis* the cheeks are chocolate below the level of the eye, the chin without any gray, while in *campestris* the sides of the head below the eye, but not including the ears, together with a narrow border of the chin, are of this color.

This species may be distinguished from *littoralis* by the less extent of ash on the cheeks. In *littoralis* it covers the whole of the ears, and extends back farther on the head all round. *L. griseinucha* is marked like *littoralis*, and is much larger than either.

The specimen described was presented to the Smithsonian Institution by Dr. Wernigk, and at the time was supposed to be *L. tephrocotis*. (Baird.)

Leucosticte tephrocotis, Swainson.

THE GRAY-CROWNED FINCH.

Linaria (Leucosticte) tephrocotis, Swainson, F. Bor. Am. II. 1831, 255; pl. 1. — *Leucosticte tephrocotis*, Swainson, Birds, II. 1837. — Baird, Stansbury's Salt Lake, 1852, 317. — Ib. P. R. Rep. IX. Birds, 430. — *Erythrospiza tephrocotis*, Bonaparte, List, 1838. — Audubon, Birds Amer. III. 1841, 176; pl. 198. — Nuttall, Man. 2d ed. I. 632. — *Fringilla tephrocotis*, Audubon, Orn. Biog. V. 1839, 232; pl. 424.

Sp. Char. Head above and nape bounded below by a line from the commissure a little below the eyes, light ashy; dusky in the loral region. Crown with a distinct patch of sooty-black, reaching nearly to the base of the bill. Lesser wing coverts and axillaries, outer edges of primaries and tail feathers, with ends of the feathers of the posterior half of body all round, pale rose-red. Rest of body dark umber-brown, tinged with dusky on

the chin and throat. Wings and tail feathers blackish. The greater coverts are tipped, and the secondaries edged, with white. Length, 7.10; wing, 4.30; tail, 2.90. Winter plumage, Salt Lake City. (Baird.)

Hab. Northern Rocky Mountains. Vicinity of Salt Lake City in winter, and Northeastern California.

I have seen a specimen brought from somewhere east of Lake Tahoe in Washoe, and presented by Mr. F. Gruber to the German Academy of Natural Sciences in San Francisco. They were said to be plentiful there in the very cold winter of 1861–62, and doubtless visited the similar country east of the northern Sierra Nevada within this State. The first specimen described was obtained on the banks of the Saskatchewan River, about lat. 51°, in May, but very little is yet known of their habits. Captain Stansbury found them common near Salt Lake City, in March, 1850.

Leucosticte arctoa, PALLAS.

THE SIBERIAN FINCH.

Passer arctous, PALLAS, Zoog. Rosso-As. II. 1831, 21. — *Fringilla (linaria) arctoa*, BRANDT, Bull. Acad. St. Pet. 1843. — MIDDENDORF, Sib. Reise, II. ii. 1852, 153.— *Montifringilla (Leucosticte) arctoa*, BONAPARTE and SCHLEGEL, Mon. Loxiens. 1850, 38; pl. 44, 45. — *Leucosticte arctoa*, BAIRD, Birds Amer. 1858, 430; pl. 74, f. 1.

SP. CHAR. (Figure, Siberia.) *Female*, body dark clear brown, with a faint suspicion of purplish; paler anteriorly, especially in the centres of the feathers of the head and breast. Nape decidedly grayish. Frontal feathers and bristles dusky. Entire tail and coverts, and the wings (except lesser coverts above and below), silvery gray; the ends of the feathers (but not the edges), especially of the quills, together with all the shafts, brownish; the

gray everywhere very finely, almost unappreciably, mottled with dusky. Bill yellowish with dusky tip; feet black. Total length, 6.50; wing, 4.70; tail, 3.40; exposed portion of first primary, 3.50. Bill: length from forehead, 0.64; from nostril, 0.37. Legs: tarsus, 0.80; middle toe and claw, 0.79; claw alone, 0.25; hind toe and claw, 0.60; claw alone, 0.36.

Hab. Siberia, the Kurile, and the Aleutian Islands.

It is not improbable that more than one species is included under the above synonymy and geographical distribution, as the specimen before me is considerably larger, and is otherwise to some extent different from those figured and described by Bonaparte and Schlegel. A *Fringilla Gebleri* of Brandt, from Siberia, (Bulletin Academy, St. Petersburg, 1843, not of 1841,) is probably a closely allied though distinct species.

No specimens from the Aleutian Islands are in American collections, and I describe the species from a Siberian skin presented to the Smithsonian Institution by Mr. John Gould. The bird is said to be very abundant in Siberia, passing southward in winter, but soon returning northward, and keeping in flocks about the villages in company with the redpolls. It is a very stupid bird; when pursued, it thrusts its head into a tuft of grass, and, imagining itself concealed, can even be taken with the hand.

Middendorf describes a male bird as having a wash of rose-red over the lower back and breast, and rose edges to feathers of hind head and cheeks. (Baird.)

Genus **CHRYSOMITRIS**, Boie.

Chrysomitris, Boie, Isis, 1828, 322. (Type, *Fringilla spinus*, Linnæus.)
Astragalinus, Cabanis, Mus. Hein. 1851, 159. (Type, *Fringilla tristis*, Linnæus.)

Gen. Char. Bill rather acutely conic, the tip not very sharp; the culmen slightly convex at the tip; the commissure gently curved. Nostrils concealed. Obsolete ridges on the upper mandible. Tarsi shorter than the middle toe; outer toe rather the longer,

C. tristis.

reaching to the base of the middle one. Claw of hind toe shorter than the digital portion. Wings very long, reaching to the middle of the tail; second quill a little longer than first and third. Tail deeply forked.

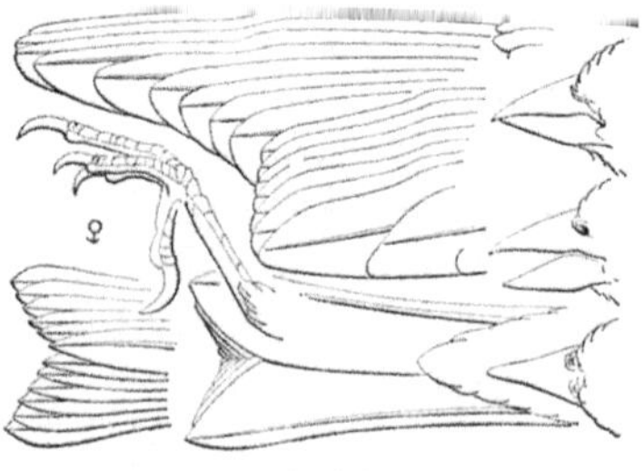

C. tristis.

The colors are generally yellow, with black on the crown, throat, back, wings, and tail varied sometimes with white.

The females are usually without yellow, or it is pale or dull, and their general hues are brown or olive.

Chrysomitris tristis, Linnæus.

THE YELLOW-BIRD; THE THISTLE-BIRD.

Fringilla tristis, Linnæus, Syst. Nat. I. 1766, 320. — *Carduelis tristis*, Audubon, Birds Amer. II. 1841, 129; pl. 181. — Heermann, P. R. Rep. X. vi. 50. — *Chrysomitris tristis*, Bonaparte, List, 1838. — Newberry, Zool. Cal. and Or. Route; Rep. P. R. R. Surv. VII. iv. 1857, 87. — Baird, P. R. Rep. IX. Birds, 421. — Cooper and Suckley, XII. iii. Zool. of W. T. 197.

Sp. Char. *Male.* Bright gamboge-yellow; crown, wings, and tail black. Lesser wing coverts, band across the end of greater ones, ends of secondaries and tertiaries, inner margins of tail feathers, upper and under tail coverts, and tibia, white. In winter the yel-

Summer.

Winter.

low becomes brownish, the black is wanting on crown, and that on wings is browner. Throat yellowish beneath, ashy-brown; abdomen white. The *female* is much like this at all times, but more olivaceous. Its bill is brownish, the tip paler. Length of male, 5.25; extent, 9.00; wing, 3.00. Iris brown; bill yellowish-white, tip black; feet pale brown.

Hab. North America generally.

The yellow-birds are constant residents in all the western parts of California, but I saw none along the Colorado. They become rare north of the Columbia along the coast, but continue in the interior up to lat. 49°. They breed near San Diego as well as northward, and seem to avoid the hot interior valleys, as well as the forest-clad mountains. Their favorite resorts are in fact those places where thistles and other composite flowers abound, and their fondness for the seeds of the former has given them, in places, the name of "thistle-bird." They are also very fond of willow-groves and cottonwoods, feeding much on their seeds, while in winter those of the sycamore (*Platanus*) supply their chief subsistence. In the Eastern States they remain throughout the snowy season, and are often seen feeding on cockle-burs and other seeds left standing above the snow.

In February, when the males acquire their yellow plumage, the flocks often collect on top of a tree, and sing in chorus for an hour, their sweet discord being particularly pleasing, the whole flock sinking and raising their voices in concert, though not keeping one time. Their song, resembling somewhat that of the canary, is well known, as they are frequently kept in cages.

At Santa Cruz I found two nests about the first of June. One was in a bush, not more than three feet from the ground; the other on the low branch of a tree, near the end, and contained young. They also build high in the forks of trees. The nest is very compactly constructed of strips of bark, roots, and fibrous plants, mixed with downy scales of leaves and catkins, and lined with thistle-down, that of the sycamore, or sometimes wool or cow-hair and fine grass, the cavity measuring 1.50 × 1.30 inches. The outside is often covered with silk of caterpillars' nests, cobwebs, or plant fibres, and seems glued smoothly together. The eggs, from three to five, are pale greenish-white, and measure 0.60 × 0.50 inch.

Being rather late in the year in building, they usually raise but one brood, though they have been known to feed their young as late as the middle of September. (Nuttall.) This lateness of incubation is observed also in the East, and supposed to depend upon the absence of some necessary kind of food in the earlier months.

Though occasionally visiting gardens for lettuce and sunflower seeds, they rarely do any mischief, not meddling with fruits, but doing much good by destroying thistle seeds.

Chrysomitris psaltria, SAY.

THE ARKANSAS FINCH.

Fringilla psaltria, SAY, Long's Exped. R. Mts. II. 1823, 40. — BONAPARTE, Am. Orn. I. 1825, 54; pl. 6, f. 3. — *Carduelis psaltria*, AUDUBON, Birds Amer. III. 1841, 134; pl. 183. — HEERMANN, P. R. Rep. X. vi. 50. — *Chrysomitris psaltria*, BONAPARTE, List, 1838. — NEWBERRY, P. R. Rep. VI. iv. 87. — BAIRD, P. R. Rep. IX. Birds, 422. — KENNERLY, X. iv. 28.

SP. CHAR. Upper parts and sides of head and neck olive-green. Hood, upper tail coverts, wings, and tail black. Beneath, bright yellow. A band across the tips of the greater coverts, the ends of nearly all the quills, the outer edges of the tertiaries, the extreme bases of all the primaries, except the outer two, and a long rectangular patch on the inner webs of the outer three tail feathers near the middle, white.

Female, with the upper parts generally, and sides, olive-green; the wings and tail brown, their white marks as in the male. Length, 4.50; extent, 8.00; wing, 2.65; tail, 1.87. Iris brown; bill horn-brown, greenish below; feet pale brown (or flesh-color in summer).

Hab. Southern Rocky Mountains to the coast of California.

I did not find this species in the Colorado Valley, though they have been obtained along Williams Fork by Dr. Kennerly. At San Diego and along the whole coast border they are rather rare, but seem to be common in the interior valleys, and breed in small numbers in the Coast Range near Santa Cruz.

They have habits very similar to those of the yellow-bird, but seem to

Male

Female.

feed more on the ground or among weeds than on trees, and are perhaps even more gregarious, keeping in flocks up to June 1st.

I have not met with their nests, or with any description of them; but they doubtless much resemble those of *C. tristis.*

Their song resembles that of the yellow-bird, but is much fainter. Probably, however, more exact observations than I have been able to make would show many differences both in habits and song, especially if they are kept in cages, in which they would doubtless do well.

Chrysomitris Mexicanus, BONAPARTE.

THE MEXICAN GOLDFINCH.

Carduelis Mexicanus, SWAINSON, Syn. Birds Mex. Phil. Mag. 1827, 435. — *Chrysomitris Mexicanus,* BONAPARTE, Consp. Av. 1850, 516. — BAIRD, Birds N. Amer. 423. — *Fringilla Texensis,* GIRAUD, Sixteen Species of Texan Birds, 1841, pl. v. f. 1.

SP. CHAR. Upper parts continuously and entirely black; the feathers of the rump white subterminally, and showing this through the black; a few of the feathers with

Var. *Arizonæ.* *Male.*

C. Mexicana. *Female.*

greenish-yellow between the white and black; a few, perhaps, without black tips. The bases of the third to seventh primaries, and the ends of the tertiaries, externally white. The tail is black, except the outer three feathers, in which the outer webs and tips only

Chrysomitris Mexicana. Male.

are this color; the rest white. Inside of wing black. Under parts of body pale yellow. Female with the black of the head and body replaced by olive-green. Length, 4 12; wing, 2.25; tail, 2.00.

Hab. Mexican side of the valley of the Rio Grande, southward.

Var. **Arizonæ**, Coues.

Chrysomitris Mexicanus, var. *Arizonæ*, Coues, Pr. A. N. Sc. 1866, 81.

Sp. Char. Differs from above in a decided gradation towards *C. psaltria.* The black of the back is mixed with about an equal amount of olive. The auriculars are black as in *Mexicanus*, but the yellow lower eyelid, like that of *psaltria*, is not disconnected with the yellow of the throat. (Coues.)

Hab. Arizona.

The Arizona variety of *C. Mexicanus*, pointed out as above by Dr. Coues, is strongly related to *C. psaltria*, in which, however, the top of the head is black, distinctly bounded on all sides with olive, and the black does not descend below the eye, as in *Mexicanus*. The back also is distinctly olive, not mixed with black. We have no special information as to its habits, but they are probably much like those of the common thistle-bird (*C. tristis*) of the United States. (Baird.)

Chrysomitris Lawrencii, Cassin.

LAWRENCE'S GOLDFINCH.

Carduelis Lawrencii, Cassin, Pr. A. N. Sc. V. Oct. 1850, 105; pl. V. (California.) — Heermann, P. R. Rep. X. vi. 50. — *Chrysomitris Lawrencii*, Bonaparte, Comptes Rendus, Dec. 1853, 913. — Baird, P. R. Rep. IX. Birds, 424.

Sp. Char. Hood, sides of head anterior to the middle of the eye, chin, and upper part of throat, black. Sides of head, neck, and body, upper part of neck and the back, and upper tail coverts, ash-color. Rump and lesser wing coverts yellowish-green. Throat below the black, breast, and outer edges of all the quills (except the first primary), bright

Female.

Male.

greenish-yellow, passing into white behind. Wings black. Tail feathers black, with a white square patch on the inner web, near the end; outer edges grayish; quills black. *Female*, similar, with the black of the head replaced by ash. Length, 4.50; extent, 8.00; wing, 3.00; tail, 2.30. Iris and feet brown; bill smoky-white (brown in female).

Hab. California.

I found a few of this species at Fort Mojave, and they are rather numerous near the coast, as far north as San Francisco at least, also in the northern mining regions.

I saw some of them in December near San Francisco, and have no doubt they remain throughout winter in all the lower country. The mountains they seem to avoid, and they have not been observed in Oregon.

Their habits and song are generally similar to those of *C. tristis*, but their voice is much weaker and more high in pitch.

Their nests I have found in a small live-oak, built very much like those of *C. tristis*, but much smaller, the cavity measuring 1.50 × 1 inch, and the eggs, four or five in number, are pure white, 0.80 × 0.46 inch. I think they usually prefer the live-oak for building, as I have not found nests in other situations.

They feed sometimes on the ground on grass-seeds, as well as on buds and seeds of various weeds and trees. They seem more of a sylvan species than the yellow-bird, and not so fond of willows and other trees growing along streams and in wet places. In the Colorado Valley they feed on seeds of an *Artemisia*. I did not see them there after April 15th.

Chrysomitris pinus, Wilson.

THE PINE FINCH.

Fringilla pinus, Wilson, Amer. Orn. II. 1810, 133; pl. xvii. f. 1. — Audubon, Orn. Biog. II. 1834, 455; V. 509; pl. 180. — *Linaria pinus*, Audubon, Birds Amer. III. 1841, 125; pl. 180. — Heermann, P. R. Rep. X. vi. 49. — *Chrysomitris pinus*, Bonaparte, Consp. 1850, 515. — Baird, P. R. Rep. IX. Birds, 425. — Cooper and Suckley, XII. iii. Zool. of W. T. 197.

Sp. Char. Tail deeply forked. Above, brownish-olive. Beneath, whitish, every feather streaked distinctly with dusky. Concealed bases of tail feathers and quills, together with their inner edges, sulphur-yellow. Outer edges of quills and tail feathers

yellowish-green Two yellowish-white bands on the wing. Length, 5.00; extent, 8.50; wing, 3.00; tail, 2.20. Iris, bill, and feet brown.

Hab. North America, from Atlantic to Pacific; chiefly northward; extending, however, along the Rocky Mountains to the table-lands of Mexico.

A common species in the far north, residing throughout the year at the Columbia River and northward, but in this State I have not met with them even in the lofty Sierra Nevada in September, though they appear along their whole western slope in winter, as far south as Fort Tejon.

They are found both among the coniferous trees and those that are deciduous, feeding on the seeds of spruces, alders, willows, and juniper-berries.

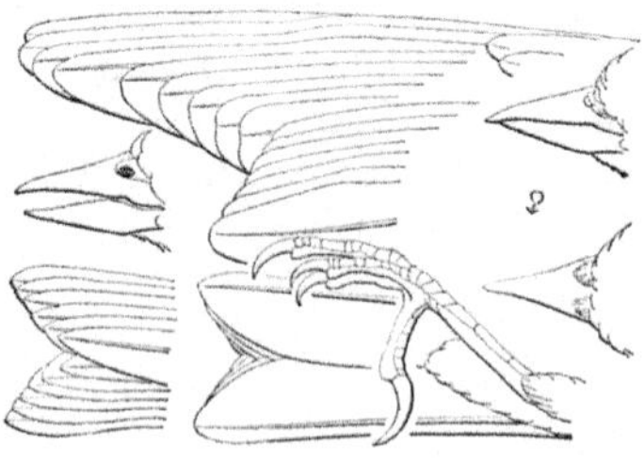

They rarely descend to the ground or lower herbage, living constantly among the trees, especially along streams. They have the same gregarious habits, and nearly the same call-notes, as the yellow-bird, but not so much song.

Genus HESPERIPHONA, Bonaparte.

Hesperiphona, Bonaparte, Comptes Rendus, XXXI. Sept. 1850, 424. (Type, *Fringilla vespertina.*)

Gen. Char. Bill largest and stoutest of all the United States fringilline birds. Upper mandible much vaulted; culmen nearly straight, but arched towards the tip; commissure curved. Lower jaw very large, but not broader than the upper, nor extending back, as in *Guiraca;* considerably lower than the upper jaw. Gonys unusually long. Feet

H. vespertina.

short, tarsi less than the middle toe; lateral toes nearly equal, and reaching to the base of the middle claw. Claws much curved, stout, and compressed. Wings very long and pointed, reaching beyond the middle of the tail. Primaries much longer than the nearly

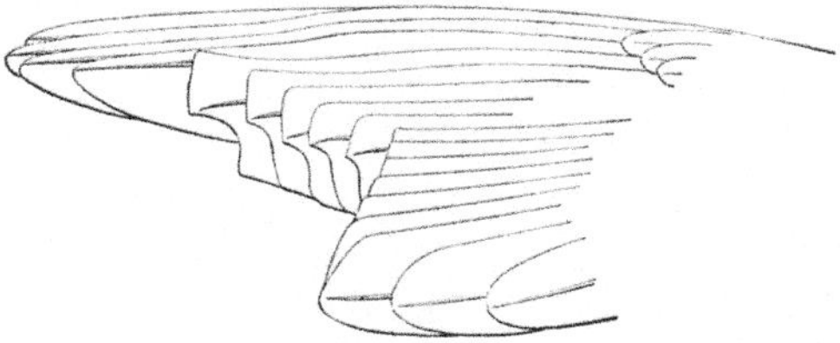

Coccothraustes vulgaris.

equal secondaries and tertials; outer two quills longest; the others rapidly graduated. Tail slightly forked; scarcely more than two thirds the length of the wings.

The genus *Hesperiphona* is very closely allied to the Old World genus *Coccothraustes.* The principal difference consists in a peculiar character of the quills, as shown in the preceding figure of *Coccothraustes,* and while *Hesperiphona* has the wing rounded off as usual in the *Fringillidæ.*

Hesperiphona vespertina, Cooper.

THE EVENING GROSBEAK.

Fringilla vespertina, Cooper, Annals New York Lyceum, N. H. I. ii. 1825, 220. (Sault St. Marie.) — *Coccothraustes vespertina,* Swainson, F. Bor. Am. II. 1831, 269. — Audubon, Syn. 134. Ib. Birds Amer. III. 1841, 217; pl. 207. — *Hesperiphona vespertina,* Bonaparte, Comptes Rendus, XXXI. Sept. 1850, 424. — Baird, P. R. Rep. IX. Birds, 409. — Cooper and Suckley, XII. iii. Zool. of W. T. 196.

Sp. Char. Bill yellowish-green, dusky at the base. Anterior half of the body dark yellowish-olive, shading into yellow to the rump above, and the under tail coverts below. Outer scapulars, a broad frontal band continued on each side over the eye, axillaries, and

middle of under wing coverts, yellow. Feathers along the extreme base of the bill, the crown, tibiæ, wings, upper tail coverts, and tail, black; inner greater wing coverts and tertiaries white. Length, 7.50; wing, 2.30; extent, 7.50; tail, 2.75. Iris brown; legs pale brown.

Hab. Lake Superior north and west; along Rocky Mountains to New Mexico; Sierra Nevada, northward. Resident in the mountains of Mexico.

This beautiful bird has been obtained at Michigan Bluffs, Placer County, near lat. 39°, by Mr. F. Gruber, and this is the lowest point at which I have heard of its occurrence in this State, though they probably go farther south along the summits of the Sierra Nevada, as they follow the Rocky Mountains down to Fort Thorn, New Mexico. I have not myself seen them in the State, but saw the feathers of one recently killed at the summit of the Sierra Nevada, lat. 39°, in September, 1863.

In the north they are not uncommon, but keep so high among the cottonwoods and pines that they are rarely obtained. They do not seem to come down near the coast, even at the Columbia River, and in this State have never been met with in the Coast Range of mountains. They feed chiefly

Mexico.

on the seeds of pines, spruces, and cottonwood poplars, occasionally seeking other seeds nearer the ground. When feeding they are very silent and difficult to perceive, but when they fly from one place to another they utter a loud call-note. In spring they have a rather short but melodious song, resembling that of the robin or black-headed grosbeak. Their nest and eggs are still undescribed. Dr. Townsend found them numerous at Fort Vancouver in May, and they were then quite tame, active, and noisy the whole day.*

In examining a large series of this species, in the Smithsonian Museum, from different localities, Mr. R. Ridgway has noted the existence of two strongly marked varieties, both represented at the beginning of this article. One of these, figured on the preceding page, has a stouter bill, broader patch of frontal yellow, and the outer tail feathers spotted with white on the inner webs at end. This style appears to belong to the northern parts of America, coming into the United States only in winter. The other, figured above, is the variety belonging to and resident on the table-lands of Mex-

* *Hesperiphona vespertina.* Since writing the previous paragraph I have met with a flock of about ten individuals that wintered near Santa Cruz, remaining until the end of April, 1866. Their favorite resort was a small grove of alders and willows, close to the town, where their loud call-note could be heard at all times of the day, though I never heard them sing. When the herbage began to grow in spring, their favorite food was the young leaves of various annual weeds that sprouted up under the shade of the trees. They then fed on the buds of the "box elder" (*Negundo*), and frequented the large pear-trees in the old mission garden, probably to eat their buds. They were generally very tame, allowing an approach to within a few yards of them when feeding. (Cooper.)

ico, and extending northward into New Mexico. It has a narrower bill, narrower frontal patch, and the tail feathers entirely black. Some specimens, again, are quite intermediate, and it is very doubtful whether the two forms can ever be considered as specifically distinct. (Baird.)

GENUS PLECTROPHANES, MEYER.

Plectrophanes, MEYER, Taschenbuch, 1810. (Type, *Emberiza nivalis.*)

GEN. CHAR. Bill variable, conical; the lower mandible higher than the upper; the sides of both mandibles (in the typical species) guarded by a closely applied brush of

P. nivalis.

stiffened bristly feathers directed forwards, and in the upper jaw concealing the nostrils; the outlines of the bill nearly straight, or slightly curved; the lower jaw considerably

P. nivalis.

broader at the base than the upper, and wider than the gonys is long. Tarsi considerably longer than the middle toe; the lateral toes nearly equal (the inner claw largest), and reaching to the base of the middle claw. The hinder claw very long; moderately curved and acute; considerably longer than its toe; the toe and claw together reaching to the middle of the middle claw, or beyond its tip. Wings very long and much pointed, reaching nearly to the end of the tail; the first quill longest, the others rapidly graduated; the tertiaries a little longer than the secondaries. Tail moderate, about two thirds as long as the wings; nearly even, or slightly emarginated.

The species of this genus are inhabitants of the northern hemisphere, mostly occurring in the highest latitudes. Two are common to the Old and New Worlds; the others peculiar to America. They all, as far as known, nest on the ground, and are strictly terrestrial in their habits.

Plectrophanes nivalis, MEYER.

THE SNOW-BUNTING.

Emberiza nivalis, LINNÆUS, Syst. Nat. 1766, 308. — *Plectrophanes nivalis*, "MEYER," AUDUBON, Birds Amer. III. 1841, 55; pl. 155. — BAIRD, Birds N. Amer. 432.

SP. CHAR. Colors, in full plumage, entirely black and white. Middle of back between scapulars, terminal half of primaries, and tertiaries, and two innermost tail feathers, black; elsewhere pure white. Legs black at all seasons. In winter dress white beneath;

the head and rump yellowish-brown, as also some blotches on the side of the breast; middle of back brown, streaked with black; white on wings and tail much more restricted. Length about 6.75; wing, 4.35; tail, 3.05; first quill longest.

Hab. Northern America, from Atlantic to Pacific; south into the United States in winter.

This species inhabits the northern parts of both hemispheres, nesting within the Arctic Circle, and going southwards in winter, usually in large flocks. In company with the skylark (*Eremophila cornuta*) they form a conspicuous feature of the winter landscapes in many parts of the United States, especially near the coast, frequenting roadsides and open fields.

They are very abundant in Greenland, breeding there as well as on the islands along the Arctic Coast of North America.

Plectrophanes Lapponicus, Selby.

THE LAPLAND LONGSPUR.

Fringilla Lapponica, Linnæus, Fauna Suecica, 1761. — *Plectrophanes Lapponica*, "Selby," Audubon, Birds Amer. IV. 50 ; pl. 152. — Baird, Birds N. Amer. 433.

Sp. Char. First quill longest. Legs black. Head all round black, this extending as a semicircular patch to the upper part of breast; sides of lower neck and under parts white, with black streaks on the sides, and spots on the side of the breast. A short, brownish-white streak back of the eye. A broad chestnut collar on the back of the neck.

Rest of upper parts brownish-yellow, streaked with dark brown. Outer tail feathers white, except on the basal portion of the inner web. Length, about 6.25 ; wing, 3.90 ; tail, 2.80.

Hab. Northern America, coming southward into the United States in winter. Not found much west of the Missouri.

This species is very seldom seen in full spring plumage in the United States. In perfect dress, the black of the throat probably extends farther down over the breast. In winter the black is more or less concealed by whitish tips to the feathers beneath, and by yellowish-brown on the crown. Some fall specimens, apparently females, show no black whatever on the throat, which, with the under parts generally, are dull white, with a short black streak on each side of the throat.

The Lapland longspur is an extremely abundant species in the far North, reaching from one ocean to another, and in winter entering the eastern portions of the United States in large flocks. It has not yet been recorded as found in California, but probably occurs there, at least as a straggler. It breeds in immense numbers on the Arctic Coast of North America, usually in company with the painted longspur (*P. pictus*), making its nest on the ground, like all the other species of the genus.

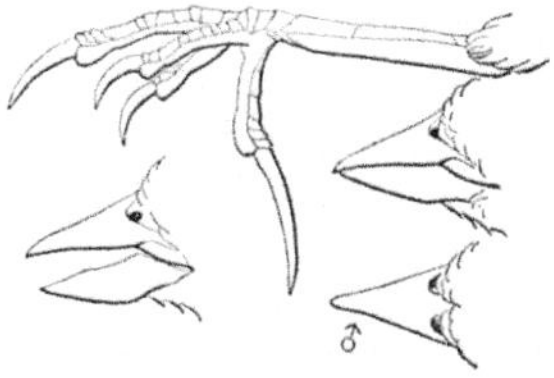

P. Lapponicus.

The special characteristics of the sub-genus, *Centrophanes*, of which this species forms the type, will be best appreciated by the above figure. (Baird.)

SUB-FAMILY SPIZELLINÆ.

CHAR. Bill variable, usually almost straight; sometimes curved. Commissure generally nearly straight, or slightly concave. Upper mandible wider than lower. Nostrils exposed. Wings moderate; the outer primaries not much rounded. Tail variable. Feet large; tarsi mostly longer than the middle toe.

The birds of this division are usually small and dull in colors, comprising most of those which are commonly called sparrows. They live mostly on or near the ground, in low bushes, thickets, and marshes, and their songs, though often sweet-toned, are neither loud nor long. The sexes are nearly similar in colors.

GENUS **PASSERCULUS**, BONAPARTE.

Passerculus, BONAPARTE, Comparative List, Birds, 1830. (Type, *Fringilla Savanna*, WILSON.)

GEN. CHAR. Bill moderately conical; the lower mandible smaller; both outlines nearly straight. Tarsus about equal to the middle toe. Lateral toes about equal, their claws falling far short of the middle one. Hind toe much longer than the lateral ones, reaching as far as the middle of the middle claw; its claws moderately curved. Wings unusually long, reaching to the middle of the tail, and almost to the end of the upper coverts. The

P. Savanna.

tertials nearly or quite as long as the primaries; the first primary longest. The tail is quite short, considerably shorter than the wings; as long as from the carpal joint to the end of the secondaries. It is emarginate, and slightly rounded; the feathers pointed and

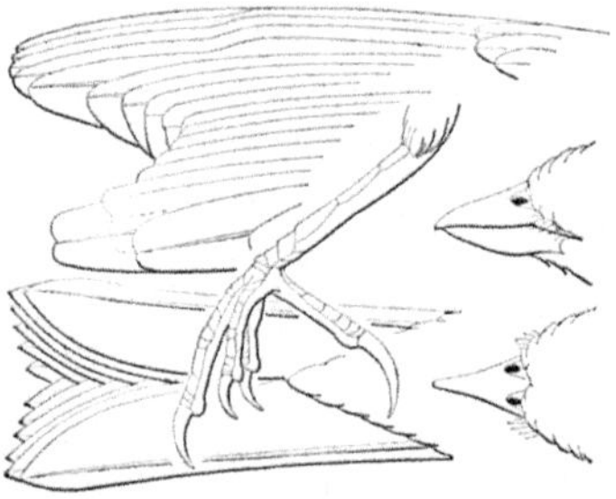

P. Savanna

narrow. Entire plumage above, head, neck, back, and rump, streaked. Thickly streaked beneath.

All the species are very similar in color, differing mainly in size and intensity of tint; scarcely at all in pattern.

Passerculus Sandwichensis, Gmelin.

THE AONALASKA SPARROW.

Emberiza Sandwichensis, Gmelin, I. 1788, 875. — *Passerculus Sandwichensis*, Baird, P. R. Rep. IX. Birds, 444. — Cooper and Suckley, XII. iii. Zool. of W. T. 199; pl. xxviii. f. 2.

Emberiza chrysops, Pallas, Zoog. Rosso-As. II. 1811, 45; tab. xlviii. fig. 1. (Aonalaska.)

Sp. Char. Feathers of the upper parts generally with a central streak of blackish-brown; the streaks of the back with a slight rufous suffusion laterally; the feathers edged with gray, which is lightest on the scapulars. Crown with a broad median stripe of yellowish-gray. A superciliary streak from the bill to the back of the head; eyelids, and edge of the carpal joint, yellow. A yellowish-white maxillary stripe curving behind the ear coverts, and margined above and below by brown. The lower margin is a series of thickly crowded spots on the sides of the throat, which are also found on the sides of the

neck, across the upper part of the breast, and on the sides of body. A few spots on the throat and chin. Rest of under parts white. Outer tail feather and primary edged with white. Length, 6.12; extent, 9.25; wing, 3.00; tail, 2.55. Iris brown; bill and feet paler.

Hab. Northwestern Coast from the Columbia River to Russian America. California in winter. (?)

This species I found in spring and fall only at the Columbia River, in their migrations to and from the north. As they were not seen there between October and April, they probably spent the winter in Southern Oregon and California, though hitherto overlooked by collectors in this State. During the warm season, between April and September, they were not to be seen near the Columbia, probably passing to the north, or possibly to the interior plains east of the Cascade Range. Their habits, as far as I observed them, did not differ from those of *P. alaudinus.* They are, however, considerably larger and darker colored.

Passerculus alaudinus, Bonaparte.

THE SKYLARK SPARROW.

Passerculus alaudinus, Bonaparte, Comptes Rendus, XXXVII. Dec. 1853, 918. (California.) — Baird, P. R. Rep. IX. Birds, 446. — *Passerculus Savanna* (and *alaudinus ?*) Heermann, X. vi. 49.

Sp. Char. Similar to *P. Sandwichensis,* but smaller; the bill rather slenderer and elongated. Little of yellow in the superciliary stripe (most distinct anteriorly); the rest of the head without any tinge of the same. General color much paler and grayer. Breast with only a few spots. Length, 6.00; extent, 9.25; wing, 3.00; tail, 2.30. Iris brown; bill brown above, flesh-color below; feet brownish-white.

Hab. California and Lower Rio Grande of Texas and Mexico.

I think it very doubtful whether these specimens (which measure larger than the dimensions given by Baird, though otherwise agreeing) are anything more than a southern form of *P. Sandwichensis*, though collected near San Diego. Whether Bonaparte's and Baird's birds are different, I cannot determine. Baird considers it almost identical with *P. Savanna* of the East, and says that *P. Sandwichensis* differs from that species only in larger size.

Spring specimens have the superciliary stripe more decidedly yellow, so that there only remains a more slender bill to distinguish this from *P. Savanna*, and the larger size (characteristic of Northern specimens generally), with darker hues, for *P. Sandwichensis*.

At the Columbia River I noticed the arrival of this species from the south in March, and that they resided there until late in October. In this State they seem chiefly winter residents, abounding on the dry interior plains as far south as San Diego, where they remain until April in large flocks. I have not seen them during the summer months, though they very probably breed in some of the higher prairies of this State, yet I saw none about the summits of the Sierra Nevada in September. They prefer the dry rolling plains to marshes, although occasionally found in the latter. Dr. Heermann's note on this species corresponds to the habits of *P. anthinus*, and was probably intended for that species, of which he collected a specimen, while he got none of this, though referring to it as "*P. Savanna*." (See lists of specimens in Baird's Rep. IX. 445, 446.)

The song of this bird is faint and lisping, delivered from the top of a tall weed, during spring. Its nest is unknown, but *P. savanna* builds in the grass, and lays pale greenish eggs, slightly spotted and splashed with pale umber. (Nuttall.)

The naturalist of the Russian Telegraph Expedition found the *P. Sandwichensis* quite common at Sitka and Kodiak, although on the Yukon the representative of the genus seemed to be a species most like the present, or even the genuine *P. Savanna*.

Passerculus anthinus, BONAPARTE.

THE TITLARK SPARROW.

Passerculus anthinus, BONAPARTE, Comptes Rendus, XXVII. Dec. 1853, 919. (Russian America.) — BAIRD, P. R. Rep. IX. Birds, 445.
Passerculus alaudinus, HEERMANN, P. R. Rep. X. vi. 49.

SP. CHAR. Similar to *P. Sandwichensis*, but smaller. Beneath tinged with reddish. Breast and upper part of belly thickly spotted with sharply defined sagittate brown spots, exhibiting a tendency to aggregation on the middle of the belly. Superciliary stripe, and

one in the middle of the crown, decided greenish-yellow, the head generally tinged with the same; as also the back and sides of the neck. Length, 5.75; extent, 9.00; wing, 2.75; tail, 2.25. Iris brown; bill brown; feet and lower mandible paler.
Hab. Coast of California; Russian America; Kodiak. (Bonaparte.)

This plain little bird is peculiarly the marsh-sparrow of this coast, as I have found them rarely out of the salt marshes, where they lie so close, and run so stealthily under the weeds, as to be flushed with some difficulty, rising only to fly a few rods and drop again into the covert. They are not very gregarious, except when migrating, and fly up singly.

They abound in winter south of San Francisco, but I am not sure that any of them spend the summer so far south, though inclined to think that they do. Near San Diego, in February, they began to utter a short but pleasant song as they perched on the top of some tall weed; and though I observed them there until April, I did not succeed in finding any nests, and have not found the species at San Pedro in summer.

This species appears better marked, as compared with *P. Savanna*, than the preceding, although I am not entirely satisfied that it is different. It may, however, constitute a race characterized by a much greater amount of spotting beneath, extending over the whole breast and upper part of jugulum. They are rather dark brown, well-defined, and unusually sagittate.

Passerculus rostratus, Cassin.

THE SEA-SHORE SPARROW.

Emberiza rostrata, Cassin, Pr. A. N. Sc. VI. 1852, 348. — *Ammodramus rostratus*, Cassin, Illust. I. 1855, 226; pl. xxxviii. — *Passerculus rostratus*, Baird, P. R. Rep. IX. Birds, 446. — Heermann, X. vi. 46.

Sp. Char. Bill very long (0.55 of an inch above). Whole upper parts and sides of head and neck pale grayish-brown, nearly every feather with a darker central blotch, darkest along the shaft. A scarcely appreciable central stripe in the crown, an obscure yellowish-white superciliary, and a whitish maxillary one. Under parts pure white;

streaked on the breast and the sides of throat and body with dark brown (streak paler externally). Under tail coverts unspotted white. Tail and wing feathers and wings margined with the color of the back; the edges of tertiaries rather paler. Length, 6.00; extent, 9.50; wing, 3.00; tail, 2.30.

Hab. Coast of California, near San Diego and San Pedro; Cape St. Lucas.

Though having some resemblance to *Passerculus*, and without the acute, graduated tail feathers of other *Ammodromi*, this bird resembles the latter genus most closely in bill and claws, as well as in habits, in which it is very different from *Passerculus*.

I found them plenty at San Pedro and San Diego at all seasons, and doubt whether they migrate at all. They frequented the shores of the bays and the sea-beach, also coming familiarly about buildings near the water, feeding on any seeds and insects they could find. On the beach they run along the sand in the rows of drift-weed, etc., seeking food, and rarely take flight unless surprised, and only fly a short distance. I have never seen them alight on bushes or heard them sing, their only note being a short chirp. At San Pedro I saw them in July feeding their young, but never found a nest that I was certain belonged to this species. This bird is a winter resident at Cape St. Lucas, where Mr. Xantus found it in abundance; but not in summer, although they probably breed near.

Passerculus guttatus, Lawrence.

THE ST. LUCAS FINCH.

Passerculus guttatus, Lawrence, Annals N. Y. Lyceum, N. H. VIII. May, 1867, 473.

Sp. Char. *Male.* The plumage above is of a dull grayish-brown, the centres of the feathers blackish; a stripe of pale yellow runs from the bill to the eye, a longer stripe of pale yellow extends from the under mandible down the side of the throat; the throat is white, and is separated from the yellow stripe by a line of dark brown spots; tail and wings umber-brown, the latter margined with dull pale fulvous; breast, upper part of abdomen, and sides conspicuously marked with elongated spots of dark brown, the lower

part of the abdomen white; the under tail coverts white, sparingly streaked with brown; bill dusky yellow, the culmen dark brown; iris dark brown; legs pale brown. Length, 5.15; wing, 2.60; tail, 2.00; tarsi, 0.80.

Hab. Lower California, San Jose. Collected by John Xantus, December, 1859. (Type, spec. in Mus. Smith. Ins., No. 26,615.)

In the size and form of its bill only, this species resembles *P. rostratus;* above it is very much darker, and differs from all its allies in the obscure grayish coloring of these parts, with no reddish-brown, and in having its under plumage more closely and fully spotted. Nothing is known of its habits, which, however, are probably much like those of *P. rostratus*, with which it was associated when killed.

Genus POŒCETES, Baird.

Poocætes, Baird, P. R. Rep. IX. Birds, 447.

Gen. Char. Bill rather large; upper outline slightly decurved towards the end, lower straight; commissure slightly concave. Tarsus about equal to the middle toe; outer toe a little longer than the inner, its claw reaching to the concealed base of the middle claw; hind toe reaching to the middle of the middle claw. Wings unusually long, reaching to

P. gramineus.

the middle of the tail, as far as the coverts, and pointed; the primaries considerably longer than the secondaries, which are not much surpassed by the tertiaries; second and third quills longest; first a little shorter, about equal to the fourth, shorter than the tail, the

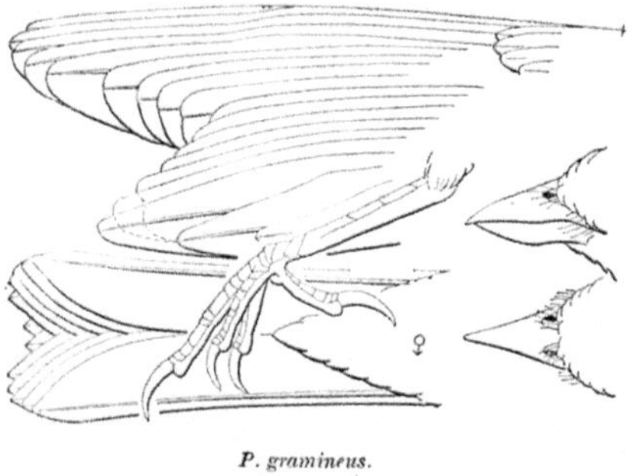

P. gramineus.

outer feathers scarcely shorter; the feathers rather stiff; each one acuminate and sharply pointed; the feathers broad nearly to the end, when they are obliquely truncate. Streaked with brown above everywhere; beneath, on the breast and sides. The lateral tail feather is white.

Poœcetes gramineus, Gmelin.

THE GRASS FINCH; BAY-WINGED BUNTING.

Fringilla graminea, Gmelin, Syst. Nat. I. 1788, 922. — *Emberiza graminea*, Wilson, Am. Orn. IV. 1811, 51; pl. xxxi. f. 5. Aud., Birds Amer. III. 1841, 65; pl. 159. — *Zonotrichia graminea*, Bonaparte, List, 1838. — Heermann, P. R. Rep. X. vi. 47. — Newberry, VI. iv. 85. — *Poocætes gramineus*, Baird, P. R. Rep. IX. Birds, 447. — Cooper and Suckley, XII. iii. Zoöl. of W. T. 200.

Sp. Char. Tail feathers rather acute. Above light yellowish-brown; the feathers everywhere streaked abruptly with dark brown, even on the sides of the neck, which are paler. Beneath yellowish-white; the breast and sides of neck and body streaked with brown. A faint light superciliary and maxillary stripe; the latter margined above and below with dark brown; the upper stripe continued around the ear coverts, which are darker than the brown color elsewhere. Wings with the shoulder light chestnut-brown, and with two dull whitish bands along the ends of the coverts; the outer edge of the

secondaries also is white. Outer tail feather, and edge and tip of the second, white. Length, about 6.25; extent, 9.75; wing, 3.50. Iris brown; bill brown, paler below; feet whitish-brown.

Hab. United States, from Atlantic to Pacific; or else one species to the high central plains, and another from this to the Pacific (var. *confinis*, Baird, differing in grayer hues, legs and wings longer, bill more slender and straighter, streaks on the breast narrower).

I found this bird wintering in the Colorado Valley in considerable numbers, but it disappeared by April. I have not seen them near the coast, and they seem to seek the interior valleys chiefly in summer. Their favorite resorts are grassy meadows and open woods or orchards, where the grass grows high; they also seek food along roads when migrating. I am not sure whether they breed in this State, but think they do towards the north. Their nests (in the East) are built on the ground under tufts of grass, and usually sunk below the surface; they are formed principally of withered wiry grass, lined with softer grass and hairs. The eggs, four or five, are white, with several shades of dark reddish-brown scattered in spots, chiefly at the larger end. They probably raise several broods annually, and do not migrate much from the Middle Atlantic States. (Nuttall.)

Their song is quite frequent, and resembles that of the canary, though less loud and varied. They also sing sometimes late in the evening. They feed much along roads, and are fond of dusting themselves in such places, running along instead of flying when followed.

According to Dr. Newberry they are common in the Sacramento Valley in summer and fall, but I have not observed them myself in the Sierra Nevada. They do not go very far north, and the only late record of their occurrence south of the United States is at Oaxaca in Western Mexico. They have not yet been observed in the West Indies.

GENUS **COTURNICULUS**, BONAPARTE.

Coturniculus, BONAPARTE, Geog. List, 1838. (Type, *Fringilla passerina*, WILSON.)

GEN. CHAR. Bill very stout and short. Tertials almost equal to the primaries. Claws small, weak; hinder one shorter than its digit. Outstretched feet not reaching

C. passerinus.

the tip of the tail. Tail feathers not stiffened, graduated, short. Upper parts generally streaked, blotches on interscapular region very wide. Breast and sides generally streaked

C. passerinus.

more or less distinctly. Edge of the wing yellow. Wings short, reaching little beyond base of tail. Primaries and secondaries nearly equal.

Two species are confined to the eastern side of the continent, and a third is found throughout the United States. There are others in South America.

Coturniculus passerinus, Wilson.

THE YELLOW-WINGED SPARROW.

Fringilla passerina, Wilson, Am. Orn. III. 1811, 76; pl. xxvi. f. 5. — *Coturniculus passerinus*, Bonaparte, List, 1838. — Baird, P. R. Rep. IX. Birds, 450. — Heermann, X. 49. — Kennerly, X. vi. 28. — *Emberiza passerinus*, Audubon, Syn. 1839. Ib. Birds Amer. III. 73; pl. 162.

Fringilla savanarum, "Gmelin," Nuttall, Man. I. 1832, 494, 2d ed. I. 1840, 570.

Sp. Char. Upper parts brownish-rufous, margined narrowly and abruptly with ash-color; reddest on the lower part of back and rump; the feathers all abruptly black in the central portion; this color visible on the interscapular region, where the rufous is more restricted. Crown blackish, with a central and superciliary stripe of yellowish tinged with brown, brightest in front of the eye. Bend of wing bright yellow, lesser coverts

tinged with greenish-yellow. Quills and tail feathers edged with whitish; tertiaries much variegated. Beneath, brownish yellow, nearly white on the middle of the belly. Feathers of the upper breast and sides of body with obsoletely darker centres. Length about 5.00; extent, 8.25; wing, 2.50; tail, 2.00. Iris brown; bill slate-color, reddish below; feet brown. Young streaked below.

Hab. Eastern United States to Central Plains, valleys of Gila and Colorado, California.

Nuttall mentions this bird as an inhabitant of "Oregon," but later collectors have not found it west of the Rocky Mountains, with the exception of Dr. Kennerly, who obtained one at Williams Fork of the Colorado, in February. He remarks that they were in some places quite numerous, going in flocks of five or six or more. I did not obtain or see the bird at Fort Mojave. It is known only as a summer visitor east of the Rocky Mountains, reaching Pennsylvania after the middle of May, and singing from trees in gardens, etc., somewhat like the purple finch, though less vigorously. Their nest is fixed on the ground among grass, composed of loose dry herbage, and lined with hair and root fibres. The eggs, five, are grayish-white, spotted with brown. (Nuttall.)

They migrate north to lat. 42°, east of the mountains, and may be found in this State west of the Sierra Nevada. Heermann mentions them as "abundant," but probably means to the eastward. I am, however, informed that Mr. Hepburn has taken several specimens in California.

Genus AMMODROMUS, Swainson.

Ammodromus, Swainson, Zool. Jour. III. 1827. (Type, *Oriolus caudacutus*, Gmelin.)

Gen. Char. Bill very long, slender, and attenuated, considerably curved towards the tip above. The gonys straight. The legs and toes are very long, and reach considerably beyond the tip of the short tail. The tarsus is about equal to the elongated middle toe;

A. caudacutus.

the lateral toes equal, their claws falling considerably short of the base of the middle one; the hind claw equal to the lateral one. Wings short, reaching to the base of the tail; much rounded; the secondaries and tertials equal, and not much shorter than the prima-

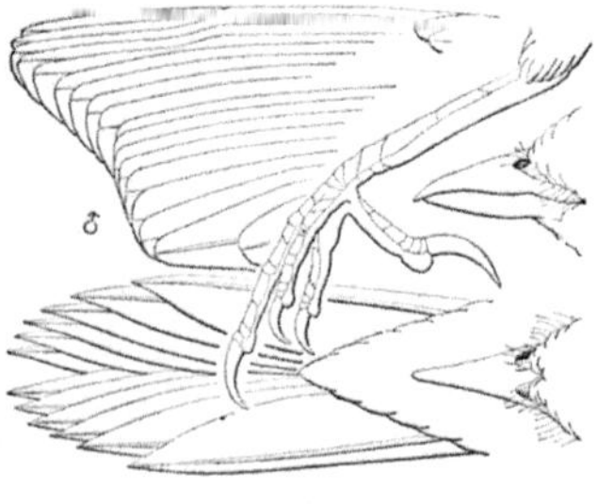

A. caudacutus.

ries. The tail is short and graduated laterally; each feather stiffened, lanceolate, and acute.

Color. Streaked above and across the breast; very faintly on the sides.

The typical species (2) are confined to the Eastern coast. *A. rostratus* differs in having a very stout bill and notched tail; *A. Samuelis* resembles *Melospiza* in color, with the bill and wings of *Ammodromus*

Ammodromus Samuelis, Baird.

Ammodromus Samuelis, Baird, Pr. Boston Soc. N. H. for June, 1858. Ib. P. R. Rep. IX. Birds, 455. Ib. Birds of N. Amer.; pl. 71, f. 1.

Sp. Char. Somewhat like *Melospiza Heermanni*, but considerably smaller. Bill slender, attenuated, and acute. Tarsus not longer than middle toe and claw. Above streaked on the head, back, and rump with dark brown, the borders of the feathers paler, but without any rufous. Beneath, bluish-white; the middle of the breast, with sides of throat and body, spotted and streaked with blackish-brown. Wings above nearly uniform

dark brownish rufous. Under tail coverts yellowish-brown, conspicuously blotched with blackish. An ashy superciliary stripe, becoming nearly white to the bill, and a whitish maxillary one; the crown with faint grayish median line. Length, 5.00; wing, 2.20; tail, 2 35.

Hab. California, lat. 38°, near the coast.

Nothing is known of the habits of this species.

Genus CHONDESTES, Swainson.

Chondestes, Swainson, Phil. Mag. I. 1827, 435. Ib. Fauna Bor. Amer. II. 1831. (Type, *Chondestes strigatus*, Swainson, equal to *Fringilla grammaca*, Say.)

Gen. Char. Bill swollen; both outlines gently curved; the lower mandible as high as the upper; the commissure angulated at the base, and then slightly sinuated. Lower mandible rather narrower at the base than the length of the gonys; broader than the upper. Tarsi moderate, about equal to the middle toe; lateral toes equal and very short, reaching but little beyond the middle of the penultimate joint of the middle toe, and fall-

C. grammaca.

ing considerably short of the base of the middle claw. Wings long, pointed, reaching nearly to the middle of the tail; the tertials not longer than the secondaries; the first quill shorter than the second and third, which are equal. The tail is moderately long,

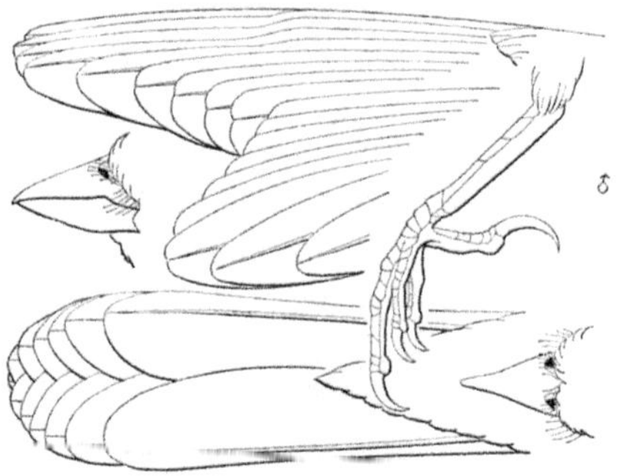

C. grammaca.

considerably graduated, the feathers rather narrow, and elliptically rounded at the end. Streaked on the back. Head with well-defined large stripes. Beneath white, with a pectoral spot.

The typical species is the only one known, and it has a wide range over the continent, from the Mississippi Valley to the shores of the Pacific east and west, and from Oregon to Mexico north and south.

Chondestes grammaca, SAY.

THE LARK FINCH.

Fringilla grammaca, SAY, Long's Exped. R. Mts. I. 1823, 139. — *Chondestes grammaca*, BONAPARTE, List, 1838. — BAIRD, P. R. Rep. IX. Birds, 456. — HEERMANN, X. vi. 48. — COOPER and SUCKLEY, XII. iii. Zool. of W. T. 200. — *Emberiza grammaca*, AUDUBON, Birds Amer. III. 1841, 63; pl. 158.
Chondestes strigatus, SWAINSON, Philos. Mag. I. 1827, 435.

SP. CHAR. Hood chestnut, tinged with black towards the forehead, with a median and a superciliary stripe of dirty whitish. Rest of upper parts pale grayish-brown, the interscapular region streaked with dark brown. Beneath white, a round spot on the upper part of the breast, a maxillary stripe, and a short line from the bill to the eye, continued

faintly behind it, black. A white crescent under the eye, bordered below by black, and behind by chestnut. Tail feathers dark brown, tipped broadly with white. Length, 7.00; extent, 11.50; wing, 3.75. Iris brown; bill horn-brown, bluish below; feet whitish.

Hab. From Wisconsin and the prairies of Illinois to the Pacific Coast; south to Texas and Mexico.

I did not observe this bird in the Colorado Valley, though one was obtained at Fort Yuma during the Mexican Boundary Survey, on the 30th of December; and, as I found large flocks of them in the valleys of San Diego County in February, I have no doubt that they winter in the southern part of the State, though not as far north as San Francisco. They reach the Columbia River east of the Cascade Mountains early in May, and breed in this State from near San Diego northward in the sheltered valleys, and at Santa Barbara. I have not found their nests in this State, but have met with many of them from Missouri west, through Kansas and Nebraska, in May and June. They build on the ground, constructing their nests chiefly of grass; the eggs are white, with scattered hair lines and spots of brown near the large end, if I remember rightly.

In the spring the males sing very sweetly, more like the canary than any other bird, but with an occasional harsh note intermingled. They frequent open plains, but usually near trees, and often alight on them in flocks. Their food consists chiefly of grass and other seeds which they find on the ground.

They are more similar in habits to the grass finch than to any other bird, and resemble this species somewhat in colors also.

Z. leucophrys.

Genus ZONOTRICHIA, Swainson.

Zonotrichia, Swainson, Fauna Bor. Am. II. 1831. (Type, *Emberiza leucophrys*, Forster.)

Gen. Char. Body rather stout. Bill conical, slightly notched, somewhat compressed, excavated inside, the lower mandible not so deep as the upper; gonys slightly convex; commissure nearly straight. Feet stout; tarsus rather longer than middle toe; the lateral

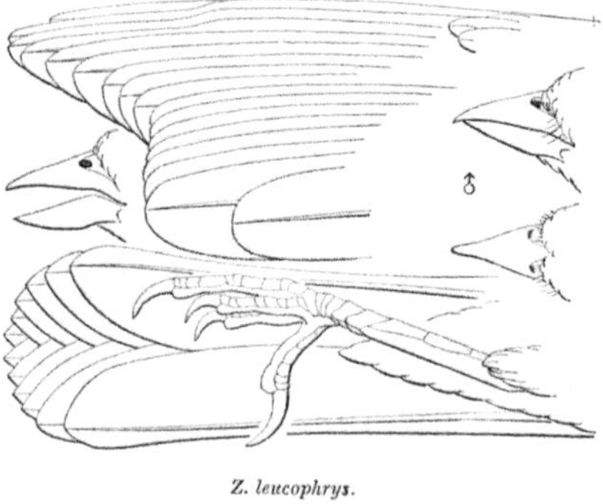

Z. leucophrys.

toes very nearly equal Hind toe longer than the lateral ones; their claws just reaching to base of middle one. Inner claw contained twice in its toe proper; claws all slender and considerably curved. Wings moderate, not reaching to the middle of the tail, but beyond the rump; secondaries and tertials equal and considerably less than longest primaries; second and third quills longest; first about equal to the fifth, much longer than tertials. Tail rather long, moderately rounded; the feathers not very broad. Back streaked. Rump and under parts immaculate. Head black, or with white streaks entirely different from the back.

Zonotrichia Gambelii, NUTTALL.

THE WESTERN WHITE-CROWNED FINCH.

Fringilla Gambelii, NUTTALL, Man. I. 2d ed. 1840, 556. (California) — *Zonotrichia Gambelli*, GAMBEL, J. A. N. Sc. 2d Series, I. Dec. 1847, 50. — BAIRD, P. R. Rep. IX. Birds, 460. IB. Birds of N. Amer.; pl. 69, f. 1. — KENNERLY, X. iv. 28. — HEERMANN, X. vi. 48. — COOPER and SUCKLEY, XII. iii. Zool. of W. T. 201.

"*Zonotrichia leucophrys*," of older authors on the West Coast birds, not of Forster (Eastern). — NEWBERRY, Zool. Cal. and Or. Route; P. R. Rep. VII. iv. 1857, 87. — NUTTALL, Man. I. 2d ed. 554.

SP. CHAR. Head above, and a narrow line behind the eye to the occiput, black; a longitudinal patch in the middle of the crown, and a line above the eye, the two confluent on the occiput, white. Lores gray. Sides of the head, fore part of breast, and lower neck all round, pale-ash, lightest beneath and shading insensibly into the whitish of the belly and

Adult.

Young.

chin; sides of belly and under tail coverts tinged with yellowish-brown. Interscapular region streaked broadly with dark chestnut brownish. Edges of the tertiaries brownish-chestnut. Two white bands on the wing.

Female similar, but smaller; immature male with the black of the head replaced by dark chestnut brown, the white tinged with brownish-yellow. Length, 6.50; extent, 9.50; wing, 3.25; tail, 3.12. Iris brown; bill brown, below yellow; feet pale brown.

Hab. Rocky Mountains to the Pacific Coast.

In the Colorado Valley, at Fort Mojave, I found this bird quite common throughout winter, and some remaining as late as May 15th, but I could find no nests, and doubt very much whether they build there, as they all seem to desert the coast south of Santa Cruz, during summer. In winter they abound through all the inhabitable country south of San Francisco, and in summer wander to the summits of the Sierra Nevada, and the regions north of the Columbia, a few remaining near San Francisco in the cold district, subject to the sea breeze at that season.

Near the mouth of the Columbia I found a nest of this bird in June, 1854. It was built in a bush, not more than a foot from the ground, formed of grasses neatly interwoven, and lined with softer materials. The eggs, if my memory of them is correct, were four or five in number, white, with thinly scattered dark spots near the large end.

The song of this species is loud but short, and remarkably melancholy. It may be heard during the whole year at intervals, and frequently at night, when its sad tone seems peculiarly suited to the darkness.

They are familiar little birds, the flocks spending most of the cool months around dwellings and barns, and I have been told that they sometimes build in gardens.

Zonotrichia leucophrys, Forster.

THE WHITE-CROWNED SPARROW.

Emberiza leucophrys, Forster, Phil. Trans. LXII. 1772, 382, 426.—*Fringilla leucophrys*, Swainson, Fauna Bor. Amer. II. 255.—Audubon, Orn. Biog. II. 1834, 88; pl. 114. *Zonotrichia leucophrys*, Swainson, Bonaparte, List, 1838.—Baird, Birds N. Amer. 458.

Sp. Char. Very similar to *Z. Gambelii*, as above described, but the white stripe from bill over eye, instead of being continuous, is interrupted by a short black line reaching

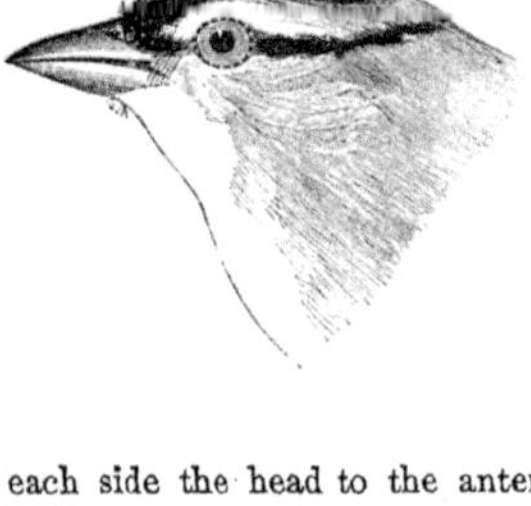

from the black stripe on each side the head to the anterior portion of the eye. Size and proportions of *Z. Gambelii.*

Hab. Eastern portion of all North America from the Rocky Mountains to the Atlantic; Cape St. Lucas in winter.

This very close ally of *Z. Gambelii* is well known in the Eastern United States as a Southern migrant; in the spring breeding abundantly in Labrador and the shores of Hudson's Bay. The habits and song are very similar to those of *Z. Gambelii*, and the mode of nesting almost identical.

The occurrence of this species at Cape St. Lucas in winter is a very curious fact in geographical distribution. (Baird.)

Zonotrichia coronata, Pallas.

THE GOLDEN-CROWNED SPARROW.

Emberiza coronata, Pallas, Zoog. Rosso-Asiat. II. 1811, 44; plate.

Emberiza atricapilla, Audubon, Orn. Biog. V. 1839, 47; pl. 394. (Not of Gmelin.) — *Fringilla atricapilla*, Audubon, Syn. 1839, 122. Ib. Birds Amer. III. 1841, 162; pl. 193.

Fringilla aurocapilla, Nuttall, Man. I. 2d ed. 1840, 555. — *Zonotrichia aurocapilla*, Bonaparte, Consp. 478. — Newberry, Zool. Cal. and Or. Route, P. R. Rep. VI. iv. 1857, 88.

Zonotrichia coronata, Baird, P. R. Rep. IX. Birds, 461. — Heermann, X. vi. 48. — Cooper and Suckley, XII. iii. Zool. of W. T. 201.

Sp. Char. Hood, from bill to upper part of nape, pure black; the middle longitudinal third occupied by yellow on the anterior half, and a pale ash on the posterior. Sides and under parts of head and neck, with upper part of breast, ash-color, passing insensibly into whitish on the middle of the body; sides and under tail coverts tinged with brownish. A yellowish spot above the eye, bounded anteriorly by a short black line from the eye to

the black of the forehead. This yellow spot, however, is reduced to a few feathers in spring dress. Interscapular region, with the feathers streaked with dark brown, suffused with dark rufous externally. Two narrow white bands on the wings. Length about 7.00; extent, 9.80; wing, 3.25. Iris brown; bill brown, paler below; feet pale brown.

Hab. Pacific Coast from Russian America to Southern California, and Cape St. Lucas; east of the Rocky Mountains (?).

I met with a few of this species wintering as far south as San Diego, associating with *Z. Gambelii*, but much less familiar, as they did not come about the house, but kept among the dense thickets. They were at that time silent, and though I have seen them north of the Columbia in May, I never heard them utter any song.

According to Heermann, they sometimes breed in California, as he mentions finding a nest near Sacramento. "It was composed of coarse stalks of weeds, and lined internally with fine roots. The eggs, four in number, are ashy-white, marked with lines of brown umber, sometimes appearing black from the depth of their shade, and covered also with a few neutral tint spots."

I saw none of this species near the summit of the Sierra Nevada.

GENUS JUNCO, WAGLER.

Junco, WAGLER, Isis, 1831. (Type, *Fringilla cinerea*, SWAINSON.)
Niphæa, AUDUBON, Syn. 1839. (Type, *Emberiza hyemalis*, GMELIN.)

GEN. CHAR. Bill small, conical; culmen curved at the tip; the lower jaw quite as high as the upper. Tarsus longer than the middle toe; outer toe longer than the inner, barely reaching to the base of the middle claw. Hind toe reaching as far as the middle of the latter; extended toes reaching about to the middle of the tail. Wings rather short, reaching over the basal fourth of the exposed surface of the tail; primaries, however, considerably longer than the nearly equal secondaries and tertials. The second

J. Oregonus

quill longest, the third to fifth successively but little shorter; first longer than sixth, much exceeding secondaries. Tail moderate, a little shorter than the wings; slightly emarginate and rounded. Feathers rather narrow, oval at the end. No streaks on the head or body; color above uniform on the head, back, or rump, separately or on all together. Belly white; outer tail feathers white.

J. hyemalis

Two other species of snow-bird, much resembling the *J. Oregonus*, inhabit the Rocky Mountains; one visits the Atlantic States in winter, and several occur in Mexico.

Junco Oregonus, TOWNSEND.

THE OREGON SNOW-BIRD.

Fringilla Oregona, TOWNSEND, J. A. N. Sc. VII. 1837, 188. — *Struthus Oregonus*, BONAPARTE, List, 1838. — NEWBERRY, Zool. Cal. and Or. Route, P. R. Rep. VI. iv. 1857, 88. — *Niphœa Oregona*, AUDUBON, Syn. 1839, 107. IB. Birds Amer. III. 1841, 91; pl. 168. — CABANIS, Mus. Hein. 1851, 134. — *Junco Oregonus*, SCLATER, Pr. Zool. Soc. 1857, 7. — BAIRD, P. R. Rep. IX. Birds, 466. — KENNERLY, X. iv. 28. — HEERMANN, X. vi. 47. — COOPER and SUCKLEY, XII. iii. Zool. of W. T. 202.

SP. CHAR. Head and neck all round sooty-black; this color extending to the upper part of the breast, but not along the sides under the wings. Interscapular region of the back and exposed surface of the wings dark rufous-brown. A lighter tint of the same on

the sides of breast and belly. Rump brownish-ash. Outer two tail feathers white; the third with only an obscure streak of white. Length, 6.00; extent, 9.00; wing, 3.00. Iris brown; bill flesh color, tip dark; feet brownish-white.

Hab. Pacific Coast of the United States to the eastern side of the Rocky Mountains. Stragglers as far east as Fort Leavenworth in winter, and Great Bend of Missouri.

This species is abundant in winter throughout most parts of the State, and resides in summer in the mountains, probably down to the 32d parallel, though I have not determined its residence along the coast farther south than Monterey. That locality is very cool, and an extensive forest of pines coming down to the coast favors the residence of several Northern birds during summer. At San Diego I observed them until April 1st, when they probably retired to the high mountains visible a few miles back from the coast. A few also visit the Colorado Valley in winter.

On the Coast Mountains south of Santa Clara I found many of them breeding in May, 1864; one nest I saw near the west base of the mountains, on the 13th, containing young just ready to fly. It was built in a cavity among the roots of a large tree on a steep bank, formed of leaves, grasses, and fine root-fibres, and covered outside with an abundant coating of green moss, raised above the general surface of the ground. The old birds showed such anxiety that I was induced to hunt for the nest nearly an hour before I found it, being satisfied from their actions that it was there, though very hard to find. The moment I saw it the young flew out in all directions, and their parents seemed more alarmed than ever.

On May 20th I discovered another nest on the very summit of the mountains, probably a second laying, as it contained but three eggs. It was slightly sunk in the ground under a fern (*Pteris*), and formed like the other, but with less moss around the edge; some cow's and horse's hair was also used in the lining. The eggs were bluish-white, with blackish and brown spots of various sizes thickly sprinkled on the larger end, measuring 0.74×0.60. According to Heermann, they build in bushes.

The only song of this species is a faint trill, much like that of the chipping sparrow (*Spizella socialis*), delivered from the top of some low tree in March and April. At other times they have merely the sharp chirp or call-note by which they are easily distinguished from most other sparrows. Though migrating so far south in winter, they also remain during winter at least as far north as the Columbia River, frequenting the vicinity of houses and barns, in great numbers, especially when the snow is on the ground, and then meriting the name of their Eastern cousin, which usually appears in the United States only in the season of snow.

They probably raise two broods in this State, and at Puget's Sound I have seen young fledged as early as May 24th.

According to Dr. Coues, this species is an exceedingly abundant winter resident in Arizona, arriving at Fort Whipple early in October, and becoming very numerous in a short time. They remain until the middle of April, and stragglers are even seen until May, keeping quietly hidden in out-of-the-way places, like the Eastern snow-bird, until cold weather sets in, when they become very familiar, and are to be seen everywhere.

Dr. Coues found numerous specimens that could only be considered as hybrids between this species and the next. Although the two in their typical dress are very appreciably different; yet individuals were met with combining the distinctive characters of both.

Junco caniceps, WOODHOUSE.

THE GREY-HEADED SNOW-BIRD.

Struthus caniceps, WOODHOUSE, Pr. A. N. Sc. Phil. 1852, 202. — *Junco caniceps*, BAIRD, Birds N. Amer. 1858, 468. — COUES, Pr. A. N. Sc. 1866, 85.
Junco dorsalis, HENRY, Pr. A. N. Sc. 1858, 117. — BAIRD, Birds N. Amer. 467.

SP. CHAR. Bill yellowish, black at the tip. Above dark plumbeous, the head and neck all round of this color, which extends (paling a little) along the sides, leaving the middle of the belly and crissum quite abruptly white. Lores conspicuously, but not very abruptly darker. Interscapular region abruptly reddish chestnut-brown, which does not

extend on the wings, except perhaps a faint tinge on some of the greater coverts. Two outer tail feathers entirely white; third with a long white terminal stripe on the inner web. Length, 6.00; wing, 3.23; tail, 3.04.

Hab. Rocky Mountains; from Black Hills to San Francisco Mountains, New Mexico.

As will be seen by the synonymy above, I combine *Junco caniceps* and *dorsalis* into one species; in this following the conclusions of Dr. Coues, after a careful investigation of the type specimens of both, and an extensive series of skins. The first-mentioned name having priority is retained for the species. It has a close relationship to a Mexican species, *J. cinereus*, as well as to one from Central Yucatan, recently described by Mr. Salvin, but appears sufficiently distinct.

As stated in the preceding article, undoubted hybrids have been met with between this species and *J. Oregonus*. One of these, described by Dr. Coues,

26

has the general appearance of *caniceps;* the head and neck being slaty-gray, not black; the lores blackish in contrast, etc. There is, however, a large dorsal area colored as in *Oregonus,* while the sides are strongly tinged with pinkish-fulvous, as in *Oregonus,* instead of being plain cinereous-gray like the throat, as in *caniceps.* Other specimens exhibit still different degrees of combination of the characters of the two species. (Baird.)

Genus **POOSPIZA**, Cabanis.

Poospiza, Cabanis, Wiegmann's Archiv, I. 1847, 349. (Type, *Emberiza nigro-rufa,* Orb., or *Pipilo personata,* Swainson.)

Gen. Char. Bill slender, conical, both outlines gently curved. Under jaw with the edges considerably inflected; not so high as the upper. Tarsi elongated, slender; considerably longer than the middle toe. Toes short, weak; the outer decidedly longer than

P. bilineata.

the inner, but not reaching to the base of the middle claw. Hind toe about equal to the middle without its claw. All the claws compressed and moderately curved. Wings rather long, reaching about over the basal fourth of the exposed portion of the rather long

P. bilineata.

tail. Tertiaries and secondaries about equal, and not much shorter than the lengthened primaries; the second to fifth about equal and longest; the first considerably shorter, and longer than the seventh. Tail long, slightly emarginate, graduated; the outer feather abruptly shorter than the others. Feathers broad, linear, and rather obliquely truncate at the ends, with the corners rounded.

Color. Uniform above, without streaks. Beneath white, with or without a black throat. Black and white stripes on the head.

Poospiza bilineata, CASSIN.

THE BLACK-THROATED SPARROW.

Emberiza bilineata, CASSIN, Pr. A. N. Sc. Phil. V. Oct. 1850, 104; pl. iii. (Texas.) IB. Illust. I. v. 1854, 150; pl. xxiii. — *Poospiza bilineata*, SCLATER, Pr. Zool. Soc. 1857, 7. — BAIRD, P. R. Rep. IX. Birds, 470. — HEERMANN, X. v. 14.

SP. CHAR. Above, uniform unspotted ashy-gray, tinged with light brown; purer and more plumbeous anteriorly. Under parts white, tinged with plumbeous on the sides, and with yellowish-brown about the thighs. A sharply defined superciliary and maxillary stripe of pure white, the former margined internally with black. Loral region black,

passing insensibly into dark slate on the ears. Chin and throat between the white maxillary stripes black, extending on the upper part of the breast in a rounded outline. Tail, black, edged externally with white. Bill blue. Length, 5.75; extent, 8.00; wing, 2.50; tail, 2.90. Iris brown; bill black, bluish below; feet brownish.

Hab. Valley of Rio Grande and of Gila, to Mojave River, California.

On the barren, treeless, and waterless mountains that border the Colorado Valley, this was one of the few birds enlivening the desolate prospect with their cheerful presence. They were nowhere numerous, but generally seen in pairs or small parties hopping along the ground under the scanty shrubbery. In winter they descended to the hills near the Colorado, where the males, perched on a low bush, sang short but lively ditties toward spring.

In crossing the Providence Range, in May, I found a nest containing white eggs, which I have little doubt belonged to this species.

Poospiza Belli, Cassin.

BELL'S FINCH.

Emberiza Belli, Cassin, Pr. A. N. Sc. Phil. V. Oct. 1850, 104; pl. iv. (San Diego, Cal.) — *Poospiza Belli*, Sclater, Pr. Zool. Soc. 1857, 7. — Baird, P. R. Rep. IX. Birds, 470. — Kennerly, X. iv. 29. — Heermann, X. vi. 46.

Sp. Char. Upper parts generally, with sides of head and neck, uniform bluish-ash, tinged with yellowish-gray on the crown and back, and with a few obsolete dusky streaks on the interscapular region. Beneath, pure white, tinged with yellowish-brown on the sides and under the tail. Eyelids, short streak from the bill to above the eye, and small median spot at the base of bill, white. A stripe on the sides of throat and spot on the

upper part of the breast, with the loral space and region round the eyes, plumbeous black. Tail feathers black; the outer edged with white. Wing feathers all broadly edged with brownish-yellow; the carpal joint tinged with yellowish-green. Bill and feet blue. Length, 6.00; extent, 9.00; wing, 3.00. Iris brown; bill brown, pale below; feet brown.

Hab. California and Valley of Gila and Colorado to Fort Thorn. North to lat. 38° 30′, in Sacramento Valley.

The extensive thickets, called chaparral, which cover barren dry tracts for miles, in all the southern half of California, are the favorite resorts of this little bird. There they pick up a living from small seeds, and probably insects, being apparently quite indifferent as to water, or depending on that dropping from the foliage after dews and fogs. They may be seen running rapidly, or rather hopping along the ground, with tail carried perfectly erect, and at the least alarm seeking the friendly thicket. They reside all the year in the same localities, and are numerous on the island of San Nicolas, eighty miles from the main-land, though I saw none on the other islands, except one on Santa Barbara.

In spring the males sing a low, monotonous ditty from the top of a favorite shrub, answering each other from long distances. Their nest, built about three feet from the ground, is composed of grasses and slender weeds, lined with hair, etc. The eggs, about four, are pale greenish, with reddish-brown dots thickly sprinkled over. At San Diego I found the young hatched on May 18th, but think they are often earlier.

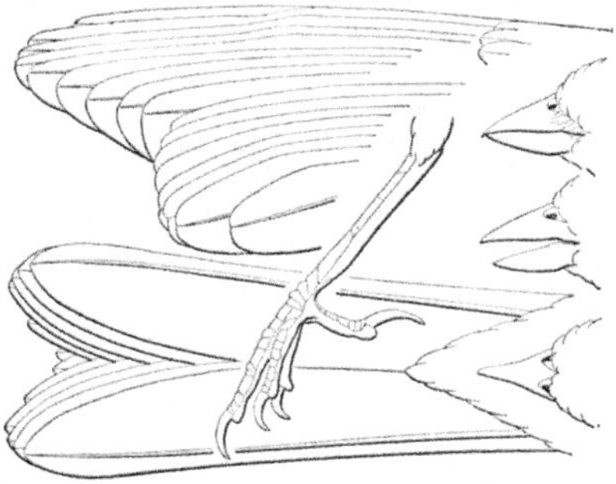

P. Belli.

It is a common bird in the chaparral of Santa Clara Valley, and, according to Heermann, along Cosumnes River.

We give above the details of external structure of *P. Belli*, to show the difference from *P. bilineata*.

Genus **SPIZELLA**, Bonaparte.

Spizella, Bonaparte, Geog. and Comp. List, 1838. (Type, *Fringilla Canadensis*, Latham.)
Spinites, Cabanis, Mus. Hein. 1851, 133. (Type, *Fringilla socialis*, Wilson.)

Gen. Char. Bill conical, the outlines slightly curved; the lower mandible not so deep as the upper, the commissure gently sinuated; the roof of the mouth not knobbed. Feet

S. monticola.

slender; tarsus rather longer than the middle toe; the hind toe a little longer than the outer lateral, which slightly exceeds the inner; the outer claw reaching the base of the middle one, and half as long as its toe. Claws moderately curved. Tertiaries and secondaries nearly equal; wing somewhat pointed, reaching not quite to the middle of the

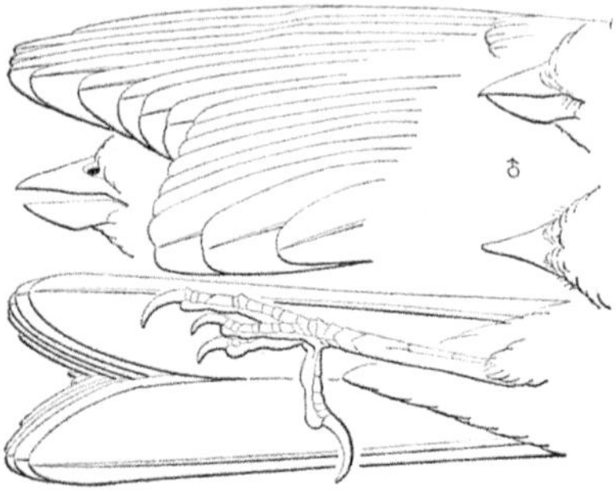

S. monticola.

tail. First quill a little shorter than second and equal to the fifth; third longest. Tail rather long, moderately forked, and divaricated at the tip; the feathers rather narrow. Back streaked; rump and beneath immaculate. Hood generally uniform.

One species of this genus (*S. pusilla*, the field sparrow), which is not found on this side of the mountains, visits the Atlantic States in summer. It is, however, closely allied to our *S. Breweri*, which replaces it in our fauna.

Spizella monticola, Gmelin.

THE MOUNTAIN OR TREE-SPARROW.

Fringilla monticola, Gmelin, Syst. Nat. I. 912, 1788. — *Spizella monticola*, Baird, P. R. Rep. IX. Birds, 472. — Kennerly, X. iv. 29. — Cooper and Suckley, XII. iii. Zool. of W. T. 203.

Fringilla Canadensis, Latham, Index Orn. I. 1790, 434. — Nuttall, Man. I. 2d ed. 572. — *Emberiza Canadensis*, Swainson, Fauna Bor. Am. II. 1831, 252. — Audubon, Birds Amer. III. 1841, 83; pl. 166.

Fringilla arborea, Wilson, Am. Orn. II. 1810, 12; pl. xii. f. 3.

Sp. Char. Middle of back with the feathers dark brown centrally, then rufous, and edged with pale fulvous (sometimes with whitish). Hood and upper part of nape continuous chestnut; a line of the same behind the eye. Sides of head and neck ashy. A broad light superciliary band. Beneath whitish, with a small circular blotch of brownish in the middle of the upper part of the breast. Edges of tail feathers, primary quills, and two bands across the tips of the secondaries, white. Tertiaries nearly black; edged externally with rufous turning to white near the tips. Length, 6.25; extent, 9.00; wing, 3.00. Iris brown; bill black; lower mandible yellow.

Hab. Eastern North America to the Missouri; Rocky Mountains to Little Colorado River, New Mexico; Sierra Nevada and Cascade Mountains (?).

Although not contained in the collections of any late explorers within this State, it is quite probable that this bird visits California in winter, and may even remain during summer about the summits of the Sierra Nevada, where I thought I saw them in September, 1863, though not successful in obtaining a specimen. I also saw them, as I thought, near the mouth of the Columbia in the winter of 1854, and Dr. Suckley obtained at the Dalles in January, 1855, what he identified with the description of this species.

In the Atlantic States it is a winter visitor, having at that season much the same habits as the chippy (*S. socialis*), but towards spring displaying considerable musical talent, singing something like the yellow-bird (*C. tristis*), though with less variety. (Nuttall.)

They retire far north in summer, breeding around Hudson's Bay, and in the Mackenzie River country to the Arctic Ocean, where, according to Hutchins, the nest is like that of *S. socialis*, and the eggs pale brown, with darker spots. They probably breed in the Rocky Mountains, lat. 39°, where they were found in August, by Lieutenant F. T. Bryan, U. S. A.

Spizella socialis, Wilson.

THE CHIPPING SPARROW.

Fringilla socialis, Wilson, Am. Orn. II. 1810, 127; pl. xvi. f. 5. — Nuttall, Man. I. 2d ed. 574. — *Spizella socialis*, Bonaparte, List, 1838. Ib. Conspectus, 1850, 480. — Baird, P. R. Rep. IX. Birds, 473. — Heermann, X. vi. 48. — Cooper and Suckley, XII. iii. Zool. of W. T. 203. — *Emberiza socialis*, Audubon, Syn. 1839. Ib. Birds Amer. III. 1841, 80; pl. 165.

Sp. Char. Rump, back of neck, and sides of head and neck, ashy. Interscapular region with black streaks, margined with pale rufous. Crown continuous and uniform chestnut. Forehead black, separated in the middle by white. A white streak over the eye, and a black one from the base of the bill through and behind the eye. Under parts unspotted whitish, tinged with ashy, especially across the upper breast. Tail feathers and primaries edged with paler, not white. Two narrow white bands across the wing coverts.

Bill black. Length, 5.60 ; extent, 8.75 ; wing, 2.87. Iris brown ; bill black, below brown : feet whitish.

Hab. North America, from the Atlantic to the Pacific.

This truly sociable little sparrow, commonly known as the "chippy," from its short and constant chirp, is quite as abundant in the northern part of California, and north at least to lat. 49°, as on the Atlantic side of the continent. I found them wintering in the Colorado Valley in large numbers, but not near San Diego. It is, however, quite probable that they winter in other parts of the interior, as they reach San Francisco by the first of April without appearing at San Diego. They spend the summer, and build in all the northern half of the State, preferring the vicinity of oak groves and gardens, coming familiarly about the doorstep to pick up crumbs, and building their nest in low branches of fruit-trees or garden shrubs. It is neatly formed of grass, rather thinly interwoven, and lined almost always with horse-hairs. The eggs are four or five, bright greenish-blue, with a few light and dark brown spots, chiefly at the larger end. They raise two or even three broods annually, in the Atlantic States.

The only song of this bird is a low trill, usually heard from the top of a tree during the still warm morning.

In autumn they collect into large flocks, and frequent open woods, pastures, etc. I found flocks of them on Catalina Island in June, but could discover no nests, and, as they were all old birds, concluded that they for some reason had forgotten to migrate.

Dr. Coues found this species breeding abundantly at Fort Whipple, Arizona, arriving there in March, and remaining until November. In a large series of skins collected by him, he found a decided difference in the young from those of Eastern specimens, the color of the crown being more like that of *S. monticola.* As, however, the adults of the two regions could not be distinguished from each other, he did not venture to give the Arizona bird a separate name.

Spizella Breweri, CASSIN.

BREWER'S SPARROW.

Emberiza pallida, AUDUBON, Orn. Biog. V. 1839, 66; pl. 398, f. 2. IB. Synopsis, 1839. — IB. Birds Amer. III. 1841, 71; pl. 161. — HEERMANN, P. R. Rep. X. vi. 48. (Not of Swainson, 1831.)

Spizella Breweri, CASSIN, Pr. A. N. Sc. VIII. Feb. 1856, 40. — BAIRD, P. R. Rep. IX. Birds, 475. — KENNERLY, X. iv. 29. — NEWBERRY, VI. iv. 88.

SP. CHAR. Smaller than *S. socialis*. Back and sides of hind neck ashy. Prevailing color above pale brownish-yellow, with a tinge of grayish. The feathers of back and crown streaked with blackish. Beneath whitish, tinged with brown on the breast and sides, and an indistinct narrow brown streak on the edge of the chin. Ear coverts brown-

ish-yellow, margined above and below by dark brown. Crown with median and superciliary ashy-white stripe. Length, 5.38; extent, 7.38; wing, 2.50. Iris and feet brown; bill black, brown below.

Hab. Rocky Mountains of United States to the Pacific Coast.

Spizella pallida, Swainson, Fauna Bor. Amer. II. 1831, 251, only differs from *S. Breweri* in having the marking more distinct, and in not having brownish shafts to the feathers of the sides of body, besides being confined, according to Baird, to the east side of the Rocky Mountains. Dr. Coues (Pr. Phil. Ac. 1866) is satisfied of their difference.

At Fort Mojave I found small flocks of this species after March 20th, frequenting grassy spots among low bushes, and a month later they were singing, much in the style of a canary, but more faintly; they probably remain in the valley all summer. Dr. Kennerly obtained one along Williams Fork of the Colorado, in February, so that they probably winter there, as do several other birds found by him, but which I did not find at Fort Mojave in winter, though situated in a lower and milder valley. Drs. Heermann and Newberry also found them common in the Sacramento and Tejon Valleys in autumn. Dr. Coues met with them in small numbers at Fort Whipple, Arizona, where they breed.

Spizella atrigularis, Cabanis.

THE BLACK-CHINNED SPARROW.

Spinites atrigularis, Cabanis, Mus. Hein. 1851, 133. — *Spizella atrigularis*, Baird, Birds N. Amer. 1858, 476. — Coues, Pr. Ac. N. Sc. 1866, 87.
Struthus atrimentalis, Couch, Pr. Ac. N. Sc. 1854, 67. — *Spizella n. s.*, Coues, Ibis. 1865, 118, 164.
Spizella cana, Baird, MSS. — Coues, Pr. Ac. N. Sc. 1866, 88.

Sp. Char. *Adult male.* Tail elongated, deeply forked, and divaricated; general color bluish-ash, paler beneath and turning to white on the middle of the belly. Interscapular region yellowish-rusty, streaked with black. Forehead, loral region, and side of head as far as eyes, chin, and upper part of throat, black. Mouth and tail feathers very dark

Adult.

brown, edged with ashy. Edges of coverts like the back. Bill red. Length, 5.50; wing, 2.50; tail, 3.00. Iris black; legs and feet brownish black.

Young birds, and perhaps the female, without the black marks on the face.

Hab. Fort Whipple, Arizona; Cape St. Lucas, and the highlands of Mexico to the Isthmus of Tehuantepec.

Young

This species is but little known, and is probably not very abundant in its area of distribution. According to Dr. Coues, it reaches the vicinity of Prescott, Arizona, in April, and remains till October, collecting towards autumn in small flocks. The note is said to be more sweet and melodious than in any other of the genus. Nothing is known of the nesting of the species, though it is quite probable that the eggs resemble those of *S. pusilla*.

GENUS **MELOSPIZA**, BAIRD.

Melospiza, BAIRD, Birds N. Amer. 1858, 476. (Type, *Fringilla melodia*.)

GEN. CHAR. Bill conical, very obsoletely notched, or smooth; somewhat compressed. Lower mandible not so deep as the upper. Commissure nearly straight. Gonys a little curved. Feet stout, not stretching beyond the tail; tarsus a little longer than the middle

M. melodia.

toe; outer toe a little longer than the inner; its claw not quite reaching to the base of the middle one. Hind toe appreciably longer than the middle one. Wings quite short and rounded, scarcely reaching beyond the base of the tail; the tertials considerably longer than the secondaries; the quills considerably graduated; the fourth longest; the first not

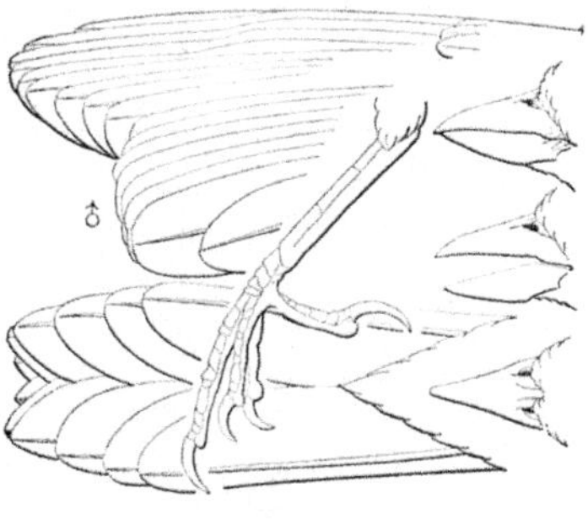

M. melodia.

longer than the tertials, and almost the shortest of the primaries. Tail moderately long, and considerably graduated; the feathers oval at the tips. Crown and back similar in color and streaked; beneath thickly streaked. Tail immaculate.

Melospiza Heermanni, BAIRD.

THE CALIFORNIAN SONG-SPARROW.

Melospiza Heermanni, BAIRD, P. R. Rep. IX. Birds, 478. IB. Birds N. Amer. pl. 70, f. 1.

SP. CHAR. General tint of upper parts olivaceous-brown, streaked with dark brown and ashy-gray. The crown is brown, with a superciliary and median stripe of dull gray, the former lighter; nearly white anteriorly; each feather of the crown with a narrow streak of dark brown. Interscapulars dark brown in the centre, grayish on the margin. Rump grayer than upper tail coverts, both with obsolete dark streaks. There is a whitish maxillary stripe, bordered above and below by one of dark brown, with a similar one from

M. Heermanni.

behind the eye. The under parts are white; the breast and sides of body and throat streaked with dark brown. On the middle of the breast these marks are rather aggregated so as to form a spot. No distinct white on tail or wings. Length, 6.50; extent, 8 50; wing, 2.75; tail, 3.00. Iris brown; bill horn-brown, bluish below, edge and angle yellow; feet pale brown.

Hab. California, from Mojave River and San Diego to San Francisco.

M. Gouldii.

Melospiza Gouldii, Baird, P. R. Rep. IX. Birds, 1858, 479, and Birds of N.

Amer. pl. 70, f. 2, is based on a specimen closely resembling *M. Heermanni*, but much smaller, the skin measuring only 4.70 inches; wing, 2.10; tail, 2.38; while the bill and feet are nearly as large as in *M. Heermanni*. It was sent from England by Mr. John Gould, labelled "California," and may possibly be a southern dwarfed specimen from the peninsula. None that I have collected, even in the most southern localities, are nearly so small, the wing being always at least half an inch longer.

Specimens from Mojave River, San Diego, Santa Barbara, S. B. Island, and San Francisco differ only in the comparative stoutness of their bills, which seems insufficient to distinguish more than one species, and may depend on age.

This species is the representative of the genus in all the southern half of California, except Colorado Valley, being found in every locality where there are thickets of low bushes and tall weeds, especially in the vicinity of water, but coming familiarly about gardens and houses if unmolested by its enemy the cat. Their usual resort is on the ground under the shade of plants, where they industriously scratch for seeds throughout the day, rarely flying more than a few yards, and never deserting their homes from one end of the year to the other.

Occasionally, especially in spring, they perch on some low bush or tree and sing their lively and pleasing melodies, for an hour at a time, each song being a complete little stanza of a dozen notes, and frequently varied or changed entirely for another of similar style, but quite distinct. There is no difficulty in distinguishing their songs when once heard, although no two birds sing precisely alike. There is a similarity of tone and style in all the species of *Melospiza* proper, that has led former observers to consider them as of only one species, when taken in connection with their similar colors and habits.

The nest of this species I cannot positively describe, though I found one at Santa Cruz in June, which I have little doubt belonged to it. It was built in a dense blackberry-bush, about three feet from the ground, formed of a thick wall of grasses and bark, lined with finer grasses. There were but two eggs, smoky white, and densely speckled with dull brown. I waited for more eggs to be laid, but on my next visit found that it had been robbed.

"*Zonotrichia guttata*," of Heermann, P. R. Rep. X. vi. 47, refers chiefly to this species, which he collected in Tejon Valley, while he did not obtain the true *guttata* (*rufina*, which see).

Though this bird was abundant around Santa Cruz, I only found two nests after much searching. The first, built on a willow, close against the tree, and three feet from the ground, contained four eggs partly hatched on May 11th. (I had seen newly fledged young on the 7th.) It was composed of coarse dry stems and leaves, lined with finer grass and horsehairs, outside five inches wide, four high, inside two and a half wide, two deep; eggs pale green, blotched and spotted with purplish-brown, chiefly at the large end; their size 0.62 × 0.82 inch. The ground color is paler and spots darker than those of *Z. Gambelii*, and the whole coloring much darker than those of *M. fallax*. This was probably an old nest used for a second brood.

I found another similar nest, also with four eggs, in a thicket, six feet up, as late as July 10th, doubtless a second brood.

Melospiza rufina, BRANDT.

THE RUSTY SONG-SPARROW.

"*Emberiza rufina*, BRANDT, Desc. Av. Rossic. 1836, tab. II. 5, Sitka." BONAPARTE.

Fringilla cinerea (GMELIN), AUDUBON, Orn. Biog. V. 1839, 22; pl. 390. IB. Syn. 1839. 119. IB. Birds Amer. III. 1841, 145; pl. 187.—(?) *Fringilla cinerea*, GMELIN, I. 1788, 922.

Fringilla (*Passerella*) *guttata*, NUTTALL, Man. I. 2d ed. 1840, 581.—*Zonotrichia guttata*, GAMBEL, J. A. N. Sc. I. Dec. 1847, 50.

Melospiza rufina, BAIRD, P. R. Rep. IX. Birds, 180.—COOPER and SUCKLEY, XII. iii. Zool. of W. T. 204.

SP. CHAR. Bill slender. Similar in general appearance to *M. Heermanni*, but much more rufous, the colors more blended. General appearance above light rufous-brown, the interscapular region streaked very obsoletely with dark brownish rufous, the feathers of the crown similar, with still darker obsolete central streaks. A superciliary and very obscure median crown stripe, ashy. Under parts brownish-white; the breast and sides of

throat and body broadly streaked with dark brownish-rufous; darker in the centre. A light maxillary stripe. Sides of the body tinged strongly with the colors of the rump, and leaving only a narrow space of the belly white. Under coverts brown. Length, 6.75; extent, 8.75; wing, 2.70; tail, 3.00. Iris, bill, and feet brown.

Hab. Russian America, to Sierra Nevada, California, lat. 35°.

This is the more northern and mountain-loving representative of the song-sparrows, being resident in the higher Sierra Nevada, and on the borders of the evergreen forests towards the Columbia and northward, where it is the only species, and common down to the level of the sea.* I there found them having habits and songs entirely similar to those of the Eastern *M. melodia,* and also of the *M. Heermanni.* I never succeeded in finding a nest, these forest birds being more artful in concealing their treasures than those that have become accustomed to the society and protection of mankind, when many species, usually wild, select the garden as the safest place for building in. In the mild winters, usual near the Columbia, these birds do not show any disposition to emigrate, but come more familiarly around the house when the snow has buried their usual supply of food.

Melospiza fallax, Baird.

THE MOUNTAIN SONG-SPARROW.

Zonotrichia fallax, Baird, Pr. A. N. Sc. Ph. VII. June, 1854, 119. (Pueblo Creek, New Mexico.)
(?) *Zonotrichia fasciata* (Gmelin), Gambel, J. A. N. Sc. Ph. 2d Series, I. 1847, 49.
Melospiza fallax, Baird, P. R. Rep. IX. Birds, 481. — Kennerly, X. iv. 29; pl. 27, f. 2.

Sp. Char. Similar to *Z. Heermanni,* but with wings longer in proportion, and bill smaller. Dark centres to the pale rufous streaks of the feathers of upper and under surfaces obsolete or wanting. Superciliary light stripe ash-color anteriorly. Length, 6.25; ex-

tent, 8.25; wing, 2.75; tail, 2.87. Iris brown; bill horn-brown, paler below; feet pale brown.

Hab. Rocky Mountain region, from Fort Thorn to the Colorado. Fort Tejon (?).

Were it not for the difference in proportions, this species might be considered one of those bleached varieties of an allied species, inhabiting the torrid and dry Colorado Valley; but the want of intermediate forms between

* I have seen a specimen obtained at Marysville, in spring, by Mr. Gruber.

it and *M. Heermanni* entitles it to distinction. It resembles most nearly the Eastern *M. melodia.* In habits and song there is no appreciable difference from those of the other species. A nest built in a willow thicket was composed of bark, fine twigs, and grass, lined with hair. The eggs are bluish-white, blotched and streaked with reddish-brown, measuring 0.74 × 0.55 inch.

Melospiza Lincolnii, AUDUBON.

LINCOLN'S FINCH.

Fringilla Lincolnii, AUDUBON, Orn. Biog. II. 1834, 539; pl. 193. — NUTTALL, Man. I. 2d ed. 1840, 569. — *Peucæa Lincolnii*, AUDUBON, Syn. 1839, 113. IB. Birds Amer. III. 1841, 116; pl. 177. — *Melospiza Lincolnii*, BAIRD, P. R. Rep. IX. Birds, 482. — KENNERLY, X. iv. 29.

SP. CHAR. Crown chestnut, with a median and two lateral or superciliary ash-colored stripes; each feather above streaked centrally with black. Back with narrow streaks of black. Beneath white, with a maxillary stripe curving round behind the ear coverts, a well-defined band across the breast, extending down the sides, and under tail coverts, brownish yellow. The maxillary stripe margined above and below with lines of black

spots. The throat, upper part of breast, and sides of body, with streaks of black, smallest in the middle of the former. There is a chestnut stripe back of the ear, streaked with black. The pectoral bands are sometimes paler. Length, 5.75; extent, 7.50; wing, 2.60. Iris and bill brown; feet brownish-white.

Hab. United States from the Atlantic to the Pacific; and south through Mexico to Guatemala.

Flocks of this species passed near San Diego on their way northward about March 25th, keeping during the day among the grass, rather shy and silent. They had indeed very much the same habits as a *Passerculus,* in which genus Bonaparte puts them. They differ much in their migratory habits, gregariousness, and general form from other *Melospizas,* and will probably merit the generic distinction suggested by Baird for this and an

Eastern species under the name of *Helospiza*, though this species does not seem, while with us, to frequent marshes like the other and typical species.

Dr. Kennerly found them in winter in New Mexico, and some probably winter in California, though I saw none in the Colorado Valley. I have not seen nor heard of them during summer in this State, and they seem to go very far north to breed, as Audubon found them in Labrador. There they had much the habits of *M. melodia*, and a similar song. The young were fledged by the 4th of July, and the nest and eggs are still undescribed.

Genus **PEUCÆA**, Audubon.

Peucæa, Audubon, Synopsis, 1839. (Type, *Fringilla æstivalis*.)

Gen. Char. Bill moderate. Upper outline and commissure decidedly curved; gonys nearly straight. Legs and feet small; the tarsus about equal to the middle toe; the lat-

P. æstivalis.

eral toes equal, their claws falling considerably short of the middle one; the hind toe reaching about to the middle of the latter. The outstretched feet reach only to the mid-

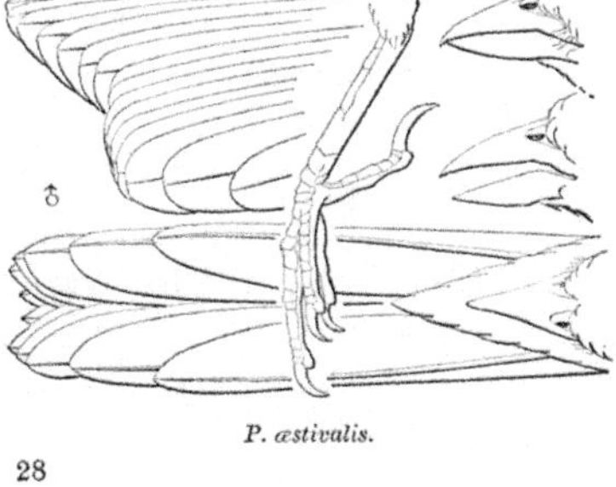

P. æstivalis.

dle of the tail. The wing is very short, reaching only to the base of the tail; the longest tertials do not exceed the secondaries, while both are not much short of the primaries; the outer three or four quills are graduated. The tail is considerably longer than the wings; it is much graduated laterally; the feathers, though long, are peculiarly narrow, linear, and elliptically rounded at the ends.

Color beneath plain whitish or brownish, with a more or less distinct dusky line each side of the chin. Above with broad obsolete brown streaks or blotches. Crown uniform, or the feathers edged with lighter. Inner tail feathers with obsolete transverse dusky bars.

A species visits the Southeastern States, and another occurs from Texas to Arizona.

Peucæa ruficeps, Cassin.

THE RED-CAPPED FINCH.

Ammodromus ruficeps, Cassin, Pr. A. N. Sc. VI. Oct. 1852, 184 (California). Ib. Illust. I. v. 1854, 135; pl. xx. — Heermann, P. R. Rep. X. vi. 49. — *Peucæa ruficeps*, Baird, P. R. Rep. IX. Birds, 486.

Sp. Char. Above brownish-ash. The crown and nape uniform brownish-chestnut. The interscapular region and neck with the feathers of this color, except around the margins. A superciliary ashy stripe, whiter at the base of the bill. Beneath, pale yellowish-brown, or brownish-yellow, darker and more ashy across the breast and on the sides of

the body; middle of belly and chin lighter; the latter with a well-marked line of black on each side. Under tail coverts more rufous. Length, 6.25; extent, 7.50; wing, 2.25; tail, 2.85. Iris brown; bill horn-brown, bluish below; feet whitish.

Hab. Coast of California to Sierra Nevada; Catalina Island; San Francisco, rare; Cosumnes River; south to Mexico.

I have only met with this species on Catalina Island in June, a few keeping about the low bushes, feeding on the ground, and very difficult even to get a sight of. I heard them sing a few musical notes that reminded me of those of the *Cyanospiza*. They flew short distances only, and in habits seemed more like the *Melospiza*. Their favorite resort, like that of the Eastern species, may, perhaps, be pine woods.

Peucæa Cassinii, Woodhouse.

CASSIN'S FINCH.

Zonotrichia Cassinii, Woodhouse, Pr. A. N. Sc. VI. April, 1852, 60. — *Passerculus Cassinii*, Woodhouse, Sitgreaves's Report, 85. — *Peucæa Cassinii*, Baird, Birds N. Amer. 1858, 485.

Sp. Char. Above light chestnut, all the feathers margined and tipped with bluish-gray. Interscapular and crown feathers with a narrow streak of brown. Beneath white, tinged with ash across the breast, and with brown towards the tail. An obsolete light

superciliary stripe, and a narrow dusky maxillary one. Tail feathers obsoletely blotched with bluish-white at the end. Bend of wing yellow; lesser coverts tinged with greenish-yellow. Length, 6.00; wing, 2.65; tail, 2.75.

Hab. Southern Texas, and west to the Gulf of California.

This species, a close relation to the Eastern *P. æstivalis,* though of paler colors, and longer wings and tail, is but little known, only a few specimens having been obtained, and nothing recorded of its habits.

Sub-Family PASSERELLINÆ.

Char. Toes and claws very stout; the lateral claws reaching beyond the middle of the middle one; all very slightly curved. Bill conical, the outlines straight; both mandibles equal; wings long, longer than the even tail, reaching nearly to the middle of its exposed portion. Hind claw longer than its digit, which is nearly as long as the middle toe; tarsus longer than the middle toe. Brown above, either uniformly so, or faintly streaked; triangular spots below.

Professor Baird has formed this sub-family to embrace the single genus

Passerella, believing that it differs essentially from all those included in the other sub-families, in the length of its lateral toes and the size of its claws, which characters are somewhat allied to those of *Pipilo* and *Xanthocephalus.*

Genus **PASSERELLA**, Swainson.

Passerella, Swainson, Class. Birds, II. 1837, 288. (Type, *Fringilla iliaca,* Merrem.)

Gen. Char. Body stout. Bill conical, not notched, the outlines straight; the two jaws of equal depth; roof of under mandible deeply excavated, and vaulted; not knobbed. Tarsus scarcely longer than the middle toe; outer toe little longer than the inner, its

P. Townsendii.

claw reaching to the middle of the central one. Hind toe about equal to the inner lateral; the claws all long, and moderately curved only; the posterior rather longer than the middle, and equal to its toe. Wings long, pointed, reaching to the middle of the tail;

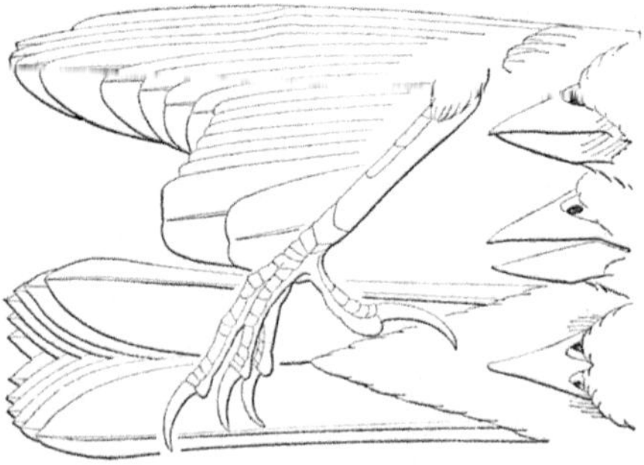

P. iliaca.

the tertials not longer than the secondaries; second and third quills longest; first equal to the fifth. Tail very nearly even, scarcely longer than the wing. Inner claw contained scarcely one and a half times in its toe proper.

Color. Rufous or slaty; obsoletely streaked or uniform above; thickly spotted with triangular blotches beneath.

Passerella Townsendii, AUDUBON.

TOWNSEND'S SPARROW.

Fringilla Townsendii, AUDUBON, Orn. Biog. V. 1839, 236; pl. 424, f. 7. — IB. Birds Amer. III. 1841, 43; pl. 187. — *Fringilla (Passerella) Townsendii,* NUTTALL, Man. I. 2d ed. 1840, 533. — *Passerella Townsendii,* BONAPARTE, Consp. 1850, 477. — BAIRD, P. R. Rep. IX. Birds, 489. — HEERMANN, X. vi. 47. — COOPER and SUCKLEY, XII. iii. Zool. of W. T. 204.

Fringilla meruloides, VIGORS, Zool. Blossom (Monterey), 1839, 19.

Emberiza Unalaschensis, GMELIN, I. 875, probably has some relation to the present species. It is based on the Aonalaska Bunting, of Pennant, Arctic Zool. II. 364.

SP. CHAR. Above, very dark olive-brown, with a tinge of rufous, the color continuous and uniform throughout, without any traces of blotches or spots; the upper tail coverts and outer edges of the wing and tail feathers rather lighter and brighter. The under parts white, but thickly covered with approximating triangular blotches like the back,

sparsest on the middle of the body and on the throat; the spots on the belly smaller. Side almost continuously like the back; tibiæ and under tail coverts similar, the latter edged with paler. Claws all very large and long; the hinder longer than its toe. First and sixth quills about equal. Length about 7.25; extent, 10.00; wing about 3.00. Iris and feet brown; bill black, yellow at base below.

Hab. Pacific Coast of the United States as far south as Monterey.

This bird is only a winter visitor in the lower country near the Columbia, but probably spends the summer in the Cascade Mountains, between April and October. Specimens have been killed near San Francisco in winter. While with us they are rather shy and silent birds, frequenting the woods and thick bushes, where they are constantly scratching among the dead

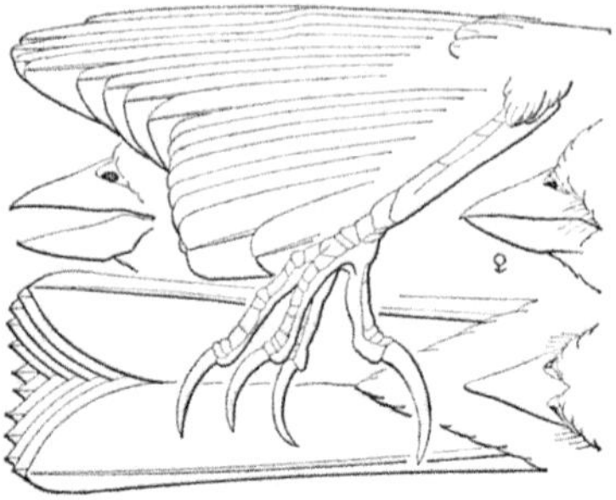

P. Townsendii.

leaves, gaining a scanty subsistence from seeds and insects. I have seen either this or the next species as far south as San Diego in winter, and noticed its arrival near San Francisco about October 20th.

Passerella megarhynchus, Baird.

THE LARGE-BEAKED SPARROW.

Passerella schistacea, Baird (in part), P. R. Rep. IX. Birds, 490. (Fort Tejon.)
Passerella megarhynchus, Baird (same vol.), 925. Ib. Birds N. Amer. pl. lxix. f. 4 (*schistacea*, f. 3).

Sp. Char. Bill very thick; the upper mandible much swollen at the base; under yellow. Above and on the sides uniform slate-gray; the upper surface of wings, tail feathers, and upper coverts dark brownish-rufous; ear coverts streaked with white. Beneath pure white, with broad triangular, arrow-shaped, and well-defined spots of slate-gray like the back everywhere, except along the middle of the belly; not numerous on the

P. megarhynchus.

throat. A hoary spot at the base of the bill above the loral region. Length, 6.80; wing, 3.08; tail, 3.40.

Hab. Fort Tejon, California, and northward in the Sierra Nevada.

P. schistacea.

I saw several of this species towards the summits of the Sierra Nevada in September, 1863, but from the difficulty of shooting them without destroy-

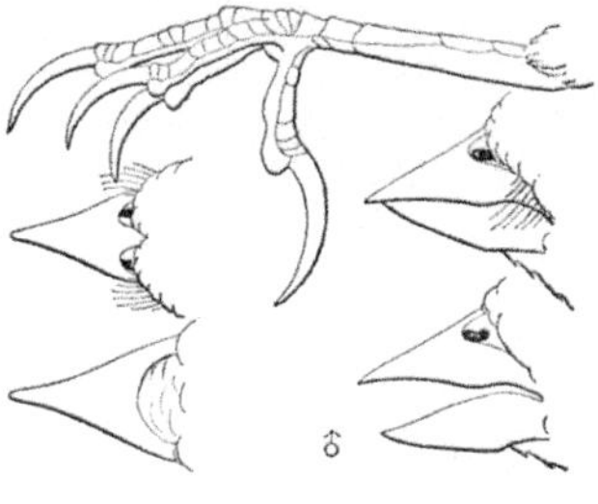

P. megarhynchus.

ing the specimens, in the dense thickets they frequented, I did not succeed in preserving any. As far as I noticed, their habits were similar to those of the *P. Townsendii*, and they had no song at that season.

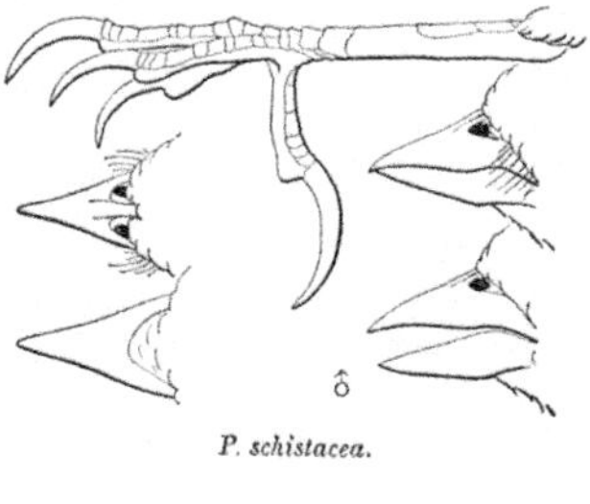

P. schistacea.

This species has usually been considered a variety of *P. schistacea*, but the figures we give on the preceding page will sufficiently show the difference. The true *schistacea* belongs to the Rocky Mountain region, and has a much slenderer bill, as indicated in the figures.

GENUS **CALAMOSPIZA** BONAPARTE.

Calamospiza, BONAPARTE, List, 1838. (Type, *Fringilla bicolor*, TOWNSEND.)

GEN. CHAR. Bill rather large, much swollen at the base; the culmen broad, gently but decidedly curved; the gonys nearly straight; the commissure much angulated near the base, then slightly sinuated; lower mandible nearly as deep as the upper, the margins much inflected, and shutting under the upper mandible. Nostrils small, strictly basal. Rictus quite stiffly bristled. Legs large and stout. Tarsi a little longer than the middle toe; outer toe rather longer than the inner, and reaching to the concealed base of

C. bicolor.

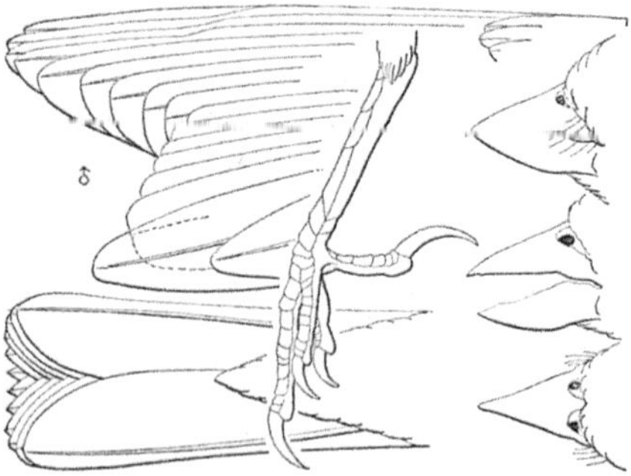

C. bicolor.

the middle claw; hind toe reaching to the base of the middle claw; hind claw about as long as its toe. Claws all strong, compressed, and considerably curved. Wings long and pointed; the first four nearly equal, and abruptly longest; the tertials much elongated, as long as the primaries. Tail a little shorter than the wings, slightly graduated, the feathers rather narrow and obliquely oval, rounded at the end.

Color. Black, with white on the wings. *Female*, varied.

This genus is well characterized by the large swollen bill, with its curved culmen; the large strong feet and claws; the long wings, a little longer than the tail, and with the tertials as long as the primaries, the first four quills equal and abruptly longest; the tail short and graduated.

The only group of North American *Spizellinæ*, with the tertials equal to the primaries in the closed wing, is *Passerculus*. This, however, has a differently formed bill, weaker feet, the inner primaries longer and more regularly graduated, the tail feathers more acute and shorter, and the plumage streaked brownish and white instead of black.

Calamospiza bicolor, Townsend.

THE WHITE-SHOULDERED BLACKBIRD.

Fringilla bicolor, Townsend, J. A. N. Sc. Phil. VII. 1837, 189. — *Calamospiza bicolor*, Bonaparte, List, 1838. — Baird, Birds N. Amer. 493. — *Corydalina bicolor*, Audubon, Birds Amer. III. 1841, 195; pl. 201.

Sp. Char. *Male*, entirely black; a broad band on the wing, with the outer edges of the quills and tail feathers, white.

Female, pale brown, streaked with darker above; beneath, white, spotted and streaked rather sparsely with black on the breast and sides. Throat nearly immaculate. A max-

Male.

illary stripe of black, bordered above by white. Region around the eye, a faint stripe above it, and an obscure crescent back of the ear coverts, whitish. A broad fulvous white band across the ends of the greater wing coverts. Tail feathers with a white spot at the end of the inner web. Length about 6.50; wing, 3.50; tail, 3.20; tarsus, 1.00; bill above, 0.60.

Hab High Central Plains to the Rocky Mountains; southwesterly to the Valley of Mimbres and Sonora. Cape St. Lucas.

Female.

This interesting species spends its time on the ground, associating in large flocks, and, according to Nuttall, is one of the sweetest songsters of the prairie. The nest is built among the grass, and the eggs are of a beautiful blue, sometimes with a few red spots.

SUB-FAMILY SPIZINÆ.

CHAR. Bill variable, always large, much arched, and with the culmen considerably curved; sometimes of enormous size, and with a great development backwards of the lower jaw, which is always appreciably, sometimes considerably, broader behind than the upper jaw at its base; nostrils exposed. Tail rather variable. Bill generally black or red. Wing shorter than in the first group. Gape almost always much more strongly bristled. Few of the species sparrow-like or plain in appearance; usually blue, red, or black and white; seldom (or never ?) streaked beneath.

This division embraces several large and gayly colored genera of sparrow-like birds, besides those here described, including the splendid cardinal birds of Lower California, Mexico, and the East. They are all frequenters of low shrubbery and the ground, unlike the brightly colored tree-finches of the first sub-family (*Coccothraustinæ*), and, besides their differences in habits, show a corresponding distinction in having short wings. Some, however, especially *Guiraca*, are intermediate in these respects.

GENUS **GUIRACA**, SWAINSON.

Guiraca, SWAINSON, Zool. Jour. III. Nov, 1827, 350. (Type, *Loxia cœrulea*, L.)
Coccoborus, SWAINSON, Class. Birds, II. 1837, 277. (Same type.)

GEN. CHAR. Bill very large, nearly as high as long; the culmen curved, with a rather sharp ridge; the commissure conspicuously angulated just below the nostril, the posterior leg of the angle nearly as long as the anterior, both nearly straight. Lower jaw deeper

G. melanocephala.

than the upper, and extending much behind the forehead; the width greater than the length of the gonys, considerable wider than the upper jaw. A prominent knob in the roof of the mouth. Tarsi shorter than the middle toe; the outer toe a little longer, reach-

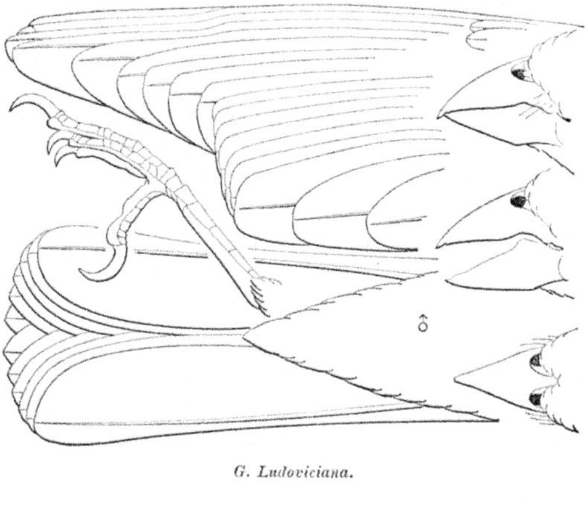

G. Ludoviciana.

ing not quite to the base of the middle claw; hind toe rather longer than to this base. Wings long, reaching the middle of the tail; the secondaries and tertials nearly equal; the second quill longest; the first less than the fourth. Tail very nearly even, shorter than the wings.

Guiraca melanocephala, Swainson.

THE BLACK-HEADED GROSBEAK.

Guiraca melanocephala, Swainson, Syn. Mex. Birds, Philos. Mag. I. 1827, 438. — Baird, P. R. Rep. IX. Birds, 498. — Cooper and Suckley, XII. iii. Zool. of W. T. 206. — *Fringilla melanocephala*, Audubon, Orn. Biog. IV. 1838, 519; pl. 373. — *Coccoborus melanocephalus*, Audubon, Synopsis, 1839, 133. — Ib. Birds Amer. III. 1841, 214; pl. 206. — Heermann, P. R. Rep. X. vi. 51.

Sp. Char. Head above and on the sides, with chin, back, wings, and tail, black. A broad median stripe on the crown, a stripe behind the eye, a well-marked collar on the

Male.

hind neck all round, edges of interscapular feathers, rump, and under parts generally pale brownish-orange, almost light cinnamon. Middle of belly, axillaries, and under wing coverts, yellow. Belly just anterior to the anus, under tail coverts, a large blotch at the end of the inner webs of first and second tail feathers, a band across the middle and greater wing coverts, some spots on the ends of the tertiaries, the basal portions of all the quills, and the outer three primaries near the tips, white.

Female, similar, with less black; wings and tail more olivaceous, the latter unspotted; the black of the head anteriorly replaced by whitish. The under wing coverts bright yellow. Length of male, 8.00; extent, 12.50; wing, 4.25; tail, 3.50. Iris brown; bill olive, fleshy white at base below; feet slate-color.

Hab. High Central Plains from the Yellowstone to the Pacific. Table-lands of Mexico.

This fine bird arrives in the State near San Diego about April 12th, and is numerous during summer throughout the mountains both of the coast and the Sierra Nevada, extending its migrations as far as Puget's Sound at least. They are often kept in cages on account of their loud and sweet song, which resembles that of the robin, but is louder and shorter. In the coast mountains in May their music is delightful, the males vying with each other from the tops of the trees, and making the hills fairly ring with their melody.

A nest I found May 12, at the eastern base of the Coast Range, was built on a low horizontal branch of an alder, consisting of a few sticks and weeds, very loosely put together, and with a lining of roots and grass. The

Female.

eggs were only three, pale bluish-white, thickly spotted with brown, densely near large end; size 0.95 × 0.70. According to Heermann, they also build in bushes.

They frequent the ground in search of food, but also live much in trees, and feed sometimes on their buds. They are not very gregarious, merely assembling in families in the autumn, and, unlike the evening grosbeak (*Hesperiphona*), to which they have much external resemblance, do not fly high, nor make any sound when flying.

This bird arrived at Santa Cruz, in 1866, about April 12th, — the same day they reached San Diego, three hundred and fifty miles farther south, in 1862; May 23d, I found a young one just fledged. No bird near the coast equals this in loudness and sweetness of song, though it is surpassed by the bow-bill thrush in variety.

Dr. Coues states that this bird is a summer visitor to Fort Whipple (Prescott), Arizona, where it is abundant, remaining until the latter part of September. It then frequents the thick brush of ravines, etc., and the cottonwood copses of the creek bottom. Its ordinary note, according to the Doctor, "resembles that of Gambel's partridge, but its song is superb, a powerful but melodious succession of clear, rich, rolling notes, somewhat like those of the Baltimore oriole."

Guiraca cærulea, Linnæus.

THE BLUE GROSBEAK.

Loxia cærulea, Linnæus, Syst. Nat. I. 1766, 306. — Wilson, Amer. Orn. III. 1811, 78; pl. xxiv. f. 6. — *Guiraca cærulea*, Swainson, Birds Mex. in Phil. Mag. I. 1827, 438. — Newberry, P. R. Rep. VI. iv. 88. — Baird, IX. Birds, 499. — *Coccoborus cæruleus*, Swainson, Birds, II. 1837, 277. — Audubon, Birds Amer. III. 1841, 204; pl. 204. — Heermann, P. R. Rep. X. vi. 51.

Sp. Char. Brilliant blue; darker across the middle of the back. Space around the base of the bill and lores, with tail feathers, black. Two bands on the wing across the

Male.

tips of the primary and secondary coverts, with outer edges of tertiaries, reddish-brown. Feathers on the posterior portion of the under surface tipped narrowly with grayish-white.

Female, yellowish-brown above, brownish-yellow beneath; darkest across the breast, and lightest on the throat. Wing coverts and tertials broadly edged with brownish-

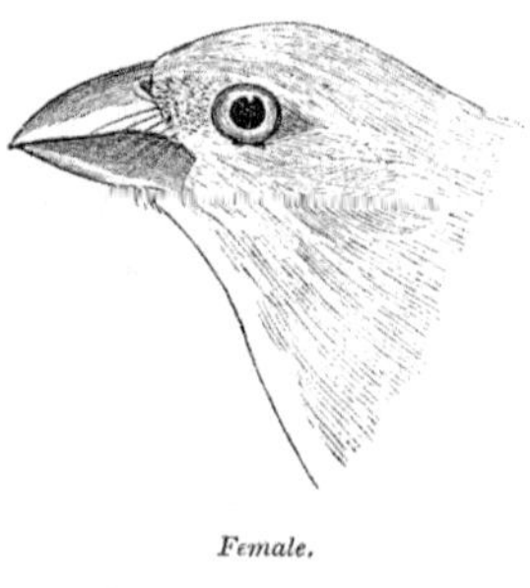

Female.

yellow. A faint trace of blue on the crown. Length of male, 7.50; extent, 11.00; wing, 3.75; tail, 2.80. Iris brown; bill black, white below; feet black.

Hab. More southern United States from the Atlantic to the Pacific, south to Mexico.

Everywhere a shy and solitary bird, this brilliant songster is rarely seen, although probably scattered throughout California in the warmer months. I noticed the first one at Fort Mojave, May 6th, and afterwards saw many more frequenting the trees and bushes along the river, and singing a lively song, resembling that of the *Carpodacus frontalis*. I have also seen them at Los Angeles and at Santa Barbara, and they were found at Pit River, in the

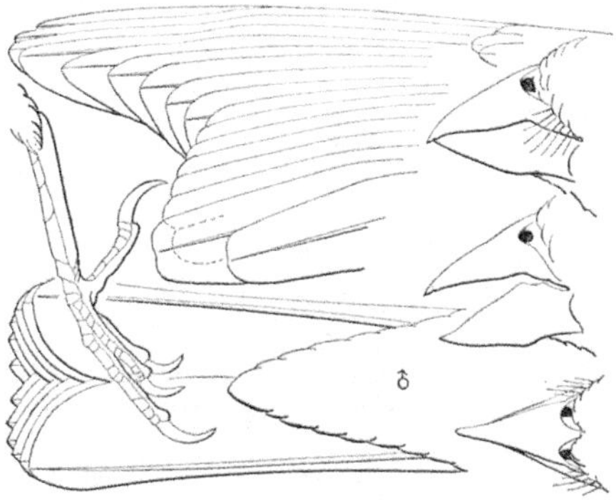

G. cærulea.

extreme northeast corner of the State, by Dr. Newberry. Their nest, as seen in the Eastern States, is made in a low bush, and composed of fine dry grass, lined with delicate root-fibres or horsehair. The eggs are about four (bluish-white), and they raise two broods in the season. (Nuttall.)

They frequent the banks of streams crossing the great interior plains and deserts, where there is little vegetation, except a few bushes, and where such brilliant birds seem quite out of place.

G. cærulea

GENUS **CYANOSPIZA**, BAIRD.

Cyanospiza, BAIRD. (Type, *Tanagra cyanea*, L.)

GEN. CHAR. Bill deep at the base, compressed; the upper outline considerably curved; the commissure rather concave, with an obtuse, shallow lobe in the middle. Gonys slightly curved. Feet moderate; tarsus about equal to middle toe; the outer lat-

C. amœna.

eral toe barely longer than the inner, its claw falling short of the base of the middle; hind toe about equal to the middle without claw. Claws all much curved, acute. Wings long and pointed, reaching nearly to the middle of the tail; the second and third quills longest. Tail appreciably shorter than the wings; rather narrow, very nearly even.

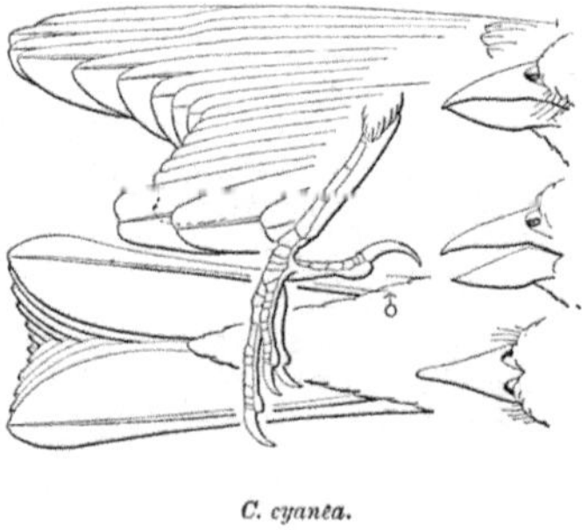

C. cyanea.

The species of this genus are all of very small size, and of showy plumage, usually blue, red, or green, in well-defined areas.

Cyanospiza amœna, Say.

THE BLUE LINNET.

Emberiza amœna, Say, Long's Exped. II. 1823, 47. — *Fringilla* (*Spiza*) *amœna*, Bonaparte, Am. Orn. I. 1825, 61; pl. vi. f. 5. — *Fringilla amœna*, Audubon, Orn. Biog. V. 1839, 64, 230; pls. 398 and 429. — Nuttall, Man. I. 2d ed. 546. — *Spiza amœna*, Bonaparte, List, 1838. — Audubon, Synopsis, 1839, 109. Ib. Birds Amer. III. 1841, 100; pl. 171. — Heermann, P. R. Rep. X. vi. 46. — *Cyanospiza amœna*, Baird, P. R. Rep. IX. Birds, 504. — Cooper and Suckley, XII. iii. Zool. of W. T. 205.

Sp. Char. *Male.* Upper parts generally, with the head and neck all round, greenish-blue; the interscapular region darker. Upper part of breast pale brownish-chestnut, separated from the blue of the throat by a faint white crescent; rest of under parts white. A white patch on the middle wing coverts, and an obscurely indicated white band across

Male.

Female.

the ends of the greater coverts. Loral region black. Length about 5.50; wing, 3.00; tail, 2.60. Iris brown; bill black, bluish below; feet black.

Female. Brown above; whitish beneath, with a trace of a buff pectoral band.

Hab. High Central Plains to the Pacific.

This is an abundant species throughout California, and north to Puget's Sound in summer, arriving at San Diego about April 22d, and remaining until October. One which I saw kept in a cage during winter retained its blue plumage, unlike the wild birds which are believed to change to the plain hues of the female in autumn. It is frequently kept in cages, and dealers, noticing its similarity of habits and song to the Eastern indigo-bird (*C. cyanea*), have absurdly given it that name, though it has not a particle of indigo in its colors. Perhaps they think it merely a faded and degenerate variety of that bird, just as Buffonian naturalists considered all American animals degenerate forms of those of Europe. Unfortunately, its name of *lazuli* finch is scarcely better understood than that of *amœna*, the

30

celebrated lazuli blue stone of Italy being known to few except foreign travellers.

During the summer there is scarcely a thicket or grove in the more open portions of the State, uninhabited by one or more pairs of this beautiful species. The male is not very timid, and frequently sings his lively notes from the top of some bush or tree, continuing musical throughout summer, and in all weathers. The song is unvaried, and rather monotonous, closely resembling that of the Eastern *C. cyanea*.

Their nest is built in a bush not more than three or four feet above the ground, formed of fibrous roots, strips of bark and grass, with a lining of plant-down or hairs, and securely bound to the surrounding branches. The eggs are four or five, white, faintly tinged with blue. At Santa Barbara I found them freshly laid on May 6th.

These birds are never very gregarious, though the males arrive in the spring in considerable flocks, travelling at night, and several days before the females. The latter are at all times very shy, and so plain in plumage that they are very difficult to obtain, unless on the nest.

They arrived at Santa Cruz in 1866, about April 12th, ten days earlier than observed at San Diego in 1862. A nest found May 7th, in a low bush close to the public road, and about three feet from the ground, was built very strongly, supported by a triple fork of the branch, composed of grass blades firmly interwoven, the inside lined with much horsehair and cobwebs. The outside measured three inches in height, three and three fourths in width; inside it was two wide, one and three fourths deep. The three eggs, partly hatched, were pale bluish-white, and measured 0.75 × 0.56 inch.

Cyanospiza versicolor, Bonaparte.

THE WESTERN NONPAREIL.

Spiza versicolor, Bonaparte, Pr. Z. S. 1837, 120. — Cabanis, Mus. Hein. 1851, 148. — *Cyanospiza versicolor*, Baird, Birds N. Amer. 503.

Sp. Char. *Male.* Posterior half of hood, with throat, dark brownish-red; interscapular region similar, but darker. Fore part of hood, lesser wing coverts, back of neck, and rump, purplish-blue; the latter purest blue; the belly reddish-purple, in places tinged with blue, more obscure posteriorly. Feathers of wing and tail dark brown, edged with dull bluish. Loral region and narrow frontal line, black. Length, 5 50; wing, 2.75; tail, 2.38.

Female. Similar to that of *C. amœna*, but distinguished by the absence of the two white bands on wings, and by the legs being black instead of dark brown. The bill is apparently more curved, and the legs stouter.

Hab. Northern Mexico, Sonora, and probably Southern Arizona. Cape St. Lucas.

Female.

Male.

This beautiful bird is not rare at Cape St. Lucas, where it breeds; the nest and eggs, and the habits of the bird itself, likewise, are probably much like those of *C. amœna.*

GENUS **PYRRHULOXIA**, BONAPARTE.

Pyrrhuloxia, BONAPARTE, Consp. 1850, 500. (Type, *Cardinalis sinuatus*, BONAPARTE.)

GEN. CHAR. The bill is very short and much curved, the culmen forming an arc of a circle of sixty degrees or more, and ending at a right angle with the straight gonys; the commissure abruptly much angulated anterior to the nostrils in its middle point; the lower jaw very much wider than the upper, and wider than the gonys is long; anterior

P. sinuata.

portion of commissure straight. Tarsus longer than middle toe; outer lateral toes longer, not reaching the base of the middle; wing considerably rounded, first quill longer than secondaries. Tail much longer than the wing, graduated; the feathers broad, truncate. Head crested.

Color, gray, with red feathers and patches.

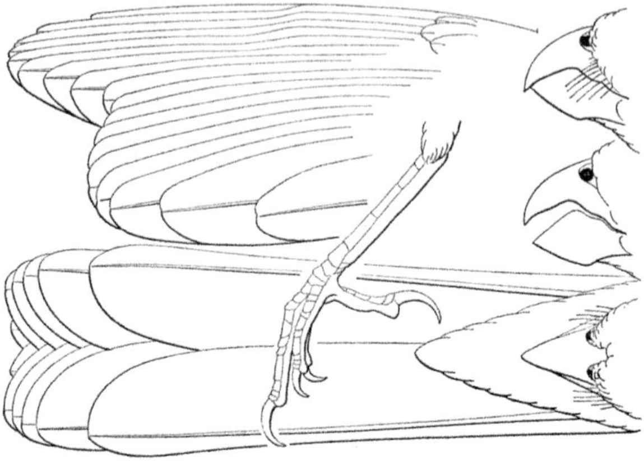

P. sinuata.

The essential character of this genus lies in the greatly curved, very short, and broad bill, something like that of Pyrrhula. In other respects like *Cardinalis*, but with less graduated wing and longer and broader tail.

Pyrrhuloxia sinuata, Bonaparte.

THE TEXAS CARDINAL.

Cardinalis sinuatus, Bonaparte, Pr. Z. S. Lond. V. 1837, 111. — Lawrence, Ann. N. Y. Lyc. 1851, 116. — Cassin, Ill. Texas Birds, I. 204; pl. 33. — *Pyrrhuloxia sinuata*, Bonaparte, Consp. 1850, 500. — Baird, Birds N. Amer. 1858, 508.

Sp. Char. Head with an elongated, pointed crest, starting from the crown. Upper parts generally pale ashy-brown; hood, sides of neck, and under parts of body rather

paler. Long crest feathers, bill all round, including lores and encircling the eye, wing and tail, dark crimson. Chin and upper part of throat, breast, and median line of the belly, under tail covers, tibiæ, edge and inner coverts of the wings, bright carmine red. Bill yellowish.

Female similar, with the under part brownish-yellow; middle of belly and throat only tinged with red. Length about 8.50; wing, 3.75; tail, 4.50.

Hab. Valley of the Rio Grande of Texas. Southern Arizona and Cape St. Lucas.

This beautiful bird is said to have much of the habits of the common cardinal of the Eastern States, and the nest and eggs are scarcely distinguishable. It appears to be very common along the Lower Rio Grande, but was observed by Dr. Heermann only once or twice westward of that river, though common in Mexico and at Cape St. Lucas. Dr. Coues does not mention in his "Prodromus" the authority for its occurrence at Fort Yuma, where none of the government collectors seem to have found it, but stragglers might reach there by way of the Gila River, deserts intervening between the fort and the wooded portions of the peninsula toward the southwest.

Captain J. P. McCown (in Cassin's Illustrations) mentions it as a gay, sprightly bird, frequenting damp bushy woods, generally in small flocks; its voice resembling that of the Virginian cardinal, which utters a loud, clear whistled note repeated several times, and varied on different occasions. Our Canada jay has at times a similar whistle. It is said to be usually very shy, as are the males of all these brilliant songsters. The nest and eggs are yet undescribed.

There does not seem to be more difference between this bird and *Cardinalis* than between *Carpodacus frontalis* and *C. purpureus*, etc., which are considered of the same genus.

C. Virginianus.

Genus CARDINALIS, Bonaparte.

Cardinalis, Bonaparte, Saggio, 1831. (Type, *Loxia Cardinalis*, Linnæus.)

Gen. Char. Bill enormously large; culmen very slightly curved, commissure sinuated; lower jaw broader than the length of the gonys, considerably wider than the upper jaw, about as deep as the latter. Tarsi longer than middle toe; outer toe rather the longer, reaching a little beyond the base of the middle one; hind toe not so long. Wings moder-

C. Virginianus.

ate, reaching over the basal third of the exposed part of the tail. Four outer quills graduated; the first equal to the secondaries. Tail long, decidedly longer than the wings, considerably graduated; feathers broad, truncated a little obliquely at the end, the corners rounded. Colors red. Head crested. (Figure on preceding page.)

The essential characters of this genus are the crested head; very large and thick bill extending far back on the forehead, and only moderately curved above; tarsus longer than middle toe; much graduated wings, the first primary equal to the secondary quills; the long tail exceeding the wings, broad and much graduated at the end.

Cardinalis igneus, Baird.

THE CAPE CARDINAL.

Cardinalis igneus, Baird, Pr. Ac. Nat. Sc. Phila. 1859, 305. — Elliot, Ill. I. plate 16.

Sp. Char. Body bright vermilion-red; darker on the back, rump, and tail. Chin and upper part of the throat black; this color extending along the base of the bill to the nostrils, but not to the forehead. Bill red. Length about 8.50; wing, 3.75; tail, 4.50. *Female* duller in color.

Hab. Cape St. Lucas and Southern Arizona.

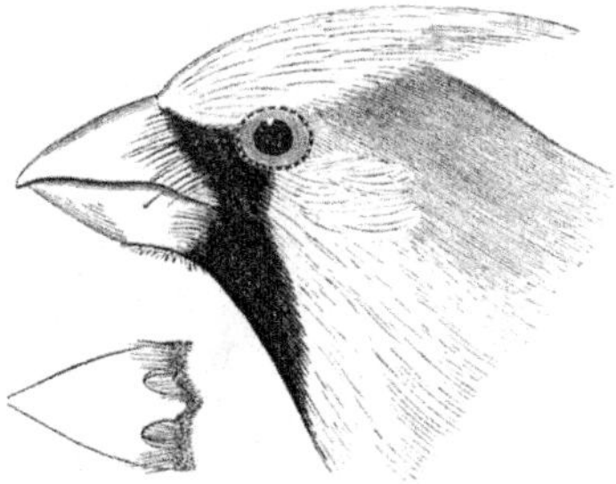

C. igneus.

This species, discovered at Cape St. Lucas by Mr. J. Xantus, is very similar to *C. Virginianus*, the well-known cardinal grosbeak or red-bird of the Eastern United States. It differs, however, in a more swollen bill, and in

C. Virginianus.

the black of the head not extending along the forehead between the eyes, as shown by the above figures. The habits are probably much the same as in *C. Virginianus;* the nest and eggs are very similar. This species is included in a collection made by Dr. Palmer at Camp Grant, east of Tucson, in Arizona.

Genus **PIPILO**, Vieillot.

Pipilo, Vieillot, Analyse, 1816 (Agassiz). (Type, *Fringilla erythrophthalma*, Linnæus.)
Kieneria, Bonaparte, Comptes Rendus, XL. 1855, 356. In part.

Gen. Char. Bill rather stout; the culmen gently curved, the gonys nearly straight; the commissure gently concave, with a decided notch near the end; the lower jaw not so deep as the upper; not as wide as the gonys is long; but wider than the base of the upper mandible. Feet large, the tarsus as long as or a little longer than the middle toe; the outer lateral toe a little the longer, and reaching a little beyond the base of the middle claw. The hind claw about equal to its toe; the two together about equal to the outer toe. Claws all stout, compressed, and moderately curved. Wings reaching about to the end of the upper tail coverts; short and rounded, though the primaries are considerably

P. erythrophthalmus.

longer than the nearly equal secondaries and tertials; the outer four quills are graduated; the first considerably shorter than the second, and about as long as the secondaries. Tail considerably longer than the wings; moderately graduated externally; the feathers rather broad; most rounded off on the inner webs at the end.

The colors vary; the upper parts are generally uniform black or brown; the under white or brown; no central streaks on the feathers. The hood sometimes differently colored.

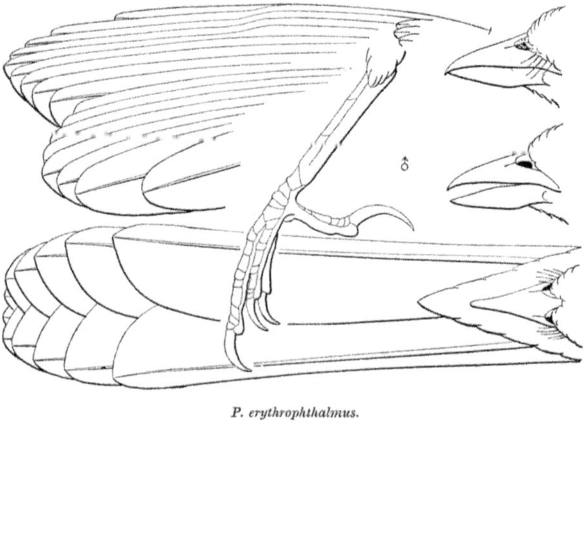

P. erythrophthalmus.

Pipilo Oregonus, Bell.

THE OREGON GROUND-ROBIN.

Pipilo Oregonus, Bell, Ann. N. Y. Lyc. V. 1852, 6 (Oregon). — Newberry, P. R. Rep. VI. iv. 89. — Baird, P. R. Rep. IX. Birds, 513. — Cooper and Suckley, XII. iii. Zool. of W. T. 206.

Fringilla arctica, Audubon, Orn. Biog. V. 1839, 49; pl. 394. — *Pipilo arctica*, Audubon, Syn. 1839, 123. Ib. Birds Amer. III. 1841, 164; pl. 194 (not of Swainson).

Sp. Char. Upper surface generally, with the head and neck all round to the upper part of the breast, deep black; the rest of the lower parts pure white, except the sides of the body and under tail coverts, which are light chestnut brown; the latter rather paler. The outer webs of scapulars (usually edged narrowly with black), and of the superincumbent feathers of the back, with a rounded white spot at the end of the outer webs of the

P. Oregonus.

greater and middle coverts; the outer edges of the innermost tertials, white; no white at the base of the primaries. Outer web of the first tail feather black, occasionally white on the extreme edge; the outer three with a white tip to the inner web.

Female, with the black replaced by brownish. Length, 8.25; extent, 10.25; wing, 3.40; tail, 4.00. Iris red; bill black; feet brown.

Hab. Coast of Oregon and Washington Territories. Northern California?

I have introduced the description of this species here, because it doubtless is found in the northern parts of the State, and the higher Sierra Nevada, where I have seen birds which I supposed to be of this species. In habits and notes I have never observed any difference between this and *P. megalonyx*, both having the complaining " mew," from which they have obtained the name of " cat-bird " on this coast, though entirely different in everything else from the Eastern cat-bird. The head of this species is scarcely distinguishable from that of *P. megalonyx*, to which we therefore refer for illustration of it.

Mr. J. K. Lord mentions finding a nest of this species (or perhaps *P. arc-*

ticus, if at Fort Colville) built "on top of a stump, round which young shoots had grown like a fringe, completely hiding it from the sharpest eye." It contained six eggs.

Pipilo megalonyx, Baird.

THE CALIFORNIA GROUND-ROBIN.

P. megalonyx, Baird, P. R. Rep. IX. Birds, 515. — Kennerly, X. iv. 30. — Heermann, X. vi. 51. — Baird, Birds N. Amer. pl. lxxiii. (feet entirely too large). — Coues, Pr. A. N. Sc. 1866, 89.

Sp. Char. Differs from *P. Oregonus* in much greater amount of white on the wings and scapulars, the spots oblong. Outer edge of outer web of external tail feathers white, sometimes confluent with that at tip of tail. Concealed white spots on feather of side of neck. Claws enormously large, the hinder longer than its digit; the hind toe and claw

reaching to the middle of the middle claw, which, with its toe, is as long as, or longer than, the tarsus. Inner lateral claw reaching nearly to the middle of the middle claw. Length, 8.00; extent, 10.50; wing, 3.40; tail, 4.25; hind toe and claw, 0.90. Iris red; bill black; feet pale brown.

Hab. California, and across through the valleys of Gila and Rio Grande, eastward.

This is a common and resident species in all the lower districts of California, and a considerable distance up in the mountains. It also inhabits Catalina Island, though it has such short wings that a flight of sixteen miles to the main-land would be a rare event for this species. I also found a few of them on San Clemente Island, twenty-two miles from Catalina, but not on the others. Though found in New Mexico, I have seen none in the barren district between the Coast Range and the Colorado, nor in the valley of that river.

Their favorite residence is in thickets and oak groves, where they live mostly on the ground, scratching among the dead leaves in the concealment of the undergrowth, and rarely venturing far from shelter. They never fly more than a few yards at a time, and only a few feet above the ground. About towns, if unmolested, they become more familiar, entering gardens and making their homes about the houses. They have little musical power, the males merely uttering a feeble, monotonous trill from the top of some low bush. The nest is made on the ground under a thicket, constructed of dry leaves, stalks, and grass mixed with fine roots. The eggs, four or five in number, are greenish-white, minutely speckled with reddish-brown. They measure 1.00 × 0.70.

When alarmed they have a note something like the "mew" of a cat, from which they are popularly known by the name of cat-bird; and I have been asked why the cat-birds of this country differ so much in color from those of the East. If observers would name from the color instead of the note, they would be more correct in comparing it with the Eastern "chewink," "towhee," or "ground-robin," — all one and the same bird, with various local names (*P. erythrophthalmus*).

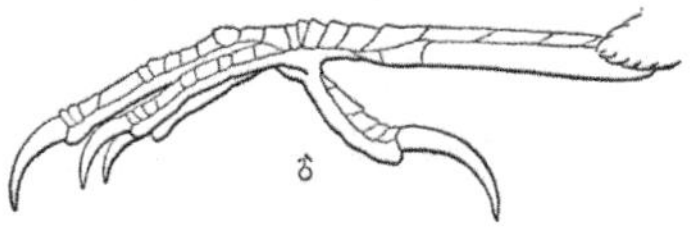

P. erythrophthalmus.

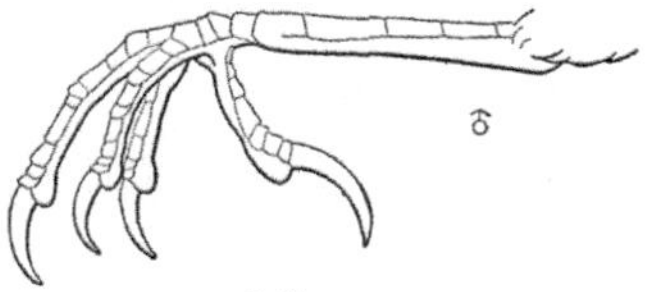

P. Oregonus.

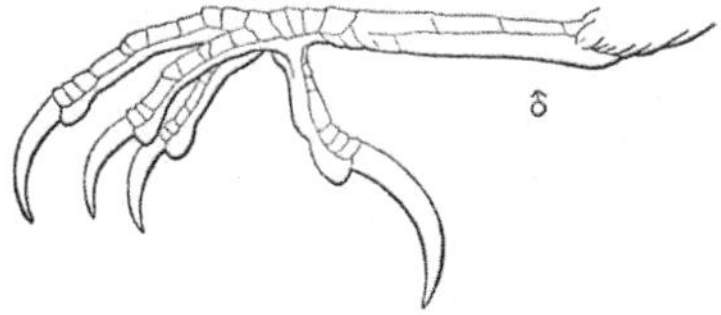

P. megalonyx.

We give above figures to illustrate the difference in the size of the claws of the feet of *Pipilo erythrophthalmus, Oregonus*, and *megalonyx*. These three species, with *P. arcticus*, form a series of which *P. erythrophthalmus* is one extreme, with no white on the wing coverts, and medium-sized claws, and *P. megalonyx*, with its much spotted coverts and long claws, the other. How far these are merely geographical varieties of one common species remains yet to be decided; here it is not necessary to settle the question, nor the relationship of these to several closely allied species from Mexico.

Pipilo Abertii, Baird.

ABERT'S FINCH.

Pipilo Abertii, Baird, Stansbury's Rep. Great Salt Lake, Zoology, June, 1852, 325 (New Mexico). Ib. P. R. Rep. IX. Birds, 516. — Kennerly, X. iv. 30; pl. 30. — Heermann, X. v. 15. — Coues, Pr. A. N. Sc. 1866, 90. — *Kieneria Abertii*, Bonaparte, Comptes Rendus, XL. 1855, 356.

Sp. Char. General color of upper parts pale brownish or yellowish-red; beneath brighter, especially on the under coverts, palest on the middle of the belly. Sides of

head anterior to eyes, and chin, dark brown. Bill yellowish. Length, 9.25; extent, 11.75; wing, 3.70; tail, 4.85. Iris chestnut; bill pale brown; feet dark brown.

Hab. Base of Rocky Mountains in New Mexico; Valleys of the Gila and Colorado.

This species is almost the exact counterpart in the Colorado Valley of the *P. fuscus*, the only difference I noticed in habits being a loud note of alarm in this species unlike that of *P. fuscus*, but remarkably similar to that of two very distinct birds of the same valley, namely, *Centurus uropygialis* and *Phænopepla nitens*. Like its congener, this species lives almost constantly on the ground, but seems rather more gregarious, especially in winter.

About April 1st I found many of their nests, generally built in thorny shrubs. They were composed of a flooring of coarse twigs or of green herbs, and strongly interwoven with strips of bark, grass, and leaves; one bird having taken advantage of the recent introduction of horses into the valley to obtain a lining of horsehair for its nest. The eggs were in all cases only three, bluish-white, with brown spots and streaks in a ring near the large end, quite variable in number, and measuring 1.00 × 0.70 inch. One nest was in a low mesquite-tree, another in a dense cluster of dead twigs hanging from a cottonwood.

The time required for hatching was twelve or thirteen days, and in a fortnight more the young left the nest. I found nests with eggs as late as May 25th, and have no doubt that they raise two or more broods. The song of the male, throughout April and May, is precisely like that of the *P. fuscus*, and reminded me of the notes of *P. Oregonus*, and of the Eastern black-throated bunting (*Euspiza Americana*).

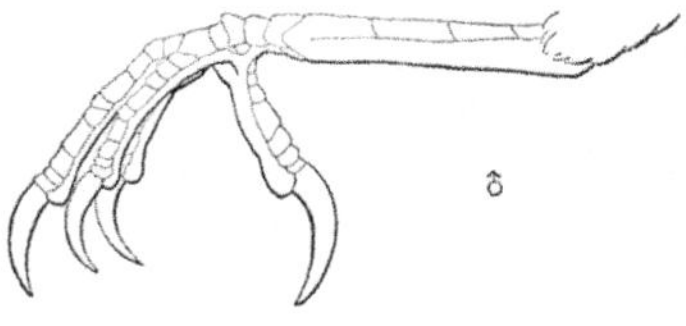

Pipilo Abertii.

According to Kennerly, they abound along the Eastern branches of the Colorado, but neither this nor any other species has been seen between that river and the coast slope, a linear distance of over one hundred miles.

Pipilo fuscus, SWAINSON.

THE BROWN FINCH.

Pipilo fusca, SWAINSON, Philos. Mag. I. 1827, 434. IB. Anim. in Menag. 1838, 347. — CASSIN, Illust. I. iv. 1853, 124; pl. xvii. — NEWBERRY, Zool. Cal. and Or. Route, P. R. R. Rep. VI. iv. 1857, 89. — BAIRD, P. R. Rep. IX. Birds, 517. — HEERMANN, X. vi. 51. — *Kieneria fusca*, BONAPARTE, Comptes Rendus, XL. 1855, 356.
Fringilla crissalis, VIGORS, Zool. Blossom, 1839, 19.
Oriturus Wrangeli, BONAPARTE, Comptes Rendus, XLIII. 1856, 413.

SP. CHAR. Above, dark olive-brown, the crown with a very slight tinge of scarcely appreciable dark rufous. Under parts with the color somewhat similar, but of a lighter shade, and washed with grayish; middle of the belly ashy-white; the under tail coverts

pale rufous, shading into lighter about the neck and sides of lower belly; chin and upper part of throat well-defined pale rufous, margined all round by brown spots, a few of them scattered within the margin. Eyelids and sides of head, anterior to the eye, rufous, like the throat. One or two feathers on the lower part of the breast with a concealed brown blotch. Outer primary not edged with white. Length, 9.00; extent, 12.00; wing, 3.75; tail, 5.00. Iris brown; feet paler; bill brown, fleshy below.

Hab. California, except the more eastern parts.

This is one of the most abundant and characteristic birds of California, residing constantly in all the lower country west of the Sierra Nevada, and up to the summits of the Coast Mountains, three thousand feet high, as well as high up the slopes of the Sierra, probably to the same elevation. The name of Cañon finch, given by most authors, seems rather inappropriate, as it is found plentifully in level districts also, wherever there are trees and shrubbery. There they have habits very similar to those of the other species, frequenting the ground, and seeking their food among the dead leaves, which they much resemble in color. This resemblance serves them as a protection from hawks, their hues corresponding as fully with that of the earth and dusty foliage, during most of the year, as the dark colors of the preceding species do with the gloomy thickets which they inhabit. Being less conspicuous in the light, they venture more fearlessly forth and feed in open grounds.

They have little power of song, merely uttering a few quaint chirping and hurried notes, as they sit perched on a low bush in spring.

At San Diego I saw the first nest with eggs on April 17th; but some laid much earlier, as I found young hatched by the 20th. I afterwards observed many more, all built in bushes, from two to four feet from the ground, and containing but three eggs, except one, which contained four. I have also seen them built in low trees, and in a vine growing over the porch of a house. The nest is formed of coarse twigs, bark, and grass, thick and large, and lined with fine grass and root-fibres. The eggs are pale blue, spotted with purplish-brown blotches mostly small and scattered, measuring 0.90 × 0.65 inch.

As remarked by Heermann, the eggs much resemble those of the redwings, and are unlike those of other *Pipilos*; I am inclined to think that they lay four oftener in the more northern parts of their range than near San Diego. I also believe that they raise two broods, as their abundance would indicate it. According to Newberry, they are common along streams throughout the Sacramento Valley (probably to near lat. 41°). At Santa Cruz, in 1866, I also found young of this species hatched as early as April 18th.

Professor Baird is strongly inclined to the opinion that the coast species of California is not the true *fusca* of Swainson, in which case the name will become *P. crissalis*, Vigors. Whether, as suggested, the true *fusca* be the *P. mesoleuca* remains yet to be decided by a critical examination of the various closely allied North American and Mexican brown *Pipilos*.

Pipilo mesoleucus, Baird.

THE CAÑON FINCH.

Pipilo mesoleucus, Baird, Pr. Ac. N. Sc. VII. 1854, 119. Ib. Birds Amer. 1858, 518; pl. xxix. — Coues, Pr. A. N. Sc. 1866, 90.

Sp. Char. Above, olivaceous-brown, with a grayish tinge; hood dull chestnut, conspicuously different from the back. Sides beyond the edge of the wing like the back, but paler; posteriorly, and about the vent and under tail coverts, pale brownish-red. The ashy olive-brown of the sides scarcely meets across the breast, the lower portion of

which, with the upper belly, is rather pure white. The chin, throat, and upper part of the breast pale yellowish-rufous, spotted on the sides and across the breast with brown; an obscure spot in the middle of the breast; edge of outer primary white. Length, 8.50; wing, 3.80; tail, 4.70.

Hab. Valley of Upper Rio Grande, and across to Gila River. East to Santa Caterina, New Leon.

This species is similar in general appearance to the *P. fuscus,* but the olive-brown and rufous are both of a lighter shade. The crown is of a decided chestnut, conspicuously different from the back, instead of nearly the same tint. The light reddish under the head is wider throughout, and extends down to the upper part of the breast, blending with the colors of the breast and belly, instead of being narrower, more sharply defined, and restricted to the chin and throat. The isolated larger spot on the breast is more conspicuous; the breast and belly are quite pure white, shaded with obsolete brownish blotches, instead of being uniform grayish-brown, with only an approach to whitish in the very middle. The edges of the wing and tail feathers are a good deal lighter, the outer web of the first primary being sharply edged with pure white, instead of obscure grayish-brown. The size generally is rather smaller.

This species is very abundant in Southern Arizona, where its habits are much like those of *P. Abertii.* The eggs resemble those of *P. fusca.*

Pipilo albigula, BAIRD.

THE WHITE-THROATED TOWHEE.

Pipilo albigula, BAIRD, Pr. Ac. N. Sc. Phil. 1859, 305. — ELLIOT, Ill. B. N. A.; pl. iv.

SP. CHAR. Similar to *P. mesoleucus,* with more white on the middle of belly; the chin and upper part of the throat with a border of dusky spots, which do not extend as far down as in *mesoleucus,* and are much better defined below. The space enclosed by this

border of spots is yellowish-brown on the chin as in *mesoleucus,* but below on the throat and in front of the spots it becomes nearly, sometimes quite white, in decided contrast to the color of the chin. Size that of *mesoleucus.*

This species is very closely related to *P. mesoleucus,* but probably distinct. It was discovered by Mr. Xantus, at Cape St. Lucas, who, however, has published no account of its habits.

Pipilo chlorura, TOWNSEND.

THE GREEN FINCH.

Fringilla chlorura, AUDUBON, Orn. Biog. V. 1839, 336 (young). — *Zonotrichia chlorura,* GAMBEL, Jour. A. N. Sc. Phil. 2d Series, I. 1847, 51; pl. ix. f. 1. — *Embernagra chlorura,* BONAPARTE, Conspectus, 1850, 483. — HEERMANN, P. R. Rep. X. vi. 46. — *Pipilo chlorurus,* BAIRD, P. R. Rep. IX. Birds, 519. — HEERMANN, X. v. 15. — COUES, Pr. A. N. Sc. 1866, 90.

Fringilla Blandingiana, GAMBEL, Pr. A. N. Sc. Phil. April, 1843, 260. — *Embernagra Blandingiana,* CASSIN, Illust. I. iii. 1853, 70; pl. xii.

SP. CHAR. Above, dull grayish olive-green. Crown uniform chestnut. Forehead with superciliary stripe, and sides of the head and neck, the upper part of the breast and sides of the body, bluish-ash. Chin and upper part of throat, abruptly defined white, the former margined by dusky, above which is a short white maxillary stripe. Under tail coverts and sides of body behind brownish-yellow. Tail feathers generally and exterior of wings

bright olive-green, the edge and under surface of the latter bright yellow; edge of first primary white. Length, 7.50; extent, 9.75; wing, 3.75; tail, 3.65. Iris and feet brown; bill black, bluish below.

Hab. Valley of the Rio Grande and Gila. Rocky Mountains, north to South Pass, south to Mexico; Sierra Nevada to lat. 40°; south to San Diego in winter.

As the other species of *Pipilo* seem suited by color to inhabit the dark thickets and dry leaves, this one is clad in a gay livery well adapted for concealment in its summer resorts, and also among the growing vegetation of the lower country during the rainy season.

I found a few of them in winter in the Colorado Valley, and rather more at San Diego, but they left both places in March. They were generally

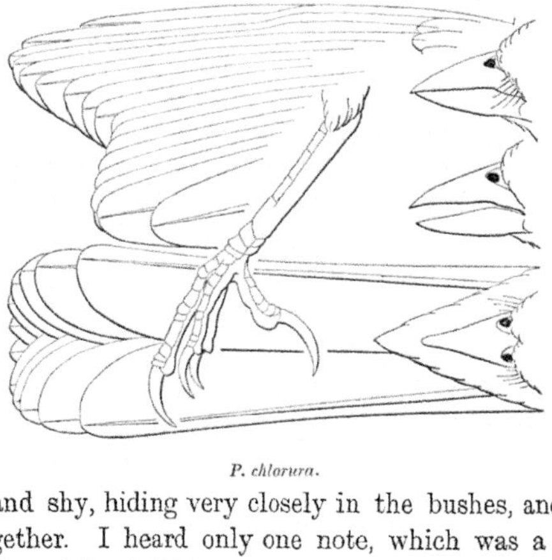

P. chlorura.

very silent and shy, hiding very closely in the bushes, and feeding on the ground altogether. I heard only one note, which was a kind of crowing resembling that of the California quail.

The nest and eggs, as well as the complete biography of this pretty bird, remain to be described. It has many points of difference from the true *Pipilos*, and will probably be yet separated from that genus, perhaps as a *Kieneria*, as suggested by Bonaparte. To illustrate its characters we give above figures of the bird and of its external anatomy.

FAMILY ALAUDIDÆ, THE LARKS.

CHAR. First primary very short or wanting. Tarsus scutellate anteriorly and posteriorly, with the plates nearly of corresponding position and number. Hind claw very long and nearly straight. Bill short, conical; frontal feathers extending along its sides; the nostrils usually concealed by a tuft of bristly feathers directed forwards. Tertials greatly elongated beyond the secondaries.

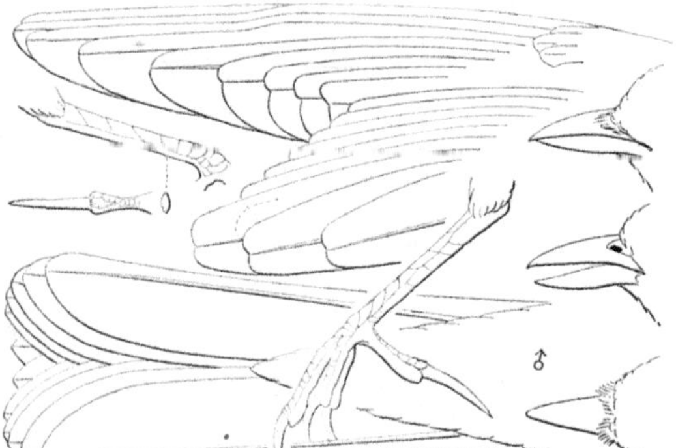

E. cornuta.

GENUS **EREMOPHILA**, BOIE.

Eremophila, BOIE, Isis, 1828, 322. (Type, *Alauda alpestris*. Not sufficiently distinct from *Eremophilus*, HUMBOLDT (Fishes), 1805.)
Phileremos, BREHM, Deutschl. Vögel, 1831.
"*Otocoris*," BONAPARTE, List, 1839. (Type, *Alauda alpestris*," GRAY.)

E. cornuta.

GEN. CHAR. First primary wanting; bill scarcely higher than broad; nostrils circular, concealed by a dense tuft of feathers; the nasal fossæ oblique. A pectoral crescent and cheek patches of black.

Eremophila cornuta, WILSON.

THE HORNED LARK.

Alauda cornuta, WILSON, Am. Orn. I. 1808, 85 (in text). — *Eremophila cornuta*, BOIE, Isis, 1828, 322. — BAIRD, P. R. Rep. IX. Birds, 403; X. iii. 13, pl. xxxii. — COOPER and SUCKLEY, XII. iii. Zool. of W. T. 195. — COUES, Pr. A. N. Sc. 1866, 79.
Alauda alpestris, FORSTER, Phil. Trans. LXII. 1772, 383. — WILSON, Am. Orn. I. 1808, 85; pl. v. f. 4. — AUDUBON, Birds Amer. III. 1841, 44; pl. 151. — *Otocoris alpestris*, NEWBERRY, P. R. Rep. VI. iv. 88. — HEERMANN, P. R. Rep. X. vi. 45.
Phileremos cornutus, BONAPARTE, List, 1838.

WESTERN AND SOUTHERN VARIETY.

Alauda chrysolœma, WAGLER, Isis, 1831, 350. — BONAPARTE, Pr. Zool. Soc. 1837, 111.
Alauda minor, GIRAUD, Sixteen Spec. Texas Birds, 1841.
Alauda rufa, AUDUBON, Birds Amer. VII. 1843, 353; pl. 497. — *Otocoris rufa*, HEERMANN, P. R. Rep. X. vi. 45.
Otocoris occidentalis, M'CALL, Pr. A. N. S. Phil. V. June, 1851, 218, Sante Fe. — BAIRD, Stansbury's Report, 1852, 318.

SP. CHAR. Above, pinkish-brown, the feathers of the back streaked with dusky. A broad band across the crown, extending backwards along the lateral tufts: a crescentic patch from the bill below the eye and along the side of the head; a jugular crescent, and the tail feathers, black. The innermost of the latter like the back. A frontal band extending backwards over the eye, and under parts, with outer edge of wings and tail,

Adult.

white. Chin and throat yellow. Length of Pennsylvania specimens, 7.75; wing, 4.50; tail, 3.25; bill above, 0.22.

Var. *chrysolæma*, smaller and lighter colored. Length, 7.25; extent, 12.75; wing, 4.25. Iris brown; bill black, bluish below; feet black.

Young.

Hab. Everywhere on the prairies and desert plains of North America. The Atlantic States in winter.

According to Professor Baird, specimens from west of the Sierra Nevada are as dark or darker colored than Eastern ones, while the difference in size is not constant. The variety is confined to the more arid and hot plains of the interior.

I found these birds in considerable flocks about Fort Mojave, about the end of February, but all of them seemed to have left the valley by the end of March. About May 29th I found numbers of them towards the summits of the Providence Range of mountains, west of the valley, and about four thousand feet above it, where I believe they had nests. They were also common on the cooler plains towards the ocean on July 1st, so that they doubtless breed in many parts of the southern portion of California, as well as up the coast to Puget's Sound, and throughout most of the Great Plains of the interior. In May or June the males rise nearly perpendicularly into the air until almost out of sight, and fly about in an irregular circle, singing a sweet and varied song for several minutes, when they descend again, nearly to the spot they started from.

The nest is made in a small depression of the ground, usually under a tuft of grass or a bush, composed of grass, sometimes lined with hairs, and the eggs are bluish-white with darker spots nearly covering them; their number four or five.

In fall they associate sometimes in very large flocks, frequenting the bare plains or prairies, but not the sea-shore. Both the name of sky-lark (be-

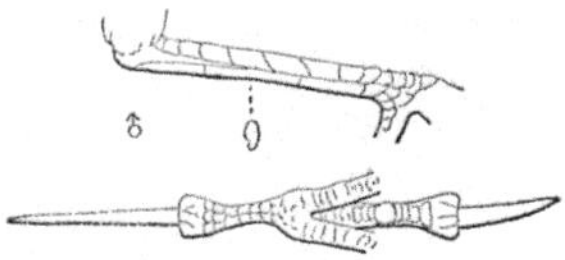

Inner face of tarsus, and under surface of foot.

longing properly to another species) and shore-lark are objectionable, and horned lark, though not strictly correct, is the most appropriate, and as applicable as that of horned owl.

Their food consists of the small seeds and insects they find among the grass. They probably never alight on a tree or bush.

Family ICTERIDÆ, The Orioles.

Char. Primaries nine. Tarsi scutellate anteriorly; plated behind. Bill long, generally equal to the head or longer, straight or gently curved, conical, without any notch, the commissure bending downward at an obtuse angle at the base. Gonys generally more than half the culmen. Basal joint of the middle toe free on the inner side; united half-way on the outer. Tail rather long, rounded. Legs stout.

Sub-Family AGELÆINÆ.

Char. Bill stout, conical, and acutely pointed, not longer than the head; the outlines nearly straight, the tip not decurved. Legs adapted for walking, longer than the head. Claws not much curved. Tail moderate, shorter than the wings, nearly even.

GENUS **DOLICHONYX**, SWAINSON.

Dolichonyx, SWAINSON, Zool. Jour. IV. 1837, 351. (Type, *Emberiza oryzivora*, LINNÆUS.)

GEN. CHAR. Bill short, stout, conical, little more than half the head; the commissure slightly sinuated; the culmen nearly straight. Middle toe considerably longer than the tarsus (which is about as long as the head); the inner lateral toe longest, but not reaching

D. oryzivorus.

the base of the middle claw. Wings long, first quill longest. Tail feathers acuminately pointed at the tip, with the shafts stiffened and rigid, as in the woodpeckers.

The peculiar characteristic of this genus is found in the rigid scansorial tail, and the very long middle toe, by means of which it is enabled to grasp the vertical stems of reeds or other slender plants. The color of the known species is black, varied with whitish patches on the upper parts.

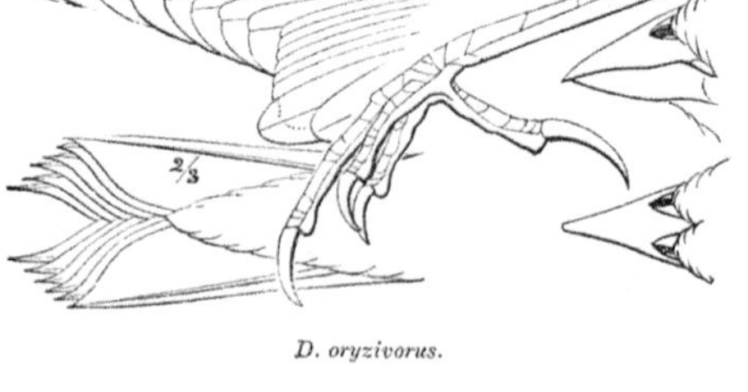

D. oryzivorus.

In coloration this genus bears a close relation to *Calamospiza*, although the other differences are very decided. Both are black, with white patches on the wings. *Dolichonyx* has, in addition, a white patch on the rump, and a yellowish one on the nape.

But one species is at present known to naturalists.

Dolichonyx oryzivorus, LINNÆUS.

THE BOBOLINK; THE RICE-BIRD.

Emberiza oryzivora, LINNÆUS, Syst. Nat. I. 1766, 311. — WILSON, II. 1810, 48; pl. xii. — *Dolichonyx oryzivora*, SWAINSON, Zool. Jour. III. 1827, 351. — AUDUBON, Birds Amer. IV. 1842, 10; pl. 211. — BAIRD, Birds N. Amer. 1858, 522.
Icterus agripennis, BONAPARTE, Obs. Wilson, 1824, 87.

SP. CHAR. General color of *male* in spring black; the nape brownish cream-color; a patch on the side of the breast, the scapulars and rump white, shading into light ash on the upper tail coverts and on the back below the interscapular region. The outer prima-

Male.

ries sharply margined with yellowish-white; the tertials less abruptly; the tail feathers margined at the tips with pale brownish-ash.

Female, yellowish beneath; two stripes on the top of the head, and the upper parts throughout, except the back of the neck and rump, and including all the wing feathers generally, dark brown, all edged with brownish-yellow, which becomes whiter near the tips of the quills. The sides sparsely streaked with dark brown, and a similar stripe behind the eye. There is a superciliary and a median band of yellow on the head. Length of male, 7.70; wing, 3.83; tail, 3 15.

Hab. Eastern United States to the high central plains, and westward to Ruby Valley, Nevada. Reaches South America in its winter migrations. Galapagos.

The introduction of the bobolink into the present work rests upon its discovery in Ruby Valley, Nevada, by Mr. Ridgway, while attached to the Geological Exploration of the 40th Parallel, where it occurs in abundance in the wheat-fields, and with habits similar to those so well known at the East. As bobolink in the North and West, reed-bird in Pennsylvania, and rice-bird in the Southern States, it is well-known to every one. It breeds

Female.

abundantly in meadows in the North and West, laying its eggs on the ground. Collecting in large flocks, after the breeding-season, it soon passes southward, and as a favorite article of food runs the gauntlet of gunners from Pennsylvania to Florida and the Gulf. (Baird.)

GENUS **MOLOTHRUS**, SWAINSON.

Molothrus, SWAINSON, F. Bor. Am. II. 1831, 277. (Type, *Fringilla pecoris*, GMELIN.)

GEN. CHAR. Bill short, stout, about two thirds the length of head; the commissure straight, culmen and gonys slightly curved, convex, the former broad, rounded, convex,

M. pecoris.

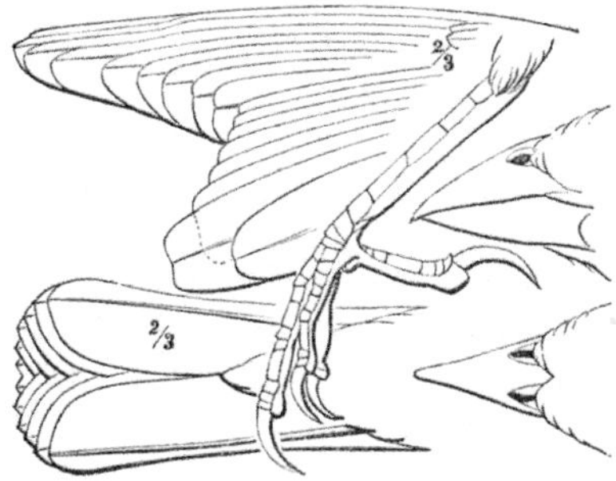

M. pecoris.

and running back on the head in a point. Lateral toes nearly equal, reaching the base of the middle one, which is shorter than tarsus; claws rather small. Tail nearly even; wings long, pointed, the first quill longest.

Molothrus pecoris, GMELIN.

THE COW BLACKBIRD; THE COW-BIRD.

Fringilla pecoris, GMELIN, Syst. Nat. I. 1788, 910 (*female*). — *Emberiza pecoris*, WILSON, Am. Orn. II. 1810, 145; pl. xviii. f. 1, 2, 3. — *Molothrus pecoris*, SWAINSON, F. Bor. Am. II. 1831, 277. — AUDUBON, Birds Amer. IV. 1842, 16; pl. 212. — BAIRD, P. R. Rep. IX. Birds, 524. — HEERMANN, X. vi. 52. — COUES, Pr. A. N. S. 1866, 90 (Arizona).

SP. CHAR. Second quill longest; first scarcely shorter. Tail nearly even, or very slightly rounded. *Male*, with the head, neck, and anterior half of the breast, light chocolate brown, rather lighter above; rest of body lustrous black, with a violet purple gloss next to the brown, of steel blue on the back, and of green elsewhere.

Male.

Female, light olivaceous-brown all over, lighter on the head and beneath. Length, 8.00; extent, 13.50; wing, 4.40; tail, 3.40. Iris brown; bill and feet black.

Hab. United States from the Atlantic to California; not found immediately on the coast of the Pacific.

I think that a few of this species pass the winter in the Colorado Valley, and, perhaps, also in the San Joaquin Valley, but I have not seen them at any time in those nearer the coast. In the country east of the Sierra Nevada they probably wander as far north as the Columbia River, as they follow wagon-trains along the Platte in great flocks, and may be expected to accompany them farther west.

The remarkable habit of this bird of laying its eggs in the nests of other birds, instead of building for itself, relieves it from the usual necessity of pairing in the spring, and it remains gregarious at all seasons, though generally

Female.

the flocks are not large, except in autumn. They seemed to be migrating northward through the Colorado Valley early in April, and on the 19th of that month I found an egg of this bird in a nest of the yellow-breasted chat (*Icteria*), showing that some of them are raised in the lat. of 35°, as well as northward.

According to Nuttall, the sexes are polygamous, not even pairing like other small birds for one year. In the East he found their eggs oftenest in the nests of the *Vireo olivaceus, Geothlypis trichas, Spizella socialis, Dendrœca æstiva, Polioptila cœrulea,* and other species corresponding to our *Vireo Huttoni, Sialia Mexicana, Cyanospiza amœna, Melospiza Heermanni, Sayornis nigricans,* and *Turdus nanus,* all of which may be supposed to act as nurses for this foundling bird in California. Though all much smaller than the cow-bird, and building in very different situations, the foster-parents usually take good care of the large egg found in their nest, especially if laid after one of their own, and frequently begin to sit immediately, although their own number is not complete. The *Vireo* even sometimes deserts the nest if the egg of the cow-bird is taken. The *Dendræca*, however, is not so easily deceived, as it sometimes builds a new floor or entire nest over the strange egg, burying it completely. Larger birds have been known to throw the egg out, but small ones are unable to do this. The *Icteria*, whose own egg is nearly as large, and quite differently colored, seemed, in the instance I saw, a willing dupe, though probably quite able to eject the parasite.

The color of the cow-bird's egg is greenish-white, thickly sprinkled with points and blotches of olive-brown, of two shades, most numerous near the large end. Sometimes the egg is nearly pure white with very dark spots. It is small for the size of the bird, obtuse, measuring about 0.71×0.56 inch, and is supposed to be hatched in about twelve days, developing sooner than the smaller eggs, perhaps because it obtains more warmth by contact with the body of the bird. When the legitimate eggs are hatched, the young are soon stifled by the larger and stronger foundling, which gets most of the food brought by the old birds, and fills up the small nest in a few days. The parents then carry off their own dead offspring, and drop them at a distance, while the foundling, receiving their whole attention, grows rapidly, and after becoming fully fledged deserts its deluded foster-parents for the society of its own species.

Another *Molothrus* in South America has the same habits, and the celebrated cuckoo of Europe, with several of its relatives in the Old World, has long been known for the same perversity, the chief difference being that the young cuckoos throw out their foster-brothers while still alive. The reason why they cannot or do not build for themselves is one of the mysteries of nature.

The name of cow-bird is derived from the partiality of this species for the society of cattle and horses. In the districts they inhabit they may almost always be found among herds of cattle, walking after them to pick up the insects disturbed by their feet, and often alighting on their backs and heads. They also associate with their relatives, the other blackbirds, especially in fall and winter.

The males, especially in spring, utter a few guttural croaking notes, either from the top of a tree, or occasionally on the ground, sounding as if they tried to imitate the more musical red-wings. They are at all times watchful and suspicious, and the female, when desirous of laying, shows much artfulness in searching for a suitable nest through the thickets, watching until the owner is absent, and then taking the opportunity to deposit her egg. Two eggs have been found in one nest, but Nuttall thinks that in these cases one is always abortive.

Why they do not approach the Pacific Coast is somewhat strange, as they are common near that of the Atlantic. The prevailing sea breeze is the only thing that may be supposed objectionable to them, and yet they do not seem elsewhere very tender birds.

Molothrus obscurus, Gmelin.

THE DWARF COW-BIRD.

Sturnus obscurus, Gmelin, Syst. Nat. I. 1788, 804. — *Molothrus obscurus*, Cassin, Pr. A. N. Sc. 1866, 18.

Sp. Char. Similar to *M. pecoris*, but considerably smaller; the bill more slender; the tail proportionally longer. Length, 6 50; wing, 4.00; tail, 2.75.

Hab. Cape St. Lucas, Southern Arizona, and Northwestern Mexico.

This bird is a dwarf representative, perhaps rare, of the cow-bird, and can hardly be considered as distinct, though worthy of note at least as a variety. The habits are doubtless the same as those of the true *M. pecoris*.

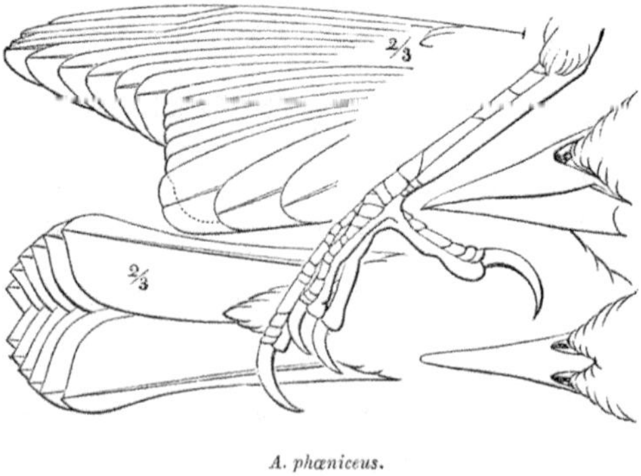

A. phœniceus.

Genus AGELAIUS, Vieillot.

Agelaius, Vieillot, Analyse, 1816. (Type, *Oriolus phœniceus*, L.)

Gen. Char. First quill shorter than second; claws short; the outer lateral scarcely reaching the base of the middle. Culmen depressed at base, parting the frontal feathers; length equal to that of the head, shorter than tarsus. Both mandibles of equal thickness and acute at tip, the edges much curved, the culmen, gonys, and commissure nearly

A. phœniceus.

straight or slightly sinuated; the length of bill about twice its height. Tail moderate, rounded, or very slightly graduated. Wings pointed, reaching to end of lower tail coverts. Colors black, with red shoulders in North American species.

Agelaius phœniceus, Linnæus.

THE SWAMP BLACKBIRD; THE RED-WING BLACKBIRD.

Oriolus phœniceus, Linnæus, Syst. Nat. I. 1766, 161. — *Agelaius phœniceus*, "Vieillot, Analyse, 1816." — Audubon, Birds Amer. IV. 1842, 31; pl. 216. — Baird, P. R. Rep. IX. Birds, 526. — Kennerly, X. iv. 30. — Cooper and Suckley, XII. iii. Zool. of W. T. 207. — Coues, Pr. A. N. S. 1866, 90.
Sturnus prædatorius, Wilson, Am. Orn. IV. 1811, 30; pl. xxx.

Sp. Char. Tail much rounded; the lateral feathers about half an inch shorter. Fourth quill longest; first about as long as the fifth. Bill large, stout; half as high or more than half as high as long.

Male. General color uniform lustrous velvet-black, with a greenish reflection. Shoulders and lesser wing coverts of a bright crimson or vermilion-red. Middle coverts brownish-yellow, and usually paler towards the tips.

Male.

Female. Brown above; the feathers streaked with rufous-brown and yellowish; beneath white, streaked with brown. Fore part of throat, superciliary, and median stripe strongly tinged with brownish-yellow. Length of male, 9.00; extent, 15.50; wing, 5.00;

Female.

tail, 4.15 Female, length, 8.25; extent, 12.00; wing, 4.00. Iris brown; bill and feet black.

Hab. United States, from the Atlantic to the Pacific.

According to Baird, many specimens of red-wings from this coast are undistinguishable from the Atlantic bird. All those I collected and saw in the Colorado Valley, and near San Diego, as well as along the coast at Santa Barbara, and many at Santa Cruz, retained the yellow edging of the lesser wing coverts throughout spring, while north of the Columbia River we found only this species. Whether the difference is specific or only an imperfectly developed state of the *A. gubernator*, has been questioned, but there are differences in the females which seem to indicate a constant distinctness. Those males which I obtained in spring have the lesser coverts entirely yellow, thus agreeing with the most essential character of plumage,

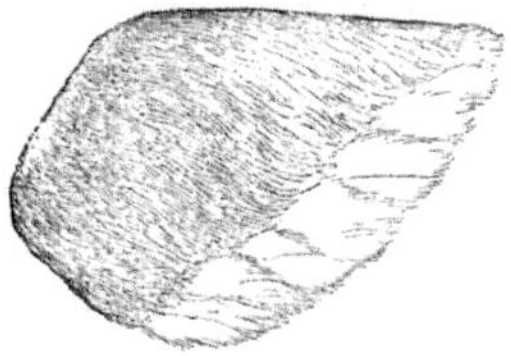

Shoulder patch.

distinguishing this from *A. gubernator*, which has them tipped with black. Eggs from Santa Barbara, measure 1.00 × 0.73 inch, and are bluish-white with blackish streaks near the large end.

Agelaius gubernator, Wagler.

THE RED-SHOULDERED BLACKBIRD.

Psarocolius gubernator, Wagler, Isis, 1832, IV. 281. — *Agelaius gubernator*, Bonaparte, List, 1838. — Audubon, Syn. 1839, 141. Ib. Birds Amer. IV. 1842, 29; pl. 215. — Newberry, P. R. Rep. VI. iv. 1857, 86. — Baird, P. R. Rep. IX. Birds, 529. — Kennerly, X. iv. 31. — Heermann, X. vi. 53.
Icterus (Xanthornus) gubernator, Nuttall, Man. I. 2d ed. 1840, 187.

Sp. Char. Bill rather shorter than the head, without any longitudinal sulci, but with faint traces of transverse ones at the base of the lower jaw. Tail rounded. First quill nearly equal to the fourth.

Male.

Male. Throughout of a lustrous velvety-black with a greenish reflection. The shoulders and lesser coverts rich crimson; the middle coverts brownish-yellow at the base, but the exposed portion black.

Female. Dusky, varied with paler. Length of male, 9.00; extent, 15.50; wing, 5.25; tail, 3.80. Female, length, 8.00; extent, 13.00; wing, 4.35. Iris brown; bill and feet black.

Hab. Colorado River (?) throughout California to the Columbia.

I did not myself see any specimens of red-wings in the Colorado Valley without the yellow edging of *A. phœniceus*, and Professor Baird mentions this species as doubtfully from there. According to my observations, this species inhabits chiefly the warmer interior of the State, Santa Cruz being the only point on the coast where I have seen them. I found them in scattered pairs in May throughout the Coast Range, even to the summits, where there are small marshes full of rushes, in which they build. I have been unable to detect any difference between the habits and notes of this and the *A. phœniceus;* and from the fact that the specimens with entirely red shoulders seem limited to the middle of the State, or are rare along the coast, while most of those on the coast closely resemble the Eastern bird, I should think it a mere local variety, though said to occur also in Mexico.

During summer this species has a variety of sweet and liquid notes, de-

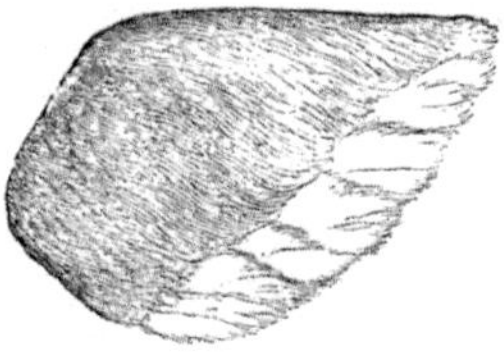

Shoulder patch.

livered from some tree near its favorite marsh, or when hovering over its nest, sometimes mingled with jingling and creaking sounds, all seeming adapted for the watery choir of frogs and aquatic birds with which they there associate.

A nest from the summit of the Coast Range is formed of grass and rushes, lined with finer grass. The eggs are pale greenish-white, with large curving streaks and spots of dark brown, mostly at the large end. They measure 1.00 × 0.75 inch.

From a comparison of three species of black *Agelaius*, belonging to Western North America, it will be seen that the differences consist mainly in the color of the middle coverts. These are entirely black in *gubernator*, entirely white in *tricolor*, and buff-yellow in *phœniceus*.

Agelaius tricolor, NUTTALL.

THE RED AND WHITE-SHOULDERED BLACKBIRD.

Icterus tricolor, "NUTTALL," AUDUBON, Orn. Biog. V. 1839, 1; pl. 388. — NUTTALL, Man. I. 2d ed. 1840, 186. — *Agelaius tricolor*, BONAPARTE, List, 1838. — AUDUBON, Syn. 1839, 141. IB. Birds Amer. IV. 1842, 27; pl. 214. — NEWBERRY, P. R. Rep. VI. iv. 86. — BAIRD, P. R. Rep. IX. Birds, 530. — HEERMANN, X. vi. 53.

SP. CHAR. Tail nearly even. Second and third quills longest; first a little shorter than the fourth. Bill slenderer, not half as high as long.

Male. General color uniform lustrous velvet black, with a decided greenish reflection.

Male.

Shoulders and lesser wing coverts brownish-red, of much the color of venous blood; the median coverts of a well-defined and nearly pure white, with sometimes a brownish tinge.

Female. Dark brown, variegated with dark grayish-ash. No median stripe on the

Female.

crown, nor any maxillary one, and scarcely a superciliary. Length, 9.50; extent, 15.25; wing, 5.25; tail, 3.90. Female, length 7.75; extent, 13.00; wing, 4.25. Iris brown; bill and feet black.

Hab. Colorado River; throughout California. Oregon.

This seems to be a rare species in the Colorado Valley, though one has been obtained there by Dr. Kennerly, according to Baird. I found them the most abundant species near San Diego and Los Angeles, and not rare at Santa Barbara, but they seem north of that point to pass more into the interior, extending up as far as Klamath Lake and Southern Oregon.

I have always seen them in considerable flocks, even in the breeding season; and according to Heermann, they build in company, selecting shrub-

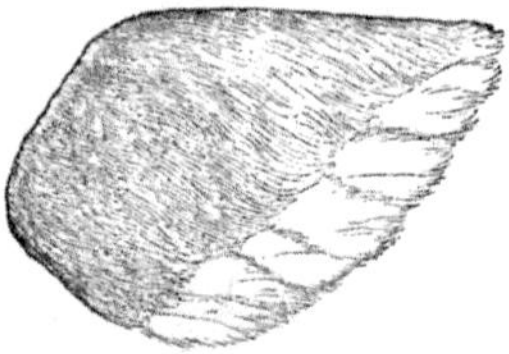

by places *near* the water, and often building four or five nests in one bush. He found one of these breeding-places in the northern part of California, covering several acres, being led to it by the flocks returning from the surrounding country, over which they had scattered for several miles in search of food. The nests were composed of mud and straw, lined with fine grasses, and the eggs were light blue, marked with lines and spots of dark umber, with a few light purple dashes. The nest and eggs are thus more like those of the *Scolecophagus* than of the other red-wings.

The song of this species is less loud and more guttural than that of the others. Their habits are otherwise very similar, and they associate in fall and winter in immense flocks in the interior, though often found separately also.

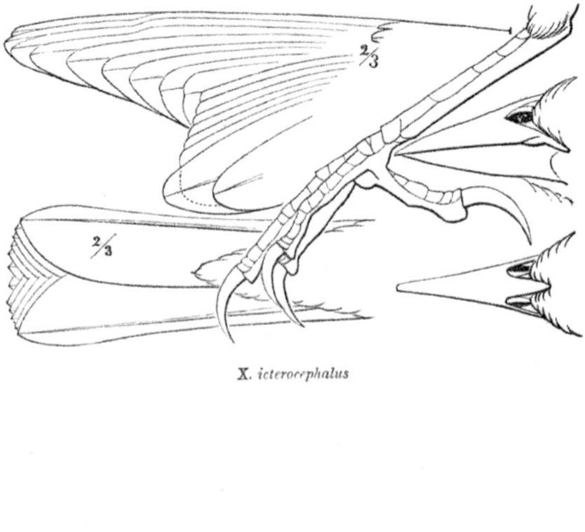

X. icterocephalus

Genus XANTHOCEPHALUS, Bonaparte.

Xanthocephalus, Bonaparte, Conspectus, 1850, 431. (Type, *Icterus icterocephalus*, Bonaparte.)

Gen. Char. Bill conical, the length about twice the height; the outlines nearly straight. Claws all very long; much curved; the inner lateral the longest, reaching be-

X. icterocephalus.

yond the middle of the middle claw. Tail narrow, nearly even, the outer web scarcely widening to the end. Wings long, much longer than the tail; the first quill longest.

There is a species in South America so closely resembling ours that they were at first considered identical.

Xanthocephalus icterocephalus, Bonaparte.

THE YELLOW-HEADED BLACKBIRD.

Icterus icterocephalus, Bonaparte, Am. Orn. I. 1825, 27; pl. iii.
Icterus xanthocephalus, Audubon, Orn. Biog. V. 1839, 6; pl. 388. — *Agelaius xanthocephalus*, Swainson, F. B. Am. II. 1831, 281. — Audubon, Birds Amer. IV. 1842, 24; pl. 213. — Newberry, Zool. Cal. & Or. Route; Rep. P. R. R. Surv. VI. iv. 86. — Heermann, X. vi. 52. — Coues, Pr. A. N. Sc. 1866, 91.
Xanthocephalus icterocephalus, Baird, P. R. Rep. IX. Birds, 531; X. iii. 13.

Sp. Char. First three quills nearly equal, the first longest and decidedly longer than the third. Tail rounded, or slightly graduated. General color black, including the inner surface of wings and axillaries, base of lower mandible all round, feathers adjacent to nostrils, lores, upper eyelids, and remaining space around the eye. The head and neck all round, the fore part of the breast, extending some distance down on the median line, and

Male.

a somewhat hidden space round the anus, yellow. A conspicuous white patch at the base of the wing formed by the spurious feathers, interrupted by the black alula.

Female. Smaller, browner; the yellow confined to the under parts and sides of the head, and a superciliary line. A dusky maxillary line. No white on the wing. Length

Female.

of male, 10.75; extent, 17.50; wing, 5.75; tail, 4.50. Female, length, 10.00; extent, 17.00; wing, 5.50. Iris brown; bill and feet black.

Hab. Western America from Texas, Illinois, Wisconsin, and North Red River, to Oregon, and California; south into Mexico. Greenland. (Reinhardt.)

This large and handsome bird is numerous in all the valleys of the State, especially where there are grassy meadows or marshes. They winter in large numbers in the middle districts, and some wander at that season to the

Colorado Valley and San Diego, though I doubt if any pass the summer so far south. They build, however, at Santa Barbara and northward, avoiding the immediate coast, but swarming about Klamath Lake. (Newberry.) Though abundant east of the Rocky Mountains up to lat. 58°, I never saw them near the Columbia.

They associate in flocks with the other blackbirds, but also keep in separate bands, and fly with such regularity that their yellow heads often show all at once as they wheel in their aerial evolutions. Sometimes also the sexes fly in separate flocks before the pairing season. They are very gregarious even in spring and summer, and seem to build in company. The only song the male attempts consists of a few hoarse chuckling notes and comical squeakings, uttered as if it was a great effort to make any noise at all. Though some kept about the marsh at Santa Barbara, in which were the nests of the red-wings, I could not find theirs. According to Heermann, the nest is composed of dry reeds and grasses, attached to the upright stalks of the reeds, and firmly fixed by pieces twisted around them. The eggs, four in number, were pale ashy-green, thickly covered and minutely dotted with points and spots of light umber brown. Nuttall describes the eggs as nearly similar, bluish-white, covered all over with minute specks of brownish-purple, largest and most numerous at the greater end. He says, however, that the nest found by Townsend near the Platte River, on the edge of a grassy marsh, was on the ground, under a tussock formed of fine grasses and canopied over like that of the meadow-lark (*Sturnella*). As there are no reeds there, the bird may vary its mode of building to suit circumstances.

S. magna.

Genus **STURNELLA**, Vieillot.

Sturnella, Vieillot, Analyse, 1816. (Type, *Alauda magna*, L.)

Gen. Char. Body thick, stout; legs large, toes reaching beyond the tail. Tail short, even, with narrow acuminate feathers. Bill slender, elongated; length about three times the height; commissure straight from the basal angle. Culmen flattened basally, extending backwards and parting the frontal feathers; longer than the head, but shorter than

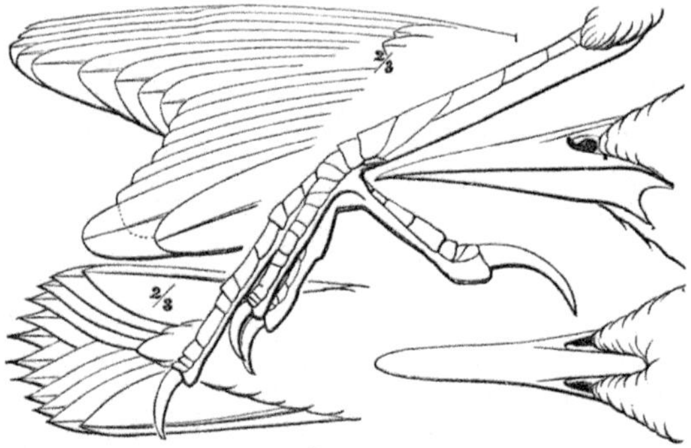

S. magna.

tarsus. Nostrils linear, covered by an incumbent membranous scale. Inner lateral toe longer than the outer, but not reaching to basal joint of middle; hind toe a little shorter than the middle, which is equal to the tarsus. Hind claw nearly twice as long as the middle. Feathers of head stiffened, and bristly; the shafts of those above extended into black setæ. Tertials nearly equal to the primaries. Feathers above all transversely banded. Beneath, yellow, with a black pectoral crescent.

Sturnella magna, Linnæus (Swainson), the Eastern meadow-lark, differs very slightly from ours, chiefly in the yellow of throat being narrower (not extending on side of face), and the bars of tertials and tail feathers being connected by a wide black stripe running along middle of feathers.

Sturnella neglecta, Audubon.

THE WESTERN LARK.

Sturnella neglecta, Audubon, Birds Amer. VII. 1843, 339; pl. 487. — Newberry, Zool. Cal. and Or. Route; Rep. P. R. R. Surv. VI. iv. 1857, 86. — Baird, P. R. Rep. IX. Birds, 537. — Kennerly, X. iv. 31. — Heermann, X. vi. 54. — Cooper and Suckley, XII. iii. Zool. of W. T. 208. — Coues, Pr. A. N. Sc. 1866, 91.

Sp. Char. Feathers above dark brown, margined with brownish-white, with a terminal blotch of pale reddish-brown. Exposed portion of wings and tail with transverse bands, which, in the latter, are completely isolated from each other, narrow and linear. Beneath, yellow, with a black pectoral crescent. The yellow of the throat extending on the side of the maxilla. Sides, crissum, and tibia very pale reddish-brown, or nearly white, streaked with blackish. Head with a light median and superciliary stripe, the lat-

Head.

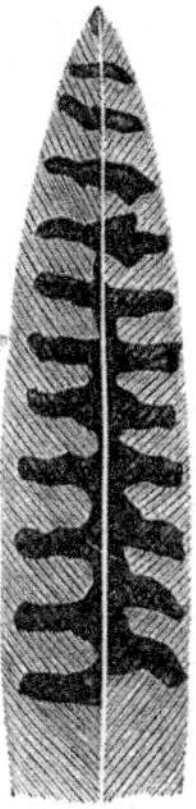

Tail feather.

ter yellow in front of the eye; a blackish line behind it. The transverse bars on the feathers above (less so on the tail), with a tendency to become confluent near the exterior margin. Length 10.50; extent, 16.50; wing, 5.00; tail, 3.25; bill, 1.25. Iris brown; bill horn-brown, bluish below; feet pale flesh-color.

Hab. Western America from the high central plains to the Pacific; east to Pembina, and perhaps to Wisconsin.

This bird is very abundant, and resident throughout nearly the entire State, though probably leaving the high mountains in winter. I think they build in the Colorado Valley, as well as all other districts not quite waterless, including all the islands except one or two.

In fact the vast grassy plains and hills of California are the most favorable nurseries of this species that could be imagined. Their abundance and large size force them on the attention of every one, while their lively, sweet, and varied songs make them general favorites. They sing at all seasons, early and late; from the ground, the tree-top, fence, or flying in the air, and when unmolested become so tame as to make the house-top a favorite perch. Their time in spring seems equally divided between an industrious search for food and musical contests with their neighbors. Even the female has considerable musical power, and cheers her mate by singing while he relieves her in sitting on the eggs. She then also has a harsh, petulant chirp, frequently repeated as if in anger.

Their flight is usually slow and laborious, partly sailing, and they furnish pretty good game for the sportsman when no other is to be found. Their flesh, though rather dry and insipid, is white and much eaten, especially by foreigners, who consider every bird, however small, a great delicacy.

Their nest is made in a slight depression under a bunch of grass, and usually more or less arched over by blades bent down. It is quite artfully concealed, and the female, if on it, generally skulks off some distance before flying. The eggs are four or five, measuring 1.15 × 0.85 inch, very obtuse, white, with a few large purplish-brown blotches and dots towards the large end.

They feed chiefly on insects, grass-seeds, and grain, but do no damage in the fields, while they destroy many noxious insects. They walk rather awkwardly but quickly, and have great ingenuity in concealing themselves when wounded.

Specimens have been shot near San Francisco and northward, which Professor Baird could not distinguish from the Eastern bird. The difference in notes, however, has always been observed, and is an important character.

Trupialis militaris, Linnæus (Bonaparte), a South American lark with a *red* breast, has been said to occur in California, but if so, must be a rare visitor, as I have never seen or heard of it here myself.

Sub-Family ICTERINÆ.

Char. Bill slender, elongated, as long as the head, generally a little decurved, and very acute. Tarsi not longer than the middle toe, nor than the head; claws short, much curved; outer lateral toe a little longer than the inner, reaching a little beyond base of middle toe. Feet adapted for perching. Tail rounded or graduated. Prevailing colors yellow or orange, and black.

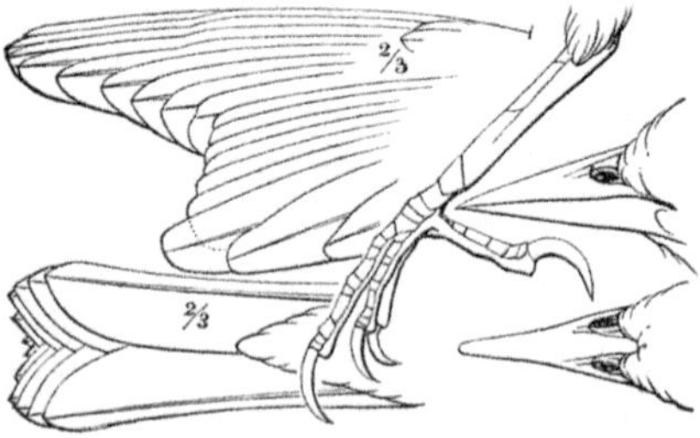

I. Baltimore.

GENUS **ICTERUS**, BRISSON.

Icterus, *Xanthornus*, *Pendulinus*, of authors, in part.

I. Bullockii

GEN. CHAR. Same as those of the sub-family, as given by Baird, being the only genus.

Icterus Bullockii, SWAINSON.

THE WESTERN ORIOLE.

Xanthornus Bullockii, SWAINSON, Syn. Mex. Birds, Taylor's Phil. Mag. I. 1827, 436.—*Yphantes Bullockii*, HEERMANN, P. R. Rep. X. vi. 52.—*Icterus Bullockii*, BONAPARTE, List, 1838.—AUDUBON, Orn. Biog. V. 1839, 9; pl. 388 and 433. IB. Birds Amer. IV. 1842, 43; pl. 218.—NEWBERRY, Rep P. R. R. VI. iv. 1857, 87.—*Hyphantes Bullockii*, CASSIN, Pr. A. N. Sc. 1867, 62.—BAIRD, P. R. Rep. Birds, IX. 549.—COOPER and SUCKLEY, XII. iii. Zool. of W. T. 209.

Psarocolius auricollis, Maxim. Reise Nordam. I. 1839, 367 (Fort Pierre, Nebraska).

SP. CHAR. Tail very slightly graduated. Upper part of the head and neck, back, wings, two central tail feathers, line from base of bill through the eye to the black of the nape, and a line from the base of the bill, running to a point on the throat, black. Under

Male.

parts generally, sides of head and neck, forehead, and line over the eye, rest of tail feathers, rump, and upper tail coverts, yellow orange. A broad band on the wings, involving the greater and middle coverts, and the outer edges of the quills, white. *Young male* with

Female.

the black replaced by greenish-yellow, that on the throat, persistent; *female* without this. Length, 7.75; extent, 12.50; wing, 4.25. Iris brown; bill black, lead-blue below; feet lead-color.

Hab. High central plains to the Pacific, south into Mexico.

This beautiful and musical bird arrives in California from the South about March 1st near San Diego; while at Fort Mojave, only one hundred and sixty miles farther north, I saw none the previous year until April 1st. They migrate north of lat. 49°, which they do not reach before the first of June, and, remaining there about three months, again retire south of the State in early autumn. They resort to the open roads, gardens, and orchards, claiming the protection of man about the towns, and repaying him for it by their sweet melody and usefulness in destroying insects. Their home is in the trees, and they rarely descend to the ground, except to pick up some bit of twine or other material for their nest. This is built near the end of a branch, often overhanging the road or house, and constructed of fibrous grasses, horse-hairs, strings, rags, down of plants, wool, and fine bark. Sometimes one or two materials alone are used, such as white horse-hair and cotton twine, which I have found in one instance. These are neatly and closely interwoven in the form of a deep bag, suspended by the edges from the forks of a branch, near its end.

The eggs, from four to six, are bluish-white, with scattered winding streaks and hair lines of black and reddish-brown near the large end, measuring 0.98 × 0.60 inch. They lay on the first or second week of May, in the southern half of California.

At Santa Cruz, in 1866, I did not observe any of this species until April 3d, which, however, was as early as they arrived at Fort Mojave in 1861.

Icterus cucullatus, Swainson.

THE HOODED ORIOLE.

Icterus cucullatus, Swainson, Phil. Mag. I. 1827, 436. — Lawrence, Ann. N. Y. Lyc. V. May, 1851, 116 (first introduced into fauna of United States). — Cassin, Ill. I. ii. 1853, 42; pl. viii. — Baird, P. R. Rep. IX. Birds, 546. — *Pendulinus cucullatus*, Bonaparte, Consp. 1850, 433. — Cassin, Pr. A. N. Sc. Phil. 1867, 60.

Sp. Char. Both mandibles much curved. Tail much graduated. Wings, a rather narrow band across the back, tail, and a patch starting as a narrow frontal band, involving the eyes, anterior half of cheek, chin, and throat, and ending as a rounded patch on

Male.

the upper part of breast, black. Rest of body orange yellow. Two bands on the wing and the edges of the quills white.

Female without the black patch of the throat; the upper parts generally yellowish-green, browner on the back. Length, 8.50; extent, 11.00; wing, 3.75. Iris brown; bill black, its base below blue; feet dark lead-color.

Hab. Valley of the Rio Grande, southward. Southern California. Abundant at Cape St. Lucas.

I found this species arriving at San Diego about April 22d, and they were not rare for a fortnight afterwards, but then retired into the warmer interior valleys, where I have seen them nearly as far north as Los Angeles. While migrating they were nearly silent.

Captain McCown, U. S. A., found them quite abundant along the Rio Grande in Texas. They were generally shy in the woods, but a pair became quite familiar around his quarters, alighting on the roof, and prying into the crevices in search of insects. (Cassin.)

Icterus parisorum, Bonaparte.

SCOTT'S ORIOLE.

Icterus parisorum, Bonaparte, Pr. Zool. Soc. V. 1837, 109. — Baird, Birds N. Amer. 544. — Cassin, Pr. A. N. Sc. 1867, 54; pl. 61, f. 1. Ib. Mex. Bound. Rep. pl. 19.
Icterus melanochrysura, Lesson, Rev. Zool. 1839, 105.
Icterus Scotti, Couch, Pr. A. N. Sc. Phil. 1854, 66.

Sp. Char. Bill attenuated; not much decurved; tail moderately graduated. Anterior half of body, or head and neck, back and breast, black. Abdomen and rump yellow, the latter generally tinged with greenish. Wings black, the greater coverts widely tipped

with white, shorter quills widely edged with white. Middle feathers of the tail black, with their bases yellow; other feathers of the tail with their basal two thirds yellow, terminal one third black. Bill dark horn-color, base of under mandible pale blue; legs bluish-brown. Bill straight, rather slender, pointed, culmen distinctly ridged; legs rather strong; wing long, third quill longest; tail moderate.

Adult male. Total length about 7.00 to 8.00 inches; wing, 4.00 to 4.25; tail, 3.50 to 3.75 inches.

Younger. Entire head and back dark brown; rump greenish-yellow, under parts of body dull pale yellow; tail olive-green; outer feathers greenish-yellow at base; wings dull brown; coverts tipped with white. Total length, 7.00.

Hab. Southern border of United States from the mouth of the Rio Grande to Cape St. Lucas. South into Mexico.

This species is abundant at Cape St. Lucas, and has been found at rare intervals all along the southern border of the United States as far east as Texas. Nothing distinctive is known of its habits. I saw a bird at Fort Mojave, in April, which I supposed to be this, but could not obtain it.

Sub-Family QUISCALINÆ.

Char. Bill rather attenuated, as long as or longer than the head. The culmen curved, the tip much bent down. The cutting edges inflected so as to impart a somewhat tubular appearance to each mandible. The commissure sinuated. Tail longer than the wings, usually much graduated. Legs longer than the head, fitted for walking. There are four species of the genus *Quiscalus* in the Eastern and Southern United States.

Genus **SCOLECOPHAGUS**, Swainson.

Scolecophagus, Swainson, F. Bor. Am. II. 1831. (Type, *Oriolus ferrugineus*, Gmelin.)

Gen. Char. Bill shorter than the head, rather slender, the edges inflexed as in *Quiscalus*, which it otherwise greatly resembles; the commissure sinuated. Culmen rounded, but not flattened. Tarsi longer than the middle toe. Tail even or slightly rounded.

S. ferrugineus.

The typical species, *S. ferrugineus,* is common in the Eastern United States, and is much like our species, which was formerly confounded with it by Nuttall, Townsend, and others.

Quiscalus major, Vieillot, the boat-tailed grakle, is said by Gambel to occur at Mazatlan, "and occasionally in California," but has not been obtained by recent collectors. The Mazatlan species is *Q. palustris,* Swainson.

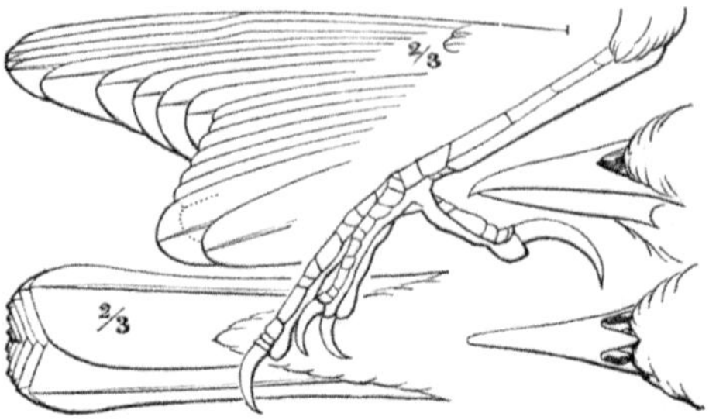

S. ferrugineus.

Scolecophagus cyanocephalus, WAGLER.

BREWER'S BLACKBIRD.

Psarocolius cyanocephalus, WAGLER, Isis, 1829, 758. — *Scolecophagus cyanocephalus*, CABANIS, Mus. Hein. 1851, 193. — BAIRD, P. R. Rep. IX. Birds, 552. — HEERMANN, X. vi. 53. COOPER and SUCKLEY, XII. iii. Zool. of W. T. 209.

Scolecophagus Mexicanus, SWAINSON, Anim. in Men. 2½ cent. 1838, 302. — NEWBERRY, Zool. Cal. and Or. Route; Rep. P. R. R. Surv. VI. iv. 1857, 86.

Quiscalus Breweri, AUDUBON, Birds Amer. VII. 1843, 345; pl. 492.

SP. CHAR. Bill stout, quiscaline, the commissure scarcely sinuated, shorter than the head and the hind toe; the height nearly half the length above. Wing nearly an inch longer than the tail; the second quill longest; the first about equal to the third. Tail

rounded and moderately graduated; the lateral feathers about 0.35 of an inch shorter. General color of male black, with lustrous green reflections everywhere, except on the head and neck, which are glossed with purplish-violet.

Female, much duller, of a light brownish anteriorly; a very faint superciliary stripe.

Length, about 10.00 : extent, 16.25 ; wing, 5.25 ; tail, 4.40. Iris of male yellow (of female brown) ; bill and feet black.

Hab. High central plains to the Pacific ; south to Mexico. Pembina, Minnesota.

An abundant species everywhere throughout California, except in the dense forests, and resident in winter as far north as the Columbia River. They frequent pastures and fields, often following cattle, like the cow-birds, for the insects which they disturb in the grass, and the undigested seeds in their ordure. They also associate much with the other blackbirds, and are fond of feeding and bathing along the edges of streams.

They have not much song, though the noise made by the associated flock, as they sit sunning themselves on the top of some tree in early spring, is quite pleasing, resembling much that of the red-wings, some of which often assist in the chorus.

Their nests are built in low trees, often several in one tree. They are large, and constructed externally of a rough frame and floor of twigs, then a thick layer of mud, and a lining of fibrous roots and grass. The eggs, laid from April 10th to May 20th, are four or five, dull greenish-white, with numerous streaks and small blotches of dark brown thickly scattered over them. They measure 1.00 × 0.72 inch. They doubtless raise two and perhaps three broods in a season.

"*Quiscalus purpureus*" (*Quiscalus versicolor*, Vieillot) is mentioned by Newberry as common near San Francisco. No collector, however, has obtained it on this coast, and I am satisfied that he mistook *S. cyanocephalus* for it, as I have myself at a distance.

At Santa Cruz these birds were more familiar than I have seen them elsewhere, frequenting the yards about houses and stables, building in the garden trees, and daily, after their hunger was satisfied, collecting on the roofs, or on convenient trees, to sing their best thanks for an hour or two at a time to their human entertainers. I have seen two of them pursue and drive away a large hawk, while the timid pigeons were wildly circling around the town in efforts to escape, even before the hawk made any approach toward them.

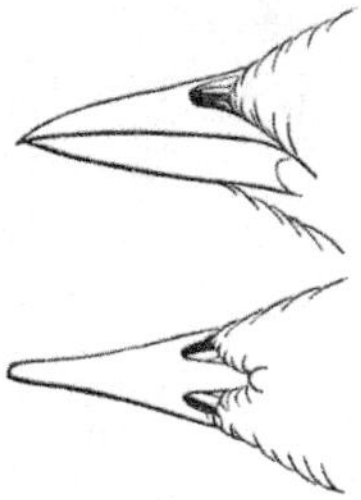

S. cyanocephalus.

FAMILY CORVIDÆ., THE CROWS.

CHAR. Primaries ten; the first short, generally about half as long as the second (or a little more); the outer four sinuated on the inner edge. The nasal fossæ and nostrils usually more or less concealed by narrow stiffened bristles (or bristly feathers), with short appressed lateral branches extending to the very tip, all directed forwards. Tarsi scutellate anteriorly, the sides undivided (except sometimes below), and separated from the anterior plates by a narrow naked strip, sometimes filled up with small scales. Basal joint of middle toe united about equally to the lateral, generally for about half the length. Bill usually notched.

SUB-FAMILY CORVINÆ, THE CROWS.

CHAR. Wings long and pointed; longer than the tail, and when closed reaching nearly to its tip, extending far beyond the under tail coverts; the third, fourth, and fifth quills forming the tip of the wing.

C. carnivorus.

GENUS **CORVUS**, LINNÆUS.

Corvus, LINNÆUS, Syst. Nat. 1735. (Type, *Corvus corax*, L.)

SP. CHAR. The nasal feathers lengthened, reaching to or beyond the middle of the bill. Nostrils large, circular, overhung behind by membrane, the edges rounded elsewhere. Rictus without bristles. Bill nearly as long as the tarsus, very stout; much higher than broad at the base; culmen much arched. Wings reaching to or nearly to the

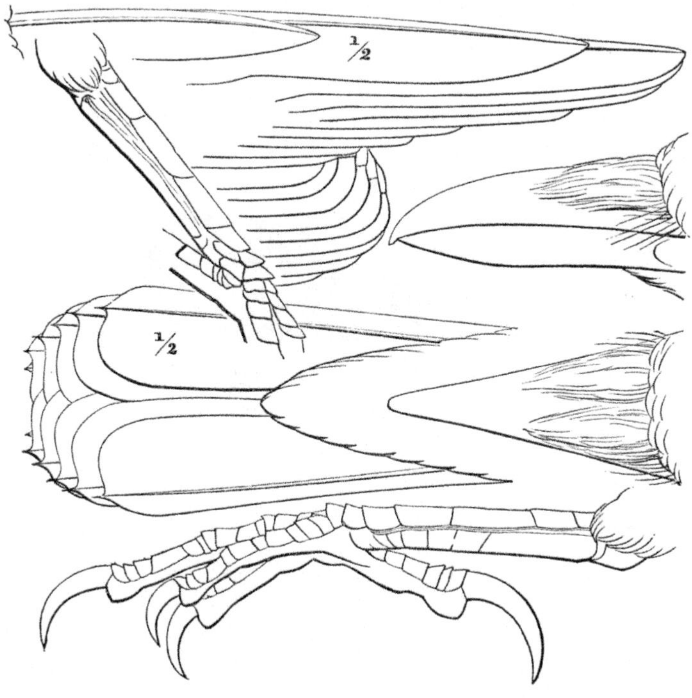

C ossifragus.

tip of the tail. Tarsi about equal to the middle toe, sometimes with a series of small scales on the middle of each side, separating the anterior scutellate portion from the posterior continuous plates. Side of the head occasionally with nearly naked patches. Tail graduated or rounded; the outer four primaries sinuated internally.

Besides our species, there is one in Texas which has the feathers of neck and breast snowy white at base (*C. cryptoleucus*, Couch); and another very large raven in Mexico (and New Mexico ?).

Corvus carnivorus, BARTRAM.

THE AMERICAN RAVEN.

Corvus carnivorus, BARTRAM, Travels in E. Florida, 1793, 290.—BAIRD, P. R. Rep. IX. Birds, 560; X. iii. 14, pl. xxi.—COOPER and SUCKLEY, XII. iii. Zool. of W. T. 210.
Corvus corax, WILSON, Am. Orn. IX. 1825, 136; pl. lxxv. f. 3.—DOUGHTY, Cab. N. H. I. 1830, 270; pl. xxiv.—NUTTALL, Man. I. 1832, 202.—AUDUBON, Orn. Biog. II. 1834, 476; pl. 101. IB. Birds Amer. IV. 1842, 78; pl. 224.—HEERMANN, P. R. Rep. X. vi. 54. (Not of Linnæus, European.)
Corvus cacalotl, NEWBERRY, P. R. Rep. VI. iv. 1857, 82.—BAIRD, P. R. Rep. IX. Birds, 563.—KENNERLY, X. iv. 31; pl. xx. (Not of Wagler ?)
Corvus lugubris, AGASSIZ, Pr. Bost. Soc. N. H. II. Dec. 1846, 188.

SP. CHAR. Fourth quill longest; third and fifth about equal; second between fifth and sixth; first nearly equal to the eighth. Tail moderately graduated; the outer about 1.60

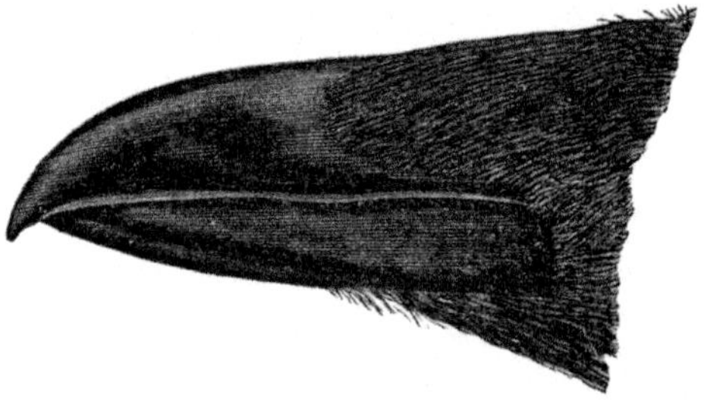

to 1.90 of an inch less than the middle. Entirely glossy black, with violet reflections. Length, about 24 00 or 25.00; extent, 50 00 to 51.00; wing, about 17.00; tail, 10.00. Iris brown; bill and feet black.

Hab. Entire continent of North America. Rare east of the Mississippi.

I obtained numbers of the raven at Fort Mojave, which, though more brilliant than usual in Northern specimens, did not seem to differ enough in size or proportions to separate them as a distinct species. The only approach to Professor Baird's "*C. cacalotl*" was in the more graduated tail feathers of one, in which the outer were 2.10 inches shorter than the middle, other specimens having them nearer alike. He gives the difference in *C. cacalotl* at 2.30, while in *C. carnivorus* it ranges from 1.60 to 1.90. Still, my specimens being intermediate, and the extremes of difference in all being only 0.07 of an inch (" the lateral feathers where shortest not having fully grown out"), I do not think there is enough reason to distinguish them as a distinct species. Their size is smaller than the average of Northern specimens. I think the greater lustre of the plumage is merely the effect of a drier and hotter climate. The differences shown in Plates XX. and XXI. are so very slight as to be of little consequence.

In cries and habits I have noticed no difference in the ravens of various parts of the coast. Being very local in their residence, and scarcely if at all migratory, there is a tendency to run into slight varieties of form, which in some other birds of similar habits we see carried so far as to produce local races or species, separated by various degrees of difference from their allies.

The American raven is found in pairs everywhere in California and the adjacent regions, being much more numerous than in the Atlantic States, where few are seen south of the Great Lakes. They frequent, as if by preference, even the most barren desert districts, but are also common near cultivated regions. Being omnivorous, and capable of enduring long fasts, they take advantage of the very barrenness and scarcity of water in the deserts, so destructive to many animals, and watch for their death, in order to support their own life by eating them. They follow trains and herds of cattle, hoping to pick up a dying straggler, and are sure to visit every deserted camp as soon as the occupants move off. When nothing better is to be had they eat lizards, snakes, birds' eggs, and in fact everything of an animal nature, however far gone to decay. They also occasionally devour grain, potatoes, etc., but do little damage in that way. They will, however, destroy young chickens, and it is said young lambs, fawns, etc., if they find them unprotected. At times they soar like the hawks to a great height, especially before storms, but also when looking out for food.

In selecting a place for nesting they show much sagacity, sometimes choosing a hole in a lofty perpendicular cliff, at others a high tree very difficult to climb. One nest I found on Santa Barbara Island, built in a cavity under a projecting ledge of rock not more than twenty feet above the water, where the waves dashed almost into it at high tide. Though a much-frequented path passed close above it, the old birds were so wary in their visits that it remained long undiscovered. The young were still unfledged in June, much later than on the main-land, where I think they often lay in March, having seen young in a nest at the Straits of Fuca in April. The nest is composed of large sticks, twigs, roots, and various soft materials inside. The eggs are said to be from three to six, of a pale muddy bluish-green, marked with numerous spots and lines of dark olive-brown. (Nuttall, of the European species ?) Those of the whole sub-family *Corvinæ* are known to be very similar in colors.

At Beales's Ranch, south of Tejon Pass, and at the eastern edge of the desert, there is a great roosting-place for ravens, which come at dusk from all directions to sleep on the oaks found up the valley.

The well-known sagacity and imitative powers of the European raven are doubtless possessed in an equal degree by ours, which differs very slightly from that species (*Corvus corax*). For full accounts of the many strange facts and superstitions recorded concerning it, I must refer to other authors.

Corvus cryptoleucus, Couch.

THE WHITE-NECKED CROW.

Sp. Char. The fourth quill is longest; the third and fifth equal; the second longer than the sixth; the first about equal to the seventh. Glossy black, with violet reflections; feathers of neck all round, back and breast, snow white at the base. Feathers of throat

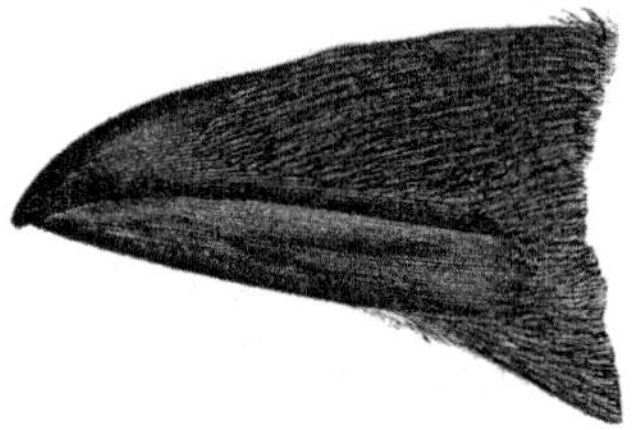

lanceolate; bristly feathers along the base of the bill, covering it for nearly two thirds its length. Length, about 21.00; wing, 14.00; tail, 8.50.

Hab. Valleys of the Rio Grande and Gila. Abundant on the Llano Estacado.

The general color of this raven is a lustrous black, with violet reflections, almost exactly as in the common species. Its most striking distinctive feature, however, is seen in the feathers of the neck all round, upper part of the back, and the whole of the breast, which are pure snowy white for about their basal half. The feathers of the head are plumbeous at the base; the greatest intensity of white is on the lower part of the neck; the color fades out on the back and belly into plumbeous. In no other North American crow is there any approach to this cottony whiteness. A species from Porto Rico (*Corvus leucognaphalus*) has the same peculiarity of color. It is of about the same size as *C. cryptoleucus*, but will be readily distinguished by the nearly naked and exposed nostrils, a character shared by most of the crows of the West Indies.

Nothing distinctive is known of the habits of this bird. The eggs have, however, been collected, and resemble those of the common raven, although perhaps of paler color. (Baird.)

Corvus caurinus, Baird.

THE WESTERN CROW.

Corvus Americanus, from West Coast. Baird, P. R. Rep. IX. Birds, 566 (in part). — Heermann, X. vi. 52. — Cooper and Suckley, XII. iii. Zool. of W. T. 211.

Corvus ossifragus, from West Coast, of authors. Nuttall, Man. I. 2d ed. 221, 227. — Newberry, P. R. Rep. VI. iv. 82.

Corvus caurinus, Baird, P. R. Rep. IX. Birds, 569. — Cooper and Suckley, XII. iii. Zool. of W. T. 211; pl. xxiv. (Pl. xxiii. *C. Americanus.*)

Sp. Char. Fourth quill longest, second longer than sixth, third and fifth about equal, first shorter than ninth. Color black, glossed with purple. Tail nearly even. Tarsus

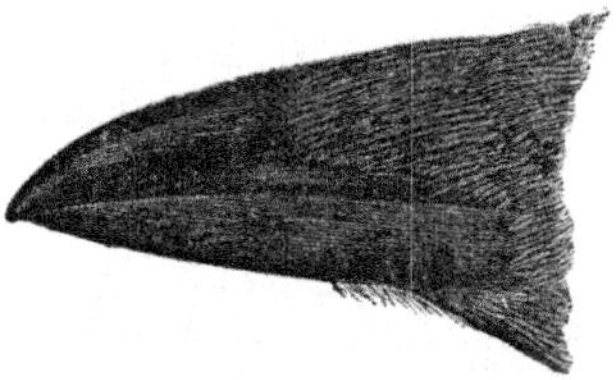

longer than middle toe and claw. *Female* the smallest. Length, 15.00 to 18.50; extent, 30.00 to 33.50; wing, 10.50 to 12.75; tail, 6.50 to 7.80. Iris brown; bill and feet black.

Hab. Pacific Coast from San Diego north; east to the Northern Rocky Mountains.

Professor Baird, in his description of *Corvus caurinus*, remarks that "but for the slight difference in size it would be difficult to tell skins" from those of *C. Americanus.* "Indeed, it is almost a question whether it be more than a dwarfed race of the other species." He, however, separates specimens from California and Fort Vancouver, "on account of their larger size," including them with *C. Americanus.* But by reference to his table of measurements, it appears that they were all below the average measurements he gives for that species (length, 19.00 to 20.00; wing, 13.00 to 13.50; tail, about 8.00), even measuring the stretched skins.

I am now satisfied that there is but one species of crow on this side of the continent, and if distinguishable at all from the Eastern, it is by its smaller size, less graduated tail, more gregarious habits, and different voice. The circumstance of their feeding much on fish towards the North is I think of secondary importance, as they are decidedly omnivorous, and feed on fish wherever they can get them more easily than other food. In this State fish are not so numerous on the sea-shore and river-banks as near the

Columbia, while the open plains offer an abundant supply of other food. Nor does the fish-eating race seem more degenerate in size on that account, since the smallest specimens I have seen were collected by Mr. F. Gruber on the west slope of the Sierra Nevada, about lat. 38° 30′, one of them in perfect condition and full grown, measuring only fifteen inches in length when fresh. Another, a male from San Diego, measured 17.50, so that the coast birds and those from the North are actually the largest, the average of Baird's *caurinus,* from Washington Territory, being 16.50 when fresh. The largest Californian from Tulare Valley measured 18.30 dry; one from the Presidio a little less, and one from Fort Vancouver 17.00, all included by Baird in *C. Americanus,* but smaller than the Eastern.

In the southern half of California this crow is rarely seen on the sea-beach, but prefers the inland districts, occasionally, however, coming to the shores of bays to feed.

They associate in large flocks during most of the year, feeding in company, and are even somewhat gregarious in the breeding-season, building in close vicinity to each other, often several nests in a tree, unlike the Eastern crows, which will not allow a strange pair in the neighborhood of their nest. The places selected for building near San Diego were the groves of evergreen oaks growing in ravines, and the nests were built from twelve to forty feet above the ground. There were indeed but few other trees in the lower country, except some few willows and sycamores, in which last they might have obtained much higher positions for their nests if desired. In the North they often build in spruces, and Dr. Suckley found one at the Dalles, Oregon, in a dense willow thicket.

The nest is strongly built of sticks, coarse on the outside, and becoming more slender toward the inside, where they are mixed with roots, grass, moss, horsehair, etc., to form a rather soft lining. The eggs, of which I have only found four in a nest, are dull green of a dark shade, thickly spotted and streaked with dark brown and olive. Size, 1.60 × 1.10. At San Diego they laid about April 15th.

Where unmolested these birds have not yet become so shy and cunning as those of the older settled districts, but they soon learn the danger of allowing a man with a gun to come near them, and show all the cunning characteristic of the tribe. They might, doubtless, be as easily tamed or raised from the nest as other species, and probably taught as many tricks, including the imitation of the human voice. They have not yet been found very destructive in this country, there being little Indian corn raised, and other crops are so prolific that the stealings of the crows are scarcely noticed. Their destruction of immense numbers of grubs, grasshoppers, and other noxious vermin counterbalances what little harm they may do. They obtain much of their food about the ranches from the offal of slaughtered cattle, and those that die of diseases, etc. In the northern half of the State they also no doubt live in great part on fish. Near the Columbia River they feed much on clams and oysters, carrying the shell high in air, and dropping it until it is broken. If unsuccessful at first they carry it higher, or to a place where the ground is harder. They show much sagacity in watching the turn of the tide, and fly in flocks at that moment for their favorite feeding-grounds, in company with the gulls and various waders.

For some reason they shun particular districts in the breeding-season. During six weeks of spring I did not see one near the east side of the Coast Range south of Santa Clara, but found them at Santa Cruz, where the only difference was a cooler climate. At the same time they avoided the coast farther north, where perhaps the wind was too strong for them. I never saw one in the Colorado Valley, nor near the summits of the Sierra Nevada. According to Dr. Newberry, they are chiefly confined to those districts where oaks flourish.

At Visalia, where there is an extensive forest of oaks, forming an oasis in the great "Tulare" plain (elsewhere either marshy or sterile gravelly prairie), I found very large flocks of this bird, with the same gregarious habits as elsewhere on the coast, and at that time (September) very noisy, as if anticipating rain, which did not come, however, for some weeks after. Near here was where Dr. Heermann obtained the large specimen called *Americanus*, by him and by Professor Baird.

During July, 1866, large numbers of this species came every evening to roost in an alder grove close to the town of Santa Cruz. They gathered in long trains from the surrounding fields, flying rather high, and suddenly pitching down toward the small trees with zigzag turns, sportively pecking and chattering at each other in the air. Dr. Suckley found this crow very abundant on Puget's Sound throughout the year, subsisting in winter on the refuse food and offal thrown out by the natives from their lodges, and attending closely upon the residences of the white settlers. He describes it as very cunning, but tame and impudent, allowing a near approach, and when closely pursued retiring but a short distance. In reference to the habit of dropping clams from a height in order to break the shell, he mentions the amusement afforded by the effort of one of these crows to get at the contents of a clam by dropping it on soft ground, carrying it over and over again to a considerable height, and letting it drop without the desired result.

A small species of crow (*C. Mexicanus*), the least of all the continental American species, is common at Mazatlan, and may yet be detected in Arizona and Southern California.

Genus **PICICORVUS**, Bonaparte.

Picicorvus, Bonaparte, Consp. Av. 1850, 384. (Type, *Corvus Columbianus*, Wilson.)

Gen. Char. Lead color, with black wings and tail. Bill longer than the head, considerably longer than the tarsus, attenuated, slightly decurved; tip without notch. Culmen and commissure curved; gonys straight or slightly concave, as long as the tarsi.

P. Columbianus.

Nostrils circular, completely covered by a full tuft of incumbent white bristly feathers. Tail much shorter than the wings, nearly even or slightly rounded. Wings pointed, reaching to the tip of tail. Third, fourth, and fifth quills longest. Tarsi short, scarcely

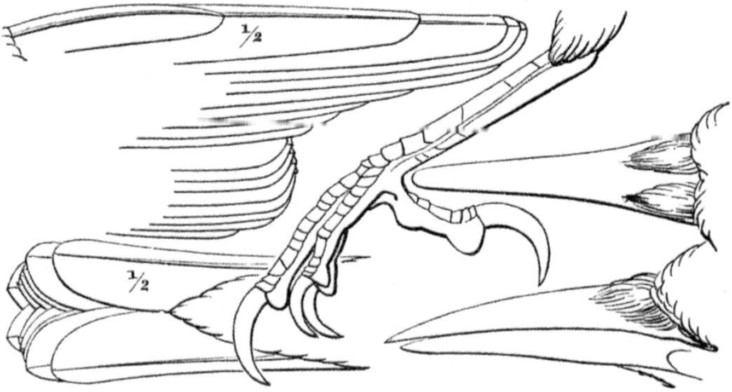

P. Columbianus.

longer than the middle toe, the hind toe and claw very large, reaching nearly to the middle of the middle claw, the lateral toe little shorter. A row of small scales on the middle of the sides of tarsus.

Picicorvus Columbianus, Wilson.

CLARKE'S CROW.

Corvus Columbianus, Wilson, Am. Orn. III. 1811, 29; pl. xx. — Nuttall, Man. I. 1832, 218. — *Nucifraga Columbiana*, Audubon, Orn. Biog. IV. 1838, 459; pl. 362. Ib. Birds Amer. IV. 1842, 127; pl. 235. — Nuttall. Man. I. 2d ed. 251. — *Picicorvus Columbianus*, Bonaparte, Consp. 1850, 384. — Newberry, P. R. Rep. VI. iv. 1837, 83. — Baird, P. R. Rep. IX. Birds, 573. — Kennerly, X. iv. 32. — Cooper and Suckley, XII. iii. Zool. of W. T. 212.

Sp. Char. Tail rounded or moderately graduated, the closed wings reaching nearly to its tip. Fourth quill longest; second considerably shorter than the sixth. General color bluish-ash, changing on the nasal feathers, the forehead, sides of head (especially around the eye), and chin to white. The wings, including their inner surface, greenish-black, the secondaries and tertials, except the innermost, broadly tipped with white; tail white,

the inner web of the fifth feather and the whole of the sixth, with the upper tail coverts, greenish-black. The axillars plumbeous black. Bill and feet black. Length of male (fresh), 12.50; extent, 24.50; wing, 8.00; tail, 4.30; tarsus, 1.20. Iris brown; bill and feet black.

Hab. From Rocky Mountains to Pacific; east to Fort Kearney; south to Fort Tejon.

This species is numerous near the summits of the Sierra Nevada, especially on their eastern side, frequenting chiefly the open forests of yellow pine (*P. ponderosa*), on the seeds of which they feed almost entirely in the autumn, swallowing them whole in such quantities as almost to suffocate themselves. I also noticed them pecking at dead bark to obtain insects, and flying short distances after them like the woodpeckers. When feeding they were very shy, scarcely allowing me to get within shooting distance, except by artifice, and, if they saw me, flying off to a long distance before alighting. I have always found them wild, except in January, 1854, when a few driven down by snow from the Cascade Mountains appeared at Fort Vancouver. They often hang head downwards while extracting the seeds from cones, reminding one of the titmice. They sometimes descend to the ground in search of seeds and insects, but probably do not eat animal food so generally as the crows and jays. They have a continuous, flapping, rapid flight, and a loud, harsh cry when flying in the manner of the crows, generally associating in flocks in the autumn. Their similarity to the woodpeckers in habits, especially to *Melanerpes torquatus*, with which they live, is sufficient to deceive most observers as to their affinities.

Of their habits in spring nothing is known, except that Townsend speaks of their building in very high pine-trees. They are in fact almost inseparable from these, and only resort to the seeds of spruces, berries, etc., when the supply of pine seeds is exhausted.

It is doubtful if they ever wander to the coast mountains south of San Francisco. Although in winter they descend to the Columbia River, they have not been seen at other seasons below an elevation of from four to ten thousand feet in California. They abound in the Rocky Mountains throughout our limits.

Near Fort Colville, Washington Territory, Mr. J. K. Lord found this species arriving in May in immense flocks, making a tremendous chattering for about a week, and then pairing off. A nest he saw was in the top of a pine-tree two hundred feet high, which was cut down on the boundary. It was composed of fir-twigs, bark, leaves of pine, and fine root-fibres, with some moss and gray lichen, — very large and shallow. The eggs were about four, of a light bluish-green.

GENUS **GYMNOKITTA**, MAXIMILIAN VON WIED.

Gymnorhinus, Pr. Max. Reise Nord Amer. II. 1841, 21. (Not of Gray.) (Type, *G. cyanocephala.*)
Gymnokitta, Pr. Max. "1850," GRAY.
Cyanocephalus, BONAPARTE, "1842." Preoccupied in Botany.

GEN. CHAR. Bill elongated, depressed, shorter than the tarsus, longer than the head, without notch, similar to that of *Sturnella* in shape. Culmen nearly straight; commissure curved; gonys ascending. Nostrils small, oval, entirely exposed, the bristly feathers at the base of the bill being very minute. Tail short, nearly even, much shorter than the

G. cyanocephala.

pointed wings, which cover three fourths of the tail. Tarsi considerably longer than the middle toe.

The species described is so far the only one of the genus known, and possesses many interesting generic peculiarities, which will be sufficiently

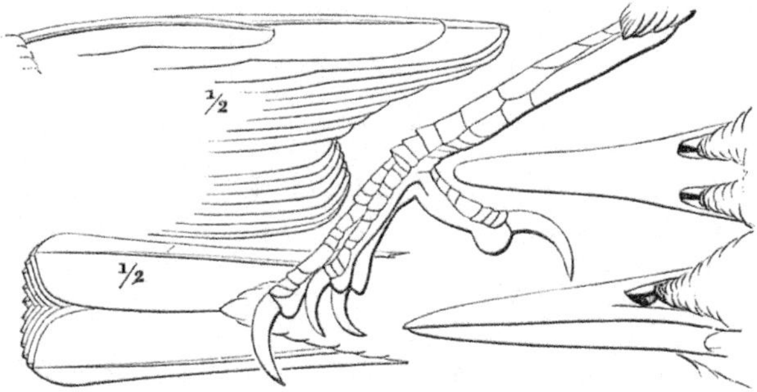

G. cyanocephala.

evident from a comparison of the above diagram with those of the other genera of *Corvidæ*. Unlike *Picicorvus*, there is no genus closely related to this in the Old World.

Gymnokitta cyanocephala, MAXIMILIAN VON WIED.

MAXIMILIAN'S JAY.

Gymnorhinus cyanocephalus, Pr. Maximilian, Reise in das innere Nord Amerika, II. 1841, 21. In. Voyage dans Am. du Nord, III. 1843, 296. — *Gymnokitta cyanocephala*, PR. MAXIMILIAN, 1850. — BONAPARTE, Conspectus, 1850, 382. — CASSIN, Illust. I. vi. 1854, 165; pl. xxviii. — NEWBERRY, Rep. P. R. R. VI. iv. 1857, 83. — BAIRD, P. R. Rep. IX. Birds, 574. — KENNERLY, X. iv. 32.
Cyanocorax Cassinii, M'CALL, Pr. A. N. Sc. V. June, 1851, 216.

SP. CHAR. Wings considerably longer than the tail, and reaching to within an inch of its tip. Tail nearly even. General color dull blue, paler on the abdomen, the middle of which is tinged with ash; the head and neck of a much deeper and more intense blue,

darker on the crown. Chin and fore part of the throat whitish, streaked with blue. Length, 10; wing, 5 90; tail, 4.50; tarsus, 1.50.

Hab. Rocky Mountains to the Sierra Nevada, California.

This bird is chiefly confined to the barren districts east of the Sierra Nevada, where occasional groves of juniper constitute its favorite resorts. According to Newberry, they were numerous in flocks of twenty or thirty among the junipers along Des Chutes River, Oregon, with the usual straggling flight of the jays, chattering as they flew from tree to tree on their way to a stream to drink. Their note when flying or feeding was a frequently repeated *că, că, că,* usually soft and agreeable, but when disturbed their cry was harsher. They were very shy, and only to be shot by lying in wait for them. They fed on the berries of the juniper, and also the seeds of the yellow-pine.

Dr. Kennerly found them in New Mexico, in November, about the watercourses. When scared they flew in circles, rising high above his head uttering their singular cry, and then suddenly descended to the top of some tree on the neighboring cliffs.

I have seen specimens from Washoe, just east of the State boundary, and am told by Mr. Clarence King that they frequent the junipers on mountains near Mariposa.

SUB-FAMILY GARRULINÆ, THE JAYS.

CHAR. Wings short, rounded; not longer, or much shorter than the tail, which is graduated, sometimes excessively so. Wings reaching not much beyond the lower tail coverts. Bristly feathers at base of bill variable. Bill nearly as long as the head, or shorter. Tarsi longer than the bill or than the middle toe. Outer lateral claws rather shorter than the inner.

GENUS **PICA**, BRISSON.

Coracias, LINNÆUS, Syst. Nat. 1735 (Gray).
Pica, BRISSON, Ornithologia, 1760, and of Cuvier (Agassiz). (Type, *Corvus pica*, L.)
Cleptes, GAMBEL, J. A. N. Sc. 2d Ser. I. 1847, 47.

GEN. CHAR. Tail very long, forming much more than half the total length; the feathers much graduated; the lateral scarcely more than half the middle. First primary

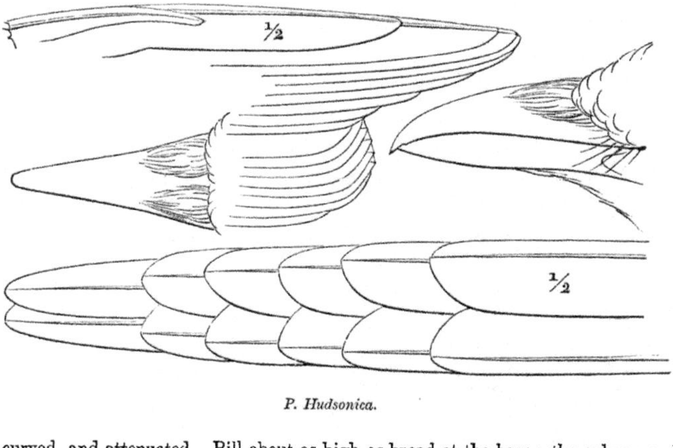

P. Hudsonica.

falcate, curved, and attenuated. Bill about as high as broad at the base; the culmen and gonys much curved, and about equal; the bristly feathers reaching nearly to the middle of the bill. Nostrils nearly circular. Tarsi very long; middle toe scarcely more than two thirds the length. A patch of naked skin beneath and behind the eye.

There are two species of magpie in the United States differing in almost nothing from each other except in the black bill of one and the yellow of the other. The latter (*P. Nuttalli*) is confined to Western California, the other (*P. Hudsonica*) is found more to the northward as well as in the mid-

P. Nuttalli.

dle region of North America. The yellow-billed bird is scarcely entitled to consideration as a distinct species, since many species of American *Corvidæ* have the bill indifferently yellow or black, or intermediate in color. The European magpie has a black bill, and we believe a variety with yellow bill is not found in the Old World.

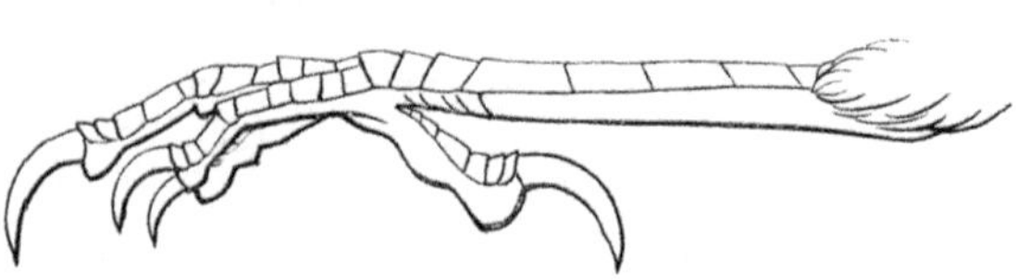

P. Hudsonica.

Pica Nuttalli, AUDUBON.

THE YELLOW-BILLED MAGPIE.

Pica Nuttalli, AUDUBON, Orn. Biog. IV. 1838, 450; pl. 362. IB. Birds Amer. IV. 1842, 104; pl. 228. — NUTTALL, Man. I. 2d ed. 1840, 236. — NEWBERRY, Rep. P. R. R. VI. iv. 1857, 84; pl. 26. — BAIRD, P. R. Rep. IX. Birds, 578. — HEERMANN, X. vi. 54. — *Cleptes Nuttalli*, GAMBEL, J. A. N. Sc. Ph. 2d Series, I. 1847, 46.

SP. CHAR. Bill, and naked skin behind the eye, bright yellow. The belly, scapulars, and inner webs of the primaries white; hind part of back grayish; exposed portion of the

tail feathers glossy green, tinged with purple and violet near the end; wings glossed with green; the secondaries and tertials blue; throat feathers spotted with white. Length, 17.00; wing, 8.00; tail, 10.00.

Hab. California, west of the Sierra Nevada.

This magpie is abundant in the valleys of California, especially near the middle of the State, except in the spring months, when I have seen none in Santa Clara Valley, and suppose they must have retired into some of the mountains eastward to build nests. There were none certainly at that time in the Coast Range west of that valley. At Santa Barbara, however, I found them numerous in April and May, and saw their nests in oak-trees, but the young were fledged by the 25th of April. The nest is composed of a large mass of coarse twigs twisted together in a spherical form, with a hole in the side. The eggs resemble those of the other species, which are whitish-green, spotted with cinereous-gray and olive-brown. They breed abundantly about Monterey.

Their food consists of almost everything animal and vegetable which they can find or eat, and they come much about farms and gardens to pick up whatever they can. Like the magpie of Europe, they probably carry off many things that can be of no use to them, out of mere curiosity or mischief.

They have a loud call like *pait, pait,* and a variety of chattering notes, in tone resembling the human voice, which they can be taught to imitate. They are said also to imitate the calls and songs of small birds, as if to attract them within their reach, as they kill them, and destroy their eggs on every opportunity. I have never seen magpies along the coast north of Monterey.

Pica Hudsonica, Sabine.

THE AMERICAN MAGPIE.

Corvus pica, Forster, Phil. Trans. LXXII. 382.
Corvus Hudsonicus, Sabine, App. Franklin's Journey, 1823, 25. — *Pica Hudsonica*, Bonaparte, List, 1838. — Newberry, P. R. Rep. VI. iv. 84. — Baird, Birds N. Amer. 576. — Cooper and Suckley, P. R. R. XII. ii. 213.
Pica melanoleuca, "Vieillot," Audubon, Birds Amer. IV. 1842, 99; pl. 227.

Sp. Char. Similar to *P. Nuttalli*, but with the bill and skin round the eye black. Throat feathers spotted with concealed white. Length, 19.00; wing, 8 50; tail, 11.00.

Hab. Northern portions of North America. Whole Rocky Mountain region and North Pacific States.

The magpie is not so familiar a bird to the inhabitants of the more settled parts of the United States as its European congener, although west of the Missouri and in the vicinity of Lake Superior it is well known. The habits are much as detailed for the yellow-billed magpie, and indeed it is almost a question whether the yellow bill of the latter is any more than a mere permanent form of a variation in this respect seen in many jays, as *Psilorhinus morio, Cyanura Beecheyi*, etc. Our bird is said not to be so much at home with man as the European species, but resembles it in most of its habits. It is a source of great annoyance to Western men from its fondness for lighting on the backs of horses and mules, and picking at any sore places it may find, causing the animals to run away or roll over to get rid of their tormentors, and often to the detriment of their load, or to their personal injury. The magpies feed largely on carrion and the refuse of camps, although nothing comes amiss to them, — eggs, young birds, fruits, etc.

Mr. J. K. Lord found this bird on Vancouver's Island, as well as at Fort Colville, and says that they build in March, in dense thickets, often very near the fish-crows, constructing a similar covered nest, and laying seven or eight eggs. Great numbers were poisoned at the camp of the Boundary Commission in winter, because they became so troublesome to sore-backed horses. They were so tame and impudent that he often gave them food from his hand!

GENUS **CYANURA**, SWAINSON.

Cyanurus, SWAINSON, Fauna Bor. Amer. II. 1831, 495, Appendix. (Type, *Corvus cristatus*, LINNÆUS.)
Cyanocitta, CABANIS, Mus. Hein. 1851. Not of Strickland, 1845.

GEN. CHAR. Head crested. Wings and tail blue, with transverse black bars; head and back of the same color. Bill rather slender, somewhat broader than high at the base;

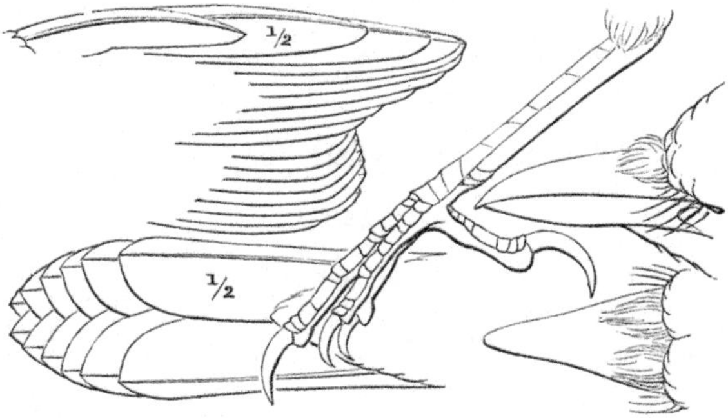

C cristata.

culmen about equal in length to the head. Nostrils large, nearly circular, concealed by bristles. Tail about as long as the wings, lengthened, graduated. Hind claw large, longer than its digit.

The Eastern blue-jay, *C. cristatus*, belongs to this genus, and is its type, and there are several others more nearly resembling ours in the Rocky Mountains, Mexico, and Central America.

Calocitta colliæi, Vigors (Gray), described and figured by Audubon as

C. cristata.

Corvus or *Pica Bullockii*, "from the Columbian River," is not found north of Mexico. The same is true of *Cyanocitta Beecheyi*, Vigors, figured in "Voyage of the Blossom," etc., as from California.

Cyanura Stelleri, Gmelin.

STELLER'S JAY.

Corvus Stelleri, Gmelin, Syst. Nat. I. 1788, 370. — *Garrulus Stelleri*, Vieillot, Dict. XII. 1817, 481. — Bonaparte, Am. Orn. II. 1828, 44; pl. xiii. — Nuttall, Man. I. 1832, 229. — Audubon, Birds Amer. IV. 1842, 107; pl. 230. — *Cyanocitta Stelleri*, Cabanis, Mus. Hein. 321. — Newberry, P. R. R. Rep. VI. iv. 85. — *Cyanurus Stelleri*, Swainson, Fauna Bor. Amer. II. 1831, 495, App. — Baird, P. R. Rep. IX. Birds, 581. — Heermann, X. vi. 55. — Cooper and Suckley, XII. iii. Zool. of W. T. 214.

Sp. Char. Crest about one third longer than the bill. Fifth quill longest; second about equal to the secondary quills. Tail graduated; lateral feathers about 0.70 of an inch shortest. Head and neck all round, and fore part of breast, dark brownish-black.

Back and lesser wing coverts blackish-brown, the scapulars glossed with blue. Under parts, rump, tail coverts, and wing, greenish-blue; exposed surfaces of lesser quills dark indigo-blue; tertials and ends of tail feathers rather obsoletely banded with black.

Feathers of the forehead streaked with greenish-blue. Length, about 12.50; extent, 18.00; wing, 6.00.

Hab. Pacific Coast of North America, to the Northern Rocky Mountains.

These birds are numerous in the mountains of California, inhabiting the whole length of the Sierra Nevada, and the Coast Range as far south as Santa Cruz at least. They show a decided preference for the coniferous forests, rarely going far from them, but sometimes in winter frequenting those of oak. Their food consists of seeds of the pines and spruces, berries, and acorns, which they crack before eating, besides insects, eggs, and any animal food they can get. They even eat potatoes in winter, and resort to the shores for dead fish. They are very noisy birds, having a variety of harsh notes, and a considerable talent for mimicry. They are sometimes very bold and prying, at others very cautious and suspicious, soon learning the effect of the gun, and showing much sagacity in their movements.

Their nests are built usually in evergreens at various heights, large, and composed of twigs and roots, with a layer of mud and a lining of root-fibres. The eggs, about four, are pale green, with small olive-brown specks, and others inclining to violet. (Nuttall.)

They lay in May near the Columbia River, and probably a month earlier in some parts of this State.

According to J. K. Lord, this species builds a nest amidst the thick foliage of a young pine-tree, of moss, small twigs, lichen, and fir leaves, lined with deer hair, in which are laid about seven bluish-green eggs. The number he gives of the eggs of several birds near the 49th parallel is so much larger than in this State, that I suspect he obtained his information from others, instead of by personal observation.

Cyanura macrolophus, Baird.

THE LONG-CRESTED JAY.

Cyanocitta macrolopha, Baird, Pr. A. N. Sc. 1854, 118. — *Cyanura macrolophus*, Baird, Birds N. Amer. 1858, 582. — Coues, Pr. A. N. S. 1866, 92.

Sp. Char. Crest nearly twice the length of the bill. Tail moderately graduated; the lateral feathers about 0.60 of an inch shorter than the middle. Fourth and fifth quills longest, second shorter than the secondaries. Head all round, throat, and fore part of the breast black, the crest with a gloss of blue; rest of back dark ashy-brown with a gloss of greenish. Under parts, rump, tail coverts, and outer surfaces of primaries, green-

ish-blue; greater coverts, secondaries, and tertials, and upper surfaces of tail feathers, bright blue, banded with black; forehead streaked with opaque-white, passing behind into pale blue; a white patch over the eye. Chin grayish. Length, 12.50; wing, 5.85; tail, 5.85; tarsus, 1.70.

Hab. Central line of Rocky Mountains to the table-lands of Mexico; Arizona.

This species is the Rocky Mountain representative of *C. Stelleri*, from which it differs mainly in the bluish-white feathers of the forehead, the white over the eye, and the longer black crest. In the northern part of the Rocky Mountains this jay has its most westward extension, and on the head-waters of the Columbia meets with *C. Stelleri*, and apparently hybridizes with it, as specimens occur there exactly intermediate in character.

The habits of *C. macrolophus* are much like those of *Stelleri*, and the nest and eggs are probably very similar.

Genus **CYANOCITTA**, Strickland.

Cyanocitta, Strickland, Annals and Mag. N. H. XV. 1845, 260. (Type, *Garrulus Californicus*, Vigors.)

Aphelocoma, Cabanis, Mus. Hein. 1851, 221. (Same type.)

Gen. Char. Head without crest. Wings and tail blue, without any bands. Back with a gray patch, different from the head. Bill about as broad as high at the base, and

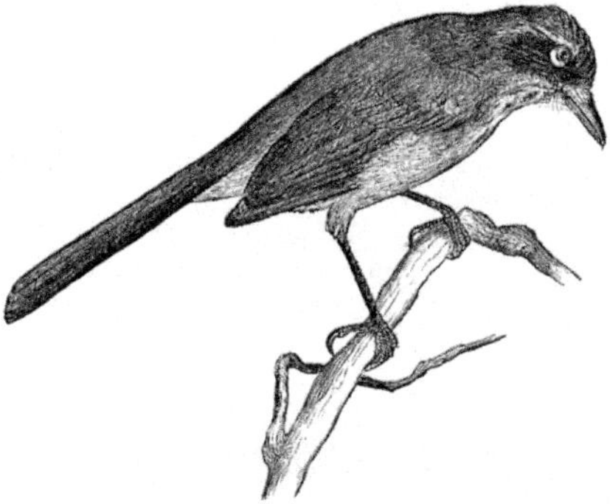

C. Californica.

the culmen a little shorter than the head. Nostrils large, nearly circular, and concealed. Tail shorter or nearly equal to the wings, lengthened, graduated.

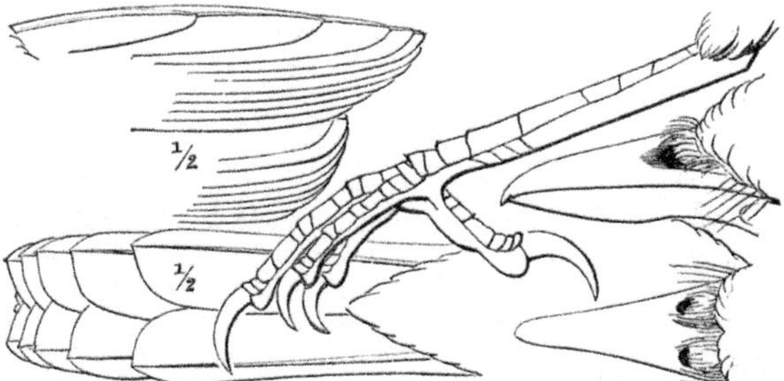

C. Californica.

Of this genus there are several species, all belonging to the southern portions of the United States and Northern Mexico; one of these, the Florida jay (*C. Floridana*), is confined at the present time to the peninsula of Florida. It is quite similar to the California jay, although smaller.

Cyanocitta Californica, Vigors.

THE CALIFORNIA JAY.

Garrulus Californicus, Vigors, Zool. Beechey's Voyage, 1839, 21 ; pl. v. — *Cyanocitta Californicus*, Strickland, Ann. Mag. XV. 342. — Gambel, J. A. N. Sc. 2d Series, I. Dec. 1847, 45. — Newberry, P. R. R. Rep. VI. iv. 1857, 85. — Baird, P. R R. Rep. IX. Birds, 554. — Kennerly, X. iv. 32. — Heermann, X. vi. 55.

Corvus ultramarinus, Audubon, Orn. Biog. IV. 1838, 456 ; pl. 362. (Not *Garrulus ultramarinus*, Bonaparte.) — *Garrulus ultramarinus*, Audubon, Syn. 1839, 154. — Audubon, Birds Amer. IV. 1842, 115 ; pl. 232. — Nuttall, Man. I. 2d ed. 245. Not of Bonaparte.

Sp. Char. Width of bill at base of lower mandible rather more than half the length of culmen. Lateral tail feathers about an inch shortest. Tail an inch longer than the wings. General color above, including the surface of the wings, bright blue, without any bars. The whole back, including to some extent the interscapulars, brownish-ash, very faintly glossed with blue in the adult. A streaked white superciliary line from a little anterior to the eye as far as the occiput. Sides of the head and neck blue, the region

around and behind the eye, including lores and most of ear coverts, black. The blue of the sides of the neck extends across the fore part of the breast, forming a crescent, interrupted in the middle. The under parts anterior to the crescent, white streaked with blue; behind it dull white; the sides tinged with brown. Length, 12.25; extent, 15.00; wing, 5.00; tail, 6.15; tarsus, 1.55. Iris brown; bill and feet black.

Hab. Pacific Coast from Columbia River south to Cape St. Lucas. Not east of the Sierra Nevada?

According to Nuttall, this species was common at Fort Vancouver in October, 1834, but it has not been seen there by recent collectors, and, according to Newberry, was not found anywhere in Oregon during his journey in 1855, its range being limited strictly by the Sacramento Valley. Nuttall found them also with newly fledged young in June, and says they were found at least as far north as Frazer's River. There is no reason to doubt his testimony, and it seems to be one of those scarcely migratory southern species which were then found north of the Columbia, but have since become nearly or quite extinct there, perhaps from the effects of uncommonly severe winters. Among these are the *Cathartes Californianus*, *Agelaius gubernator*, *Oreortyx pictus*, and other more migratory species elsewhere mentioned.

In California this jay is one of the most common and conspicuous birds, frequenting every locality where oaks grow, even close to the towns, entering gardens and audaciously pilfering fruits, etc., before the owner's eyes. They share the usual cunning of the tribe, and if alarmed become very quiet, concealing themselves in the thick foliage, so as to be found with difficulty. They are usually, however, noisy and fearless, their odd cries, grotesque actions, and bright plumage making them rather favorite guests, in spite of their petty depredations. They live chiefly on small acorns and insects, but, like other jays, are decidedly omnivorous.

Their cries are less harsh and loud than those of Steller's jay, and they have also some talent for mimicry, besides notes to express their various wants and ideas.

They build throughout the western parts of California, constructing a large and strong nest of twigs, roots, grass, etc., in a low tree or bush, and laying about five eggs, dark green, with numerous pale brown blotches and spots, measuring 1.04 × 1.80 inch. At San Diego I found eggs laid about April 5th.

These birds inhabit the Coast Range to its summits, south of San Francisco, and the Sierra Nevada, probably up to the commencement of dense pine woods and the limits of oaks, or from one thousand to five thousand feet, according to latitude. They may feed on the pine seeds, as observed by Dr. Kennerly of *C. Woodhousii* in New Mexico. I saw none on the east side of the Sierra Nevada in lat. 39°.

Their flight is slow and laborious on account of their short wings, and they never fly far at a time. They are very active, continually hopping about and jerking their long tails. They are very destructive to bird's eggs, making a business of hunting for them in the spring, and are then justly detested and attacked by birds much smaller than themselves. They watch the movements of small birds with great attention, in order to discover their nests, and if possible make a meal of the proprietor; but they are very cowardly, and never attack a bird that is prepared for them.

This bird is very abundant at Cape St. Lucas, and, like all the species resident there, is much smaller than in more northern regions; in size, indeed, not exceeding, if equalling, that of the diminutive Florida jay. The habits, however, are said to be much the same.

Cyanocitta Woodhousii, Baird.

WOODHOUSE'S JAY.

Cyanocitta Woodhousii, Baird, Birds N. Amer. 1858, 585; pl. 59. — Coues, Pr. A. N. Sc. 1866, 92.

Cyanocorax Californicus, Woodhouse, Sitgreave's Report, 1853, 77. (San Francisco Mountains.)

Sp. Char. Size and general appearance of *C. Californica*. Graduation of tail one inch. Blue, with a very obscure ashy patch on the back. Sides of the head and neck and incomplete pectoral collar blue; throat streaked with the same. Breast and belly uniform

brownish-ash glossed with blue; under tail coverts bright blue. Sides of head, including lores, black, glossed with blue below; a streaked white superciliary line. Length, 11.50; wing, 5.35; tail, 6.10; tarsus, 1.60.

Hab. Central line of Rocky Mountains to the table-lands of Mexico. Southern Arizona.

This species is distinguishable from *C. Californica* by the characters given above, especially by the grayish under parts, and the greater or less amount of blue on the under tail coverts. There is no white whatever in the under parts, except immediately around the anus, and there is a gloss of blue distinctly appreciable, especially along the middle of the body. The back is more blue, and the lores are quite black, without any hoary, as in *Californica*. According to Dr. Coues, it is very abundant in Arizona, being found everywhere, but preferring open hillsides among the scrub-oaks. In winter it collects in flocks, and, like most jays, it is shy, restless, and noisy.

Cyanocitta sordida, Swainson.

SIEBER'S JAY.

Garrulus sordidus, Swainson, Philos. Mag. June, 1827, 437. Ib. Zool. Ill. III. n. s. tab. 86. — *Aphelocoma sordida*, Cabanis, Mus. Hein. 1851, 221. — *Cyanocitta sordida*, Baird, Birds N. Amer. 1858, 587; pl. lx. f. 1. Ib. Mex. Bound. Rep. II. 21. — Coues, Pr. A. N. Sc. 1866, 92.

Pica Sieberi, Wagler, Syst. Av. 1827, No. 23.

Sp. Char. Bill short, thick; half as high as long. Wings about as long or but little longer than the tail, which is graduated 0.85 of an inch. Above and on sides of head and neck bright blue, scarcely duller in the middle of the back. Beneath white; the throat and breast tinged with very faint bluish, especially across the latter. Tibial feathers dull

bluish-ash; crissum white, the tips of posterior feathers very faintly tinged with bluish-gray. Length, 13.00; wing, 6.60; tail, 6.60; tarsus, 1.65; culmen, 1.00; height of bill at base, 0.45.

Hab. Mimbres region of Rocky Mountains, and south to table-lands of Mexico. Southern Arizona (Fort Buchanan).

This bird is found along the southern border of Arizona and New Mexico, but in what degree of abundance cannot now be stated. Nothing is known of its habits.

It is not at all improbable that another species of *Cyanocitta, C. ultramarina* (Bonaparte) (Baird, Birds N. Amer. 588) may yet be found in Arizona or New Mexico, as it has been taken very near the southern boundaries of those territories. It is allied to *C. Californica*, but has a shorter, more even tail, much longer wings, and stouter feet. There are no dark marks beneath, and the ash of the back is scarcely appreciable.

Genus **PERISOREUS**, Bonaparte.

Perisoreus, Bonaparte, Saggio di una dist. met. 1831. (Type *Corvus Canadensis*?)
Dysornithia, Swainson, Fauna Bor. Amer. II. 1831, 495. (Same type.)

Gen. Char. Feathers lax and full, especially on the back, and of very dull colors, without any blue. Head without distinct crest. Bill very short; broader than high. Culmen scarcely half the length of the head; straight to near the tip, then slightly

P. Canadensis

curved; gonys more curved than culmen. Bill notched at tip. Nostrils round, covered by bristly feathers. Tail about equal to the wings; graduated. Tarsi rather short; but little longer than the middle toe.

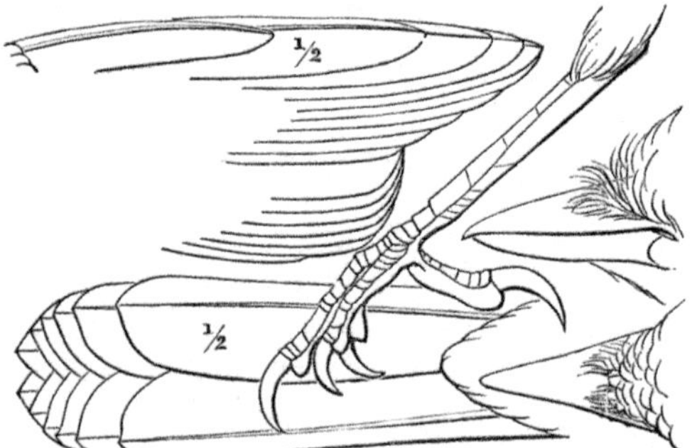

P. Canadensis.

Perisoreus Canadensis, LINNÆUS.

THE CANADA JAY.

Corvus Canadensis, LINNÆUS, Syst. Nat. I. 1766, 158. — WILSON, Am. Orn. III. 1811, 33; pl. xxi. — *Garrulus Canadensis,* BONAPARTE, (Saggio, 1831?) Syn. 1828, 58 — SWAINSON, Fauna Bor. Amer. II. 1831, 295. — NUTTALL, Man. I. 1832, 232 — AUDUBON, Birds Amer. IV. 1842, 121; pl. 234. — *Perisoreus Canadensis,* BONAPARTE, List, 1838. — NEWBERRY, Rep. P. R. R. Surv. VI. iv. 1857, 85. — BAIRD, P. R. Rep. IX. Birds, 590. — COOPER and SUCKLEY, XII. iii. Zool. of W. T. 216.

Garrulus brachyrhynchus, SWAINSON, Fauna Bor. Amer. II. 1831, 296; pl. 55 (young).

SP. CHAR. Tail graduated; lateral feathers about one inch shortest. Wings a little shorter than the tail. Head and neck and fore part of breast white. A plumbeous brown

Adult.

occipital patch, becoming darker behind, from the middle of the crown to the back, from which it is separated by an interrupted whitish collar. Rest of upper parts ashy-plumbeous; the outer primaries margined, the secondaries, tertials, and tail feathers obscurely

Young.

tipped with white. Beneath smoky gray. Crissum whitish. Length, 10.50; extent, 17 00; wing, 5.50; tail, 6.00; tarsus, 1.40. Iris brown; bill and feet black.

Hab. Northern America into the northern parts of the United States from the Atlantic to the Pacific; more south in Rocky Mountains.

Dr. Newberry found this bird common in California north of lat. 40° in autumn, and they probably descend much farther south in winter, if not resident in the highest Sierra Nevada, for nearly their whole length. I did not see them about lat. 39° in September, but have everywhere towards the north found them so scattered in their range that they may easily be overlooked in a short visit to a district, though common there at times.

They are generally rather shy birds, migrating in small families through the woods, occasionally whistling in a loud and clear tone quite unlike other jays. They have indeed much the habits and appearance of the titmice, though so much larger, searching closely among the evergreens for seeds, insects, etc., hanging head downwards and uttering a variety of quaint and musical notes. At times, especially in winter, they become very bold, entering cabins in the woods, and following hunters to obtain scraps of meat and fat of the game he may hang up, from which they have obtained one of their names of "meat-bird." They are also called in the far North "Whiskey Jack," not from any fondness for liquor, but from a corruption of their Chippewa name of *Wiskachon*. (Suckley.)

They are most numerous in the cold regions of the fur countries north of lat. 49°, and do not migrate much even in that climate. According to Hutchins, they in winter become so bold as to steal from the very dishes in the hunter's camp. They lay up stores of berries, etc., in hollow trees, and even are said to eat lichens. They are considered mockers and birds of ill omen by the Indians, and are very noisy about the commencement of storms.

Their nests are built in pine-trees of twigs and grass, and their eggs, four to six, are light grayish, with faint brown spots. (Nuttall.) The young for the first autumn are nearly as black as crows.

ORDER CLAMATORES.

CHAR. Toes, three anterior and one posterior, not versatile. Tail feathers usually twelve. Primaries always ten, the first nearly as long as the second. Tarsal scales generally passing entirely around.

But one family of this order is found within our limits, though there are numerous others in the tropics, presenting great variety of form.

FAMILY TYRANNIDÆ, THE TYRANT FLYCATCHERS.

CHAR. Bill broader than high at the base, much depressed, more or less triangular. Culmen nearly as long as the head or shorter; straight to near the tip, then suddenly bent down into a conspicuous hook, with a notch behind it; tip of lower jaw also notched. Commissure straight to near the notch; gonys slightly convex. Nostrils oval or rounded, in the anterior extremity of the nasal groove, and more or less concealed by long bristles which extend from the posterior angle of the jaws along the base of the bill, becoming smaller, but reaching nearly to the median line of the forehead. These bristles with lateral branches at the base. Similar bristles mixed in the loral feathers and margining the chin. Tarsi short, generally less than the middle toe, completely enveloped by a series of large scales, which meet near the posterior edge of the inner side, and are separated either by naked skin or by a row of small scales. Sometimes a second series of rather large plates is seen on the posterior face of the tarsus, these, however, usually on the upper extremity only. Basal joint of middle toe united almost throughout to that of the outer toe, but more than half free on the inner side; outer lateral toe rather the longer. Wings and tail variable; first quill always more than three fourths the second. The outer primaries sometimes attenuated near the tip.

Genus TYRANNUS, Cuvier.

Tyrannus, Cuvier, Leçons Anat. Comp. 1799, 1800 (Agassiz).

Gen. Char. Tail nearly even, or moderately forked; rather shorter than the wings; the feathers broad, and widening somewhat at the ends. Wings long and pointed; the

T. Carolinensis.

T. Carolinensis.

outer primaries rather abruptly attenuated near the end, the attenuated portion not linear, however. Head with a concealed patch of red on the crown.

This is exclusively an American genus, of which there are many species in warm and tropical regions.

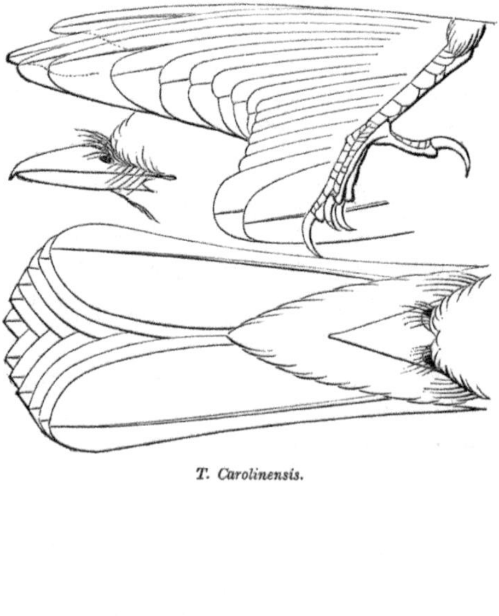

T. Carolinensis.

Tyrannus Carolinensis, BAIRD.

THE KING-BIRD; BEE-MARTIN.

Lanius tyrannus, LINNÆUS, Syst. Nat. I. 1766, 136. (This belongs to the Cuban *T. matutinus*, according to Bonaparte.) — *Muscicapa tyrannus*, (BRISSON ?) WILSON, Am. Orn. I. 1808, 66; pl. xiii. — AUDUBON, Orn. Biog. I. 1832, 403; V. 1839, 420; pl. 79. IB. Birds Amer. I. 1840, 204; pl. 56. — *Lanius tyrannus*, var. *Carolinensis*, *Ludovicianus*, GMELIN, Syst. Nat. I. 1788, 302.

Muscicapa rex, BARTON, Fragments, N. H. Penna. 1799, 18.

Tyrannus Carolinensis, BAIRD, P. R. Rep. IX. Birds, 171. — COOPER and SUCKLEY, XII. iii. Zool. of W. T. 167. — LORD, Pr. R. A. Inst. Woolwich, 1864, 113. British Columbia.

SP. CHAR. Two, sometimes three, outer primaries abruptly attenuated at the end. Second quill longest; third little shorter; first rather longer than fourth, or nearly equal. Tail slightly rounded. Above dark bluish-ash, the top and sides of the head to beneath the eyes bluish-black. A concealed crest on the crown, vermilion in the centre, white behind, and before partially mixed with orange. Lower parts pure white, tinged with pale bluish-ash on the sides of the throat and across the breast; sides of the breast and

under the wings similar to, but rather lighter than, the back. Axillaries pale grayish-brown tipped with lighter. The wings dark brown, darkest towards the ends of the quills; the greater coverts and quills edged with white, most so on the tertials; the lesser coverts edged with paler. Upper tail coverts and upper surface of the tail glossy black, the latter very dark brown beneath; all the feathers tipped and the exterior margined externally with white, forming a conspicuous terminal band about 0.25 of an inch broad. Length, 8.75; extent, 15.00; wing, 4.65; tail, 3.70; tarsus, 0.75. Iris brown; bill and feet black.

Hab. Eastern North America to Rocky Mountains. West of this seen only in Washington Territory. California? South to Panama and the Upper Amazon, in winter.

Although this bird has never yet been collected in California, it is almost impossible that one of its migratory habits should be abundant at the

Columbia River, as I found it, without passing through California in the spring and fall. It is scarcely probable that they prefer a route of travel through the barren regions towards the Rocky Mountains, when they can find abundance of food on the route through California, and a country quite congenial to their habits. I have found them, however, all along the Upper Missouri, and across the Rocky Mountains northward.

Their habits are well known to settlers from the Atlantic States. Their name of king-bird is derived from their quarrelsome disposition during the breeding season, when they drive every bird, even to the size of the eagle, away from their nest. This is built on a low branch, often in the garden or orchard, constructed firmly of roots, grass, and a lining of soft materials, horse-hair, etc. The eggs are three to five, yellowish-white, and with a few large well-defined brown spots. Their notes consist of harsh guttural twitters and chirps, not worthy to be called a song. Their food is almost wholly of insects, for which they watch from some prominent perch; also of berries, and they sometimes watch around hives for bees. They leave the Atlantic States in October, to winter in Tropical America. (Nuttall.)

Tyrannus verticalis, Say.

THE ARKANSAS FLYCATCHER.

Tyrannus verticalis, Say, Long's Exped. II. 1823, 60. — Nuttall, Man. II. 2d ed. 1840, 306. Baird, P. R. Rep. IX. Birds, 173. — Heermann, X. vi. 37. — Cooper and Suckley, XII. iii. Zool. of W. T. 168. — Lord, Pr. R. A. Inst. Woolwich, 1864, 113. British Columbia. — *Muscicapa verticalis*, Bonaparte, Am. Orn. I. 1825, 18; pl. xi. — Audubon, Orn. Biog. IV. 1838, 422; pl. 359. Ib. Birds Amer. I. 1840, 199; pl. 54.

Sp. Char. The four exterior quills attenuated very gently at the end, the first most so; third and fourth quill longest, second and fifth successively a little shorter. Tail slightly forked; bill shorter than the head. Crown, sides of head above the eyes, nape, and sides of neck, pale lead-color or ash-gray; a concealed crest in the crown, vermilion in the centre, and yellowish before and behind. Hind neck and back ashy-gray, strongly tinged with light olivaceous-green, the gray turning to brown on the rump; upper tail coverts nearly black, lower dusky; chin and part of ear coverts dull white; throat and upper part of breast similar to head, but lighter, and but slightly contrasted with the chin; rest of the lower parts, with the under wing coverts and axillars, yellow deepening to gamboge on the belly, tinged with olivaceous on the breast. Wing brown, the coverts with indistinct ashy margins; secondaries and tertials edged with whitish; inner webs of primaries whitish towards the base. Tail nearly black above, and glossy, duller brownish beneath; without olivaceous edgings. Exterior feather, with the outer web and the shaft, yellowish-white; inner edge of latter brown. Tips of remaining feathers paler. Bill and feet dark brown.

Female rather smaller and colors less bright. Length of male, 8.50; extent, 13.25; wing, about 4.50.

Hab. Western North America, from the High Central Plains to the Pacific. Accidental in the Eastern States.

This beautiful and abundant species arrives in the State about March 20th, and I think none remain within our limits during winter. Small parties of males come first, which are constantly quarrelling until the victors have decided the contest and selected their mates. This does not seem to take place for several weeks, as I have not found a nest with eggs earlier than May 12th at Santa Barbara. This nest was built on a branch of a low oak near the town, was five inches wide, and constructed of lichens, twigs,

coarse grass, and wool, lined with hair, and the four eggs it contained were creamy-white spotted with purple of two shades near large end, measuring 0.94 × 0.70 inch. They exhibit similar courage in defence of their nest, and are in fact almost the exact counterpart of the king-bird in habits. Their notes are rather more varied and noisy, and they utter them almost constantly during the spring, often when flying and fighting. As "bee-birds" they are fully as destructive, but compensate for the damage by destroying great numbers of noxious insects, and therefore ought to be driven away from the neighborhood of the hives instead of shot. They leave the State, I think, in October. With this species as with the king-bird, it is not uncommonly the case that smaller birds take advantage of its courageous behavior in driving away predatory birds from its nest, by constructing their own in the same neighborhood, so as to enjoy a certain amount of protection. This reliance of one bird on another is shown still more strikingly in regard to a Jamaican species of allied family, the *Hadrostomus niger*, of about the size of the king-bird, which builds a nest as large as a bushel measure, in the substance of which half a dozen birds of other species sometimes make their nest likewise.

Tyrannus vociferans, Swainson.

CASSIN'S FLYCATCHER.

Tyrannus vociferans, Swainson, Mon. Tyrant Shrikes in Quarterly Journal Sc. XX. Jan. 1826, 273. Ib. Philos. Mag. I. 1827, 368. — Baird, P. R. Rep. IX. Birds, 174; U. S. and Mex. Bound. Rep. II. iii. 8; pl. x. — Coues, Pr. Phil. Ac. 1866, 59.
Tyrannus Cassinii, Lawrence, Ann. N. Y. Lyceum, V. 1852, 39; pl. iii. f. 2. (Texas.)

Sp. Char. Bill from the forehead, about as long as the head. Tail even or slightly rounded. Outer five primaries attenuated; the first four abruptly and deeply emarginated; third quill longest, second and fourth a little less, first shorter than the sixth, and half an inch less than the longest. Head and neck above and on the sides rather dark bluish-ash; the throat and breast similar, and only a little paler. Rest of upper parts olive-green tinged with gray, mixed with brown on the rump; the upper tail coverts and surface of the tail nearly black; the outer web of the external feather and tips of all pale

brown. The chin is white, in strong contrast to the dark ash of the throat; the rest of the under parts bright sulphur yellow (the sides olivaceous), palest on the under tail coverts and inside of wings. A concealed vermilion patch in the crown, bordered by straw yellow. Wing feathers brown, tinged with olive, becoming paler towards the edge. Length, 9.00 to 10.00; extent, 15.00 to 17.00; wing, 4.25 to 5.50; tail, 4.25. Iris brown; bill black; feet grayish.

Hab. Pecos River, Texas, and into Mexico, and Guatemala on table-lands. West to Coast of California.

Quite common in the southern half of California, and resident as far north as Los Angeles during winter. They much resemble the *T. verticalis* in color, but are less lively and quarrelsome in habits. During the early part of the year they begin to sing at daylight, generally from the top of a sycamore-tree, their notes being loud and much more musical than those of the other species, with considerable variety for a bird of this family. During the middle of the day they are rather quiet, and sit much of the time on their perch, occasionally catching an insect that comes very near, but I think their feeding is done mostly in the very early morning.

I found them breeding at San Diego as early as March 28th, as well as subsequently. The nest is much larger and more firmly built than that of the others, being 5.50 × 2.50 inches externally, and 3.00 across the cavity, the eggs 0.96 × 0.70 inch, white with large scattered reddish-brown and umber blotches. I found some of these birds in Santa Clara Valley in May, 1864, which were smaller and greener on the back than those from the south.

They winter in small numbers at Santa Cruz, lat. 37°, together with *Troglodytes Parkmanni* and *Hirundo bicolor*, indicating a much milder climate than anywhere else in the State.

Genus MYIARCHUS, Cabanis.

Myiarchus, Cabanis, Fauna Peruana, 1844-1846, 152. — Burmeister, Thiere Brasiliens, II. Vögel, 1856, 469.

Gen. Char. Tarsus equal to or not longer than the middle toe, which is decidedly longer than the hinder one. Bill wider at base than half the culmen. Tail broad, long,

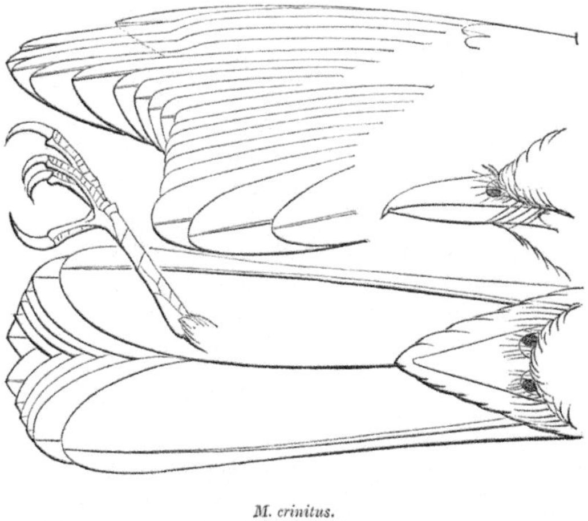

M. crinitus.

even, or slightly rounded, about equal to the wings, which scarcely reach the middle of the tail; the first primary shorter than the sixth. Head with elongated lanceolate dis-

M. crinitus.

tinct feathers. Above brownish-olive, throat ash, belly yellow. Tail and wing feathers varied with rufous

Myiarchus Mexicanus, KAUP.

THE ASH-THROATED FLYCATCHER.

Tyrannula Mexicana, KAUP, Pr. Zool. Soc. Feb. 1851, 51.—*Myiarchus Mexicanus*, BAIRD, P. R. Rep. IX. Birds, 179.—HEERMANN, X. vi. 37; pl. iv.—COUES, Pr. Phil. Ac. 1866, 59.

Tyrannula cinerascens, LAWRENCE, Annals N. Y. Lyc. N. Hist. V. Sept. 1851, 109.—NEWBERRY, P. R. Rep. VI. iv. 81.

SP. CHAR. Bill black, the width opposite the nostrils not half the length of the culmen. Head crested. Tail even, the lateral feathers slightly shorter. Second, third, and fourth quills longest; first rather shorter than the seventh. Above a dull grayish-olive; the

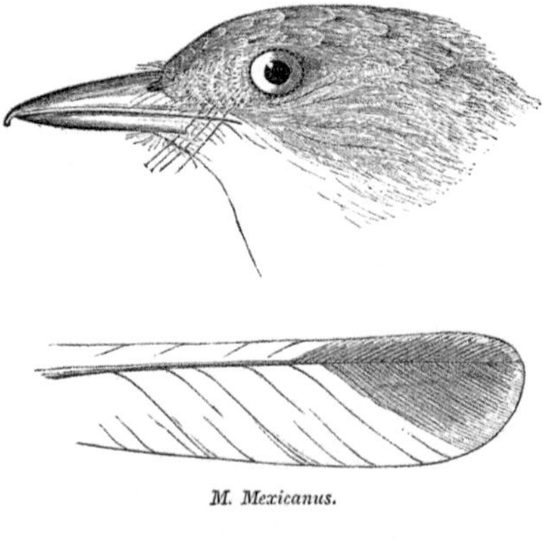

M. Mexicanus.

centres of the feathers rather darker; the crown, rump, and upper tail coverts tinged with brownish. The forehead and sides of the head grayish-ash; the chin, throat, and fore part of the breast ashy-white; the middle of the breast white; the rest of the under parts very pale sulphur yellow; wings and tail brown. Two bands across the wing, with outer edges of the secondaries and tertials dull white; the outer edges of the primaries light chestnut brown (except towards the tip and on the outer feather); the inner edges tinged with the same. Whole of the middle tail feathers, with the outer webs (only), and the ends of the others brown; the rest of the inner webs reddish-chestnut, the outer web of exterior feather yellowish-white. Legs and bill black; lower mandible brownish at the base. Length, about 8.00; wing, 4.00; tail, 4.10; tarsus, 90.

Hab. Coast of California, and across by Valley of Gila and Rio Grande to Northeastern Mexico, and south to Guatemala.

I found one of this species at Fort Mojave on January 15th, and think a few may habitually winter in the Colorado Valley. They began to arrive from the south about March 10th, and extend their range through nearly the whole State or its lower portions. Their notes are few, loud, and harsh, little varied, and uttered from time to time as they fly after an insect from their accustomed perch (which is usually on the lower dead limbs of the forest trees), preferring shady situations, and feeding late in the evening. According to Heermann, the nest, found in a hollow tree, or a deserted burrow of a woodpecker, is composed of grasses lined with feathers. The eggs, five in number, are cream-color, marked and speckled with purplish-red dashes and faint neutral tint blotches.

M. crinitus.

A variety (*Myiarchus Mexicanus*, var. *pertinax*, Baird, Pr. Phil. Acad. 1859, 303) of this species occurs at St. Lucas, where it was collected by Mr. Xantus, characterized by the possession of a decidedly stouter bill and stronger feet than usual. The size of the bird itself is perhaps less than the average.

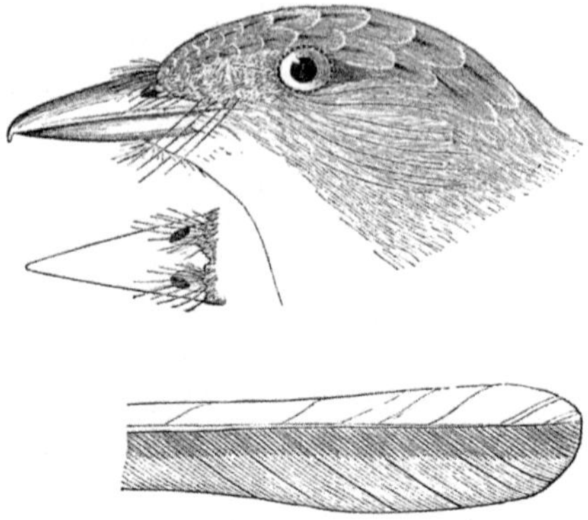

M. Cooperi.

This species is closely allied to the *M. crinitus* of the Eastern States, and *M. Cooperi* of Mexico; the difference will be shown in the preceding figures. The principal characters will be found in the markings of the tail.

GENUS **SAYORNIS**, BONAPARTE.

Sayornis, BONAPARTE? Ateneo Italiano, 1854. IB. Comptes Rendus, 1854; Notes Orn. Delattre.

Aulanax, CABANIS, Journal für Orn. 1856, 1. (Type, *Tyrannula nigricans.*)

GEN. CHAR. Head with a blended depressed moderate crest. Tarsus decidedly longer than middle toe, which is scarcely longer than the hind toe. Bill rather narrow; width at

S. nigricans.

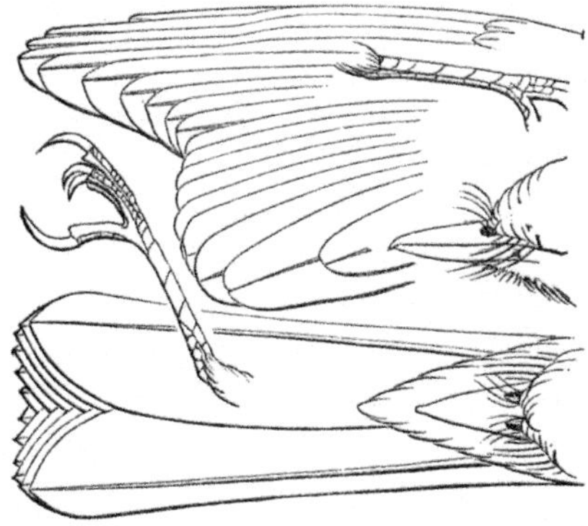

S. nigricans.

base about half the culmen. Tail broad, long, slightly forked; equal to the wings, which are moderately pointed, and reach the middle of the tail. First primary shorter than the sixth.

Sayornis nigricans, Swainson.

THE BLACK FLYCATCHER.

Tyrannula nigricans, Swainson, Syn. Birds Mex. Taylor's Phil. Mag. I. 1827, 367. — Newberry, Zool. Cal. and Or. Route, P. R. Rep. VI. iv. 1857, 81. — *Muscicapa nigricans*, Audubon, Orn. Biog. V. 1839, 302; pl. 474. Ib. Birds Amer. I. 1840, 218; pl. 60. — *Tyrannus nigricans*, Nuttall, Man. I. 2d ed. 1840, 326. — *Sayornis nigricans*, Bonaparte, Comptes Rendus, XXVIII. 1854, notes Orn. 87. — Baird, P. R. Rep. IX. Birds, 183. — Kennerly, X. iv. 23. — Heermann, X. vi. 38. — Coues, Pr. Phil. Ac. 1866, 60.

Sp. Char. Wings rounded; second, third, and fourth quills longest; first shorter than sixth. Tarsi with a second row of scales behind. The head and neck all round,

fore part and sides of the breast dark sooty-brown; the rest of the upper parts similar, but lighter; faintly tinged with lead-color towards the tail. The middle of the breast, abdomen, and lower tail coverts white; some of the latter with the shafts and the centres brown. The lower wing coverts grayish-brown, edged with white. Wings dark brown; the edges of the secondary coverts rather lighter; of primary coverts dull white. Edge of the exterior vane of the first primary, and of the secondaries, white. Tail dark brown, with the greater part of the outer vane of the exterior tail feather white; this color narrowing from the base to the tip. Bill and feet black. The tail rounded; rather emarginate; feathers broad; more obliquely truncate than in *S. sayus*. The bill slender; similar to that of *S. fuscus*. Length, nearly 7.00; wing, 3.60; tail, 3.45.

Hab. California to Texas, and south into Mexico.

An abundant and resident species in all the lower parts of California, except the Colorado Valley, where I saw none later than March 25, 1861, as they had gone north. At San Diego they had nests and eggs at that date (1862), and are elsewhere among the earliest birds to build. The nest is formed of an outer wall of mud, about 5.25 inches wide, and 3.50 high, in little pellets piled on successively as they dry, in the shape of half a cup, stuck against a wall or sometimes on a shelf, beam, or ledge of rock, but always under some protecting roof, often under a bridge. It is lined with fine grass or moss, and horse or cow hair, and the eggs, four or five in number, are pure white; measuring 0.74 × 0.55 inch. They prefer the vicinity of human habitations, and also to keep around water, on account of the numerous flies they find in such situations.

They often sit for hours on the end of a barn or other perch, uttering their monotonous but not unpleasing ditty, which sounds like "Pittic, pittit," alternately repeated, much like the cry of the Eastern "Pewee" or "Phœbe-bird" (*S. fuscus*), which is their exact analogue in habits. They fly only short distances at a time, turning and dodging quickly in pursuit of their prey, which they capture with a sharp snap of the bill.

Sayornis Sayus, Bonaparte.

SAY'S FLYCATCHER.

Muscicapa Saya, Bonaparte, Am. Orn. I. 1825, 20; pl. xi. f. 3. — Audubon, Orn. Biog. IV. 1838, 428; pl. 359. Ib. Birds Amer. I. 1840, 217; pl. 59. — *Tyrannus Saya*, Nuttall, Man. I. 2d ed. 1840, 311. — Maxim. Cab. Jour. 1858, 183. — *Tyrannula Saya*, Bonaparte, Conspectus, 1850. — Newberry, P. R. Rep. VI. iv. 81. — *Sayornis Sayus*, Baird, P. R. Rep. IX. Birds, 185. — Kennerly, X. iv. 24. — Heermann, X. vi. 37. — Coues, Pr. Phil. Ac. 1866, 60.

Sp. Char. Above and on the sides of the head, neck, and breast, grayish-brown, darker on the crown; region about the eye dusky. The chin, throat, and upper part of the breast, similar to the back, but rather lighter and tinged with the color of the rest of the lower parts, which are pale cinnamon. Under wing coverts pale rusty white. The wings of a rather deeper tint than the back, with the exterior vanes and the tips of the quills darker. Edges of the greater and secondary coverts, of the outer vane of the outer primary, and of the secondaries and tertials, dull white. The upper tail coverts and tail nearly black. Edge of outer vane of exterior tail feather white. Bill dark brown; rather

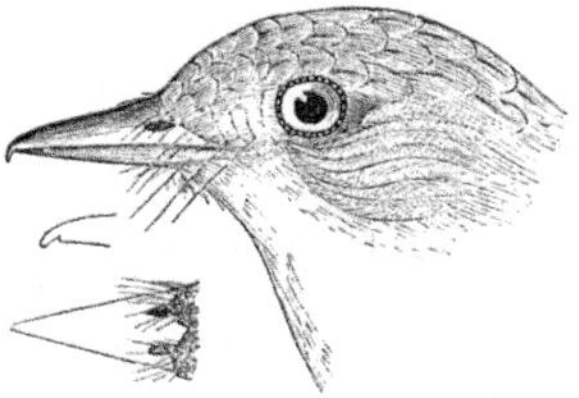

paler beneath. The feet brown. Second, third, and fourth quills nearly equal; fifth nearly equal to sixth; sixth much shorter than the first. Tail broad, emarginate. Tarsi with a posterior row of scales. Length, 7.00; wing, 4.30; tail, 3.35.

Hab. Missouri and central high plains westward to the Pacific, and south to Mexico.

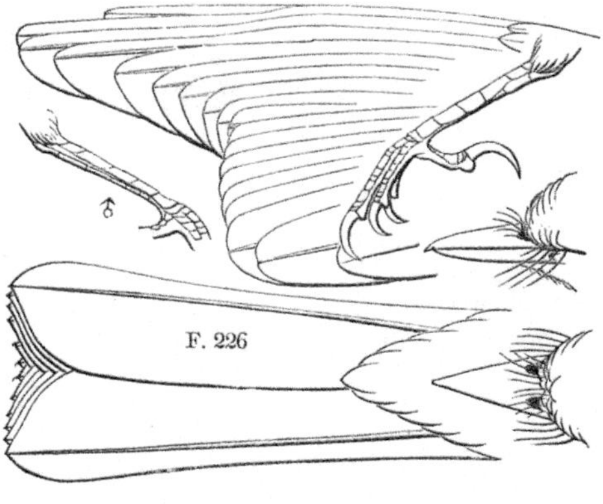

F. 226

This seems to be mostly a winter visitor in the southern and western parts of California, where I have seen none after March. They migrate in summer throughout the great interior plains, and, according to Richardson, as far north as lat. 60° in the Hudson's Bay region, where they seem to take the place of the other two species. Their nest was found there in a tree, resembling that of the *S. nigricans*. They probably, however, build more frequently among the cliffs, which are the only suitable situations in the barren, treeless regions they inhabit. Their habits are similar to those of the other species, and the few notes they utter while with us are also similar.

I first noticed the arrival of this bird from the north at the end of September, near Los Angeles, but it probably arrives much earlier in the northern part of the State, if indeed it does not breed there, east of the Sierra.

GENUS CONTOPUS, CABANIS.

Contopus, CABANIS, Journal für Orn. III. Nov. 1855, 479. (Type, *Muscicapa virens*, L.)

GEN. CHAR. Tarsus very short, but stout; less than the middle toe and scarcely longer than the hinder. Bill quite broad at the base; wider than half the culmen. Tail mod-

C. borealis.

erately forked, much shorter than the wing (rather more than three fourths). Wings very long and much pointed, reaching beyond the middle of the tail; the first primary

C. borealis.

about equal to the fourth. All the primaries slender and rather acute, but not attenuated. Head moderately crested. Color, olive above, pale yellowish beneath, with a darker patch on the sides of the breast. Under tail coverts streaked.

There is one Eastern species not found on this coast, though it has been attributed to it (*C. virens*). It differs from *C. Richardsonii* in smaller size, outer primary edged with whitish, the breast paler than the back, and with a median white line.

Contopus borealis, Swainson.

THE OLIVE-SIDED FLYCATCHER.

Tyrannus borealis, Sw. and Rich., Fauna Bor. Amer. II. 1831, 141; plate.
Muscicapa inornata, Nuttall, Man. I. 1832, 282.
Muscicapa Cooperi, Nuttall, Man. I. 1832, 282. — Audubon, Orn. Biog. II. 1834, 422; V. 1839, 422; pl. 174. Ib. Birds Amer. I. 1840, 212; pl. 58. — *Tyrannus Cooperi*, Bonaparte, List, 1838. — Nuttall, Man. I. 2d ed. 1840, 298. — *Contopus Cooperi*, Cabanis, Jour. für Ornithol. III. Nov. 1855, 479.
Contopus borealis, Baird, P. R. Rep. IX. Birds, 188. — Cooper and Suckley, XII. iii. Zool. of W. T. 169.
Contopus mesoleucus, Sclater, Pr. Z. S. 1859, 43. (Mexico.)

Sp. Char. Wings long, much pointed; the second quill longest; the first longer than the third. Tail deeply forked. Tarsi short. The upper parts ashy brown, showing darker brown centres of the feathers; this is eminently the case on the top of the head; the sides of the head and neck, of the breast and body resembling the back, but with the edges of the feathers tinged with gray, leaving a darker central streak. The chin, throat, narrow line down the middle of the breast and body, abdomen, and lower tail coverts

white, or sometimes with a faint tinge of yellow. The lower tail coverts somewhat streaked with brown in the centre. On each side of the rump, generally concealed by the wings, is an elongated bunch of white silky feathers. The wings and tail very dark brown, the former with the edges of the secondaries and tertials edged with dull white. The lower wing coverts and axillaries grayish-brown. The tips of the primaries and tail feathers rather paler. Feet and upper mandible black, lower mandible brown. The young of the year similar, but the color duller; the feet light brown. Length, 7.25; extent, 12.75; wing, 4.25; tail, 3.30; tarsus, 0.60. Iris brown; bill black, yellow at base; feet black.

Hab. Atlantic and Pacific Coasts of the United States, and interior towards the north. Greenland (Reinhardt). South to Costa Rica.

This is a Northern flycatcher, which seems to be resident in most parts of this State where it is found, but not occurring south of Monterey, where Dr. Gambel found young in July. I found them rather common in the Coast Range toward Santa Cruz, where they had nests in May, but I could not examine any of these, their location being generally on a high inaccessible branch. I also found this bird at Lake Tahoe in September. It is rather silent, keeping mostly on tops of the trees, and catching passing insects. In spring it utters a loud, monotonous chirping note, frequently repeated, sometimes for hours together, and sometimes a call resembling "phēe-bēē." Nuttall found a nest, about fifty feet from the ground, in Massachusetts, built much like that of the king-bird. The eggs were four, yellowish cream-color, with dark brown and lavender purple spots thinly dispersed.

It is a much more abundant bird near the Columbia River and throughout the Northern Rocky Mountains. There they migrate, remaining at the Lower Columbia from May to October.

Contopus pertinax, Cabanis.

COUES FLYCATCHER.

Contopus pertinax, Cabanis, Mus. Hein. II. 1859, 72. (Mexico.) — Coues, Pr. Phil. Acad. 1866, 60. (Fort Whipple.) — Elliot, B. N. A. I.; pl. 8.

Sp. Char. Second, third, and fourth quills nearly equal; first intermediate between fifth and sixth. Tail slightly forked. Upper parts uniform olive-green, rather darker on the

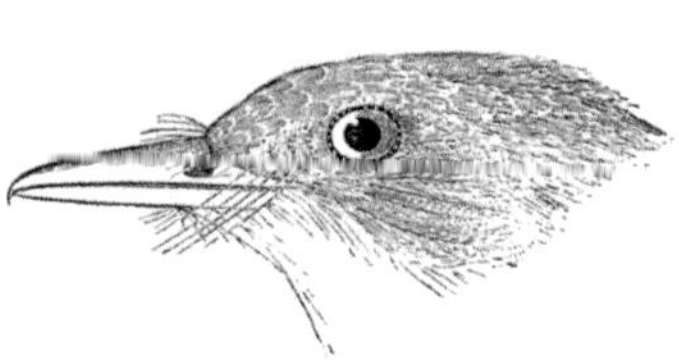

head. Beneath, the olive is a little lighter, and covers the whole under surfaces in slightly varying shade, except the chin, middle of belly, and crissum. Axillars, and inside and edge of wings, yellowish with tinge of buff. Middle and greater wing coverts rather paler at ends, without forming a distinct band. Secondaries edged externally with lighter olive. A concealed white tuft on flanks. Bill black above; yellow beneath. Young with two pale buff bands on the wings. Length, about 7.50; wing, 4.40; tail, 4.00; bill above, 0.70; tarsus, 0.70; middle toe and claw, 0.65.

Hab. Mexico to Fort Whipple, Arizona.

This species was first detected within the limits of the United States by Dr. Coues, who found a young bird at Fort Whipple. He noticed nothing peculiar in its habits. Of much the same size as *C. borealis*, the tail is disproportionately longer, and the wing formula and colors quite different. The species is not uncommon in Mexico, and will probably be yet found more abundantly in the United States, both in Arizona and New Mexico, as its inconspicuous plumage renders it very liable to be overlooked by collectors.

Contopus Richardsonii, Swainson.

THE SHORT-LEGGED PEWEE.

Tyrannula Richardsonii, Swainson, Fauna Bor. Amer. II. 1831, 146; plate. — *Muscicapa Richardsonii*, Audubon, Orn. Biog. V. 1839, 299; pl. 434. — *Contopus Richardsonii*, Baird, P. R. Rep. IX. Birds, 189. — Coues, Pr. Ph. Ac. 1866, 61.

Muscicapa phœbe, Audubon, Syn. 1839, 42. Ib. Birds Amer. I. 1840, 219; pl. 61 (not of Latham). — *Tyrannus phœbe*, Nuttall, Man. I. 2d ed. 1840, 319.

Contopus sordidulus, Sclater.

Contopus plebeius, Cabanis.

Sp. Char General appearance of *C. virens.* Bill broad. Wings very long and much pointed, considerably exceeding the tail; second quill longest; third a little shorter; first shorter than fourth, and about midway between distance from second to fifth (0.60 of an inch). Primaries 1.20 inches longer than secondaries. Tail moderately forked. Above, dark olive-brown (the head darker), the entire breast and sides of head, neck, and body

of a paler shade of the same, tinging strongly also the dull whitish throat and chin. Abdomen and under tail coverts dirty pale yellowish. Quills and tail dark blackish-brown; the secondaries narrowly, the tertials more broadly edged with whitish. Two quite indistinct bands of brownish-white across the wings. Lower mandible yellow; the tip brown. Length, 6.20; extent, 10.50; wing, 3.50; tail, 3.10. Iris brown; bill black, brown below; feet black.

Hab. High central plains to the Pacific; Rio Grande Valley, southward to Mexico and Costa Rica; Labrador (Audubon).

This bird arrives in California from the south about April 15th, and spends the summer in most mountainous parts of the State. It perches mainly on lower dead limbs, and watches for passing insects, uttering occasionally a plaintive "pé-āh," but is usually very silent, and prefers the dark, solitary recesses of the forests.

I have not discovered their nest, but Audubon describes one found in Labrador as built in a bush, of a large size, chiefly composed of dry moss, and lined with feathers, being almost suspended like that of an oriole. The eggs were five to seven, white, and minutely speckled nearly all over with brown.

If Audubon made no mistake in this, the nest was not like that of most birds of this family, but may have been modified to suit the cold climate of Labrador.

GENUS **EMPIDONAX**, CABANIS.

Empidonax, CABANIS, Jour. für Ornithologie, III. Nov. 1855, 480. (Type, *Tyrannula pusilla.*)

Tyrannula of most authors.

GEN. CHAR. Tarsus lengthened, considerably longer than the middle toe, which is decidedly longer than the hind toe. Bill variable. Tail very slightly forked, even or

E. pusillus.

rounded; a little shorter only than the wings, which are considerably rounded; the first primary much shorter than the fourth. Head moderately crested. *Color*, olivaceous above, yellowish beneath; throat generally gray.

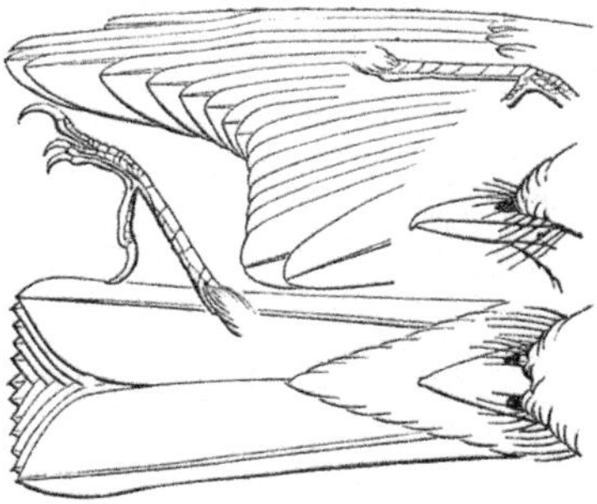

E. acadicus.

Empidonax Traillii, Audubon.

TRAILL'S FLYCATCHER.

Muscicapa Traillii, Audubon, Orn. Biog. I. 1832, 236; V. 1839, 426; pl. 45. Ib. Birds Amer. I. 1840, 234; pl. 65. — *Tyrannus Traillii*, Nuttall, Man. I. 2d ed. 1840, 323. — *Empidonax Traillii*, Baird, P. R. Rep. IX. Birds, 193.

Sp. Char. Third quill longest; second scarcely shorter than fourth; first shorter than fifth, about 0.35 shorter than the longest. Primaries about 0.75 of an inch longer than the secondaries. Tail even. Upper parts dark olive-green; lighter under the wings, and duller and more tinged with ash on nape and sides of the neck. Centre of the crown feathers brown. A pale yellowish-white ring (in some specimens altogether white) round

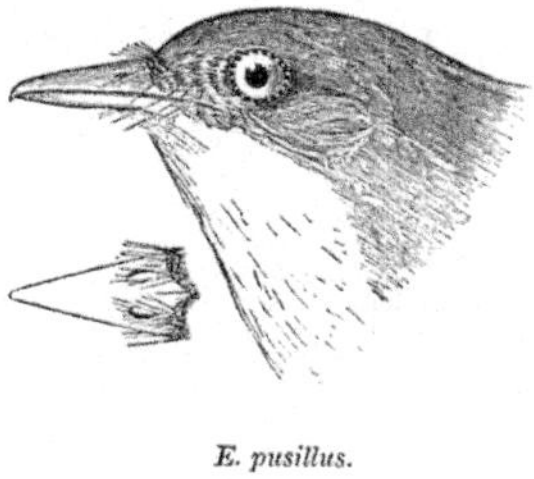

E. pusillus.

the eye. Loral feathers mixed with white. Chin and throat white; the breast and sides of throat light ash, tinged with olive, its intensity varying in individuals, the former sometimes faintly tinged with olive. Sides of the breast much like the back. Middle of the belly nearly white; sides of the belly, abdomen, and the lower tail coverts sulphur-yellow. The quills and tail feathers dark brown. Two olivaceous yellow white bands on the wing, formed by the tips of the first and second coverts, succeeded by a brown one; the edge of the first primary and of the secondaries and tertials a little lighter shade of the same.

The outer edge of the tail feathers like the back; that of the lateral one rather lighter. Length, 6.00; extent, 8.50; wing, 2.90; tail, 2.60. Iris brown; bill black, pale brown below; feet black.

Hab. Eastern United States and south to Mexico. Colorado River, California.

Although Nuttall thought he saw this species on the banks of the Columbia, no specimens were obtained west of the Rocky Mountains by any of the later explorers.

At Fort Mojave on the first of May, I found several of them inhabiting a very dark dense thicket, being attracted by their note, which sounded like *quèt-quéah.* They were very shy; and though afterwards more common there, I obtained but three. They differed from the description given by Baird only in having the sides of body *more* yellow than the back. I afterwards heard their peculiar note in the thickets along Mojave River, near Los Angeles, and in May, 1863, at Santa Barbara.

Heerman, though he found no specimens in the State, mentions it as "abundant," but whether in California or Texas is uncertain.

[I do not agree with Dr. Cooper in considering the California species to be *E. Traillii*, as all the specimens examined by me from the Middle and Western Provinces appear to belong to *E. pusillus.* This differs in having the tarsus decidedly longer than middle toe; the first quill shorter than sixth, not longer; the color is also appreciably distinct. The synonymy is as follows.* S. F. B.]

Empidonax flaviventris, Baird.

THE YELLOW-BELLIED FLYCATCHER.

Tyrannula flaviventris, Wm. M. and S. F. Baird, Pr. Ac. Nat. Sc. Phila. I. July, 1843, 283. Ib. Am. Journ. Science, April, 1844. — Audubon, Birds Amer. IV. 1844, 341; pl. 490.

Tyrannula pusilla (Swainson), Reinhardt, Vidensk. Meddel. for 1853, 1854, 82. — Gloger, Cab. Journ. 1854, 426.

Empidonax flaviventris, Baird, P. R. Rep. IX. Birds, 198.

Empidonax difficilis, Baird (provisional name for Western specimens), Birds N. Amer. pl. 76, f. 2. — Cooper and Suckley, XII. iii. errata, p. xv. Zool. of W. T. 170.

Sp. Char. Second, third, and fourth quills nearly equal; first intermediate between fifth and sixth. Tail nearly even, slightly rounded. Tarsi long. Above bright olive-green; crown rather darker. A broad yellow ring round the eye. The sides of the head, neck, breast, and body, and a band across the breast like the back, but lighter; the rest

* Empidonax pusillus. — ? *Platyrhynchus pusillus*, Swainson, Phil. Mag. 1827, 366. — *Tyrannula pusilla*, Swainson, F. Bor. Am. II. 1831, 144; plate. — *Muscicapa pusilla*, Audubon, Orn. Biog. V. 1839, 288; pl. 434. — *Tyrannus pusilla*, Nuttall, Man. I. 2d ed. 1840. — *Empidonax pusillus*, Baird, Birds N. Amer. 1858, 194. — Coues, Pr. Phil. Acad. 1866, 61.

Hab. Middle and Western Provinces of United States; north to Saskatchewan.

of the lower parts bright sulphur-yellow; no white or ashy anywhere on the body. Quills dark brown; two bands on the wing formed by the tips of the primary and secondary coverts, the outer edge of the first primary and of the secondaries and tertials pale yellow, or greenish-yellow. The tail feathers brown, with the exterior edges like the back. The

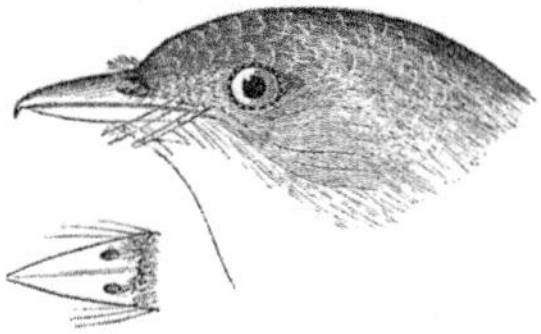

bill dark brown above, yellow beneath. The feet black. In the autumn the colors are purer, the yellow is deeper, and the markings on the wings of an ochry tint. Length, 5.15 to 5.75; extent, 8.50; wing, 2.83; tail, 2.45. Iris brown; bill dark brown, below flesh-color, edge orange; feet black.

Hab. United States generally.

At San Diego I observed the arrival of this species on April 15th, and preserved specimens which agree exactly with the description of Eastern specimens by Baird, as does one collected at Cape St. Lucas by Xantus. I obtained one also at Lake Tahoe, which is like the Eastern bird, while one obtained at Monterey has the darker markings on the wings, characterizing the variety *difficilis*. This, however, seems scarcely more than the young plumage; and though Baird mentions the difference in proportions of the quills also (first shorter than sixth, second considerably shorter than fourth), this does not characterize spring specimens, as far as I have obtained them, and I cannot divide the species.

This bird frequents the woods, chiefly those of *Coniferæ*, and is very silent, having in summer only a lisping song of three notes. I have never seen their nest or eggs.

Empidonax obscurus, SWAINSON.

THE GRAYISH FLYCATCHER.

? *Tyrannula obscura*, SWAINSON, Syn. Mex. Birds, in Philos. Mag. I. 1827, 367. — *Empidonax obscurus*, BAIRD, P. R. Rep. IX. Birds, 200. IB. U. S. and Mex. Bound. Survey, II. iii. 9; pl. xi. f. 3. — COUES, Pr. Phil. Ac. 1866, 62.

SP. CHAR. Bill very narrow. Tarsi long. Second, third, and fourth quills longest; first shorter than sixth. Tail rounded. Above dull brownish-olive, paler on the rump, tinged with gray on the head. Loral region and space round the eye whitish. Throat and fore part of the breast grayish-white, slightly tinged with olive across the latter; the rest of the under parts pale yellowish. Wings and tail brown, the former with two conspicuous bands of brownish-white; the outer primary edged, the secondaries and tertials edged and tipped with the same. The outer web of the external tail feather white, in

strong contrast. Length, 5.75; extent, 9.25; wing, 2.75; tail, 2.55; tarsus, 0.70. Iris brown; bill black, and yellow below; feet black.

Hab. Rocky Mountains of Texas, to Colorado River, California; south to Mexico.

I first observed this species at Fort Mojave about April 1st, and a few afterwards until May 25th. They kept about low bushes, generally silent or with only a single lisping chirp, occasionally flying a short distance after insects, like other species of the genus.

Empidonax Hammondii, XANTUS.

HAMMOND'S FLYCATCHER.

Tyrannula Hammondii, XANTUS, Pr. Phil. Ac. May, 1858. — *Empidonax Hammondii*, BAIRD, P. R. Rep. IX. Birds, 199. IB. Birds N. Amer. pl. 76, f. 1.

SP. CHAR. Tail moderately forked; the feathers acutely pointed. Third quill longest; second and then fourth a little shorter. First much shorter than fifth, a little longer than sixth. Bill very slender, dark brown. Above dark olive-green, considerably darker

on the head. Breast and sides of the body light olive-green, the throat grayish-white; the rest of under parts bright sulphur-yellow. A whitish ring round the eye. Wings and tail dark brown; the former with two olivaceous gray bands across the coverts; the latter with the outer edge a little paler than elsewhere, but not at all white. Length, 5.75; extent, 8.50; wing, 2.80; tail, 2.50; tarsus, 0.67. Iris brown; bill black, flesh-color below; feet black.

Hab. Vicinity of Fort Tejon to Los Angeles. Colorado Valley; south to Mexico.

I obtained but one specimen of this species at Fort Mojave on May 20th. This closely resembled in habits, at that time, *E. obscurus*, for which I mistook it. I afterwards saw on Catalina Island, in June, several birds which I think were of this species, but did not succeed in getting one. They kept on low trees, and uttered a few faint lisping notes.

The first of this species arrived at Santa Cruz, March 13, 1866, and they were numerous during the summer, disappearing in September. April 27th, I found the first nest built on a horizontal branch of a "box-elder" (*Negundo*), about eighteen feet from the ground, but in pulling down the branch the eggs were broken. I found four others afterwards from four to ten feet high, either on horizontal branches or in forks of small trees, and containing three or four eggs or young. The last found with eggs was as late as June 29th, probably a second attempt of a pair before robbed. All were thick walled, composed externally of dry moss and downy buds, with a few leaves and strips of bark, then slender fibres of bark, often a few hairs and feathers, lining the inside. The size outside was about four inches wide, two and a half high, the cavity two inches wide, one and a half deep, the walls nearly one and a half thick. The eggs were white with brown blotches and specks near the large end, mostly in a circle. They measured 0.68×0.52 inch.

These birds frequented only the darkest groves along the river, had very few simple calls of two or three monotonous notes, and were so shy that I did not get near enough to determine the species positively.

This bird is among the many species of birds that, while breeding, perhaps even as far north as the higher latitudes of North America, occur at the same time on the table-lands of Mexico, from which country numerous summer specimens have been sent to the Smithsonian Institution by Dr. Sartorius, Professor Sumichrast, and other correspondents. As in other instances, these are rather smaller than our birds, but no other difference is appreciable.

There is considerable resemblance in immature specimens between *E. Hammondii* and *obscurus*; the former species is the smaller bird, though with nearly as long a wing. The tail is decidedly shorter, being appreciably less than the wing, instead of about equal to it, as in *obscurus*. The first quill in *Hammondii* is rather longer than the sixth, while in *obscurus* it is shorter, sometimes shorter than the eighth or ninth. The feet and bill of *obscurus* are considerably the larger.

GENUS **PYROCEPHALUS**, GOULD.

Pyrocephalus, GOULD, Zool. of Beagle, 1838, 44.

GEN. CHAR. Tarsus moderate, very little longer than the middle toe; hind toe not longer than the lateral. Bill slender; very narrow at the base. Tail broad, even, about

P. Mexicanus.

four fifths the length of the wings, which reach beyond the middle of the tail. First quill shorter than the fifth. Head with a conspicuous rounded crest. Sexes dissimilar. *Male* with the crown and beneath red; tail, back, and wings brown.

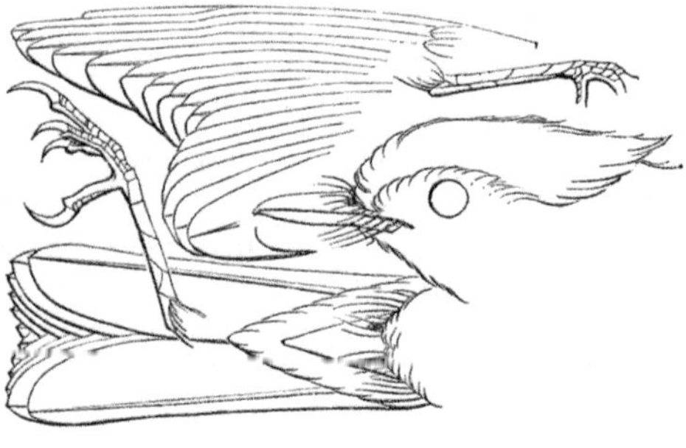

P. Mexicanus.

Of this genus several other species are known to naturalists, all, with perhaps a single exception, of much the same colors. In the red of the under parts they are perhaps unique in their family, where gray, white, or yellow are so much more frequently met with.

Pyrocephalus Mexicanus, Sclater.

THE RED FLYCATCHER.

Pyrocephalus Mexicanus, Sclater, Pr. Zool. Soc. 1859, 45.—Coues, Pr. Phil. Ac. 1866, 64.
Pyrocephalus rubineus, Lawrence, Ann. N. Y. Lyc. V. May, 1851, 115.—Cassin, Illust. I. iv. 1853, 127; pl. xvii.—Baird, P. R. Rep. IX. Birds, 201.—Heermann, X. vi. 38.
Pyrocephalus nanus, Woodhouse, Sitgreave's Report, 1853, 75.

Sp. Char. Head with a full rounded or globular crest. Tail even. Crown and whole under parts bright carmine red; rest of upper parts, including the cheeks as far as the bill, dull dark brown; the upper tail coverts darker; the tail almost black; greater and middle wing coverts and edges of secondaries and tertials dull white towards the edges. Outer web of exterior tail feather and tips of all the tail feathers whitish.

Female, similar, without the crest; the crown brown, like the back; the under parts whitish anteriorly, streaked with brown; behind white, tinged with red or ochraceous. Length of male, about 5.50; wing, 3 25; tail, 2.75. (Baird.)

Hab. Valleys of Rio Grande and Gila southward. Colorado Valley, California, south to Honduras.

On the 24th of May, 1861, I saw at Fort Mojave a small brilliant red bird, which I supposed must have been of the above species. It perched on the top of bushes, and would not allow me to approach within shooting distance, always flying off for several hundred yards before alighting again.

According to Dr. Heermann, they are quite common in spring at Fort Yuma, and he noted exactly the same wild habits above mentioned. Lieutenant Couch describes its habits as like those of a flycatcher, and its note as a low chirp.

One has since been obtained by Mr. W. W. Holder in Colorado Valley, lat. 34°, on April 18th.

The North American species, originally supposed to be identical with that inhabiting South America, has lately been described by Dr. Sclater as distinct. It may, however, perhaps be fairly questioned whether here, as in many other instances, there is anything more than a slight difference produced by peculiarities of climate, etc.

GENUS **MITREPHORUS**, SCLATER.

Mitrephorus, SCLATER, Pr. Zool. Soc. 1859, 45. (Type, *M. phæocercus*.) — COUES, Pr. Phil. Ac. 1866, 63.

SP. CHAR. Similar in general character to *Empidonax*, but with fulvous, fulvous-olive, and rufous tints, instead of clear olive, gray-white, and sulphur or olive-yellow. Head crested. Bristles of gape reaching nearly to tip of bill.

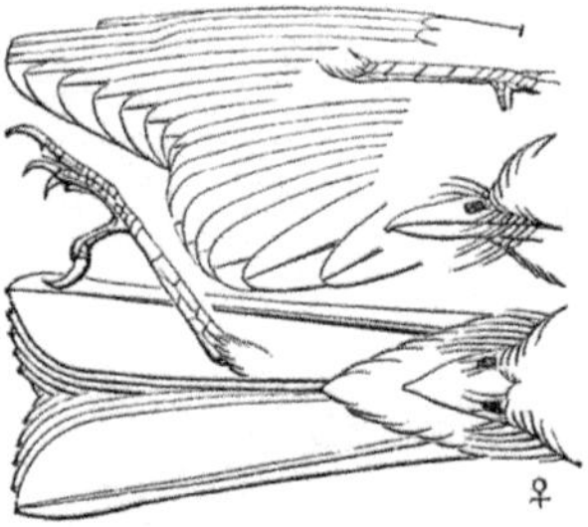

M. pallescens.

The single North American species exhibits a strong resemblance in form to *Empidonax*, although the type is decidedly different.

Mitrephorus pallescens, COUES.

THE BUFF-BREASTED FLYCATCHER.

Mitrephorus pallescens, COUES, Pr. Phil. Ac. 1866, 63. (Fort Whipple.)
Mitrephorus fulvifrous, ELLIOT, Illus. B. N. A. I. pl. 19.

SP. CHAR. Wing short, much rounded; first quill considerably shorter than sixth; the second, third, and fifth little shorter than the fourth. Above plain dull grayish-brown,

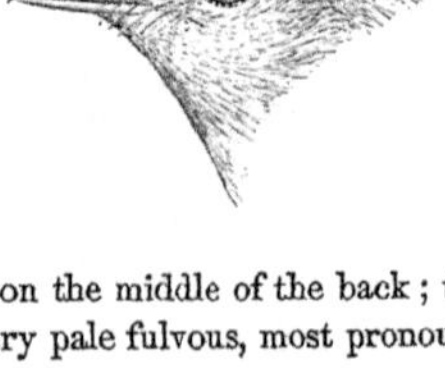

tinged with olive, particularly on the middle of the back; the head and rump hardly appreciably thus tinged. Below very pale fulvous, most pronounced across the breast, the chin and throat being much lighter, and the abdomen almost white. No fulvous suffusion about the forehead; the dark feathers of the crown reaching to the bill; the space between eye and bill, the auriculars and sides of the head generally light brownish-olive, with no trace of fulvous. Wings and tail plain dusky; the outer web of the external rectrices, the margins of the inner primaries, except just at their base, and the tips of greater and median coverts, dull white, with no tint of olive or ferruginous. Iris brown; upper mandible and feet black: lower mandible and mouth bright yellow. Length, 4.75; extent, 7.30; wing from the carpus, 2.15: tail, 2.00; tarsus, 0.55; middle toe and claw, 0.45 · bill above, 0.40.

Hab. Fort Whipple, Arizona.

Since this species was described by Dr. Coues, Mr. Giraud has presented to the Smithsonian Institution his type of *Muscicapa fulvifrons,* which, though much like the *pallescens,* seem yet sufficiently distinct. The coloration is similar but darker, and more ochraceous throughout, with less olive above. The wing bands are fulvous (the edges of secondaries more yellowish), instead of grayish-white. While the bill seems actually smaller, the other dimensions of *M. fulvifrons* are considerably larger (length, 4.80; wing, 2.65; tail, 2.40; tarsus, 0.61). The wing is considerably more pointed; the first quill equal to, not shorter than, the sixth; the third, not fourth, quill longest. (Baird.)

ORDER STRISORES.

CHAR. Toes variable, generally three anterior, and one posterior, sometimes in pairs. Primaries ten. Tail feathers generally ten. Tarsi with small scales or with naked skin, without the broad encircling plates of the preceding order.

This order is represented in North America by a greater number and variety of families than the preceding. Several others, however, are found in Middle and South America.

FAMILY ALCEDINIDÆ, THE KINGFISHERS.

CHAR. Head large; bill long, strong, straight, and sub-pyramidal, usually longer than the head. Tongue very small. Wings short; legs small; the outer and middle toes united to their middle. Toes with the usual number of joints (2, 3, 4, 5).

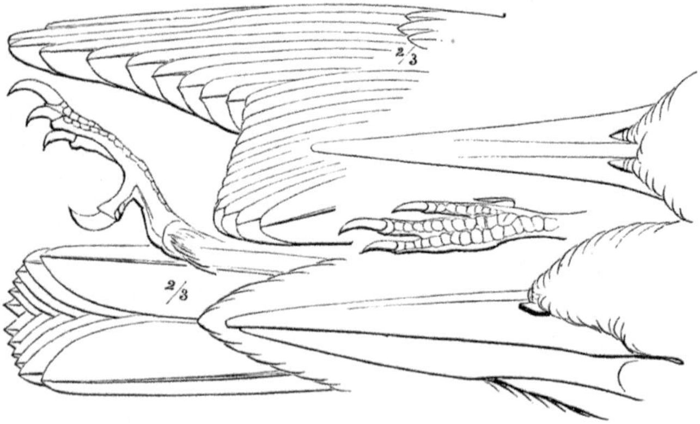

C. alcyon.

Genus CERYLE, Boie.

Ceryle, Boie, Isis, 1828, 316. (Type, *C. rudis ?*)
Spida, Swainson, Birds, II. 1837, 336. (Type, *C. alcyon.*)

Gen. Char. Bill long, straight, and strong, the culmen slightly advancing on the forehead and sloping to the acute tip; the sides much compressed; the lateral margins

C. alcyon.

rather dilated at the base, and straight to the tip; the gonys long and ascending. Tail rather long and broad. Tarsi short and stout.

Ceryle alcyon, Linnæus.

THE BELTED KINGFISHER.

Alcedo alcyon, Linnæus, Syst. Nat. I. 1766, 180. — Wilson, Am. Orn. III. 1811, 59. — Audubon, Orn. Biog. I. 1831, 394; pl. 77. — *Ceryle alcyon*, Boie, Isis, 1828, 316. — Cassin, Illust. I. 1855, 254. — Brewer, N. Am. Ool. I. 1857, 110; pl. iv. f. 52 (egg). — Newberry, P. R. Rep. VI. iv. 79. — Baird, P. R. Rep. IX. Birds, 158. — Heermann, X. vi. 57. — Cooper and Suckley, XII. iii. Zool. of W. T. 167. — Coues, Pr. Phil. Ac. 1866, 59.

Sp. Char. Head with a long crest. Above blue, without metallic lustre. Beneath, with a concealed band across the occiput, and a spot anterior to the eye, pure white. A band across the breast, and the sides of the body under the wings, like the back. Primaries white on the basal half, the terminal unspotted. Tail with transverse bands and spots of white.

Half natural size.

Young, with the sides of body and a transverse band across the belly below the pectoral one, light chestnut; the pectoral band more or less tinged with the same. Length of adult, 12.00 to 14.00; extent, 20.00 to 23.00; wing, 6.00 to 7.00. Bill slate-blue; feet reddish-white.

Hab. The entire continent of North America.

This bird is common along the coast, and about every clear stream and lake throughout California, but rare about those which, like the Colorado, are constantly muddy. They are usually to be seen perched on some dead branch, stake, or other object overhanging the water, watching for their prey, which consists wholly of small fish. When one is seen, the bird plunges suddenly headlong into the water, and usually catches the victim. If unsuccessful, he sometimes flies off and searches for fish, flying rapidly along near the shores, and if any are seen, he stops and hovers for a moment over them, diving with a somewhat spiral motion. When alarmed he utters a harsh rattling sound, as he flies off, and is usually rather wary.

The hole for a nest is burrowed in a sandy bank, to the depth of five or six feet, and usually not far from the top of the ground. The inner end of the burrow is enlarged into a rounded chamber, and the same hole is used for several years. The eggs are white, six in number, and laid on a nest composed of a few twigs, grass, and feathers. According to Nuttall, both parents sit on them, and they are hatched in sixteen days.

This bird is not migratory in California, unless it is from the highest mountain streams to the valleys and coast. Pacific Coast specimens are larger than Eastern.

Ceryle Americana, GMELIN.

THE TEXAS KINGFISHER.

Alcedo Americana, GMELIN, Syst. Nat. I. 1788, 451. — *Ceryle Americana*, BOIE, Isis, 1828, 316. — LAWRENCE, Annals N. Y. Lyceum, V. 1851, 118. (First introduction into the fauna of the United States.) — CASSIN, Illust. I. 1855, 255. — BREWER, N. Am. Ool. I. 1857, 3; pl. iv. f. 53 (egg). — BAIRD, Birds N. Amer. 1858, 159; pl. 45. IB. Rep. Mex. Bound. II. 7; pl. vii. — COUES, Pr. Phil. Ac. 1866, 59. — *Chloroceryle Americana* REICHENB. Handb. Sp. Orn. I. ii. 1851, 27; pl. 413, f. 3112 - 3115.
Alcedo viridis, VIEILLOT, Nouv. Dict. XIX. 1818, 413 (Cassin).
Alcedo Cabanisii, TSCHUDI, Fauna Per. 253.

SP. CHAR. Head slightly crested. Upper parts, with a pectoral and abdominal band of blotches, glossy green, as also a line on each side the throat. Under parts generally,

Natural size.

a collar on the back of the neck, and a double series of spots on the quills, white; a chestnut band across the breast in some skins. Length, about 8.00; wing, 3.25.

Hab. Rio Grande and the Lower Colorado, southward.

This diminutive species, scarcely one sixth the weight of the belted kingfisher, is not at all rare along the Rio Grande, but has been seen more seldom in Southern Arizona. Dr. Coues observed it at several points on the Colorado River between Forts Mojave and Yuma.

No detailed account of the habits of this species has been published, but it is probable that they resemble those of *C. alcyon*. The eggs are white, as in other species of the family. (Baird.)

FAMILY CAPRIMULGIDÆ, THE GOAT-SUCKERS.

CHAR. Bill very short, triangular, the culmen less than one sixth the gape. The anterior toes united at the base by a membrane. The inner anterior toe with three joints, the others with four; all with distinct scutellæ above. The toe much elongated, its middle claw pectinated on the inner edge. Hind toe directed a little more than half forwards. Tarsi partly feathered superiorly. The bill more or less bristled; the nostrils separated, rather nearer the commissure than the culmen. Plumage soft, lax, and owl-like. (Baird.)

A. Carolinensis.

The chuck-wills-widow (*A. Carolinensis*) is the only species of the family in which the bristles of the mouth have lateral filaments, as shown in the figure above.

Genus **ANTROSTOMUS**, Gould.

Antrostomus, Gould, Icones Avium, 1838 (Agassiz).

Gen. Char. Bill remarkably small, with tubular nostrils, and the gape with long, stiff,

A. Nuttalli.

sometimes pectinated bristles. Wings long, somewhat rounded, second quill longest, the primaries emarginated. Tail rounded. Plumage loose and soft.

Antrostomus Nuttalli, Audubon.

NUTTALL'S WHIPPOORWILL.

Caprimulgus Nuttalli, Audubon, Birds Amer. VII. 1843; pl. 495, Appendix. — *Antrostomus Nuttalli*, Cassin, J. A. N. Sc. Phila. 2d Series, II. 1852, 123. Ib. Illus. I. 1855, 237. — Newberry, Zool. Cal. and Or. Route, 77; P. R. Rep. VI. iv. 77. — Baird, P. R. Rep. IX. Birds, 149. — Kennerly, X. iv. 23. — Heermann, X. vi. 35. — Cooper and Suckley, XII. iii. Zool. of W. T. 166. — Coues, Pr. Phil. Ac. 1866, 58.

Sp. Char. *Male.* General color of upper parts dark brownish-gray, lighter on the head and medial tail feathers, which extend a half-inch beyond the others, all which are minutely streaked and sprinkled with brownish-black and ash gray. Quills and coverts dull cinnamon color, spotted in bars with brownish-black; tips of former mottled with light and dark brown; three lateral tail feathers barred with dark brown and cinnamon, and tipped with white. Throat brown, annulated with black; a band of white across fore neck; beneath the latter black, mixed with bars of light yellowish-gray and black lines. Under tail coverts dull yellow. Length, 7.25 to 8.00; extent, 17.50; wing, 5.75; bill, edge, 0.19; second and third quills nearly equal. Tail to end of upper feathers, 3.50; tarsus, 0.63; middle toe, 0.63; claw, 0.25; strongly pectinated. Iris brown; bill and feet gray.

Hab. High central plains, and mountains to the Pacific Coast.

I did not myself meet with this bird in the Colorado Valley, where, however, Dr. Kennerly obtained one on February 23, 1854, indicating that they may perhaps winter there. I heard their call of "poor-will" on the barren mountains west of the valley in May, but have neither heard nor seen any west of the Coast Range, nor in the Santa Clara Valley in spring, though their habits of concealment during the day might lead to mistakes as to their

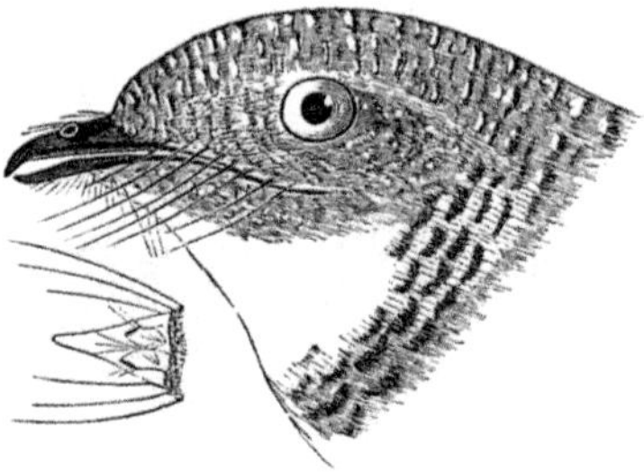

real occurrence. They are, however, common in the hot interior valleys, and remain near San Francisco as late as November, hiding usually on the ground under a low bush or tuft of grass, and flying at dusk in short fitful courses in pursuit of insects. They inhabit the almost bare and barren "Sage-plains," east of the Sierra Nevada, and their rather sad whistle is heard all night during spring, like an echoing answer to the cry of their cousins, the Eastern "whip-poor-will."

Their eggs are probably laid on the ground.

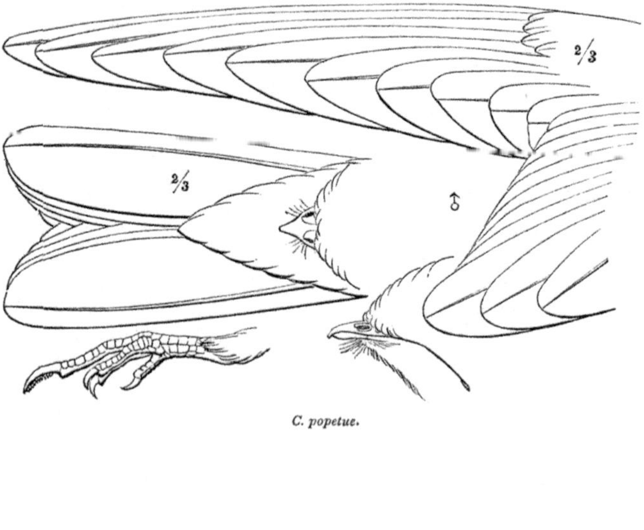

C. popetue.

Genus CHORDEILES, Swainson.

Chordeiles, Swainson, Fauna Bor. Amer. II. 1831, 496.

Gen. Char. Bill very small, the gape with very short feeble bristles. Wings very long and pointed, with the first quill nearly or quite equal to the second, and the prima-

C. popetue.

ries not emarginated on the inner edge. Tail long; slightly forked in North American species; plumage rather compact.

Chordeiles popetue, Vieillot.

THE NIGHT-HAWK; THE BULL-BAT.

Caprimulgus popetue, Vieillot, Ois. Am. Sept. I. 1807, 56; pl. xxiv. (♀). — Bonaparte, Obs. Wilson, 1825, 177; J. A. N. Sc. Phil. VI. — *Chordeiles popetue*, Baird, P. R. Rep. IX. Birds, 151. — Heermann, X. vi. 35. — Cooper and Suckley, XII. iii. Zool. of W. T. 166.

Caprimulgus Americanus, Wilson, Am. Orn. V. 1812, 65; pl. cxl. f. 1, 2. — *Chordeiles Americanus*, De Kay, N. Y. Zool. II. 1844, 34; pl. xxvii.

Caprimulgus Virginianus, Brisson, II. 1760, 477. (In part only.) Audubon, Orn. Biog. II. 1834, 273; pl. 147. — Swainson, Fauna Bor. Amer. II. 1831, 62.

Chordeiles Virginianus, Bon. List, 1838. — Aud. Birds Amer. I. 1840, 159; pl. 43. — Cassin, Illus. I. 1855, 238. — Newberry, Zool. Cal. and Or. Route, 79; P. R. Rep. VI. 1857.

Chordeiles Henryi, Cassin, Illus. I. Jan. 1855, 233. — Baird, P. R. Rep. IX. Birds, 153, 922; X. iii. 13; pl. xvii. — Coues, Pr. Phil. Ac. 1866, 58. (Arizona.)

Sp. Char. *Male*, above greenish-black, with but little mottling on the head and back. Wing coverts varied with grayish; scapulars with yellowish-rufous. A nuchal band of fine gray mottling, behind which is another coarser one of rufous spots. A white V-shaped mark on the throat; behind this a collar of pale rufous blotches, and another on the breast of grayish mottling. Under parts banded transversely with dull yellowish or

reddish-white and brown. Wing quills quite uniformly brown. The five outer primaries with a white blotch midway between the tip and carpal joint, not extending on the outer web of the outer quill. Tail with a terminal white patch.

Female without the caudal white patch, the white of the throat mixed with reddish. Length of male, 9.50; wing, 8.20.

Hab. North America generally.

This species shuns the coast border of this State, probably on account of the cold winds, though it is common near the mouth of the Columbia River. In Santa Clara Valley and the Coast Range I have seen none, but, according to Newberry, they are quite common in the Sacramento Valley in summer. I saw but one in the Sierra Nevada in September, the month in which they depart for the South. At the Columbia River they are very numerous, and during June, even on clear days, may be seen pursuing their insect prey high in the air, uttering their monotonous croaking at short intervals. I saw one, probably of this species, migrating north high above the town of Santa Barbara on the 27th of April, 1863, and this is probably about the time they arrive from Mexico. Their flight is easy and swift, resembling that of a swallow, and they often sweep down close to a person's head, especially in the twilight. When they have a nest, they attempt to drive away intruders by pitching down perpendicularly towards them, with their mouth wide open, producing a singular hollow sound like that made by blowing into a large bottle. They sometimes alight on the ground in pursuit of insects, and also lay two whitish eggs, speckled with brown, in a slight cavity among the dead leaves on the ground, or sometimes among gravel or on the bare sand. They generally roost during the day on a horizontal branch, sitting lengthwise. The female if found on the nest feigns lameness to draw away the intruder.

Chordeiles Henryi, described by Mr. Cassin as distinct, is probably only a lighter race of this species, in which the markings are more distinctly appreciable, owing to the lighter color.

Chordeiles Texensis, LAWRENCE.

THE TEXAS NIGHT-HAWK.

Chordeiles Brasilianus, LAWRENCE, Ann. N. Y. Lyceum, V. May, 1851, 114. (Texas.) (Not of Gmelin.) — CASSIN, Illus. I. 1855, 238.

Chordeiles Texensis, LAWRENCE, Ann. N. Y. Lyceum, VI. Dec. 1856, 167. — BAIRD, P. R. Rep. IX. Birds, 154. — U. S. and Mex. Bound. Rep. II. iii. 7; pl. vi. — COUES, Pr. Phil. Ac. 1866, 58. (Arizona.)

SP. CHAR. Much smaller than *C. popetue*, but similar. White on the wing extending over only four outer primaries, the bases of which, as well as the remaining ones, with other quills, have round rufous spots on both webs. Under tail coverts and abdomen with a strong yellowish-rufous tinge. *Female*, more rufous, and without the spot of the tail. Length, 8.50; extent, 21.00; wing, 7.30; tail, 3.40.

Hab. Rio Grande Valley and south; west to Southern California.

On the 17th of April I saw the first of this species at Fort Mojave, and soon after they became quite numerous, hunting in company after sunset, and hiding during the day on the ground under low bushes. About the

25th of May they were paired, but continued nearly silent, making only a low croaking when approached. They flew like *C. popetue,* but sailed rather more in small circles. I found them common as far west as the Coast Mountains. The eggs are said to be much paler than those of the other species.

FAMILY CYPSELIDÆ, THE SWIFTS.

CHAR. Bill very small, without notch, triangular, much broader than high; the culmen not one sixth the gape. Anterior toes cleft to the base, each with three joints (in the typical species), and covered with skin; the middle claw without any serrations; the lateral toes nearly equal to the middle. Bill without bristles, but with minute feathers extending along the under margin of the nostrils. Nostrils elongated, superior, and very close together. Plumage compact. Primaries ten, elongated, falcate. Species of this family are found abundantly both in the Old and the New World.

GENUS **PANYPTILA**, CABANIS.

Panyptila, CABANIS, Wiegm. Archiv, 1847, 1, 345.
Pseudoprocne, STREUBEL, Isis, 1848, 357.

GEN. CHAR. Tail half as long as the wings; moderately forked; the feathers rather lanceolate, rounded at tip, the shafts stiffened but not projecting. First primary shorter

P. melanoleuca.

than the second. Tarsi, toes, and claws very thick and stout; the former shorter than the middle toe and claw, which is rather longer than the lateral one; middle claw longer

P. melanoleuca.

than its digit. Hind toe very short; half versatile, or inserted on the side of the tarsus. Tarsi and toes feathered to the claws, except on the under surfaces.

Panyptila melanoleuca, Baird.

THE WHITE-THROATED SWIFT.

Cypselus melanoleucus, Baird, Pr. A. N. Sc. Phil. VII. June, 1854, 118. (San Francisco Mountains, N. M.) — Cassin, Illus. I. 1855, 248. — *Panyptila melanoleucus*, Baird, P. R. Rep. IX. Birds, 141. — Kennerly, X. iv. 23. — Heermann, X. v. 10; X. vi. 35; pl. xviii. f. 1 (lower figure). — Coues, Pr. Phil. Acad. 1866, 57.

Sp. Char. Wings very long; tail forked; tarsi and feet covered with feathers. Black all over, except the chin, throat, middle of the belly as far as the vent, a patch on each side of the rump, the edge of the outer primary, and blotches on the inner webs of the

median tail feathers, near the base, which are white, as is also a band across the ends of the secondaries. Length, 5.50; wing, 5 50; tail, 2.70. Iris brown; bill black.

Hab. Colorado Basin, New Mexico to San Diego, California.

Dr. Kennerly discovered this species along Williams Fork of the Colorado, on February 16, 1854, and remarks that large flocks could be seen at any time in the vicinity of the cañons, flying and circling about very high, and far beyond the reach of shot, occasionally descending lower towards sunset. It is very probable that they winter about the cañons, though I saw none at Fort Mojave until May, probably stragglers from their favorite mountains. On the 7th of June, near the head of the Mojave River, I found a few of them about some lofty granite cliffs, and by watching them closely succeeded in shooting one as it came sweeping towards me about sunset as if about to fly in my face. Their flight is exceedingly swift and changeable, resembling that of the *Chætura,* and they are probably the most difficult of all our birds to shoot. Their cry when flying is low but harsh, and sounds much farther off than the actual distance of the bird. They also utter a harsh croaking noise.

About twelve miles north of San Diego, I again found them rather numerous about some high rocky bluffs close to the sea-shore. They were there March 22d, and may have been about for a month previously, but generally fly so high during the day that they are first betrayed by their harsh twitter when scarcely perceptible in the zenith. Occasionally they dart off like lightning for several miles and back, as if it were only a slight curve in their course. Sometimes they sweep for a moment near the ground, and the next disappear in the sky above. They doubtless build in the crevices of the lofty rocks about which they live, but, though I have looked for their nests, I have not been able to find them.

Genus **NEPHŒCETES**, Baird.

Nephœcetes, Baird, P. Rep. IX. Birds, 1858, 142.

Gen. Char. Tail rather less than half the wings; quite deeply forked; the feathers obtusely acuminate; the shafts scarcely stiffened. First quill longest. Tarsi and toes completely bare, and covered with naked skin, without distinct indications of scutellæ.

N. niger.

Tarsus rather longer than middle toe; the three anterior toes about equal, with moderately stout claws. Claw of middle toe much shorter than its digit. Hind toe not versatile,

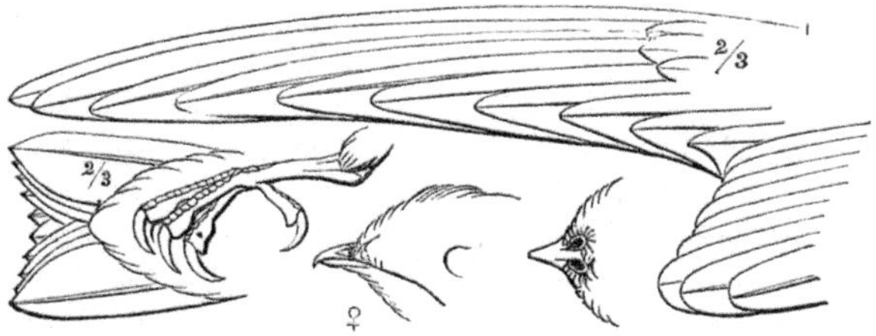

N. niger.

but truly posterior and apposite, with its claw rather longer than the middle toe without it. Toes all slender; claws moderate. Nostrils widely ovate, the feathers margining its entire lower edge.

Nephœcetes niger, Gmelin.

THE NORTHERN SWIFT.

? *Hirundo nigra*, Gmelin, Syst. Nat. I. 1788, 1025. — ? *Cypselus niger*, Gosse, Birds Jamaica, 1847, 63. Ib. Illus. Birds Jamaica. — Gundlach and Lawrence, Ann. N. Y. Lyceum, VI. 1858, 268. — *Nephœcœtes niger*, Baird, P. R. Rep. IX. Birds, 142.
Cypselus borealis, Kennerly, Pr. Ac. Nat. Sc. Phil. IX. Nov. 1857, 202.

Sp. Char. Wing the length of the body. General color rather lustrous dark sooty-brown, with a greenish gloss, becoming a very little lighter from the breast anteriorly below, but rather more so on the neck and head above. The feathers on the top of the head edged with light gray, which forms a continuous wash on each side the forehead

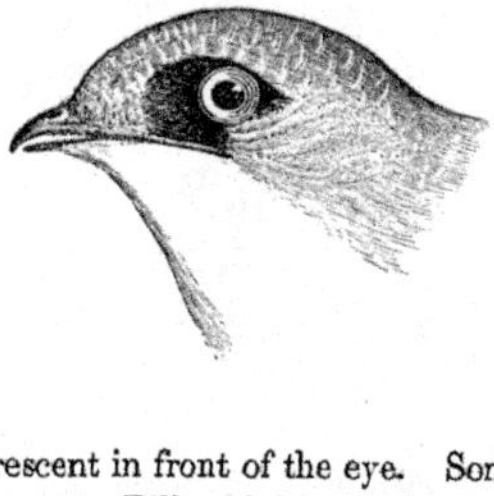

anterior to the usual black crescent in front of the eye. Some feathers of the under parts behind narrowly edged with gray. Bill and feet black. Length, 6.75; wing, 6.75; tail, 3.00; depth of fork, 45.

Hab. Northwestern North America to Jamaica and Cuba.

This bird has not yet been obtained within this State; yet it must undoubtedly migrate through some part of it to reach its summer residence near the Straits of Fuca. I have seen a black swift, without any white apparent, flying about, very high above Pah-Ute cañon, west of Fort Mojave, on May 29th, 1861. I also saw an apparently black swallow at Santa Barbara in May, 1863, but this may have been the *Chætura Vauxii.* As mentioned above, it is found in Jamaica and Cuba, where it appears to be resident, and of smaller size. It is also occasionally seen on the table-lands of Mexico, according to Professor Sumichrast; but its movements otherwise are but little known. Like other swifts, flying habitually almost out of sight, and rarely coming near the ground, except in cloudy weather, it would very readily escape the attention of travellers in the West, however abundant it might be.

It is by no means improbable that other species of the present family will yet be discovered in New Mexico and Arizona, as there are several kinds belonging to Mexico and Guatemala, that might readily extend their spring migrations across our boundary without detection.

Of its habits no account has been published, and yet it is one of the most interesting additions lately made to the North American fauna.

Genus **CHÆTURA**, Stephens.

Chætura, Stephens, Shaw's Gen. Zool. Birds, XIII. ii. 1825, 76. (Type, *C. pelasgia.*)
Acanthylis, Boie, Isis, 1826, 971. (*A. spinicauda.*)

Gen. Char. Tail very short, scarcely more than two fifths the wings, slightly rounded; the shafts stiffened and extending some distance beyond the feathers in a rigid spine.

C. Vauxii.

First primary longest. Legs covered by a naked skin, without scutellæ or feathers. Tarsus longer than middle toe. Lateral toes equal, nearly as long as the middle. Hind toe scarcely versatile, or quite posterior; with the claw, shorter than the middle toe with-

C. pelasgia

out it. Toes slender; claws moderate. Feathers of the base of the bill not extending beyond the beginning of the nostrils.

Chætura Vauxii, Townsend.

THE OREGON SWIFT.

Cypselus Vauxii, Townsend, J. A. N. Sc. VIII. 1839, 148. (Oregon.) Ib. Narrative, 1839. — Nuttall, Man. I. 2d ed. 738. — *Chætura Vauxii*, De Kay, N. Y. Zool. II. 1844, 36. — Baird, P. R. Rep. IX. Birds, 145; upper fig. pl. 18. — Cooper and Suckley, XII. iii. Zool. of W. T. 165. — *Acanthylis Vauxii*, Bonaparte, Comptes Rendus, XXVIII. 1854; notes Delattre, 90. — Cassin, Illus. I. 1855, 250. — Newberry, Zool. Cal. and Or. Route, 78; P. R. R. Surv. VI. 1857.

Sp. Char. Light sooty-brown; rump and under parts paler; lightest on the chin and throat. Length, 4.25; extent, 11.50; wing, 4.50; tail, 1.90. Iris brown.

Hab. Pacific Coast, from Puget's Sound to California.

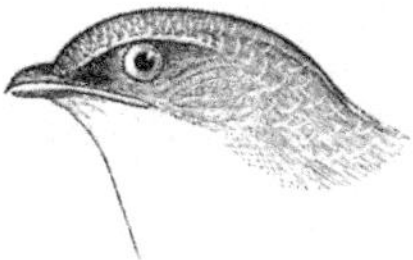

I first met with these birds on May 4, 1864, in the Coast Range, twelve miles south of Santa Clara. They may have arrived some time earlier. I was satisfied that they had nests in hollow trees at the summit of the Coast Range, but did not succeed in finding any. Dr. Townsend, in 1834, found them at the Columbia River, breeding in hollow trees, forming a nest like that of the Eastern chimney-swallow, of small sticks glued together, and stuck against the wall of the cavity by the saliva of the bird, containing four pure white eggs. I did not, however, observe them in the summer of 1853, at Fort Vancouver.

Not having been in the interior at the proper season, I do not know whether they occur in the central valleys, but Heermann does not mention seeing them there. They may become common where high chimneys are built. Like the Eastern species they probably go south early.

In 1866 their first appearance at Santa Cruz was May 4th, though as they rarely descend to the town, they may have been among the hills for some time previously.

October 5th, I observed five of them in company with a large flock of *Hirundo bicolor*, which spent the morning hunting insects near town, apparently delayed in their migration southward by a thick fog and cold south wind. As this is two months later than the Eastern species departs south, these may have been a late brood from the far north.

Family TROCHILIDÆ, The Humming-Birds.

Char. Bill long and thin, forming a sheath for the very long forked thread-like tongue. Secondaries six. Wings falcate, shafts of the primaries strong, the first always longest. Tail of ten feathers. Feet very small, claws sharp and strong.

There are over four hundred species of these splendidly brilliant birds known to naturalists, all inhabiting America; but only nine of them occur within the limits of the United States, and but one is found in the Atlantic States, the *T. colubris*, or ruby-throated hummer.

T. colubris.

Genus **TROCHILUS**, Linnæus.

Trochilus, Linnæus, Syst. Nat. 1748 (Agassiz).

T. Alexandri.

Gen. Char. Feathers of throat but little elongated laterally. Lateral tail feathers but little narrower than the others, and lanceolate acute. Tail forked.

Trochilus Alexandri, Bourcier and Mulsant.

THE BLACK-CHINNED HUMMING-BIRD.

Trochilus Alexandri, Bourcier and Mulsant, Ann. de Soc. d'Agric. de Lyons, IX. 1846, 330. — Heermann, Jour. A. N. Sc. Phil. 2d Series, II. 1853, 269. — Cassin, Illus. N. Amer. Birds, I. v. 1854, 141; pl. xxii. — Baird, P. R. Rep. IX. Birds, 133. Ib. U. S. and Mex. Bound. Report, II. iii. 6; pl. v. f. 3. — Heermann, P. R. Rep. X. vi. 56.

Sp. Char. Tail slightly forked; the chin and upper part of the throat opaque velvety-black, without metallic reflections, which are confined to the posterior border of the gorget, and are violet, changing to steel-blue or green, instead of coppery-red.

Female without the metallic scales; the tail feathers tipped with white, the tail graduated, not emarginated; the innermost feather among the longest. Length of male, 3.30; wing, 1.70; tail, 1.26; bill, 0.75.

Hab. California southward.

I observed none of this species in the Colorado Valley, and in coming westward first saw them along the Mojave River on the 3d of June. I also found one of their nests there, built in a dark willow thicket in a fork of a tree about eight feet from the ground. I have since found several more nests near Santa Barbara, all of them built near the end of hanging branches of the sycamore (*Platanus*), constructed entirely of white down from the willow or sycamore catkins, agglutinated by the bird's saliva, and attached in the same way to the branch on which they rested. These were built in April, and early in May I found several containing two white eggs, like those laid by all humming-birds, oblong in shape, and alike at each end; size 0.51 × 0.32. Dr. Heermann found their nests as far north as Sacramento, and south to Guaymas. I have never seen the species in places exposed to the cold sea-winds, where others are found. It is a less interesting and conspicuous bird than the larger species found in this State, and probably not often recognized, though its small size is alone sufficient to distinguish it. I believe they go entirely beyond this State in winter, as I have seen none at that season. Mr. W. W. Holder has since obtained a specimen in Colorado Valley, March 20.

During the progress of the Northwestern Boundary Survey, Mr. J. K. Lord, of the British Commission, was so fortunate as to find this species between the Cascade and Rocky Mountains, near lat. 49°, where they arrived

T. colubris. Male.

towards the end of May, and frequented the vicinity of lakes, pools, and swamps where the birch-tree grew. The sap exuding from the bark of this tree attracted numbers of insects, on which this humming-bird chiefly fed. He found the nests in high forks of branches of the birch or alder.

This species is very closely allied to *T. colubris*, the common humming-bird of the Eastern States, and the only one found east of the Rocky Mountains. The difference, consisting in the color of the chin and the shape of the tail, will be best expressed by the above figure. As will be seen, the tail in the male is nearly even, or slightly rounded, instead of being decidedly forked. The females of the two species are very similar, and can scarcely be distinguished. The locality will be the best indication, as the one is confined to the region west of the Rocky Mountains, and the other to that east of the same Range.

GENUS **SELASPHORUS**, SWAINSON.

Selasphorus, SWAINSON, F. Bor. Am. II. 1831.

S. rufus.

GEN. CHAR. Feathers of the throat laterally much elongated, forming a ruff. Lateral tail feathers much narrower than the middle, linear, or with the sides parallel to the

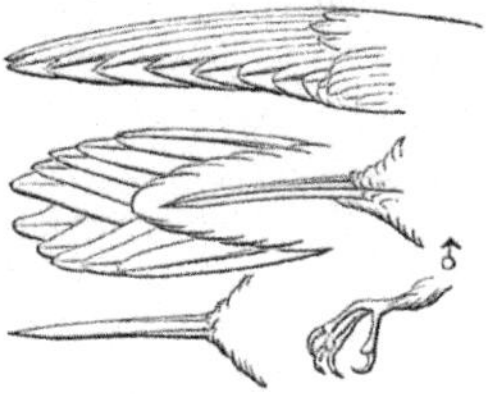

S. rufus.

rounded end. Tail graduated or cuneate. Outer primary attenuated at the tip. Crown without red metallic scales.

Selasphorus rufus, GMELIN.

THE RED-BACKED HUMMING-BIRD.

Trochilus rufus, GMELIN, Syst. Nat. I. 1788, 497. — AUDUBON, Orn. Biog. IV. 1838, 555; pl. 372. — NUTTALL, Man. I. 2d ed. 1840, 714. — *Selasphorus rufus*, SWAINSON, F. Bor. Am. II. 1831, 324. IB. Birds Amer. IV. 1842, 200; pl. 264. — BAIRD, P. R. Rep. IX. Birds, 134. — HEERMANN, X. vi. 57. — COOPER and SUCKLEY, XII. iii. Zool. of W. T. 164.

SP. CHAR. Tail strongly cuneate and wedge-shaped. Upper parts, lower tail coverts, and breast, cinnamon. A trace of metallic green on the crown, which sometimes extends over the back; never on the belly. Throat coppery-red, with a well-developed ruff of the same; below this a white collar. Tail feathers cinnamon, edged or streaked at the end with purplish-brown.

Female with the rufous of the back covered or replaced with green; less cinnamon on the breast. Traces only of metallic feathers on the throat. Tail rufous, banded with

black and tipped with white; middle feathers glossed with green at the end. Tail cuneate. Length of male, 3.50 to 4.00; extent, 4.40; wing, 1.55 to 1.75; tail, 1.30. Iris brown; bill and feet black.

Hab. West coast of North America to highlands of Mexico.

Although the most northern in its range of any North American species, this humming-bird does not appear to be so hardy as the *S. Anna,* since I have found none of them during winter, even at San Diego, where they first arrived on the 5th of February, 1862. Still, some may remain in the warmer interior valleys, as I saw several on the 22d of February, a few miles back from the coast, feeding among the flowers of the evergreen gooseberry. The blooming of this plant may indeed be the guide to their migration northward, as that of another species (*Ribes sanguineum*) is for their arrival at the Columbia River about March 10th. By the first week of April they were swarming about San Diego.

I found a nest of this species on Catalina Island, in which young had already been raised by the middle of June. Some also build about San Francisco, and probably also on the higher Sierra Nevada, as well as more abundantly near the Columbia River, where this is the only species I have seen. The nest is made as usual of moss, and the down of plants, covered with lichens, and the two small eggs can scarcely be distinguished from those of the other species.

When perching the males have a shrill wiry call, like the highest note of a violin. When sporting in the air, or trying to drive an enemy from their nest, they have a curious habit of rising to a considerable height, then plunging perpendicularly downward with a hollow whirring sound, perhaps produced by their wings. They also produce a kind of a bleating sound.

They are probably the most lively and noisy species in the country, being very quarrelsome, chasing each other away from favorite flowers, and rising into the air until out of sight, buzzing and chirping in an angry manner as they go.

Though found on the interior table-lands of Mexico and the Rio Grande

Valley, I did not see this species along the Colorado, and am inclined to think that it is replaced in most parts of the interior by the next species.

Mr. J. K. Lord, in his interesting "Naturalist in British Columbia," etc., published 1866, mentions the first arrival of this species at the "Little Spokan River," near lat. 49°, early in May, together with the *Calliope*, and confirms the statement that they come simultaneously with the flowering of the *Ribes sanguineum*, which seems to be two months later east of the Cascade Range than on Whidby's Island. He found the nests usually in a low shrub, and close to a rippling stream. The females of all the species arrived about a week behind the males.

Selasphorus platycercus, Swainson.

THE BROAD-TAILED HUMMING-BIRD.

Trochilus platycercus, Swainson, Phil. Mag. I. 1827, 441. — *Selasphorus platycercus*, Gould, Mon. Troch. 1852. — Baird, P. R. Rep. IX. Birds, 135. Ib. U. S. and Mex. Bound. Surv. II. iii. 6; pl. v. f. 1, 2. (El Paso, Texas.)
Ornismia tricolor, Lesson, Colibris, 125; pl. xiv. — Jardine, Nat. Lib. II. 77; pl. xiii. — *Ornismia montana*, Lesson, Troch. 161; pls. 63 and 163, 64 (young), 1831.

Sp. Char. Outer primaries greatly attenuated at the end. Outer tail feather nearly linear, but widening a little from the base; its width 0.20 of an inch. Tail slightly graduated and emarginated. Male above and on the sides metallic green; chin and throat light reddish-purple, behind which, and along the belly to the tail, is a good deal of white.

Wings and tail dusky purplish; the tail feathers, excepting the internal and external ones, edged towards the base with light cinnamon. Length, 3.75; extent, 4.75; wing, 1.90; tail, 1.40. Bill, gape, 0.80. Iris brown; bill and feet black.

Hab. Rocky Mountains, to lat. 42°; west to Sierra Nevada, and south to Mexico.

At Lake Tahoe, over six thousand feet above the sea, I found the young of this species quite common near the middle of September, but supposing them to be the *S. rufus*, obtained only one specimen. It probably extends north to the Blue Mountains, near Snake River, Oregon, and is referred to by Nuttall as the *S. rufus* seen there in autumn. Their habits at that season do not exhibit anything peculiar.

GENUS **CALYPTE**, GOULD.

Calypte, GOULD, Introd. Trochilidæ, 1861, 87. (Type, *Ornismya costæ*.)

GEN. CHAR. Bill longer than the head; straight or slightly curved; tail rather short; forked, the outer feather abruptly narrower, linear and incurved (the second and third

sometimes so to a less degree). Outer primary not attenuated at end. Top of head as well as throat with metallic scale-like feathers. A decided and elongated ruff on each side the neck.

The species of this genus will be readily distinguished from their allies in North America by the metallic feathers of top of head.

Calypte Anna, LESSON.

THE ANNA HUMMING-BIRD.

Ornismya Anna, LESSON, Oiseaux Mouches, 1830 (?); pl. cxxiv. — *Trochilus Anna*, JARDINE, Nat. Lib. Humming-Birds, I. 93; pl. vI. — AUDUBON, Orn. Biog. V. 1839, 428; pl. 428. IB. Birds Amer. IV. 1842, 188; pl. 252. — NEWBERRY, P. R. Rep. VI. ii. 79. — HEERMANN, X. vi. 56. — *Calliphlox Anna*, GAMBEL, Pr. A. N. Sc. Phil. III. 1846, 3. IB. Journ. 2d Series, I. 1847, 32. — *Atthis Anna* (REICHENBACH), BAIRD, P. R. Rep. IX. Birds, 137. — *Calypte Anna*, GOULD, Introd. Trochilidæ.

Trochilus icterocephalus, NUTTALL, Man. I. 2d ed. 1840, 712. (Male with forehead covered with yellow pollen.)

SP. CHAR. Tail deeply forked; external feather narrow, linear. Top of the head, throat, and a moderate ruff, metallic red, with purple reflections. Rest of upper parts and a band across the breast, green. Tail feathers purplish-brown, darkest centrally. In the *female* the tail is slightly rounded, not emarginate; the scales of the head and throat are wanting. Tail barred with black, and tipped with white. Length, 3.80 to 4.00; extent, 5.00; wing, 2.00; tail, 1.45. Iris brown; bill and feet black.

Hab. California.

This large species seems to be peculiar to California, and constantly resident south of San Francisco, in mild winters remaining among the foot-hills of the Sierra Nevada, at least fifteen hundred feet above the sea, where I have found them common in February. They find flowers more abundant at that season than in the dry summer, and many insects. The males are in

fine plumage very early in January, though the brilliancy of their ruby head increases towards spring. The nest described by Nuttall seems to be rather that of *T. Alexandri*, as all those of this species which I have seen are about twice as large, and covered externally with lichens, even when on branches not having these parasites. One I found in a thicket of wild gooseberry, about two feet above the ground, and built in a triple fork. It was on top of an old nest of the same kind, and seemed to have been deserted before I found it, perhaps robbed of its eggs. Others are built at various heights and positions, often in gardens, and sometimes on dead branches, without any attempt at concealment except the outside covering of lichens. They are often almost wholly of moss, with only a lining of feathers and down from plants. The eggs are, as usual, two, and white. About San Francisco the young are sometimes hatched as early as March 15th, when the climate is really warmer than it becomes in summer, after the cold sea-winds begin to blow. This species is, however, more hardy than others, being common along the coast border, though I did not observe them near the summits of the Sierra Nevada.

The notes of the male sound very much like those produced by filing a saw, and sometimes like the sound of whetting a scythe. They come familiarly about the city, and often into rooms, in search of flowers. Like all humming-birds, they are so confident of their power of flight as to alight within a few feet of a man, eying him curiously, and uttering their harsh cry; but on the least motion darting away like a meteor.

Calypte Costæ, Bourcier.

COSTA'S HUMMING-BIRD.

Ornismya Costæ, Bourcier, Rev. Zool. Oct. 1839, 294. (Lower California.) — *Atthis Costæ*, Reichenbach, Journ. für Orn. Extraheft, 1853, 1854. — Baird, P. R. Rep. IX. Birds, 138. — Kennerly, X. iv. 22; pl. xix.
Calypte Costæ, Gould, Mon. Humming-Birds.

Sp. Char. Tail very slightly emarginated and rounded; exterior feather very narrow, and linear. A very long ruff on each side of the throat. Head above and below, with the ruff, covered with metallic-red, purple, violet, and steel-green. Remaining upper parts and sides of the body green. Throat under and between the ruffs, side of head be-

hind the eye, anal region, and under tail coverts, whitish. *Female* with the tail rounded, scarcely emarginate; barred with black, and tipped with white. The metallic colors of the head wanting. Length, 3.30 to 3.50 inches; extent, 4.40; wing, 1.75; tail, 1.10; bill, 0.68. Iris brown; bill and feet black.

Hab. Southern California and Colorado Basin. Cape St. Lucas.

This species, distinguished from *A. Anna* by its smaller size, and purple instead of ruby-red head, probably winters in the Colorado Valley in small numbers, as Dr. Kennerly found them along William's Fork on the 9th of February. I did not, however, observe any at Fort Mojave until March 5th, and they were not numerous afterwards. At San Diego, in the backward spring of 1862, I first saw them April 22d, and have since found them north to San Francisco, where, however, they are rare.

Their notes are faint chirps, and sometimes a sound like the highest and sharpest note that can be made on a violin, uttered as a song, when the male is resting on a dead twig.

Their habits of building have not yet been observed, but doubtless resemble those of the Anna humming-bird, to which this is so nearly related.

This species is extremely abundant at Cape St. Lucas, where many specimens, mostly in immature plumage, were collected by Mr. Xantus. It appears also to occur on the eastern side of the Gulf of California, at Guaymas and Mazatlan, as well as on the table-lands of Mexico, and perhaps of Guatemala.

Genus ATTHIS, Reichenbach.

Atthis, Reichenbach, Cabanis, Journ. für Orn.; Extraheft für 1853, 1854, Appendix B. (Type, *Ornysmya Heloisæ*.)

Gen. Char. Size very diminutive; bill short, scarcely longer than the head. Outer primary attenuated as in *Selasphorus*; the tail graduated; the feathers, however, not lanceolate-acute, but rounded at end, and tipped with white in the male.

This genus seems closely related to *Selasphorus*, agreeing in character of throat, the curious attenuation of outer primary, and the general shape of the tail, with its rufous base and edging. The feathers, however, are not lanceolate or pointed, either sharply as in *S. rufus*, or obtusely as in *platycercus*, but are more equal to near the end, where they round off. The white tips of the tail in the male seem to be the principal reason why Mr. Gould removes the single species from *Selasphorus*, where it was previously placed by him, and where perhaps it might have not inappropriately remained.

Atthis Heloisæ, Lesson.

HELOISA'S HUMMING-BIRD.

Ornysmya Heloisæ, Lesson, Delattre, Rev. Zool. 1839, 15. (Xalapa.) — *Mellisuga Heloisæ*, Gray and Mitchell, Gen. Birds, I. 113. — *Tryphæna Heloisæ*, Bonaparte, Consp. Troch. Rev. Mag. Zool. 1854, 257. — *Selasphorus Heloisæ*, Gould, Mon. Trochil. III.; pl. 141. — *Atthis Heloisæ*, Reichenbach, Jour. für Orn. Extraheft, 1853, App. 12. — Gould, Introd. Trochil. 1861, 89. — Elliot, B. N. A. I. pl. 21.

Sp. Char. Above, metallic green with golden reflections; beneath white, the sides of breast glossed with green, the flanks with rufous, which tinges very faintly the crissum. Gorget brilliant violet or light purplish-red, bordered behind by clear white. All the tail

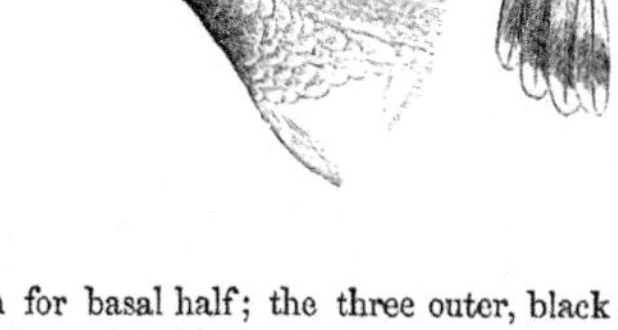

feathers rufous-cinnamon for basal half; the three outer, black centrally and tipped with white (mixed with reddish on the third); the fourth, green centrally tipped with black; the central, entirely green for the exposed portion, perhaps glossed with blackish at the end. Length, 2.70; wing, 1.35; tail, 1.00; exposed part of bill above, 0.45.

Female. Outer primary not attenuated; colors similar to male, wanting the metallic gorget; the feathers spotted with dusky; crissum and flanks more rufous; innermost tail feathers entirely green; other feathers as in male, but with the central black encroaching on the basal rufous; third and fourth feathers tipped with reddish-white.

Hab. Southern New Mexico to Guatemala.

The introduction of this species into the fauna of the United States is based on a female specimen collected by Mr. J. H. Clark at El Paso, Texas, and for a time supposed to be *Selasphorus rufus*, but, after a careful examination by Mr. Lawrence, has been pronounced to belong to this species. Its range is southward along the highlands of Mexico to Guatemala.

The species is very much like *Selasphorus* in shape, and hardly differs more than *S. rufus* and *platycercus* do from each other. The male is easily distinguished from its allies; the females are closely related to those of *rufus*, differing in much shorter bill (0.45 to 0.65), much less rufous on the more nearly even tail, with broader feathers, etc.

GENUS **STELLULA**, GOULD.

Stellula, GOULD, Introd. Trochil. 1861, 90. (Type, *Trochilus calliope*, GOULD.)

GEN. CHAR. Bill rather longer than the head, straight; wings much developed, reaching beyond the tail, which is short, nearly even or slightly rounded; and with the innermost feathers abruptly shorter; the outer feather rather narrower and more linear than

the others. Metallic throat feathers elongated and rather linear and loose, not forming a continuous metallic surface. Central tail feathers without green.

This genus established by Gould has a slight resemblance to *Atthis*, but differs in absence of the attenuated tip of outer primary. The outer three tail feathers are longest and nearly even (the second rather longest), the fourth and fifth equal and abruptly a little shorter; the latter without any green. The feathers are rather broad (the outermost least so), and are obtusely rounded at end. The tail of the female is quite similar. The absence of green on the tail seems a good character. But one species is known of the genus.

Calothorax is a closely allied genus, in which the tail is considerably longer. One species, *C. cyanopogon,* will probably be yet detected in New Mexico. (Baird.)

Stellula calliope, GOULD.

THE CALLIOPE HUMMING-BIRD.

Trochilus calliope, GOULD, Pr. Z. S. 1847, 11. (Mexico.) — *Calothorax calliope,* GRAY, Genera, I. 100. — BONAPARTE, Rev. Mag. Zool. 1854, 257. — GOULD, Mon. Troch. III.; pl. 142. — XANTUS, Phil. Ac. 1859, 190. — *Stellula calliope,* GOULD, Introd. Troch. 1861, 90.

SP. CHAR. *Male.* Above, except on tail, golden green; beneath white, the sides glossed with green; the flanks somewhat with rusty; crissum pure white. Throat feathers pure white at base, terminal half violet-red, as is *Atthis Heloisæ;* the sides of neck pure white. Tail feathers brown, edged at base, especially on inner webs (but inconspicuously) with rufous; the ends paler as if faded; central feathers like the rest. Under mandible yellow. Length, 2.75; wing, 1.60; tail, 1.00; bill above to base of feathers, 0.55.

Female, without the metallic gorget (replaced by a few dusky specks), and the throat

feathers not elongated; no green on side, and more tinged with rufous beneath. A white crescent under the eye. Tail more rounded and less emarginate than in the male; the outer three feathers green at base, then black, and tipped with white; the fourth green and black; the fifth green, with a dusky shade at end; all except central, edged internally at base with rufous. The under mandible is paler at base than elsewhere, but not yellowish-white as in the male.

Hab. Mountains of Washington, Oregon, California, to Northern Mexico.

The male bird is easily distinguished from other North American species by its very small size, the snowy-white bases of the elongated loose throat feathers, and by the shape of the tail, as also the absence, at least in the only male before me, of decided metallic green on the central tail feathers. The females resemble those of *A. Heloisæ* most closely, but have longer bills and wings, broader tail feathers, and their rufous confined to the edges, instead of crossing the entire basal portion. *Selasphorus platycercus* and *rufus* are much larger, and have tails marked more as in *A. Heloisæ.* (Baird.)

Mr. Lord found this species numerous near the northwest boundary line, east of the Cascade Mountains in May, arriving with *S. rufus*, and frequenting the *Ribes* flowers also. During the summer he observed them chiefly about rocky hillsides at great altitudes, frequently above the line of perpetual snow, and discovered the nest, usually on the branch of a young pine, artfully concealed among the leaves at the very end so as to be rocked by the breeze.

I probably saw the same species in August, 1853, on the summit of the Cascade Mountains, but mistook them in their moulting plumage for the young of *S. rufus*. (Cooper.)

GENUS **HELIOPÆDICA**, GOULD.

Heliopædica, GOULD, Mon. Trochil. II.; Introd. Trochil. 1861, 60. (Type, *Trochilus melanotis*, SWAINSON.)

GEN. CHAR. Bill longer than head, depressed; broad at its exposed base; the frontal feathers not advancing forward beyond the beginning of the nostrils, nor so far as those of the chin. Hind toe shorter than the lateral; tarsi feathered. Outer primary not attenuated. Tail nearly even, slightly rounded and emarginate, the feathers broad, the webs nearly even. Metallic feathers of throat not elongated. *Female* quite similar in form.

This genus is quite peculiar among those of North America, in the exposure of the base of bill, which is entirely bare between the lengthened nostrils, instead of covered by the frontal feathers. This makes the bill appear very broad, although it really is more so than in our other genera. The feathers on the chin extend considerably beyond those of the forehead, instead of to about the same line. The tail and its feathers are much broader than in the other genera.

Of *Heliopædica* there are two species, both with green throats and white stripe in the black of the side of the head; the top of head blue; they differ, however, in decided characters. (Baird.)

Heliopædica Xantusii, Lawrence.

XANTUS'S HUMMING-BIRD.

Amazilia Xantusii, Lawrence, Ann. N. Y. Lyc. VII. April, 1860, 109. — *Heliopædica Xantusii*, Gould, Mon. Troch. II.; pl. 65. Ib. Introd. Troch. 61. — Elliot, Ill. B. N. A. I. pl. 23.

Heliopædica castaneocauda, Lawrence, Ann. N. Y. Lyc. 1860, 145 (female).

Sp. Char. *Male.* Above metallic green; the forehead, cheeks, and chin, velvety-black (the former perhaps deep blue when in perfect plumage). A distinct white stripe from bill, through and behind the eye. Throat and fore part of breast, brilliant metallic green; rest of under parts cinnamon-rufous. All the tail feathers purplish-rufous; the central glossed with green above near the edges; the others obscurely edged with blackish

along ends. Bill red at base; black at ends. Length, 3.50; wing, 2.10; tail, 1.40; exposed portion of bill about 0.65.

Female. Forehead and all under parts light cinnamon beneath, without any green or any dusky specks on throat; white cheek stripe appreciable, but tinged with rufous. Tail as in male, but the central feathers entirely green above; the others, except the outer, with a dusky greenish or purplish spot on each web near the end. Whole upper mandible apparently dusky; base of lower, red.

Hab. Cape St. Lucas.

This well-marked and interesting species we owe to Mr. Xantus, with many other birds of the West Coast. It is sufficiently distinct to require no comparison than that given under the general head; it can be separated from *H. melanotis* in all stages of plumage by the rufous tail.

Specimens vary sometimes in the intensity of the rufous shade; and, as stated, it is probable that the forehead, instead of being black in full plumage, is deep blue, as in *melanotis*. (Baird.)

ORDER SCANSORES.

CHAR. Toes in pairs, two in front and two behind, the outer anterior being usually directed backwards. Tail feathers eight to twelve. Primaries ten. Tarsi (in our species), with broad plates anteriorly, small scales behind.

In this order we find the most perfect development of the bird type, analogous to *Quadrumana*.

The parrots and trogons of the tropics are members of this order, as usually understood. There are, however, none of these two families found within our State.

The term *Zygodactyle*, or cross-toed, is also applied to these birds.

FAMILY CUCULIDÆ, THE CUCKOOS.

CHAR. Bill compressed laterally, gently curved, sometimes attenuated and generally lengthened. A few bristles at the base of bill or none. Tarsi lengthened, toes rather short. Tail long and soft, of ten feathers in our genera.

This family is numerously represented in warm countries. The genus *Cuculus*, and allied forms of the Old World, are celebrated for never building a nest of their own, but laying eggs in those of others birds, like our *Molothrus*.

GENUS **GEOCOCCYX**, WAGLER.

Geococcyx, WAGLER, Isis, 1831, 524.
Leptostoma, SWAINSON, Classification Birds, II. 1837, 325.

GEN. CHAR. Bill long and strong, slightly compressed, and at least as long as the head; head crested; loral feathers and those at base of bill stiffened and bristly. Nostrils

G. Californianus.

elongated, linear. A naked colored skin around and behind the eye; the eyelids ciliated. Tarsi longer than the toes; very stout. Wings very short and concave; the tertials as long as the primaries. Tail longer than the head and body; composed of ten narrow, much graduated feathers.

Another species inhabits Mexico.

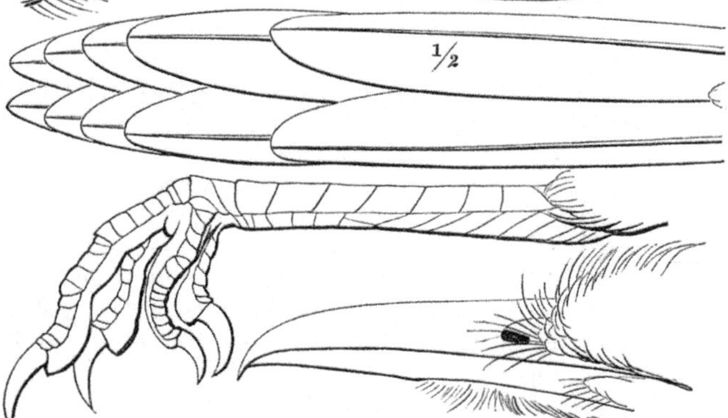

G. Californianus.

Geococcyx Californianus, Lesson.

THE PAISANO; THE ROAD-RUNNER; THE CHAPPARAL COCK.

Saurothera Californiana, "Lesson, Complem. Buff. VI. 1829, 420. Ib. Ann. du Mus. 1835, 121 ; pl. ix."

Geococcyx viaticus, Hartlaub, Rev. Zool. 1844, 215. — M'Call, Pr. A. N. Sc. III. July, 1847, 234. — Heermann, J. A. N. Sc. Phil. 2d Series, II. 1853, 270. — Newberry, Zool. Cal. and Or. Route, 91 ; P. R. R. Rep. VI. 1857.

Leptostoma longicauda, Swainson, Birds, II. 1837, 325. — Gambel, Pr. A. N. S. I. 1843, 263.

Geococcyx Mexicanus, Gambel, Jour. A. N. Sc. Phil. 2d Series, I. 1849, 215 (not of Gmelin). — Cassin, Ill. I. 1855, 213 ; pl. xxxvi. — Heermann, P. R. Rep. X. vi. 59.

G. Californianus (Lesson), Baird, P. R. Rep. IX. Birds, 73. — Kennerly, X. iv. 21.

Sp. Char. Tail very long ; the lateral feathers much shortest. An erectile crest on the head. A bare skin around and behind the eye, colored bluish-white and orange when alive. Legs very long and stout. All the feathers of the upper parts and wings of a dull metallic olivaceous-green, broadly edged with white near the end. There is, however, a tinge of black in the green along the line of white, which itself is suffused with brown. On the neck the black preponderates. The sides and under surface of the neck have the white feathers streaked centrally with black, next to which is a brownish suffusion. The remaining under parts are whitish, immaculate. Primary quills tipped with white, and with a median band across the outer webs. Central tail feathers olive-brown ; remaining ones clear dark green, all edged, and (except the central two) broadly tipped with white. Top of the head dark blackish blue. Length, 20.00 to 23.00 ; extent, 21.00 ; wing, 6.00 to 7.00 ; tail, 12.00 to 13.00. Iris brown and yellow ; bill olive ; feet black.

Hab. Middle Texas, New Mexico, and California to Central Mexico. Seen as far north as Fort Reading, California.

This remarkable bird is abundant in the southern part of the State, and follows the valleys northward nearly to the head of the Sacramento, being resident everywhere south of San Francisco at least. I observed them first at Fort Mojave, Colorado Valley, on the 25th of February, when I heard one making a low cooing noise like that of a dove, for which I at first mistook it. After much watching I surprised it perched on top of a low tree, and shot it. On April 3d the Indians brought me a female specimen, probably caught on the nest, as it laid an egg in the cage in which I put it, which unfortunately got broken. This bird was timid, and made no attempts to defend itself, running off on the ground when I released it, instead of attempting to fly. I afterwards saw this one or others several times sitting on trees, where it allowed me to approach within ten feet, cooing harshly, and chattering its mandibles together, at the same time jerking up its tail and erecting its crest.

I saw none in the barren mountains and plains towards the coast, until I reached San Diego, where, a little after sunrise, I saw a pair sitting on a fence in the cemetery, close to the town, where they sat still while several wagons passed by.

Where not molested, they become quite tame, and seem to have a preference for the vicinity of towns and houses. At Santa Barbara I observed a

young one nearly fledged in May. I have seen several nests, but never discovered one with young or eggs in it. The nest is built in a low thicket, composed mostly of sticks, rather loosely put together, and very shallow. The eggs are two, larger than a pigeon's egg, white, and nearly equal in size at each end ; size, 1.65 × 1.22.

The food of this bird consists of insects, lizards, and snakes, probably also of any small or young animal it can destroy. It is said to hedge in the rattlesnake with a circle of cactus-joints, until the reptile, becoming enraged, bites itself and falls an easy prey to the bird. This I have never seen, nor can I understand how such a *hard-shelled* reptile could be enclosed by even a spiny hedge so as to be unable to escape. It is, however, undoubtedly true that the bird kills these, and other snakes also. Its fleetness on foot, when in an open plain, is well known, a fast horse being scarcely able to overtake it. On such occasions it never flies, unless downwards from a height, and its wings seem scarcely ever of much use to it, as it probably cannot fly upwards at all.

Mr. H. E. Dresser, in an article published in the London Ibis, (1865, 466,) speaks of it as abundant in the mezquite region of Texas, especially near the Rio Grande. He found its eggs near San Antonio as late as September, laid usually in a clumsy nest of mezquite twigs, and from two to four in number, as ascertained by him. Its food consists of small snakes, lizards, ticks, and large insects, and when tamed it catches and devours mice with great avidity. The practice of keeping them tamed he found quite general, the object being apparently to have them in readiness in case of sickness, for some forms of which their flesh is considered a perfect cure. One in his possession became very familiar and mischievous, stealing and hiding anything it could carry off, and being particularly fond of tearing up letters and spilling the ink. It was not confined in any way, and frequently visited the neighbors, always, however, returning before evening.

Mr. A. J. Grayson, of San Jose, in an article published in Hutchings's Magazine for November, 1856, gives the following additional particulars, together with a good figure: —

"The road-runner is seldom seen on trees, unless pursued very closely, when it has been seen to spring from the ground to the branches, at a height of ten or fifteen feet at a single bound; but it prefers running along the road or path, from which habit it derives its name.

"I have now in my possession one of these birds, which is becoming quite tame, and readily feeds upon any kind of raw meat, but prefers lizards and small birds, which it swallows whole, feathers and all. If given to him alive, he will play with them awhile before swallowing them, just as a cat will do with a mouse. I have seen him devour three sparrows, one lizard, and a portion of the breast of a coot for his breakfast, without experiencing any apparent inconvenience.

"Although it cannot fly well, by its activity and quickness it easily catches small birds, whether on the ground or in the thicket."

I have heard of their being tamed and kept about gardens to kill mice, insects, etc.

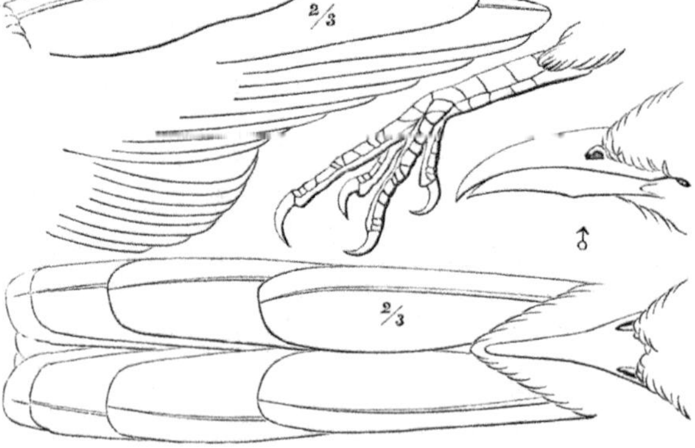

C. Americanus.

GENUS **COCCYGUS**, VIEILLOT.

Coccygus, VIEILLOT, Analyse, 1816.
Erythrophrys, SWAINSON, Class. Birds, II. 1837, 322.

GEN. CHAR. Head without crest; feathers about base of bill soft; bill nearly as long as the head, decurved, slender, and attenuated towards the end. Nostrils linear. Wings

C. Americanus.

lengthened, reaching the middle of the tail; the tertials short. Tail of ten graduated feathers. Feet weak; tarsi shorter than the middle toe.

Several species inhabit Tropical America.

Coccygus Americanus, LINNÆUS.

THE YELLOW-BILLED CUCKOO.

Cuculus Americanus, LINNÆUS, Syst. Nat. I. 1766, 170, 10. — *Coccygus Americanus*, BONAPARTE, Obs. Wilson, 1825, No. 47. — AUDUBON, Orn. Biog. I. 1832, 18; V. 520; pl. 2. IB. Birds Amer. IV. 1842, 293; pl. 275. — NUTTALL, Man. I. 2d ed. 652. — *Coccygus Americanus*, BAIRD, P. R. Rep. Birds, IX. 76.

SP. CHAR. Upper mandible and tip of lower, black; rest of lower mandible and cutting edges of the upper yellow. Upper parts of a metallic greenish-olive, slightly tinged with ash towards the bill; beneath, white. Tail feathers (except the median, which are like the back) black, tipped with white for about an inch on the outer feathers, the external one with the outer edge almost entirely white. Quills orange-cinnamon; the terminal portion and a gloss on the outer webs olive; iris brown. Length, 12.00; wing, 5.95; tail, 6.35. (Baird.)

Hab. Eastern United States to the Missouri Plains. Very rare in California. Oregon. (Nuttall.)

Although Nuttall and Townsend mentioned seeing this bird in "Oregon," no one has since confirmed its occurrence on any part of the West Coast until the summer of 1862, when Mr. Gruber obtained two specimens, shot in Napa Valley. Dr. Newberry, in P. R. Rep. VI. iv. 92, does indeed mention *hearing* what he thought was the note of a cuckoo, but did not obtain a specimen, and referred it to the other Atlantic Coast species (*Coccygus erythropthalmus*). He might easily have been mistaken about the cry of the bird, as I was myself at San Diego, until I ascertained that the noise was made by a kind of toad (*Scaphiopus*), and produced the same deceptive effect as that of the bird does, sounding as if in a different direction from the real one. I have not heard this bird in Santa Clara Valley, and think its migration, if any more than an occasional visitor, must be by way of the interior valleys of the State. In the East it is called "koubird" from its note, like *kou-kou-kou,* often repeated, and also rain-crow.

Though nearly allied to the cuckoo of Europe, it builds a nest and hatches its own eggs. The nest is usually in a low tree loosely constructed of small twigs, almost without a cavity, and the eggs are bluish-green, without spots, the ends nearly equal, and rather large. According to Audubon, they begin to sit on one egg, but lay another soon after, the first being hatched so much earlier that the young are often of quite different sizes. When one is fledged, another egg is laid, and the young assists by its warmth in hatching it. This occurs successively through the summer, until as many as eleven have been raised from one nest.

This remarkable habit is not known to characterize any other bird, except perhaps some eagles. They feed, like the *Geococcyx,* at times on small birds and their eggs, as well as insects. In "Hutchings's California Magazine," an account of the habits of one of the North American species, extracted from

"Wilson's Ornithology," is published, with a figure of some very distinct foreign species.

While stationed at Sacramento in 1865, I found these birds quite common in the large cottonwood-trees about the city, from about May 1st to September 1st, but as usual very shy, and I could not find any nests. While travelling south through the San Joaquin Valley, in September, I neither saw nor heard them, though there are groves along the river-banks suited for them; and as they migrate south in that month from the Atlantic States, they must pass very silently and rapidly into Mexico, as I have not seen them in the south parts of California.

FAMILY PICIDÆ, THE WOODPECKERS.

CHAR. Bill straight, rigid, and chisel-shaped at the tip, the base without rictal bristles. Feet stout, covered anteriorly with broad plates. Tail feathers twelve, the exterior very small and concealed. Primaries ten, the first very short, secondaries nine to twelve. Tongue elongated and acute, with short spines or barbs on each side near the point.

Of the eight genera found in Northern America, California possesses representatives of all, and more species than any other district of equal ex-

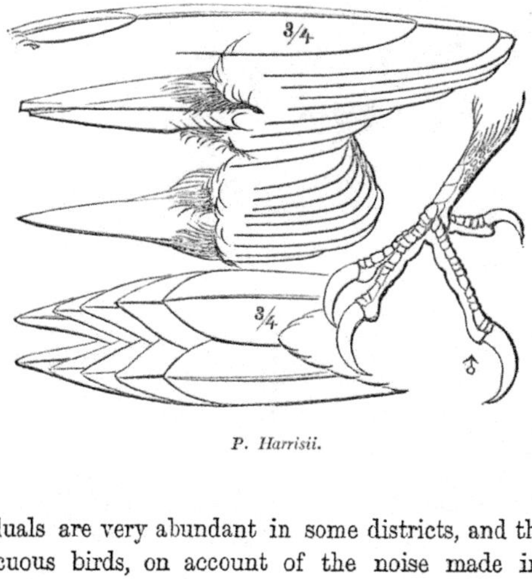

P. Harrisii.

tent. Individuals are very abundant in some districts, and they are everywhere conspicuous birds, on account of the noise made in hammering against trees, and the size of many of the species. All make their nests in holes of trees, usually excavated by themselves with considerable effort; they lay four or five white eggs, with crystalline shell. Their food consists in great part of insects, though fruit, berries, and even the bark of trees are occasionally devoured.

Genus **PICUS**, Linnæus.

Picus, Linnæus, Syst. Nat. 1748.

Gen. Char. Bill equal to the head, or a little longer; the lateral ridges conspicuous, starting about the middle of the base of the bill; the basal, elongated, oval nostrils nearer the commissure; the ridges of the culmen and gonys acute, and very nearly straight, or slightly convex towards the tip; the bill but little broader than high at the base, becoming compressed considerably before the middle. The outer posterior toe

P. villosus.

longest; the outer anterior about intermediate between it and the inner anterior, the inner posterior reaching to the base of the claw of the inner anterior. Tarsus about equal to the inner anterior toe; shorter than the two other long toes. Wings rather long, reaching the middle of the tail, rather rounded; the fourth and fifth quills longest, the quills rather broad and rounded.

The species are numerous, and found in nearly all parts of the world. Probably none of the genus, as now restricted, are to be found on the West Coast, except those here mentioned.

Picus Harrisii, AUDUBON.

HARRIS'S WOODPECKER.

Picus Harrisii, AUDUBON, Orn. Biog. V. 1839, 191; pl. 417. IB. Syn. 1839, 178. IB. Birds Amer. II. 1842, 242; pl. 261 (dark-bellied variety). — NUTTALL, Man. I. 2d ed. 1840, 627. — BAIRD, P. R. Rep. IX. Birds, 87. — KENNERLY, X. iv. 21. — HEERMANN, X. vi. 57. — COOPER and SUCKLEY, XII. iii. Zool. of W. T. 159.
Picus hyloscopus, CABANIS and HEINE.

SP. CHAR. Above black, a white stripe down the back. The only white spots on the surface of the folded wings are seen on the outer webs of the primaries and outer secon-

Male.

daries (none on tertials). Beneath whitish, with faint streaks on the side of the body. Two white and two black stripes on each side of the head; the latter confluent with the black of the neck, the upper white stripe nearly confluent. Three outer tail

Female.

feathers with the exposed portions white. Length, 9.75; extent, 16.00 to 17.00; wing, 5.75.

Male, with a nuchal scarlet crest covering the white of the back of the head; wanting in the female. Iris brownish-red; bill slate-colored; feet dark gray.

Young.

Young, with whole top of head red.
Hab. From the Pacific Coast to the eastern slope of the Rocky Mountains.

This is chiefly a Northern bird, frequenting the forests of all kinds up to the summits of the Sierra Nevada, but also resident as far south as Santa Barbara, descending to the eastern branches of the Colorado River in winter (Kennerly), and to Tejon Pass.

I found a pair excavating a burrow in a stump not more than four feet above the ground, May 4, 1863. I enlarged the opening, hoping to find eggs, as the old birds were very solicitous, but was disappointed, and did not examine it again, supposing that they would certainly desert the place. I found them more common in the higher Coast Range near Santa Cruz, but they become still more so towards the Columbia River.

The cry of this species is louder than that of most of the small woodpeckers, and it is rather shy, especially if it thinks itself pursued. Like most other species, it feeds at times on fruits and berries, and sometimes visits gardens. It is one of the "sap-suckers" also, though, like *P. Gairdneri* it probably does more good than harm in the orchard by destroying the larvæ of insects burrowing in the bark, and also the perfect insects that are depositing the eggs from which the larvæ are hatched.

It is much to be questioned whether this species is any more than a Western variety of *Picus villosus*, the common hairy woodpecker of the Eastern States, differing from it in the absence of many of the white spots of the wings.

Picus Gairdneri, AUDUBON.

GAIRDNER'S WOODPECKER.

Picus Gairdneri, AUDUBON, Orn. Biog. V. 1839, 317. IB. Syn. 1839, 180. IB. Birds Amer. IV. 1842, 252. — BAIRD, P. R. Rep. IX. Birds, 91. IB. Birds N. Amer.; pl. 85. — HEERMANN, X. vi. 57. — COOPER and SUCKLEY, XII. iii. Zool. of W. T. 159.
Picus meridionalis, NUTTALL, Man. I. 2d ed. 1840, 690 (not of Swainson).
Picus turati, MALHERBE, — *Dryobates homorus*, CABANIS and HEINE.

SP. CHAR. Very similar in color to *P. Harrisii*, but smaller. Larger wing coverts, and more exposed tertials, either pure black, or with but occasional spots on the outer web in the latter. Back with a white median stripe. Side of head with two white and two

black stripes. Two outer tail feathers white, with two bands of black at the end. Length, 6.75 to 7.00; extent, 12.00; wing, 3.75, generally rather less.

Male, with a scarlet occipital band. Iris brown; bill slate-color; feet gray.

Hab. With *P. Harrisii*, from the Pacific Coast to the eastern base of the Rocky Mountains, northward.

This species is apparently confined chiefly to the northern parts of the State, as I have not met with it south of the Santa Clara Valley. I there found one of their nests containing young on the 24th of May, 1864. It was burrowed in a small, partly rotten tree about five feet from the ground. From the fact of their breeding so far south, we may suppose that they occur much farther south on the mountains, like almost all the northern birds.

This woodpecker frequents chiefly the smaller trees in the vicinity of the evergreen woods, and may be found at all seasons industriously tapping their bark to obtain insects. It is so closely analogous to *P. pubescens*, the "little sap-sucker" of the Atlantic States, that we may expect it to show a similar habit of perforating apple-tree bark to suck the sap. It is, however, very doubtful whether the injury thus caused counterbalances the benefit which the bird effects in the destruction of insects, especially as no one has ever observed trees to be killed in this way, and the holes made are so shallow as not to reach the principal sap-wood or inner bark.

As this is a familiar, industrious little bird, we may hope to see it encouraged in the vicinity of orchards, at least until the true extent of benefit or damage it produces shall be fully ascertained.

This species, like *P. Harrisii*, will probably ultimately be considered as a mere local variety of another (in this case *P. pubescens*).

Picus Nuttalli, Gambel.

NUTTALL'S WOODPECKER.

Picus Nuttalli, Gambel, Pr. A. N. Sc. I. April, 1843, 259. (Los Angeles, Cal.) — Baird, P. R. Rep. IX. Birds, 93. — Newberry, VI. 89. — Heermann, X. vi. 57.

Picus scalaris, Gambel, J. A. N. Sc. Phil. 2d Series, I. Dec. 1847, 55; pl. 9, f. 2, 3 (not of Wagler).

Picus Wilsonii, Malherbe, Rev. Zool. 1849, 529.

Sp. Char. Back black, banded transversely with white; not, however, as far forward as the neck. Crown black, with white spots. Occiput and nape crimson. Tufts of

Male.

feathers at the base of the bill white. Sides of the head black, with two white stripes, one above the eye and passing down on the side of the neck, the other below and interrupted by the black. Under parts smoky yellowish-white, spotted on the sides of the head with black. Predominant color of the three outer tail feathers white, with three, two, or one spot on the outer web near the end. Length, 8.00; wing, 4.50.

Female, with the top of the head uniform black. Iris brown; bill black; feet gray.

Hab. Coast region of California.

This little woodpecker is abundant towards the coast of California, and among the foothills of the interior, west of the Sierra Nevada. It frequents the oaks and smaller trees almost exclusively, avoiding the coniferous forests. It is very industrious, and not easily frightened when engaged in its laborious hammering on the bark of trees, allowing a very near approach, though, like all woodpeckers, somewhat suspicious and wary when pursued. On April 20, 1862, I discovered a burrow of this bird near San Diego, in

Female.

a rotten stump only about four feet above the ground, and captured the female on the nest. It contained five pure and pearly white eggs, which is the number and character of those laid by all of the small woodpeckers, as far as known.

This species apparently remains throughout the year in the valleys, and migrates little if at all. I have not observed it west of the Coast Range, except near Santa Barbara, nor have I seen any around gardens or orchards. None have been observed north or east of this State, the *P. scalaris* taking its place east of the mountains.

Picus scalaris, WAGLER.

THE ARIZONA WOODPECKER.

Picus scalaris, WAGLER, Isis, 1829, V. 511. (Mexico.) — BONAPARTE, Consp. 1850, 138.
Picus (*Dyctiopicus*) *scalaris*, BONAPARTE, Consp. Zygod. Aten. Ital. 1854, 8. — BAIRD, P. R. Rep. IX. Birds, 94. — HEERMANN, X. v. 18; VI. 57. — U. S. and Mex. Bound. Rep. II. iii. 5; pl. iii.
Picus parvus, CABOT. — *Picus Bairdii*, SCLATER. — *Picus vagatus*, and *Orizabæ*, CASSIN.

SP. CHAR. Back banded transversely with black and white to the neck. Black and white bands of back about equal. Crown crimson spotted with white, from the bill to the nape; tuft of feathers at the base of the bill brown. A white stripe above the eye, continued on the side of the neck; another under the eye, interrupted by the black of the side of the head. Under parts smoky brownish-white, spotted on the sides of the breast, and banded on the flank with black. Predominating color of the three outer tail feathers

Male.

black, with white bands chiefly on the outer webs. Length, about 7.50; extent, 13.00; wing, 4.25.

Female, without red on the head. Iris dark brown; bill horn-color; feet gray.

Female.

Hab. Rocky Mountains and its slopes, west to San Bernadino Mountains of California.

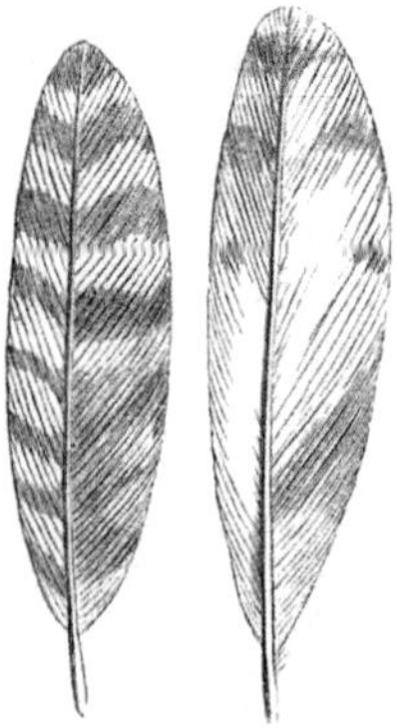

P scalaris *P. Nuttalli.*

Abundant in the Colorado Valley, and sometimes seen on the bushes covering the neighboring mountains. In habits they are the exact representatives of *P. Nuttalli,* which they so much resemble.

The relationships of color between this species and *Nuttalli* are very close. The differences are, however, well illustrated by the figures of the head and by the diagram, on the preceding page, of the tails of the two.

Picus scalaris, var. **lucasanus**, Xantus.

THE CAPE WOODPECKER.

Picus lucasanus, Xantus, Pr. A. N. S. 1859, 298, 302. — Malherbe, Mon. *Picidæ,* I. 166. — Cassin, Pr. A. N. Sc. 1863, 195.

Sp. Char. General appearance that of *Picus Nuttalli* and *scalaris*. Bill stout, as long as or longer than the head. Above black, banded transversely with white on the back and scapulars to the nape, the white band narrower, the rump and inner tail feathers entirely black; quills with a row of white spots on each web, the outer square, the inner rounded, these spots on the tertials becoming transversely quadrangular. Beneath brownish-white, with rounded black spots on the sides of the breast, passing behind on the flanks

and under tail coverts into transverse bars. Greater inner wing coverts transversely barred. Outer two tail feathers white, with one, sometimes two terminal bars, next to which are one or two bars on the inner web only; third feather black, the outer web mostly white, with traces of a terminal black bar; sometimes there is a greater predominance of black on the inner web. Two white stripes on side of head, one starting above, the other below the eye, with a tendency to meet behind and form a whitish collar on the nape.

Male, with the entire top of the head streaked with red, becoming more conspicuous behind; each red streak with a white spot at base. Feathers covering the nostrils smoky-brown. Length, 7.15; extent, 12.15; wing, 4.00; bill above, 1.00; middle toe and claw, 0.80; tarsus, 0.76.

Hab. Cape St. Lucas.

Mr. Xantus makes no mention of any peculiar habits of this variety, which are probably much like those of *scalaris*. Of the distinctness of this bird as a species from *P. Nuttalli* and *scalaris* I had at one time no doubt, but the discovery that otherwise typical *scalaris* from Mazatlan and Western Mexico generally have the same markings on the tail, has induced me to consider it as a kind of connecting link. I have, however, thought it best to give a detailed description for comparison. Of about the same size with *Nuttalli*, the bill and feet are much larger. The legs, indeed, are nearly if not quite as large as those of a male *P. villosus*, from Pennsylvania; the bill, however, is somewhat less. The relationship to typical *P. scalaris* is seen in the dorsal bands extending to the nape; the smoky-brown feathers of the nostrils; the whole top of head red (scattering anteriorly); the brownish shade beneath; the width of the white cheek bands. On the other hand, it has the black bands of the back rather wider than the white, as in *Nuttalli;* and the white outer tail feathers even less banded with black. The two outer are entirely white, with one terminal black bar; one or two spots on the outer web; and two or three bands on the inner, with a sub-basal patch on the inner web even smaller than in *Nuttalli*. It is rarely that even two continuous transverse bands can be seen to cross both webs of the tail. The bill and feet are much larger.

The following measurements, taken from the largest specimens before me of *Dyctiopicus*, and one of *P. villosus*, will illustrate what has been said of the size of the bill and feet of *P. lucasanus*. (Baird.)

	P. villosus.	*P. lucasanus.*	*P. Nuttalli.*	*P. scalaris.*
	884	♂ 12939	♂ 4482	♂ 6105
Bill from forehead,	1.26	1.10	.90	.99
Tarsus,	.76	.76	.70	.68
Middle toe and claw,	.87	.84	.75	.65
Claw alone,	.39	.34	.32	.31
Outer hind toe and claw,	.95	.84	.79	.80
Claw alone,	.40	.32	.31	.31

Picus albolarvatus, Cassin.

THE WHITE-HEADED WOODPECKER.

Leuconerpes albolarvatus, Cassin, Pr. A. N. Sc. V. Oct. 1850, 106. (California.)—*Melanerpes albolarvatus*, Cassin, Jour. A. N. Sc. 2d Series, II. Jan. 1853, 257; pl. 22.—Newberry, Zool. Cal. & Or. Route; P. R. Rep. VI. 1857.—Heermann, X. vi. 59.—*Picus (Xenopicus) albolarvatus*, Baird, P. R. Rep. IX. Birds, 96.—Cooper and Suckley, XII. iii. Zool. of W. T. 160.

Sp. Char. Fourth and fifth quills equal and longest. Entirely bluish-black, excepting the head and outer edges, with the entire basal portion, of the primaries, which are white.

Length, about 9.50; extent, 16.50; wing, 5.25. *Male*, with a narrow line of red on the nape
Hab. Cascade Mountains of Oregon, and southward into California.

I found this beautiful species quite common near the summits of the Sierra Nevada, lat. 39°, in September, 1863, and obtained three specimens. In 1860 I found it also common at Fort Dalles, Columbia River, its chief range of distribution being between these two points, though I have also found it as far north as Fort Colville, Washington Territory, near lat. 49°. It is rather a silent bird, and little known.

Genus **PICOIDES**, Lacep.

Picoides, Lacep. Mem. Inst. 1799.
Tridactylia, Stephens, Shaw, Gen. Zool. 1815.
Apternus, Swainson, Fauna Bor. Amer. II. 1831, 311.

Gen. Char. Bill about as long as the head, very much depressed at the base; the outlines nearly straight; the lateral ridge at its base much nearer the commissure than

P. Arcticus.

the culmen, so as to bring the large rather linear nostrils close to the edge of the commissure. The gonys very long, equal to the distance from the nostrils to the tip of the bill. Feet with only three toes; the outer lateral a little longer than the inner, but slightly ex-

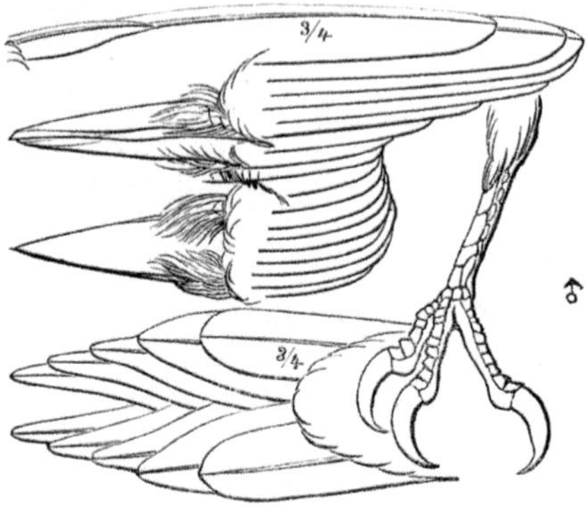

P. hirsutus.

ceeded by the hind toe, which is about equal to the tarsus. Wings very long, reaching beyond the middle of the tail; fourth and fifth quills longest. Color black with a broad patch of yellow on the crown; body transversely banded on the sides. Quills with round spots.

Three species inhabit North America, and others are found on the Old Continent. *P. dorsalis*, Baird, with a white stripe down the middle of back, may perhaps stray as far to the west as the Northern Sierra Nevada or the Cascade Mountains, as it inhabits the Rocky Mountains at least as far south as lat. 40°.

Picoides arcticus, SWAINSON.

THE ARCTIC THREE-TOED WOODPECKER.

Picus (Apternus) arcticus, SWAINSON, Fauna Bor. Am. II. 1831, 313. — *Apternus arcticus*, BONAPARTE, List, 1838. IB. Consp. 1850, 139. — NEWBERRY, Zool. Cal. and Or. Route, 91; Rep. P. R. Surv. VI. 1857. — *Picus arcticus*, AUDUBON, Syn. 1839, 182. IB. Birds Amer. IV. 1842, 266; pl. 268. — NUTTALL, Man. I. 2d ed. 1840, 691. *Picoides arcticus*, GRAY, BAIRD, P. R. Rep. Birds, IX. 98.

Picus tridactylus, BONAPARTE, Am. Orn. II. 1828, 14; pl. xiv. f. 2. — AUDUBON, Orn. Biog. II. 1834, 198; pl. 132.

SP. CHAR. Above entirely uniform glossy bluish-black; a square patch on the middle of the crown, saffron yellow, and a few white spots on the outer edges of both webs of the primary and secondary quills. Beneath white, on. the sides of the breast longitudinally striped, and on the sides of the belly, flanks, and tibial region banded transversely with black. A narrow concealed white line from the eye a short distance backwards, and a white stripe from the extreme forehead (meeting anteriorly) under the eye, and down the sides of the neck. Bristly feathers of the base of the bill brown. Exposed portion

Male.

of two outer tail feathers (first and second) white. Iris brown; bill and feet black. Length, 9.00 to 10.00; extent, 16.00 to 17.00; wing, 5.25; tail, 3.85. *Female*, without yellow on the head.

Hab. Northern portions of the United States to the Arctic regions, from the Atlantic to the Pacific.

I found this bird quite numerous about Lake Tahoe, and the summits of the Sierra Nevada above six thousand feet altitude, in September, and it extends thence northward, chiefly on the east side of these and the Cascade Mountains, as I never saw it near the Lower Columbia. At the lake they were quite fearless, coming close to the hotel, and industriously rapping the trees in the early morning and evening. In the North I found them very wild, probably because the Indians pursue them for their scalps, which they consider very valuable. I noticed their burrows in low pine-trees near the lake, where they had doubtless raised their young. According to Nuttall, they lay four or five white eggs. I have found them silent birds, though probably in the spring they have more variety of calls. The only note I heard was a shrill, harsh, rattling cry, sufficiently distinct from that of any other woodpecker.

Picoides Americanus, Brehm.

THE STRIPED THREE-TOED WOODPECKER.

Var. **fasciatus.** Banded back.

Picus hirsutus, Vieillot, Orn. Am. Sept. II. 1807, 68; pl. cxxiv. (European specimen). — Wagler, Syst. Av. 1827, No. 27 (mixed with *undulatus*). — Audubon, Orn. Biog. V.

1839, 184; pl. 417. IB. Birds Amer. IV. 1842; pl. 269. — NUTTALL, Man. I. 2d ed. 1840, 692. — *Apternus hirsutus*, BONAPARTE, List. — *Picoides hirsutus*, BAIRD, Birds N. Amer. 1858, 98.

? *Picus undulatus*, VIEILLOT, Orn. Am. Sept. II. 1807, 69 (based on Pl. enl. 553, a fictitious species ?).

Picus undatus, TEMM.

Picus undosus, CUV. R. A. 1829, 457 (all based on same figure).

Tridactylia undulata, CABANIS and HEINE, Mus. Hein. IV. 2, 1863, 28.

Picus tridactylus, SWAINSON, Fauna Bor. Am. 1831, 311; pl. lvi.

Picoides Americanus, BREHM, Vögel Deutschlands, 1831, 195. — MALHERBE, Mon. Picidæ, I. 176; pl. 17, 36. — SCLATER, Catal. GRAY, Cat. Br. Mus. III. 3, 4, 1868, 30. — *Apternus Americanus*, SWAINSON, Class. II. 1837, 306.—*Picus Americanus*, SUNDEVALL, Consp. Av. Picin. 1866, 15.

Var. **dorsalis.** Striped back.

Picoides dorsalis, BAIRD, Birds N. Amer. 1858, 100; pl. 85, f. 1. — *Tridactylia dorsalis*, CABANIS and HEINE. — *Picus dorsalis*, SUNDEVALL, Consp. 1866, 14.

SP. CHAR. Black above; the back with transverse bands of white to the rump. A white line from behind the eye, widening on the nape, and a broader one under the eye from the loral region, but not extending on the forehead; occiput and sides of head uni-

P. Americanus.

form black. Quills, but not coverts, spotted on both webs with white, seen on inner webs of inner secondaries. Under parts, including crissum, white; the sides, including axillars and lining of wing, banded transversely with black. Exposed portion of outer three tail feathers white; that of third much less, and sometimes with a narrow tip of black.

P. dorsalis.

Upper tail coverts sometimes tipped with white. Top of the head spotted with white; the crown of the male with a yellow patch. Nasal bristles black, mixed with gray.

Female, with the whole top of head spotted with white; very rarely entirely black.

Hab. Arctic regions of North America.

Var. *dorsalis.* The black transverse bands on the ends of the central dorsal feathers wanting, partly or entirely, leaving the white more or less continuous as in *Picus villosus.*

Hab. Rocky Mountains of the United States and British America.

This species varies considerably in its markings, especially in the amount of white above. The head is sometimes more coarsely spotted with white than in the average; very rarely are the white spots wanting, leaving merely the broad malar- and interrupted post-ocular stripe. The rictal black stripe is sometimes much obscured by white. In typical specimens from the Hudson Bay and Labrador provinces, which seem to be darkest, the feathers of the centre of the back have three transverse bars of white (one of them terminal), and rather narrower than the intermediate black bars; the basal white one disappearing both anteriorly and posteriorly, leaving but two. In specimens from the Mackenzie River district, there is a greater development of white, the white bands being broader than the black, and sometimes extending along the shafts so as to reduce the black bars to pairs of spots. The next step is the disappearance of these spots on one side or the other of the feather, or on both, leaving the end entirely white, especially anteriorly, where the back may have a longitudinal stripe of white as in *Picus villosus.* Usually, however, in this extreme the upper tail coverts remain banded transversely.

In all the specimens from the Rocky Mountains of the United States, especially Laramie Peak, this white back unbarred, except on the rump, is a constant character, and added to it we have a broad nuchal patch of white running into that of the back, and connected with the white post ocular stripe. The bands, too, on the sides are less distinct. It was to this state of plumage that I applied the name of *P. dorsalis,* in 1858, and although, in view of the connecting links, it may not be entitled to consideration as a distinct species, it yet appears to be a well-marked geographical variety.

This same character prevails in all the Rocky Mountain specimens (from more northern regions) before me, including those from Fort Liard, and in only one not found in that region, named No. 49,905, collected at Nulato by Mr. Dall. Here the middle of the back is very white, although the nuchal band is less distinct. Other specimens from that locality and the Yukon River generally, as also from Kodiak, show the transverse bars distinctly.

In one specimen, 29,126, from the Mackenzie River, all the upper tail coverts are banded decidedly with white, and the wing coverts spotted with the same. Even the central tail feathers show white scollops. The back is, however, very distinctly banded transversely, not longitudinally.

P. Americanus, in all stages of color, is distinguished from *arcticus* by the white along the middle of the back, the absence of distinct frontal white and black bands, more numerous spots of white on the head, etc. The inner webs of inner secondaries are banded with white, not uniform black. The maxillary black stripe is rather larger than the rictal white one, not smaller; the size is decidedly smaller. Females almost always have the top of head spotted with white, instead of uniform black, which is the rule in *arcticus*. Still, the relationship is very close; and from the New Brunswick specimens of *P. Americanus*, with very little white on the back, to typical *arcticus*, without any, is but a single step. In reality, if *dorsalis* and *Americanus* are one, *Americanus* and *arcticus* would scarcely seem entitled to separation.

It is possible that the difference in the amount of white on the upper parts of this species is to some extent due to age and season, the winter specimens and the young showing it to the greatest degree. Still, however, there is a decided geographical relationship, as already indicated.

This species can be easily distinguished from the *Picoides tridactylus* of Northern and Alpine Europe, by the tail feathers; of these, the outer three are white (the rest black) as far as exposed, without any bands, the tip of the third being white only at the end. The supra-ocular white stripe is very narrow and scarcely appreciable; the crissum white and unbanded. The back is banded transversely in one variety, striped longitudinally in the other. In *P. tridactylus*, the two outer feathers on each side are white, banded with black, the outer with the bands regular and equal from base; the second black, except one or two terminal bands. The crissum is well banded with black; the back striped longitudinally with white; the supra-ocular white stripe almost as broad as the infra-ocular. *P. crissoleucus*, of Siberia, is similar to the last, but differs in white crissum, and from both species in the almost entire absence of dark bands on the sides.

I follow Sundevall in using the specific name *Americanus*, Brehm, for this species as being the first legitimately belonging to it. *P. hirsutus*, of Vieillot, usually adopted, is based on a European bird, and agrees with it, though referred by the author to the American. The name of *undulatus*, Vieillot, selected by Cabanis, is based on Buffon's figure (pl. enl. 553) of a bird said to be from Cayenne, with four toes; the whole top of the head red from base of bill to end of occiput, with the edges of the dorsal feathers narrowly white, and with the three lateral tail feathers regularly banded with black, tipped with red; the fourth banded white and black on outer web, tipped with black. None of these features belong to the bird of Arctic America, and the markings answer, if to any, better to the European. (Baird.)

GENUS SPHYROPICUS, BAIRD.

Pilumnus, BONAPARTE, Consp. Zygod. Ateneo Italiano, May, 1854. (*P. thyroideus.*) Name preoccupied.

Sphyrapicus, BAIRD, P. R. Rep. IX. Birds, 1858, 102.

GEN. CHAR. Bill as in *Picus*, but the lateral ridge, which is very prominent, running out distinctly to the commissure at about its middle, beyond which the bill is rounded

P. borealis.

without any angles at all. The culmen and gonys are very nearly straight, but slightly convex, the bill tapering rapidly to a point; the lateral outline concave to very near the slightly bevelled tip. Outer pair of toes longest; the hinder exterior rather longest; the

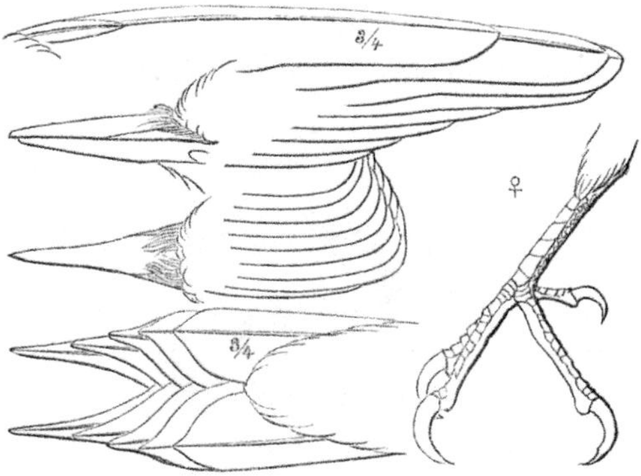

S. varius.

inner posterior toe very short; less than the inner anterior without its claw. Wings long and pointed; the fourth quill longest. Tail feathers very broad, abruptly accuminate, with a very long linear point.

Professor Baird includes five species in this new division of *Picus*, all but one inhabiting this coast.

Sphyropicus nuchalis, Baird.

THE RED-NECKED WOODPECKER.

Picus varius, Heermann, P. R. Rep. X. vi. 58 (not of Linnæus). — *Sphyrapicus varius*, var. *nuchalis*, Baird, P. R. Rep. IX. Birds, 103.
Sphyrapicus nuchalis, Baird, P. R. Rep. IX. Birds, 921. Ib. Birds of N. Amer. pl. xxxv.

Sp. Char. General color above black, variegated with white, the back and rump brownish-white, spotted with black. Crown scarlet, bordered behind by a black occipital band, and behind this a scarlet nuchal band or crescent curving forward to the eye. A

S. nuchalis.

triangular broad patch of scarlet on the chin and front of throat. A stripe from above the eye, and another from the bristles of the bill passing below the eye, and into the fairest yellowish of the belly, white. Between the lower stripe and the red of chin, a black

S varius.

stripe, not extending to the large quadrate black patch on the breast. A white stripe along the edges of wing coverts, outer webs of secondaries almost entirely black, the rest of the quills spotted with white on the edges of both webs. Tail feathers black, except the inner webs of the innermost, which are white banded with black, the others occasionally edged with yellowish.

Female with the throat red, the chin more or less white.

Length, 8 50; extent, 15.50; wing, 5.12. Iris brown; bill horn-color; feet gray.

Hab. Rocky Mountains near lat. 48°, south to Colorado Valley.

At Fort Mojave, on the 20th of February, 1861, I shot a female specimen of this beautiful woodpecker, which may have wandered in a storm from the mountains, being the only one I saw, though Dr. Heermann states that they are not rare at Fort Yuma. It was silent and seemed inactive, as if exhausted by a long flight, keeping concealed in the tall poplars.

I found this species rather common in crossing the Rocky Mountains near lat. 48°, in September, 1860, and noticed a great similarity in habits to

S. nuchalis, ♀

those of the Eastern *P. varius*. They frequent chiefly small deciduous trees, and feed in the same manner as other woodpeckers, having also a shrill, unvaried call, or note of alarm.

It is very likely that this species occurs on the eastern slope of the Sierra Nevada, although I saw none there in September, 1863. Its nest has not yet been discovered.

We have placed figures of the heads of this species and the Eastern *P. varius* on the opposite page, as the best mode of exhibiting the differences.

The female of *S. nuchalis* is shown in the above figure.

Sphyropicus ruber, Gmelin.

THE RED-BREASTED WOODPECKER.

Picus ruber, Gmelin, Syst. Nat. I. 1788, 429. — Wagler, Syst. Av. 1827, No. 151. — Audubon, Orn. Biog. V. 1839, 179; pl. 416. Ib. Birds Amer. IV. 1842, 261; pl. 266. — Heermann, P. R. Rep. X. vi. 57. — *Pilumnus ruber*, Bonaparte, Consp. Zyg. Aten. Ital. 1854, 8.

Picus flaviventris, Vieillot, Ois. Am. Sept. II. 1807, 67.

Sphyropicus ruber, Baird, P. R. Rep. IX. Birds, 104. — Cooper and Suckley, XII. iii. Zool. of W. T. 160.

Sp. Char. Fourth quill longest; third intermediate between fourth and fifth. Bill brown wax-color. Head and neck all round, and breast, carmine red. Above black; central line of back from nape to rump spotted with whitish; rump, wing coverts, and inner web of the inner tail feathers white, the latter with a series of round black spots. Belly

sulphur-yellow, streaked with brown on the sides. Narrow space around and a little in front of the eye black. A narrow yellowish stripe from the nostrils, a short distance below and behind the eye. Length, about 9.50; extent, 15.75; wing, 5.00; tail, 3.40. Iris, bill, and feet, pale brown.

Hab. Pacific slope of the United States.

This is rather a Northern species, as I have not seen any south of Santa Clara, and there only in the mountains of the Coast Range in early spring. According to Dr. Heermann, they are not uncommon in the Sierra Nevada, but I did not observe them near the summits, and even at the Columbia River found them rather scarce, solitary, and shy, keeping chiefly in the high coniferous trees. Nuttall states that he saw one of their burrows containing young in a tall fir-tree in Oregon. During some weeks' stay in the coast mountains towards Santa Cruz, I saw none of them, and think that most of them go North in summer. Their cry is compared by Heermann to that of a young child in distress.

Sphyropicus Williamsonii, Newberry.

WILLIAMSON'S WOODPECKER.

Picus Williamsonii, Newberry, Zool. Cal. and Or. Route, 89; P. R. Rep. VI. iv. 1857; pl. xxxiv. fig. 1. — *Sphyrapicus Williamsonii*, Baird, P. R. Rep. IX. Birds, 105.
Melanerpes rubrigularis, Sclater, Annals and Mag. N. H. 3d Series, I. Feb. 1858, 127.

Sp. Char. Black; middle line of belly yellow; central line of chin and throat above red. A large patch on the wing, rump, and upper tail coverts, a line from the forehead

beneath the eye, and another from its upper border, white. Tail entirely black. Exposed surface of wing without any white, except on the outer primaries.

Female with the chin white instead of red. (?) Length, 9.00; wing, 5.00; tail, 4.70.

Hab. Rocky Mountains to the Cascade Mountains, and Sierra Nevada, California.

The first specimen of this beautiful bird that I met with was a straggler in winter to the Colorado Valley, which I shot on the 12th of March, 1861, but unfortunately did not find it until three days after, when it was nearly destroyed by ants. In September, 1863, I found them rather common near the summit of the Sierra Nevada in lat. 39°, where I shot two, both of which stuck in the trees. Dr. Newberry saw but one near Klamath Lake, but it has since been found at Laramie Peak, and towards the mouth of Klamath River.

Of its habits I observed nothing peculiar on the few occasions when I saw them. They seemed to keep high in the trees, shy and silent, like others of this group, less noisy in their hammering than other woodpeckers.

The male of this species is very remarkable in the absence of any red on the upper part of the head, a character which, shared by the *S. thyroideus*, is scarcely to be found on any other North American woodpecker. It, however, appears to be strictly congeneric with *S. nuchalis*, *varius*, etc.

The first indication of this species is to be found in the report of Dr. Newberry on the birds collected by Lieutenant Williamson's party, and published in 1857. Early in 1858 Mr. Sclater renamed and described it, as stated above, Dr. Newberry's report not having reached him at the time.

Sphyropicus thyroideus, Cassin.

THE ROUND-HEADED WOODPECKER.

Picus thyroideus, Cassin, Pr. A. N. Sc. V. Dec. 1851, 349. (California.) — Heermann, J. A. N. Sc. Phil. 2d Series, II. 1853, 270. Ib. P. R. Rep. X. vi. 58. — *Melanerpes thyroideus*, Cassin, Illust. I. 1854, 201; pl. xxxii. — *Pilumnus thyroideus*, Bonaparte, Consp. Zygod. Aten. Ital. 1854, 8. — *Sphyropicus thyroideus*, Baird, P. R. Rep. IX. Birds, 106.

Sp. Char. About the size of *P. ruber*. Head dark ashy-brown; rest of body apparently encircled by narrow transverse and continuous bands crossing the wings, of black and brownish-white, except a large, round, black patch on the breast; and the central

line of the body from the crest to the vent, which is the color of roll-sulphur. No red on top of the head, but the chin and throat tinged with this color in adult males.

Female with rather duller colors. Length, about 9.00; wing, 5.00; tail. 4.10.

Hab. Interior mountain ranges of California and Oregon, to mountains of New Mexico.

This seems to be a very rare bird, as I have never met with it, and doubt if it is ever found in the Coast Range south of San Francisco. Mr. Bell of New York first found it in the Lower Sierra Nevada, Dr. Heermann near the Colorado River, and it has since been found near Fort Crook. It may be more common in the mountains of Eastern Oregon and Central Utah.

Genus HYLOTOMUS, Baird.

Dryotomus, Malherbe, Mem. Ac. Metz. 1849, 322. (Not of Swainson, 1831.)
Dryopicus, Bonaparte, Consp. Zygod. in Aten. Ital. May, 1854. (Not of Malherbe.)
Hylatomus, Baird, P. R. Rep. IX. Birds, 1858, 107.

Gen. Char. Bill a little longer than the head; considerably depressed or broader than high at the base; shaped much as in *Campephilus*, except shorter, and without the bristly feathers directed forwards at the base of the lower jaw. Gonys about half the length of the commissure. Tarsus shorter than any toe, except the inner posterior. Outer posterior toe shorter than the outer anterior, and a little longer than the inner anterior.

H. pileatus.

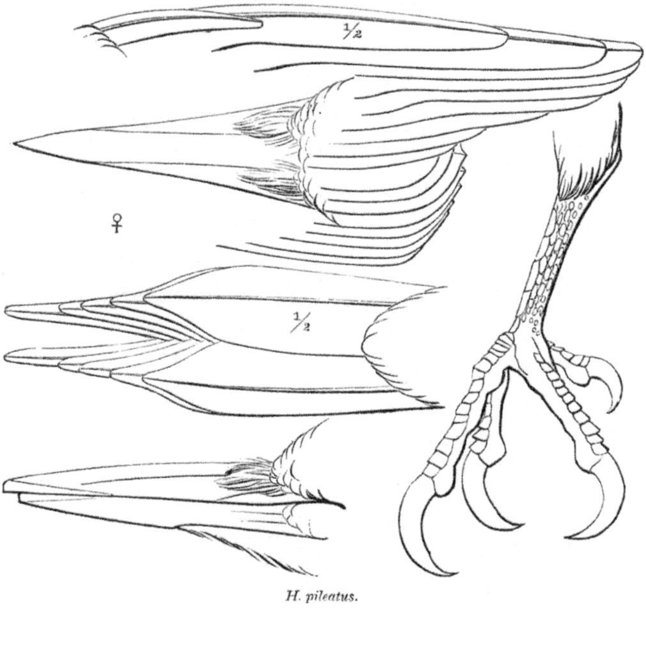

H. pileatus.

Inner posterior very short; not half the outer anterior; about half the inner anterior one. Tail long, graduated; the long feathers much incurved at the tip. Wing longer than the tail, reaching to the middle of the exposed surface of tail; considerably graduated, though pointed; the fourth and fifth quills longest.

Color, uniform black, with white patches on the side of the head. Head with pointed crest.

Hylotomus pileatus, Linnæus.

THE BLACK WOODCOCK; THE LOG-COCK.

Picus pileatus, Linnæus, Syst. Nat. I. 1766, 173. — Vieillot, Ois. Am. Sept. II. 1807, 58; pl. cx. — Wilson, Am. Orn. IV. 1811, 27; pl. xxix. f. 2. — Wagler, Syst. Av. 1827, No. 2. — Audubon, Orn. Biog. II. 1834, 74; V. 533; pl. 111. Ib. Birds Amer. IV. 1842, 266; pl. 257. — *Picus (Dryotomus) pileatus*, Swainson, Fauna Bor. Amer. II. 1831, 304. — *Hylatomus pileatus*, Baird, P. R. Rep. IX. Birds, 107. — Cooper and Suckley, XII. iii. Zool. of W. T. 161.

Sp. Char. Fourth and fifth quills equal and longest; third intermediate between sixth and seventh. Bill blue-back. General color of body, wings, and tail dull greenish-black. A narrow white streak from just above the eye to the occiput; a wider one from the nostril feathers (inclusive) under the eye and along the side of the head and neck; sides of

Male.

the breast (concealed by the wings), axillaries, and under wing coverts, and concealed bases of all the quills, with chin and beneath the head, white, tinged with sulphur-yellow. Entire crown from the base of the bill to a well-developed occipital crest, as also a patch on the ramus of the lower jaw, scarlet red. A few faint white crescents on the sides of the body and on the abdomen. Length, about 18 00; extent, 29.00; wing, 9.50.

Female.

Female without the red on the cheek and the anterior half of that on the top of the head replaced by black.

Hab. North America from the Atlantic to the Pacific.

This large species is rarely if ever seen to the south of lat. 38°, on this coast. I have never met with them, or heard of them, except one killed near Mount Diablo. Even on the summits of the Sierra Nevada I saw none, though they doubtless wander occasionally through the whole extent of that range.

They are abundant near the Columbia River, and their loud call resembling that of the *Colaptes*, but much louder, often indicates their presence for miles away. They are usually very shy and difficult to kill. Their burrow for a nest I have seen dug in the side of a tree about forty feet above the ground. According to Nuttall, they lay about six snowy white eggs, and are said sometimes to raise two broods in a season.

Specimens from Washington Territory are much larger than those of the Middle Atlantic States.

GENUS **CENTURUS**, SWAINSON.

Centurus, SWAINSON, Class Birds, II. 1837, 310. (Type, *C. Carolinus.*)
Zebrapicus, Malh. Mem. Acad. Metz. 1849, 360. (Type, *C. Carolinus.*)

GEN. CHAR. Bill about the length of the head, or a little longer; decidedly compressed, except at the extreme base. A lateral ridge starting a little below the culmen at the base of the bill, and angular for half the length of the bill, then becoming obsolete though traceable nearly to the tip. Culmen considerably curved from the base; gonys nearly straight. Nostrils very broad, elliptical; situated about midway on the side of

C. Carolinus

the mandible, near the base; partly concealed. Outer pairs of toes unequal; the anterior longest. Wings long, broad; third to fifth primaries equal and longest. Tail feathers rather narrow, stiffened.

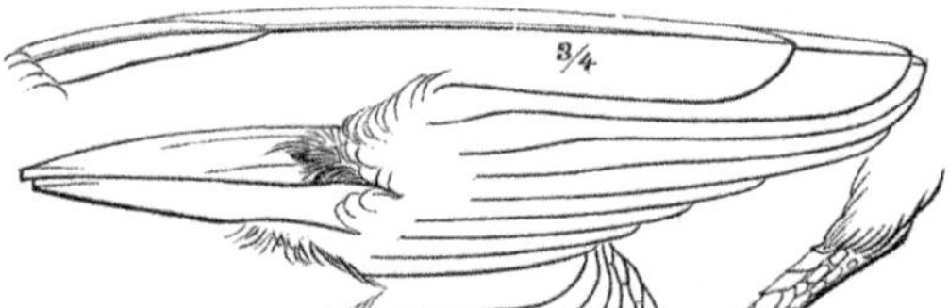

C. Carolinus.

The species are all banded above transversely with black and white; the rump white. The head and under parts are brown; the belly with a red or yellow tinge.

Centurus aurifrons, Gray.

THE YELLOW-BELLIED WOODPECKER.

Centurus flaviventris, Swainson, Anim. in Menag. 1838 (2½ centenaries), 354. — Baird, Birds N. Amer. 1858, 110; pl. 42. Ib. Rep. Mex. Bound. U. S.; pl. 4. — Heermann, P. R. Rep. X. c. 18. — Dresser, Ibis. 1865, 469. (Resident in Texas.)
Centurus elegans, Lawrence, Ann. N. Y. Lyc. V. May, 1851, 116.
Centurus Santacruzii, Lawrence, Ann. N. Y. Lyc. V. 1851, 123. (Not of Bonaparte.)
Picus aurifrons, Wagler, Isis, 1829, 512. — Sundevall, Consp. Pic. 53. — *Centuris aurifrons*, Gray, Genera. — Cabanis, Jour. 1862, 323.
Picus ornatus, Less. Rev. Zool. 1839, 102.

Sp. Char. Fourth and fifth quills nearly equal; third a little shorter; longer than the fourth. Back banded transversely with black and white; rump and upper tail coverts pure white. Crown with a sub-quadrate spot of crimson, about half an inch wide and long, and separated from the gamboge-yellow at the base of the bill by dirty white; from the orbit and occiput by brownish-ash. Nape half-way round the neck orange-yellow. Under parts generally and sides of head dirty white. Middle of belly gamboge-yellow.

Tail feathers all entirely black, except the outer, which has some obscure bars of white. Length, about 9 50; wing, 5.00.

Female without the red of the crown.

Hab. Rio Grande region of the United States; south into Mexico. Probably Arizona.

Young birds are not different from adults, except in showing indication of dark shaft-lines beneath, becoming broader behind on the sides. The yellow of the nape extends over the whole side of the head.

Centurus uropygialis, Baird.

THE GILA WOODPECKER.

Centurus uropygialis, Baird, Pr. A. N. Sc. Ph. VII. June, 1854, 120. (Bill Williams's River, N. M.) Ib. P. R. Rep. IX. Birds, 111. — Kennerly, X. iv. 22; pl. 36. — Heermann, X. v. 17; X. vi. 58.

Sp. Char. Third, fourth, and fifth quills longest, and about equal. Back, rump, and upper tail coverts barred transversely with black and white, purest on the two latter.

Head and neck all round pale dirty brown, or brownish-ash, darkest above. A small subquadrate patch of red on the middle of the crown, separated from the bill by dirty white. Middle of the abdomen gamboge-yellow; under tail coverts and anal region strongly barred with black. First and second outer tail feathers banded black and white, as is

also the inner web of the inner tail feather; the outer web of the latter with a white stripe. Length, about 9.50 to 10.00; wing, 5.50. Iris red; bill horn-black; feet gray.

Female with the head uniform brownish-ash, without any red or yellow.

Hab. Lower Colorado River of the West, and along the Gila.

At Fort Mojave I found this woodpecker abundant in winter, when they fed chiefly on the berries of the mistletoe (*Viscum*), and were rather shy. I rarely saw them pecking at the trees, but they seemed to depend for a living on catching insects, which were numerous on the foliage during the spring. They had a loud note of alarm, strikingly similar to that of *Phainopepla nitens*, which associated with them in the mistletoe boughs.

About March 25th they were preparing their nests in burrows near the dead tops of trees, none of those which I saw being accessible. By the 25th

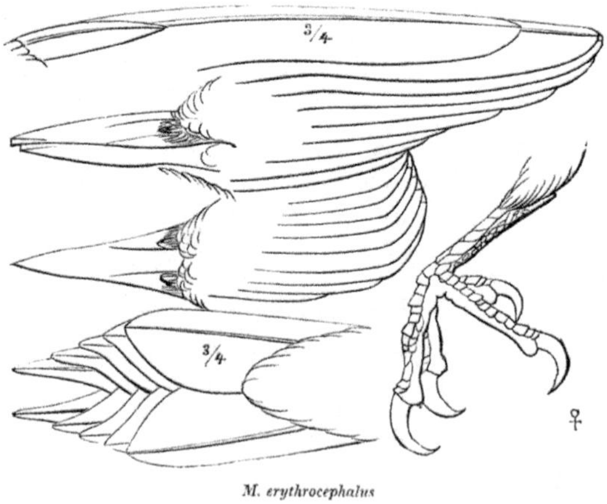

M. erythrocephalus

of May they had deserted the mistletoe entirely, and were probably feeding their young on insects. According to Heermann, they burrow holes in the giant cactus (*Cereus giganteus*), which often grows forty feet high along the Gila, and at Tucson they frequent the corn-fields.

GENUS **MELANERPES**, SWAINSON.

Melanerpes, SWAINSON, Fauna Bor. Amer. II. 1831. (Type, *M. erythrocephalus.*)
Melampicus (Section 3), MALHERBE, Mém. Ac. Metz, 1849, 365.

GEN. CHAR. Bill about equal to the head; broader than high at the base, but becoming compressed immediately anterior to the commencement of the gonys. Culmen and gonys with a moderately decided angular ridge; both decidedly curved from the very base. A rather prominent acute ridge commences at the base of the mandible, a little below the ridge of the culmen, and proceeds but a short distance anterior to the nostrils

M. formicivorus.

(about one third of the way), when it sinks down, and the bill is then smooth. The lateral outlines are gently concave from the basal two thirds; then gently convex to the tip, which does not exhibit any abrupt bevelling. Nostrils open, broadly oval; not concealed by the feathers, nor entirely basal. The outer pair of toes equal. Wings long, broad; third and fourth quills longest. Tail feathers broad. Back plain black.

Species of this genus are quite numerous in America, including the West Indies, but do not occur in the Old World. They are among the most brightly colored of woodpeckers, and usually present themselves in considerable abundance where found at all.

Melanerpes erythrocephalus, Swainson.

THE RED-HEADED WOODPECKER.

Picus erythrocephalus, Linnæus, Syst. Nat. I. 1766, 174. — Vieillot, Ois. Am. Sept. II. 1807, 60; pl. cxii., cxiii. — Wilson, Am. Orn. I. 1810, 142; pl. ix. f. 1. — Wagler, Syst. Av. 1827, No. 14. Ib. Isis, 1829, 518 (young). — Audubon, Orn. Biog. I. 1832, 141: V. 536; pl. 27. Ib. Birds Amer. IV. 1842, 274; pl. 271. — Max. Cab. Jour. VI. 1858, 419. (Upper Missouri.) — Sundevall, Consp. 50. — *Melanerpes erythrocephalus*, Swainson, Fauna Bor. Amer. II. 1831, 316. — Bonaparte, List, 1838. Ib. Conspectus, 1850, 115. — Gambel, J. Ac. Nat. Sc. Ph. 2d Series, I. 1847, 55. — Baird, Birds N. Amer. 1858. — Dresser, Ibis, 1865, 469. (Resident from Nueces to Brazos, Texas.)

Picus obscurus, Gmelin, I. 1788, 429 (young).

Red-headed woodpecker, Pennant, Kalm, Latham.

White-rumped woodpecker, Latham.

Sp. Char. Head and neck all round crimson red, margined by a narrow crescent of black on the upper part of the breast. Back, primary quills, and tail, bluish-black. Under parts generally, a broad band across the middle of the wing (exposed portion of sec-

ondaries), and the rump, white. The female is not different. Length, about 9.75; wing, 5.50.

Hab. North America, from the Atlantic Coast to the eastern slope of the Rocky Mountains. (Coast of California, Gambel.)

The crimson feathers on the head and neck all round have the same bristly texture as described under *M. torquatus.* The red descends much lower below than above; its posterior outline well defined and semicircular. The white on the wing involves the whole of the secondaries, except the extreme bases; the shafts are black. There is a yellowish tinge to the white on the middle of the belly, and the exterior tail feathers are tipped with whitish. The inside of the wing is white.

I can detect no difference in Western specimens. Occasionally the secondaries are blotched or barred with black near the end (587). Immature specimens almost always have this character. The young lack the red of the head, which is replaced by brown obscurely spotted and streaked. The white of the breast too is duller, and also streaked. Dr. Gambel speaks of this species as common in oak timber near the Mission of San Gabriel, California, but none have been noticed west of the mountains by any one else. As, however, it extends to the Rocky Mountains, and perhaps west of them, we introduce the species into the present report.

Melanerpes formicivorus, Swainson.

THE CALIFORNIA WOODPECKER.

Picus formicivorus, Swainson, Birds Mex. in Philos. Mag. I. 1827, 439. (Mexico.) — Vigors, Zool. Blossom, 1839, 23. (Monterey.) — Nuttall, Man. I. 2d ed. 1840. — *Melanerpes formicivorus*, Bonaparte, Conspectus, 1850, 115. — Heermann, J. A. N. Sc. Phil. 2d Series, II. 1853, 270. — Cassin, Illust. II. 1853, 11; pl. ii. — Newberry, Zool. Cal. and Or. Route, 90; P. R. R. Surv. VI. 1857. — Baird, P. R. Rep. IX. Birds, 114. — Heermann, X. vi. 58.

Sp. Char. Fourth quill longest, third a little shorter. Above on the anterior half of the body glossy bluish or greenish black; the top of the head and a short occipital crest red. A white patch on the forehead, connected with a broad crescentic collar on the

Male.

upper part of the neck by a narrow isthmus, white tinged with sulphur-yellow. Belly, rump, bases of primaries, and inner edges of the outer quills, white. Tail feathers uniform black.

Female with the red confined to the occipital crest, the rest replaced by greenish-black; the three patches white, black, and red, very sharply defined. Length, about 9.50; extent, 18.00; wing, about 6.00. Iris brown; bill black; feet gray.

Female.

Hab. Coast region of California and south; in Northern Mexico, eastward almost to the Gulf of Mexico; also on Upper Rio Grande.

This beautiful bird is one of the commonest species in all the lower regions of California, frequenting chiefly the oaks, and extending up as far as they grow on the mountains. Its brilliant plumage, lively and familiar habits, and loud notes make it a very conspicuous inhabitant of the woods, and it will, if unmolested, become quite familiar around dwellings.

Their usual resorts are among the topmost and decayed branches, where they seek their insect food; but they also feed in great part on insects caught among the leaves and on the bark, as well as on fruits, being less industrious in hammering for a subsistence than the *Pici.*

They burrow out the cavity for a nest in a dead branch, making it, according to Heermann, from six inches to two feet deep, and laying four or five pure white eggs, on the dust and chips at the bottom, like nearly all woodpeckers.

They are fond of playing together around the branches, uttering their rattling calls, and often darting off to take a short sail in the air, returning to the same spot. They have a habit, peculiar to them, of drilling small holes in the bark of trees, and fitting acorns tightly into them, each one being carefully adapted, and driven tight. The bark is often so full of these holes as to leave scarcely room to crowd in another without destroying the bark entirely. These are generally considered as laid up for a winter supply of food; but while in this climate no such provision is necessary, it is also very improbable that birds of this family would feed on hard nuts or seeds of any kind. The more probable explanation is that they are preserved for the sake of the grubs they contain so frequently, which, being very small when the acorn falls, grow until they eat the whole interior, when they are a welcome delicacy for the bird. Whether they select only those containing grubs, or put away all they meet with, is uncertain; but as they leave great numbers in the tree untouched, it is probable that these are sound acorns, and often become a supply to the squirrels and jays.

From this strange habit the bird has received the name of "carpintero," and this is also adopted by many Americans.

Melanerpes formicivorus, var. **angustifrons**, Baird.

THE NARROW-FRONTED WOODPECKER.

Sp. Char. Compared with *M. formicivorus*, the size is smaller, the light frontal bar is much narrower; in the female scarcely more than half the black one behind it, and not

Adult.

reaching anything like as far back as the anterior border of the eye, instead of exceeding this limit. The light frontal and the black bars together are only about two thirds the

Female.

length of the occipital red, instead of exceeding it in length; the red patch reaches forward nearly or quite to the posterior border of the eye, instead of falling a considerable distance behind it, and being much broader posteriorly. The frontal band, too, is gamboge-yellow, much like the throat, and not white; the connection with the yellow throat patch much broader. The white upper tail coverts show a tendency to a black edge. Length, 8.00; wing, 5.20; tail, 3.20.

Hab. Cape St. Lucas.

As the differences mentioned are constant, I consider the Cape St. Lucas bird as forming at least a permanent variety, and indicate it as above. A single specimen from the Sierra Madre of Colima is very similar.

Melanerpes torquatus, Wilson.

LEWIS'S WOODPECKER.

Picus torquatus, Wilson, Amer. Orn. III. 1811, 31; pl. xx. — Wagler, Syst. Av. 1827, No. 82. — Audubon, Orn. Biog. V. 1839, 176; pl. 416. Ib. Birds Amer. IV. 1842, 280; pl. 272. — *Melanerpes torquatus*, Bonaparte, Conspectus, 1850, 115. — Heermann, J. A. N. Sc. Phil. 2d Series, II. 1853, 270. — Newberry, Zool. Cal. and Or. Route, 90; P. R. R. Surv. VI. 1857. — Baird, P. R. Rep. IX. Birds, 115. — Heermann, X. vi. 58. — Cooper and Suckley, XII. iii. Zool. of W. T. 161.

Sp. Char. Feathers on the under parts bristle-like. Fourth quill longest; then third and fifth. Above dark glossy green. Breast, lower part of the neck, and a narrow collar all round hoary grayish-white. Around the base of the bill and sides of the head to behind the eyes, dark crimson. Belly blood red, streaked finely with hoary-whitish. Wings and tail entirely uniform dark glossy green.

Female with the markings more obscure. Length, about 10.75; extent, 21.00; wing, 6.50. Female smaller.

Hab. Western America from the Black Hills to the Pacific.

I have found this species quite common near New Almaden, but not elsewhere in the Coast Range southward, during summer. They are, however, numerous in the foothills of the Sierra Nevada, and doubtless also in the more northern Coast Range. They occur also on the higher parts of the Sierra, frequenting chiefly the coniferous trees, and having very much the same habits and notes as the *M. formicivorus.* They keep much about the higher parts of the trees, circling around them in pursuit of insects, and not troubling themselves much to hammer the bark for food. They feed in great part also on fruits, and the flocks of black young, associating together in autumn, might be taken for something very different from woodpeckers. Their flight is rather slow and flapping, causing them to look like crows at a distance, while they frequently sail in circles around the tree-tops like hawks. In fact, they are less like ordinary woodpeckers than any other species we have.

Their nest is burrowed out at a considerable height, but no description of the eggs has yet been given, though they doubtless resemble those of other woodpeckers.

Genus **COLAPTES**, Swainson.

Colaptes, Swainson, Zool. Jour. III. Dec. 1827, 353. (Type, *C. auratus.*)

Gen. Char. Bill slender, depressed at the base, then compressed. Culmen much curved; gonys straight, both with acute ridges, and coming to quite a sharp point with

C. Mexicanus.

the commissure at the end. No ridges on the bill. Nostrils basal, median, oval, and exposed. Gonys very short; about half the culmen. Feet large; the anterior outer toe

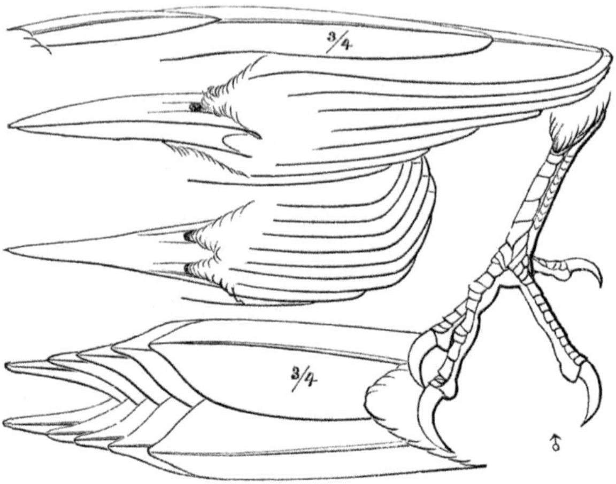

C. auratus.

considerably longer than the posterior. Tail long, exceeding the secondaries, the feathers suddenly acuminate, with elongated points.

The species of the Atlantic States, known as golden-winged woodpecker, flicker, high-holder, and yellow-hammer (*C. auratus*), closely resembles *C. chrysoides*; and in fact the hybrids between the former and *C. Mexicanus* are often marked so much like *C. chrysoides* that, were not the plumage of the latter constant, its validity as a species would be doubtful.

Colaptes Mexicanus, Swainson.

THE RED-SHAFTED FLICKER.

Colaptes Mexicanus, Swainson, Syn. Mex. Birds, Philos. Mag. I. 1827, 440. Ib. Fauna Bor. Amer. II. 1831, 315. — Newberry, Zool. Cal. and Or. Route, 91; P. R. R. Rep. VI. 1857. — Baird, P. R. Rep. IX. Birds, 120. — Kennerly, X. iv. 22. — Heermann, X. v. 59. — Cooper and Suckley, XII. iii. Zool. of W. T. 163. — *Picus Mexicanus*, Audubon, Orn. Biog. V. 1839, 174; pl. 416. Ib. Birds Amer. IV. 1842, 295; pl. 274.

Colaptes collaris, Vigors, Zool. Jour. IV. Jan. 1829, 353. Ib. Zool. Beechey's Voy. 1839, 24; pl. ix.

Sp. Char. Shafts and under surfaces of wing and tail feathers orange-red. A red patch on each side the cheek; nape without red crescent; sometimes very faint indications laterally. Throat and stripe beneath the eye bluish-ash. Back glossed with purplish-brown. Spots on the belly, a crescent on the breast, and interrupted transverse bands on the back, black.

Female without the red cheek patch. Length, 12.50 to 14.00; extent, 21.00; wing, 6 00 to 7.00.

Hab. Western North America from the Black Hills to the Pacific.

This beautiful and common species is found in every part of the State, except the barest plains, frequenting even the low bushes, where no trees are to be seen for miles. They are much more abundant, however, in the middle wooded districts, and towards the north.

The burrows of this species are made at all heights above the ground, and about a foot in depth; the eggs are five or six, and white. Towards the south they are laid early in April, and in May at Puget Sound.

Along the Colorado I found this species very shy, probably on account of their being much hunted by the Indians for their bright feathers. In most other places they are, however, rather tame, and their interesting habits may be watched without difficulty. In these they are the exact counterpart of the Eastern golden-wing, whose history is so completely given by American ornithologists. They do not depend much on hard work for their subsistence, but live in great part on ants, for which they visit the large ant-hills so common in many localities, and, like the four-footed ant-eater, catch them by means of their glutinous tongue. They also feed much on berries during the season; and if they do destroy decayed wood in search of insects, it is usually the softest branches, their curved bill not being very well suited for cutting into harder materials. Of the Eastern species, however, Nuttall remarks that they have been known to make a winding burrow for their nest through solid oak, for fifteen inches in depth, requiring several weeks for its completion. The young of our species I have found when nearly fledged, and at that time they hiss like snakes when a hand is inserted in the burrow, — a sufficient warning for those who do not know what is in it, and doubtless intended as such by nature. They are fond of playing, and chasing each other around the trunks of trees, always uttering at that time a peculiar note of recognition, like " whittoo, whittoo, whittoo."

Of the various names given to it by immigrants from various parts of the

C. auratus.

East, that of " flicker " seems the most appropriate, while " yellow-hammer " is quite unsuitable for a bird without any yellow on it, and was besides originally the English name of a small yellow bird not found in America.

The relationship of this species to the Eastern *C. auratus* will be best exhibited by a comparison of the preceding figures.

Colaptes chrysoides, MALHERBE.

MALHERBE'S FLICKER.

Colaptes chrysoides, MALHERBE, Rev. et Mag. de Zool. IV. 1852, 553. — BAIRD, P. R. Rep. IX. Birds, 1858, 125. IB. Proc. Ac. N. Sc. Phil. XI. 1859, 302.
? *Colaptes Ayresii*, HEERMANN, P. R. Rep. X. vi. 59. (Not of Audubon ?)

SP. CHAR. Above yellowish-ash, transversely barred with black; chin, throat, and sides of head clear ash; under parts white; a broad pectoral crescent, and rounded spots on remaining under parts, black. Top of head light brown. Shafts of wings and tail feathers gamboge-yellow. Tail black, the basal portion yellow; the outer feathers uniformly black on the exposed terminal half, including the shafts. No red on the nape. Bill blackish horn-color; iris blood red; feet lead-gray. Length, male, 11.75; extent, 19.25; wing, 6.25. Female little smaller. Smaller southward.

C. chrysoides.

Hab. Colorado Valley, lat. 35°, to Cape St. Lucas and Mexico. Cosumnes River, California? (Heermann.)

I found only two pairs of this species at Fort Mojave, after February 20th, which were then mated, and at that time fed like many other insectivorous birds on the insects they found among the blossoms of the poplars. They had precisely the same habits, flight, and cries as the *C. Mexicanus*, but were somewhat smaller than specimens of that species killed at the same place. I had great difficulty in obtaining them on account of their wariness, and shot only three. The mate of one of these was still about there in May when I left, but no others had arrived. As they seemed somewhat migratory, coming from the south, it is possible that they do not go so far to the north as the Cosumnes River, where Dr. Heermann mentions finding *C. Ayresii.* His specimens may have been hybrids between this and *C. Mexicanus,* analogous to the forms found in the Rocky Mountains, and described by Baird as *C. hybridus,* — a mixture of *C. Mexicanus* and *C. auratus,* including as one variety the form described by Audubon as *C. Ayresii.* The combinations of colors found in those specimens are almost endless, and furnish one of the most interesting objects of study in ornithology. Whether they are all males or productive, and how far these distinctions of color can be relied on to distinguish species, and to determine their "origin," will be for future Darwins to examine into.*

A female bird from Cape St. Lucas is represented in the accompanying figure.

NOTE. — Since writing the above, two or more specimens have been obtained at Oakland, opposite San Francisco, which are evidently of the form *hybridus*, Baird. One, sent to Mr. H. P. Carlton fresh, differs from the *auratus* only in having the head grayish like *Mexicanus*, and the *black* cheek feathers tipped with *red*. These must be stragglers from the Rocky Mountains.

Order RAPTORES.

Char. Base of upper mandible with a soft skin or cere. Upper mandible compressed; its point curving down over that of the lower, forming a strong sharp hook. Toes four, one behind. Size usually large, and frame powerful. Female bird the larger, except in the vultures (and *Polyborus?*).

These birds may be divided into two sub-orders, namely: —

A. True birds of prey, which catch their victims alive, and are distinguished by powerful bills, claws, and vigorous swift flight.

B. Carrion-eaters, feeding on dead animals, having rather weak bills, long, straight claws, and heavy flight. The gradation of links between these two is, however, very close.

Species are found in all parts of the world, the carrion-eaters, however, being limited to warm climates. They are analogous to the *Rapacia* among mammals, and their office in the economy of nature is to keep in check the excessive increase of the smaller kinds of other animals. Some foreign snake-eating species have the legs very long like the cranes, to protect their body from the bite of venomous serpents.

Family STRIGIDÆ, The Owls.

Char. Form usually short and heavy, the head very large, round, and frequently with tufts of feathers resembling ears. Eyes usually very large, directed forwards, surrounded by short bristles and radiating feathers, which form a more or less perfect disk around the face. Bill rather strong, much curved from the base, nearly concealed by projecting bristles; wings generally long; outer edges of primaries fringed with soft ends; whole plumage very loose and soft, legs rather short, and in all American species feathered or bristly. Cavity of the ear very large.

About one hundred and forty species of owls have been described, some inhabiting every country. (Cassin.)

SUB-FAMILY STRIGINÆ.

CHAR. Of medium size; facial disk perfect; bill rather long; eyes rather small for owls; somewhat lateral; legs rather long, fully feathered to the toes.

GENUS **STRIX**, LINNÆUS.

Strix, LINNÆUS, Syst. Nat. I. 1766, 131.

GEN. CHAR. Head large, without ear tufts; wings long; tail rather short; toes and claws rather long; tarsi thinly covered with small feathers; middle claw serrated. There are about twelve species known.

S. flammea.

NOTE. — We give a figure of the European race of the barn owl, as type of the genus, one of a series of cuts kindly furnished by the Society for the Diffusion of Christian Knowledge, of London, from a work published by it, entitled "British Birds in their Haunts," by St. John. The illustrations of this work are of remarkable excellence, having been drawn by John Wolf, and engraved by F. Whymper.

Strix pratincola, Bonaparte.

THE BARN-OWL.

Strix flammea, Wilson, Am. Orn. VI. 57; pl. 50, f. 2. — Audubon, Birds Amer. pl. 171 (not of Linnæus, which is an European species).
Strix pratincola, Bonaparte, Comp. List, 1838, 7. — De Kay, Nat. Hist. N. Y. I. 31; pl. 13, f. 28. — Newberry, P. R. Rep. VI. ii. 76. — Cassin, Birds, IX. 47. — Heermann, X. vi. 34. — Coues, Pr. Acad. Phil. 1866, 49.
Strix Americana, Audubon, Orn. Biog. II. 1834, 421; oct. ed. I. 127; pl. 34. — Nuttall, Man. I. 2d ed. 149 (not of Gmelin, which is *Otus Wilsonianus*, according to Bonaparte ?).
Strix perlata, Kaup, Trans. Zool. Soc. Lond. V. 1859, 247.

Sp. Char. Above pale fawn-color, or tawny brownish-yellow, frequently very pale, nearly every feather with a small subterminal black spot, succeeded by a white one. Be-

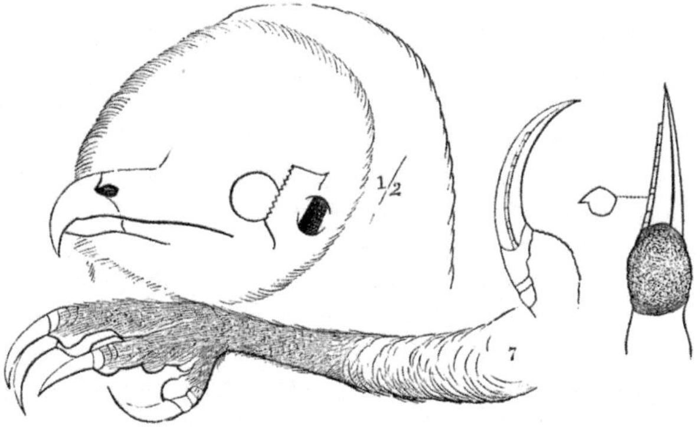

S. pratincola

neath, generally pale fawn-color, but frequently pure white, with small lanceolate and circular spots of brownish-black; under coverts of wings and tail white; quills fawn-

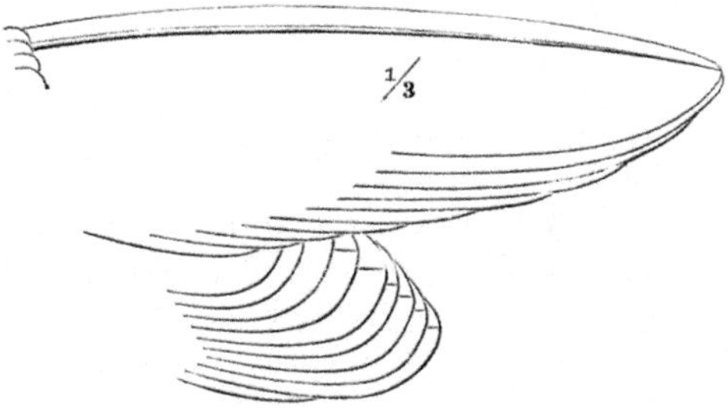

colored; primaries with about five irregular transverse bars of brownish-black; tail with about four or five bands of dark brown. Face white, spots of dark chestnut-brown around the eyes. Length, 15.50 to 17.00; extent, 40.00 to 45.00; wing, 12.75 to 14.00; tail, 5.00 to 5.50. Iris brownish-black; bill, toes, and claws yellowish.

Hab. The United States as far north as Long Island, New York, and the Columbia River, lat. 46°. The West Indies; Central and South America generally.

Abundant throughout the southern part of California, especially near the coast, frequenting chiefly old buildings, barns, etc., but often found hid in dark thickets and hollow trees. It closely resembles the European barn-owl, and others of almost every part of the world, and its habits seem to resemble closely those of its Old World relative. In the Atlantic States it is said rather to avoid human habitations, but this is probably on account of the thoughtless persecution too much practised among our countrymen against all owls, under the impression that they destroy fowls. Careful observations of the habits of the European species have shown that they very rarely if ever do so, and that, on the contrary, they destroy an incredible number of rats and mice, — in fact, more than they and their young can eat; a pair of old ones being watched and seen to arrive at the nest every few minutes with a rat or mouse, during the early night. When flying about at dusk they utter a variety of loud, harsh, and rather strange cries, which are sometimes heard throughout the night. Their nest is merely the natural floor of the cavity in which they live, and their eggs are said by Nuttall to be three to five, of a whitish color.

Audubon also found them nesting among high grass on the ground in Texas, on the 3d of May. They had avenues leading under the grass and bushes for some distance from the nest. The eggs were ovate, 1.79 by 1.25 inches, white and rough.

The figure of the European *Strix flammea*, under the general head, will illustrate the American bird sufficiently, as they are scarcely, if at all, to be distinguished.

Specimens from different regions of the earth exhibit certain peculiarities which have been made the basis of specific distinctions, but it is questionable whether these are more than climatic variations.

Sub-Family BUBONINÆ, The Horned Owls.

Char. Head with erectile and prominent ear-like tufts. Eyes large; facial disk not complete above the eyes, and bills, legs, feet, and claws usually very strong. Size various.

The species are numerous, and found almost everywhere throughout the globe.

They include some of the largest and most powerful of the owls, among them, species nearly equal to the eagle in strength.

Genus BUBO, Cuvier.

Bubo, Cuvier, Règne Anim. I. 1817, 331.

Gen. Char. Size large; general form very robust and powerful Head with long tufts; eyes very large; wings long; tail short; legs and toes densely feathered; bill rather

B. Virginianus.

short, strong, covered at base by projecting feathers. Concha of ear moderate, oval, with an operculum.

About fifteen species are known, some of them the largest of the family. Their feathers are quite compact, in this respect resembling somewhat the *Falconidæ*, and some of the species are quite diurnal in their habits.

Bubo Virginianus, Gmelin.

THE GREAT HORNED OWL.

Strix Virginiana, Gmelin, Syst. Nat. I. 1788, 287. — Wilson, Am. Orn. VI. 52; pl. 50, f. 1. — Audubon, Birds Amer.; pl. 61. — Nuttall, Man. I. 129. — *Bubo Virginianus*, Bonaparte, Comp. List, 6. — Audubon, Birds Amer.; pl. 61; oct. ed. I. 143, pl. 39. — Nuttall, Man. I. 124. — Vars. *arcticus, Atlanticus, Pacificus*, Cassin, Birds of Cal. and Texas, I. 1853, 178. Ib. P. R. Rep. Birds, IX. 49. — Newberry, VI. 76. — Kennerly, X. iv. 20. — Heermann, X. vi. 35. — Cooper and Suckley, XII. iii. Zool. of W. T. 154. — Coues, Pr. Phil. Ac. 1866, 49.

Sp. Char. Upper parts usually dark brown, each feather mottled and barred with irregular lines of pale ashy and reddish, the latter color prevailing at the base of feathers. Ear tufts dark brown, nearly black, their inner webs edged with dark fulvous; a black spot above the eye, posterior feathers of disk pale or dark reddish; tips of all black.

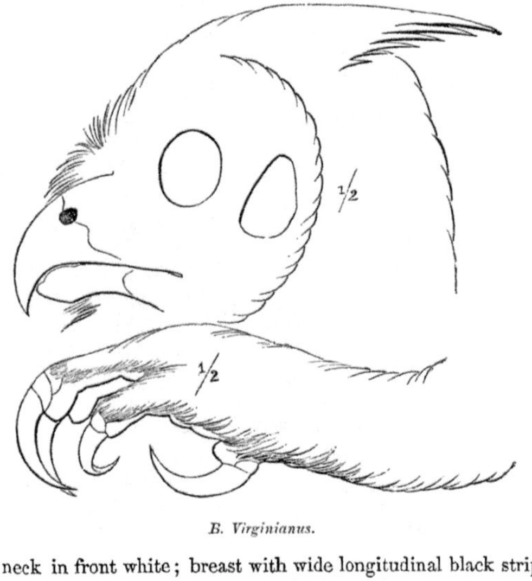

B. Virginianus.

Throat and neck in front white; breast with wide longitudinal black stripes, other under parts variegated with white and fulvous, each feather with narrow transverse brown lines. Middle of abdomen frequently white. Feathers of legs and toes white or fulvous, frequently with narrow dark-brown bars. Quills brown, with wide transverse cinereous bands; usually tinged, on inner webs, with fulvous; tail similar, with much fulvous on outer feathers.

Var. *Pacificus* is less reddish, and the posterior part of facial disk is pale ashy; the general color darker.

Var. *arcticus*, pale yellowish, face and legs nearly white, markings indistinct.

Length, 18.00 to 23.00 ; extent, 35.00 to 52.00 ; wing, 14.00 to 16.00 ; tail, 9.00 to 10.00. Iris yellow ; bill and claws brownish-black.

Hab. All of North America. Var. *arcticus*, in the most northern and central regions ; also found in the Alpine Sierra Nevada, about lat. 36°, by Mr. C. King. Found also, in varying characters, through Mexico and Central America into South America. Not recorded from the West Indies.

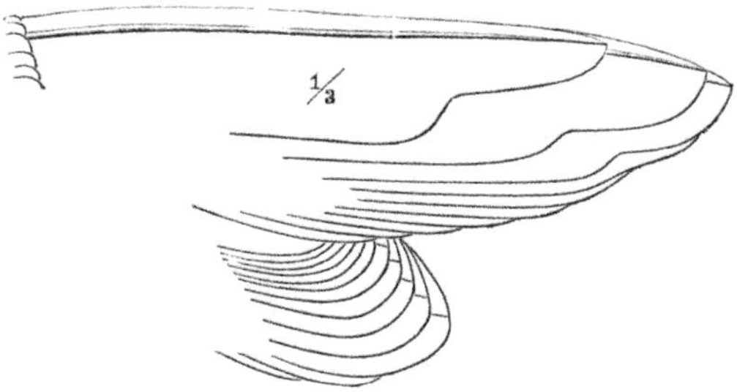

Next to the burrowing owl, this species is probably the most common in California, or the most frequently seen, on account of its large size, and the difficulty of its finding concealment during the day, where large hollow trees are scarce. They are often started from the evergreen oaks, where they sit, scarcely hid at all, and indeed seem able to bear a great deal of light. I found a nest of this species near San Diego, apparently an old crow's-nest, of which there were several in the same tree, about sixty feet from the ground. On climbing to it, the female flew off into the bright sunlight, and we were surprised to see two young, lately hatched and covered with white down. I preserved one of them, and afterwards found an egg-shell in a brook under the tree, white and larger than that of a hen when perfect. They also build sometimes in hollow trees, and in Florida I have seen a nest built on the top of a lonely column, which was left standing after the ruin of a fine building, almost in the broad sunlight. Eggs three to six, 2.38 by 2.00 inches, the ends alike. (Audubon.) This is probably the only species of owl that preys much on domestic fowls, as all the other owls common in California are rather too small to kill them. Their loud hooting is well known, and not like the cry of any other owl that I know of; and travellers' tales are filled with allusions to the terror inspired in the uninitiated by hearing their sudden and unexpected note, when camped about a fire by night in the woods.

GENUS **SCOPS**, SAVIGNY.

Scops, SAVIGNY, Hist. Nat. d'Egypte, I. 1809, 105.

GEN. CHAR. Size small, ear-tufts conspicuous. Facial disk imperfect in front; bill short, nearly covered by projecting feathers; wings long; tail rather short, frequently

S. asio.

curved inwards; tarsi rather long, more or less fully covered with short feathers; those on the toes hair-like.

About twenty-five species of *Scops* are known.

Scops asio, LINNÆUS.

THE MOTTLED SCREECH-OWL.

Strix asio, LINNÆUS, Syst. Nat. I. 1766, 132 (red form). — *Bubo asio*, AUDUBON, Birds Amer. oct. ed. I. 147, pl. 40. — PR. MAX. Cab. Jour. 1858, 23. — *Scops asio*, BONAPARTE, Comp. List, 6. — CASSIN, P. R. Rep. Birds, IX. 51. — HEERMANN, X. vi. 35. — COOPER and SUCKLEY, XII. iii. Zool. of W. T. 155.

Strix nævia, GMELIN, Syst. Nat. I. 1788, 287. — WILSON, Am. Orn. V. 83; pl. 42, f. 1; III. 17; pl. 19, f. 1. — AUDUBON, Birds Amer. pl. 97. — NUTTALL, Man. I. 125, 127.

? *Scops McCallii*, CASSIN, Birds of Cal. and Texas, I. 1854, 180. IB. P. R. Rep. Birds, IX. 52. IB. Mex. Bound. Rep. II. iii. 4; pl. 1.

SP. CHAR. Above pale ashy-brown, with streak of brownish-black, and irregular mottlings of the same mixed with cinereous. Beneath ashy-white with brownish-black stripes, and transverse lines of the same; face, throat, and tarsi ashy-white, irregularly lined and mottled with pale brownish; quills brown, with transverse bands, nearly white on the outer webs; tail pale ashy-brown, with about ten transverse narrow bands of pale cinereous; under wing coverts white, the larger tipped with black. (*Scops nævia.*)

Younger. Entire upper parts pale brownish-red with streaks of brownish-black, especially on the head and scapulars; face, throat, under wing coverts, and tarsi, reddish-white; quills reddish-brown; tail rufous with bands of brown, darker on the inner webs. (*Scops asio.*)

S. asio.

Young. Entire plumage banded with ashy-white and pale brown; wings and tail pale rufous.

S. McCallii does not seem to have tangible differences in plumage, and its smaller size is according to the usual rule in Southern specimens of widely distributed birds. Length, 8.00 to 10.75; extent, 18.50 to 22.50; wing, 6.00 to 7.00; tail, 3.00 to 3.50. Iris yellow; bill and claws horn-color; cere greenish; toes whitish-gray.

Hab. All temperate North America. Cuba?

According to Dr. Bachman, the young are *red* for two years, when they change to gray. I have never seen a red specimen in California.

This little owl is quite common in the wooded parts of the State, and often captured in houses. At Fort Mojave one was brought to me from a hollow tree, which differed from one I found at Santa Barbara only in being rather smaller and much paler gray, as if its hue was affected by the hot, dry climate, although it never exposed itself to the sunshine. It did not, therefore, present the "darker" color and more numerous better defined transverse lines below, being scarcely more than light ash-color in any part. Being thus unlike Cassin's *S. McCallii*, I have considered it only a pale variety of *S. asio*, although its southern locality would lead us to expect to find it to be *S. McCallii*, if indeed that be a distinct species.

At Santa Barbara I found a young one half fledged on May 4th, and in another tree a fine full-grown bird. The young was gray, but, according to Cassin, they become red when the feathers are fully grown, and afterwards gray again, specimens of both colors being found breeding together in the East, showing probably that it requires some years to effect the change. They have been, however, often taken for distinct species. Their food is entirely composed of small birds, mice, and insects. I have seen one living harmlessly in a pigeon-house.

Scops flammeola, LICHTENSTEIN.

THE FLAMMULATED OWLET.

Scops flammeola, LICHT. Nomenclator Mus. Berol. 1831. — KAUP, Trans. Zool. Soc. IV. 226. SCHLEGEL, Mus. Pays Bas, Ois. 27. — SCLATER, P. Z. S. 1858, 96. — SCLATER and SALVIN, Pr. Z. S. 1868, 57. IB. Exotic Ornithol. VII. 68, 99; pl. 50.

SP. CHAR. Above grayish-brown, streaked and vermiculated with black; the margins of the scapulars and the plumes of the head partially varied with rufous, the former forming a rufous line between the wing and the back. Outer webs of the primaries with quadrate white spots. Beneath grayish-white, all the feathers with a shaft streak, and numerous fine transverse lines of black. Throat and breast slightly tinged with rufous. Tarsi en-

S. flammeola.

tirely feathered, white slightly varied with black; toes entirely naked and brown. Length, 7.00; wing, 5.50; tail, 1.60; tarsi, 0.90. Iris golden yellow; bill pale horn-color, yellowish at tip. (Sclater and Salvin.)

Hab. Fort Crook, California, and the mountains of Mexico (Orizaba) to Guatemala.

The introduction of this species into the fauna of California rests upon a specimen collected at Fort Crook, by the late Captain John Feilner, and transmitted by him to the Smithsonian Institution. Though not entirely mature, it is unmistakably this species. It is much the smallest of the American eared owls, and readily distinguished by the naked toes and the rufous scapular stripe. It is well figured by Sclater and Salvin in the work quoted above, the Californian skin, and another from Orizaba, having been submitted to those gentlemen for their inspection. (Baird.)

Scops Kennicotti, Elliot.

KENNICOTT'S OWL.

Scops Kennicotti, Elliot, Pr. Acad. Nat. Sc. Phil. 1867, 99. Ib. Illust. Birds N. A. I. pl. 27.

Sp. Char. Head and upper parts light rufous brown, each feather having a central streak of brownish-black, and also barred with the same color. The rufous-brown hue is

S. Kennicotti.

lightest on the lower part of the neck, where it is almost a buff. Outer feathers of interscapulars with outer webs light buff, forming a distinct bar. Wings same color as the back, but the central streak broader. Primaries dark brown, outer webs marked with distinct spots of light buff, slightly discernible on the inner. Secondaries blackish-brown, outer webs distinctly spotted with dark buff. Tertials mottled with light buff and black. Ear-tufts light buff, with a central streak of black, and barred with the same, broadest on outer webs. Feathers round the eye reddish-brown; those covering the nostrils soiled white, with black shafts. Concealed patches of white feathers equidistant between the ear-tufts and the ears. Upper part of breast light buff, several feathers on each side having very broad central streaks of black, forming together a conspicuous spot; the rest have this mark much narrower, and the black bars either nearly obsolete or mere wavy lines. Flanks light buff, with a broad line of black in the middle, and a conspicuous bar of pale yellowish-white near the tips. Centre of abdomen and under tail coverts yellowish-white, a few indistinct brown bars on the latter. Length, 11.00; wing, 7.25; tail, 4.00; bill, 0.87 along the curve; claws, 0.50. Tarsi reddish-brown; feet yellowish-white; bill black, white at tip.

The general color is reddish-brown, mottled and blotched with black. In size it is between *Scops asio* and *Otus Wilsonianus*. (Elliot.)

Hab. Sitka.

This species was described by Mr. Elliot, from the single specimen taken in Sitka by Mr. Bischoff. It was called after Mr. Kennicott, with a well-deserved tribute of respect to his memory. Nothing is known of its habits.

It is not a little remarkable that our knowledge of four species of owls of Western North America, so far, rests upon single specimens. Three of these, *Syrnium occidentale, Scops Kennicotti*, and *Micrathene Whitneyi*, are types; of the fourth, — *Scops flammeola*, — the type was taken in Mexico. (Baird.)

Otus Wilsonianus.

GENUS **OTUS**, CUVIER.

Otus, CUVIER, Règne Anim. I. 1817, 327.

GEN. CHAR. Form somewhat elongated. Head moderate; ear-tufts long, erectile; bill rather short, curved from the base; facial disk nearly perfect. Wings long, tail mod-

O. vulgaris.

erate; tarsi and toes covered with short feathers; claws long, curved. Eyes rather small. Concha of ear semicircular, extending from the bill towards summit of head, anteriorly operculate. (Cassin.)

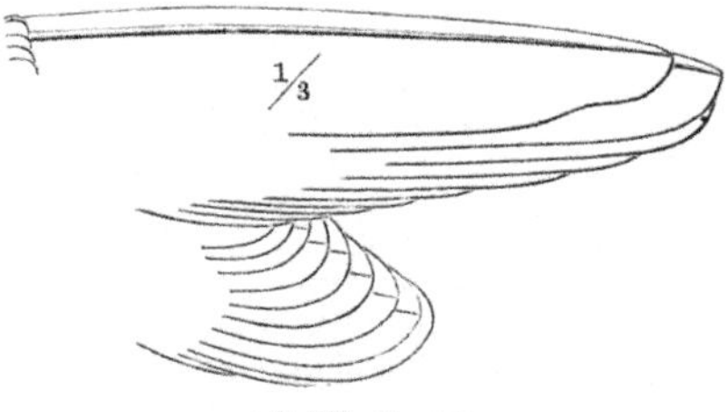

*O. Wilsonianus.**

* This figure is taken from the European bird, and not distinguishable in a woodcut from the American.

54

Otus Wilsonianus, Lesson.

THE LONG-EARED OWL.

Strix otus, Wilson, Am. Orn. VI. 73; pl. 51, f. 3. — Audubon, Birds Amer. pl. 383; oct. ed. I.; pl. 37. — Nuttall, Man. I. 139. (Not of Linnæus, the European species.)
Strix Americana, Gmelin, Syst. Nat. I. 1788, 288. — *Otus Americanus,* Bonaparte, Comp. List, 1838, 7. — Pr. Max. Cab. Jour. 1858, 25.
Otus Wilsonianus, Lesson, Traité d'Orn. I. 1831, 110. — Cassin, P. R. Rep. Birds, IX. 53. — Cooper and Suckley, XII. iii. Zool. of W. T. 155. — Coues, Pr. Phil. Acad. 1866, 50.

Sp. Char. Above mottled with brownish-black, fulvous, and ashy-white, the former predominating. Breast pale fulvous, with stripes of brownish-black; abdomen white, every feather with a wide stripe crossed by others of brownish-black; legs and toes pale fulvous, usually unspotted, frequently with irregular dark-brown bars. Eye nearly encircled by black; other feathers of the face ashy-white, with minute lines of black; ear-tufts brownish-black, edged with fulvous and ashy-white; quills pale fulvous at their bases, with irregular transverse brown bands; lower wing coverts pale fulvous, frequently nearly white; the larger widely tipped with black; tail brown, with several irregular transverse bands of ashy-fulvous, mottled as on the quills Length, 13.00 to 15.00; extent, 35 00 to 38.00; wing, 11.00 to 12.00; tail, 5.00 to 6.00. Iris yellow; bill and claws horn-black.
Hab. All of temperate North America.

I found this owl quite common near San Diego, and in March observed them sitting in pairs in the evergreen oaks, apparently not much troubled by the light. One which I shot was much infested by insects, which may have caused them to leave the hollow trees. On the 27th of March I found a nest, perhaps that of a crow, built in a low evergreen oak, in which a female owl was sitting on five eggs, then half hatched. The bird was quite bold, flying round and snapping her bill at me. I took one egg, and on the 23d of April found the rest hatched. In trying to draw me away from the nest, the female imitated the cries of wounded birds in a remarkably accurate manner, showing a power of voice not heretofore attributed to the owls, but indicating their affinity to the parrots.

The eggs are white, measuring 1.60 × 1.36 inches, and rounded at both ends about equally, like those of most other owls.

According to Nuttall, the cry of this owl is a plaintive and hollow moaning, sounding like "clōw clōud," and incessantly repeated during the night. I have not heard this in California, but it is doubtless correct.

This, as well as the great horned owls, wanders into the barren, treeless deserts east of the Sierra Nevada, and may be found quite frequently in autumn hiding in the thickets of willows along the streams, where the sun shines with scarcely diminished brightness. They also resort to caves, where there are any to be found.

Their food consists of small animals entirely, and I think it very doubtful if they ever attack poultry. By many persons they are, however, mistaken for young horned owls, and ruthlessly slain. I need scarcely remark that a bird which has become able to fly never grows afterwards to any very appreciable degree.

Genus **BRACHYOTUS**, Gould.

Brachyotus, Gould, Proc. Zool. Soc. Lond. 1837, 10.

Gen. Char. Ear-tufts very short and inconspicuous. General form rather strong; wings long; tail moderate; legs rather long, and with the toes fully covered by short

B. Cassinii.

feathers; claws long, very sharp and rather slender. Head moderate; eyes rather small; facial disk imperfect on the forehead and above eyes; tail moderate. (Cassin.)

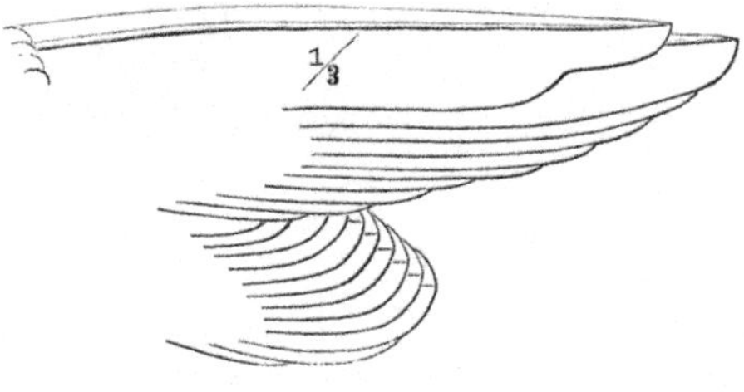

B. Cassini.

Of several species referred to this genus, and representing the different quarters of the globe, naturalists are now inclined to make but one, varying somewhat with the region, but scarcely differing in essential character.

Brachyotus Cassinii, Brewer.

THE SHORT-EARED OWL.

Strix brachyotus, Forster, Phil. Trans. LXII. 1772, 384. — Wilson, Am. Orn. IV. 64; pl. 33, f. 3. — Audubon, Birds Amer.; pl. 410, oct. ed. I. 140; pl. 38 (Otus). — Nuttall, Man. I. 141.

Brachyotus palustris Americanus, Bonaparte, Consp. Av. 1849, 51. — *Otus brachyotus Americanus*, Pr. Max. Cab. Jour. 1858, 27.

Brachyotus Cassinii, Brewer, Proc. Bost. Soc. Nat. Hist. 1855? — Newberry, P. R. Rep. VI. iv. 1857, 76. — Cassin, P. R. Rep. Birds, IX. 54. — Heermann, X. vi. 34. — Cooper and Suckley, XII. iii. Zool. of W. T. 155.

Sp. Char. Ear-tufts very short. Entire plumage buff or pale fulvous; each feather above with a wide dark brown stripe, this color predominating on the back. Under parts paler, frequently nearly white on the abdomen, with brownish-black stripes, most numer-

*B. palustris.**

ous on the breast, very narrow and less numerous on the abdomen and flanks; legs and toes usually like the abdomen. Quills pale reddish-fulvous at base; brown at their ends, with wide bands and large spots of darker shade; tail similar, with about five irregular transverse dark brown bands, this color predominating on the central feathers; under tail

* The figure is taken from the European bird. The American is not distinguishable in a cut.

coverts nearly white. Throat white; eyes enclosed by large brownish-black spots; ear-tufts brown, edged with fulvous. Iris, cere, and toes yellow; bill and claws horn-color. Length, 14 to 15; wing, 11 to 12; tail, $5\frac{1}{2}$ to 6.

Hab. All of temperate North America and South America; Greenland; Cuba.

This is a more northern species than the *Otus*, visiting the United States only in winter, and then appearing in considerable numbers or flocks. They generally sit in long grass or bushes during the day, little disturbed by the sunlight, and on cloudy days even hunt over the prairies, on which they usually obtain their prey. I have not seen them south of Santa Clara Valley; but east of the Sierra they have been seen along the Mojave River by Dr. Heermann.

Audubon found a nest of this species in the pine forest on the mountains of Pennsylvania. It was built on the ground under a bush, hidden among long grass, of which it was roughly formed. This nest was flat, and much spread out, and contained, on the 17th of June, four bluish-white eggs, 1.50 × 1.12.

Sub-Family SYRNIINÆ.

Char. Head large, with concealed ear-tufts or none. Facial disk nearly perfect; eyes small for owls; wings rather short; tarsi and toes generally fully feathered. Size moderate or small. Concha of ear semicircular, operculate.

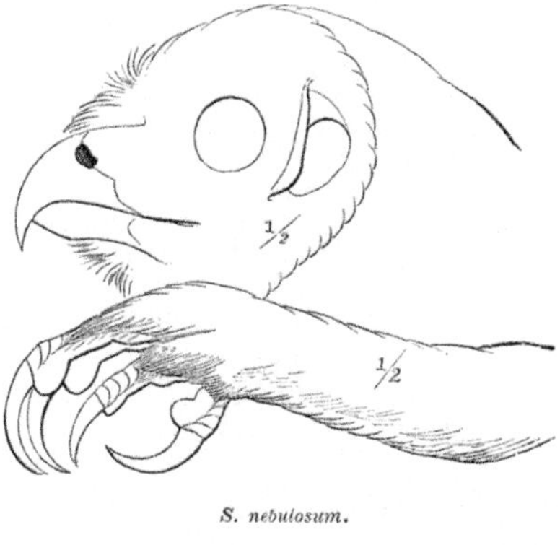

S. nebulosum.

This sub-family embraces species differing greatly in size. Some are among the largest, and others among the smallest of the owls, and they inhabit extremes of latitude. The plumage is generally very lax and soft.

Genus SYRNIUM, Savigny.

Syrnium, Savigny, Hist. Nat. d'Egypte, I. 1809, 112.

Gen. Char. Size usually large; head large, without ear-tufts; eyes rather small; bill strong, curved from the base; fourth and fifth quills longest; tail rather long, wide, rounded; legs moderate or rather long; claws long, strong, very sharp. Conch of the ear a simple oval cavity only half the height of the cranium.

A genus containing fifteen or twenty species, chiefly from the northern parts of the world.

Syrnium occidentale, Xantus.

THE WESTERN BARRED OWL.

Syrnium occidentale, Xantus, Proc. A. N. Sc. Phil. XI. July, 1859, 193.—Baird, Cassin, and Lawrence, Birds N. Amer. II. v.; pl. lxvi. 1860. Index, p. v. descr.

Sp. Char. General color liver-brown, the feathers barred everywhere, even on the flanks. Axillars and under wing and tail coverts banded transversely with white, the

S. occidentale

bands towards and on the head contracted into rounded spots. Bars of white on feathers of body, two; several on the scapulars, axillars, and other long feathers. Quills and tail with 7 – 9 bars, one terminal. Legs dirty-yellowish, with obscure transverse brown mottlings. Toes thickly feathered, exposing only two scutellæ at the base of claws. Fourth

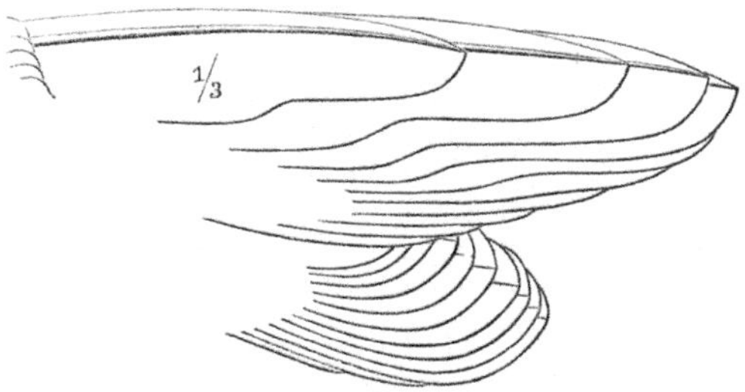

quill longest, fifth and third shorter, second between sixth and seventh, first shorter than eighth. Length of male, 18.00; extent, 40.00; wing, 13.00; tail, 8.50; tarsus about 2.00. Bill greenish-yellow; iris gamboge; claws horn-color.

Hab. Fort Tejon, California (and northward?).

This species, of the history of which nothing is yet known, probably resembles the Eastern barred owl (*S. nebulosum*) in habits, as it does in plumage. It was discovered by Mr. J. Xantus, at Fort Tejon, March 6, 1858; but only one specimen was obtained. The allied Eastern species is widely spread, and this will probably be found common in some parts of the State.

Syrnium nebulosum, FORSTER.

THE BARRED OWL.

Strix nebulosa, FORSTER, Trans. Philos. Soc. London, LXII. 1772, 386, 424.
Strix varius, BARTON, Frag. Nat. Hist. Penn. 1799, 11.
Syrnium nebulosum, CASSIN, Birds N. Amer. 1858, 56.

FIGURES. — Vieillot, Ois. d'Am. Sept. pl. 17; Wilson, Am. Orn. IV. pl. 33, fig. 2; Audubon, Birds Amer. pl. 46: oct. ed. I. pl. 36; Nat. Hist. New York, Birds, pl. 10, fig. 21; Gould, Birds of Eur. I. pl. 46.

SP. CHAR. Smaller than the following; head large, without ear-tufts; tail rather long. Upper parts light ashy-brown, frequently tinged with dull yellow; with transverse narrow bands of white, most numerous on the head and neck behind, broader on the back. Breast with transverse bands of brown and white; abdomen ashy-white, with longitudinal stripes of brown; tarsi and toes ashy-white, tinged with fulvous, generally without spots, but frequently mottled and banded with dark brown. Quills brown, with six or seven transverse bars nearly pure white on the outer webs, and ashy-fulvous on the inner webs; tail light brown, with about five bands of white, generally tinged with reddish-yellow. Feathers of disk tipped with white; face ashy-white, with lines of brown, and a spot of

S. nebulosum.

black in front of the eye; throat dark brown; claws horn-color; bill pale yellow; irides bluish-black. Sexes alike. Total length about 20.00; wing, 13.00 to 14.00; tail, 9.00. Sexes nearly of the same size.

Hab. Eastern North America. South to Mexico.

No specimens of this well-known species have yet been found west of the Rocky Mountains of the United States. As it is, however, exceedingly improbable that a bird of such wide range in other parts of North America should not cross the mountains, we introduce it here, in anticipation of the period of its detection in the mountains of California, which will undoubtedly occur sooner or later. The nocturnal habits of the owls naturally keep them from the notice of observers, and even in the Eastern States some species are but rarely seen or killed, although perhaps known to be actually abundant.

Messrs. Sclater and Salvin have lately described *S. fulvescens* as a very closely allied Mexican species. (Baird.)

Syrnium cinereum, Gmelin.

THE GREAT GRAY OWL.

Strix cinerea, Gmelin, Syst. Nat. I. 1788, 291. — Bonaparte, Am. Orn.; pl. 23, f. 2. — Audubon, Orn. Biog.; pl. 353. — Nuttall, Man. I. 134.

Syrnium cinereum, Bonaparte, B. of Eur. and Am. 6. — Audubon, Birds Amer.; pl. 58. — Cassin, P. R. Rep. Birds IX. 56. — Newberry, VI. iv. 77. — Cooper, XII. iv. Zool. of W. T. 156. — Kaup, Trans. Zool. Soc. IV. 1859, 256.

Sp. Char. The largest North American owl. Above smoky or ashy-brown, mottled and barred with ashy-white; beneath ashy-white, with numerous dark ashy-brown stripes, especially on the breast, and with bars of the same on the abdomen, legs and lower tail

S. cinereum.

coverts. Quills brown, with about five wide irregular bands of ashy-white; tail similar, with five or six bands, and mottled with dark brown. Feathers of the disk, on the neck tipped with white; eye nearly encircled by a black spot; radiating feathers round the eye, with regular transverse narrow bars of dark brown and ashy-white. Length, 25.00; extent, 56.00; wing, 18.00; tail, 12.00 to 15.00. Iris yellow; bill and claws paler.

Hab. Northern North America. Sacramento Valley, according to Newberry.

This owl is common in the dense spruce forests near the Columbia River, and northward. It is probably resident throughout the year in that latitude, and perhaps in the high mountains of California.

Its habits are to some extent diurnal, or it is active towards sunset, and at times utters a low laughing cry, said to resemble that of the mottled owl. Richardson found a nest near lat. 60°, built of sticks in the top of a lofty balsam poplar, and lined with feathers. It contained three young, covered with whitish down.

GENUS **NYCTALE**, BREHM.

Nyctale, BREHM, Isis, 1828, 1271.

GEN. CHAR. Size small. Head with very small ear-tufts, only seen when erected; eyes small; bill moderate; facial disk nearly perfect. Wings rather long; tail short; toes densely feathered. Conch of ear very large, with an operculum.

N. acadica.

Only four or five species of *Nyctale* are known, most of them in America. They are all of diminutive size, and rarely met with living, owing to the close concealment which they practise during the day, in this respect quite different from some of our American species.

Nyctale albifrons, SHAW.

KIRTLAND'S OWL.

Strix albifrons, SHAW, Nat. Misc. V. 1794. — *Nyctale albifrons*, CASSIN, B. of Cal. and Tex. 187. IB. P. R. Rep. Birds, IX. 57. — COUES, Pr. A. N. Sc. 1866, 50.
Strix frontalis, LICHTENSTEIN, Trans. Berlin Ac. 1838, 340.
Nyctale Kirtlandii, HOY, Pr. A. N. Sc. Phil. VI. 1852, 210. — CASSIN, B. of Cal. and Tex. I. 1855, 63 ; pl. xi.

SP. CHAR. Head, upper part of breast, and entire upper parts, dark chocolate-brown ; forehead and eyebrows white. Throat, and a line running down on each side from base of under mandible, white ; other under parts of body reddish ochre-yellow. Quills dark brown, with small white spots on their outer edges, and large white spots on their inner

N. albifrons.

webs. Tail dark brown, with two transverse bands of white, and a narrow white tip. Length about 8.00 ; wing, 5.25 ; tail, 3.00. Iris yellow ; bill and claws dark. (Cassin.)

Hab. Northern North America to Canada and Wisconsin. California (Lichtenstein).

I have seen in the museum of the German Academy of Natural Sciences, of San Francisco, a specimen of this owl, brought from Nevada, close to the boundary of California, about lat. 39°, altitude seven thousand feet.

Cassin's beautiful plate, and interesting description above referred to, include all at present known of its habits. On the eastern side of the continent it seems limited to the northern regions, and on this coast probably confines its range to the highest mountains, unless driven down to the valleys in winter by cold.

Dr. Hoy remarks that one flew into an open shop in Racine, Wisconsin, in July, 1852. It was strictly nocturnal, uttered a low tremulous note, and was an active, efficient mouser.

Nyctale Acadica, GMELIN.

THE ACADIAN OWL.

Strix Acadica, GMELIN, Syst. Nat. I. 1788, 296. — NUTTALL, Man. I. 145. — AUDUBON, Birds Amer.; pl. 199. — *Nyctale Acadica*, BONAPARTE, Eur. and N. Amer. Birds, 7. — CASSIN, P. R. Rep. Birds, IX. 58. — KAUP, Trans. Zool. Soc. IV. 206. — COOPER and SUCKLEY, XII. iii. Zool. of W. T. 156. — SCLATER, Pr. Zool. Soc. 1858, 295. (Oaxaca.) — LORD, Pr. R. Art. Inst. Woolwich, IV. 1863, 111.
Strix passerina, "LINNÆUS," WILSON, Am. Orn. IV. 66; pl. 34.

SP. CHAR. Above reddish-brown, tinged with olive; head in front with fine white lines, and large partly concealed white spots on the neck behind, rump, and scapulars. Face ashy-white, throat white, under parts ashy-white, with pale reddish-brown stripes;

N. Acadica.

under wing and tail coverts white. Quills brown, with small white spots on their outer edges, and large white spots on their inner webs; tail brown, each feather with about three pairs of white spots. Iris yellow; bill and claws dark.

Hab. Northern parts of North America. South to Oaxaca, Mexico.

Found at Fort Tejon by Mr. J. Xantus de Vesey. It is probably more common northward, as I obtained it at Vancouver, Washington Territory, and in the Eastern States it does not seem to be common south of Pennsylvania. It is very nocturnal, and its habits are therefore but little known. It is said to have a cry like the whetting of a saw. Audubon found their eggs in Maryland, deposited in crow's nests or holes of trees, numbering three to six, white, and elliptical.

SUB-FAMILY ATHENINÆ.

CHAR. Size small; facial disk very imperfect, nearly obsolete; tarsi partially or but thinly feathered; head untufted.

The species of this group are numerous, and less strictly nocturnal than others.

GENUS **ATHENE**, BOIE.

Athene, BOIE, Isis, 1822, 549.

GEN. CHAR. Wings rather long; tail rather short; bill short; legs rather long, thinly covered with short bristly feathers; toes naked, or with a few hair-like feathers.

Contains forty species, generally distributed.

Athene cunicularia, MOLINA.

THE BURROWING OWL.

Strix cunicularia, MOLINA, Sagg. Stor. Nat. Chili, 1782. — NUTTALL, Man. I. 123, and other authors on the Californian species. — *Athene cunicularia*, BONAPARTE, Eur. and Am. Birds, 6. — CASSIN, P. R. Rep. Birds, IX. 60. — BAIRD, X. iii. 13. (Utah.) — KENNERLY, X. iv. 20. — KAUP, Trans. Z. S. iv. 201. — HEERMANN, X. vi. 33. — CANFIELD, Amer. Nat. 1869, 583 (habits). — COOPER and SUCKLEY, XII. iii. Zool. of W. T. 157.

Strix Californica, AUDUBON, Birds Amer.; pl. 432, f. 2 (name). IB. oct. ed. I.; pl. 31. (Lower figure.)

Athene hypugæa, NEWBERRY, P. R. Rep. VI. ii. 17, but not of Bonaparte; the species found east of the Rocky Mountains, according to Cassin.

SP. CHAR. Upper parts light ashy-brown, with large spots of dull white enclosed in edgings of brownish-black. Throat white; a transverse band of brownish-black and red-

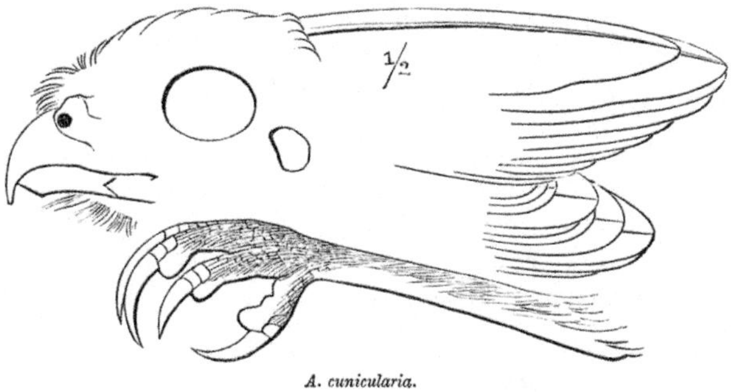

A. cunicularia.

dish-white feathers across the neck in front, succeeded by a large patch of white. Breast light brown with large spots of white as above; abdomen yellowish-white, with hastate or crescentic spots of reddish-brown, in transverse bands. Under tail coverts, tibiæ, tarsi, and under wing coverts yellowish-white; quills and tail light brown, with spots of reddish-white, which are edged with brownish-black. Tail with about six transverse bands, or

A. cunicularia.

pairs of spots, of reddish-white edged with dark brown. Other specimens are much lighter and tinged with dull yellow, having a faded appearance.

Younger. Above with light grayish-brown spots irregular in shape and confused; frequently predominating on head. Abdomen yellowish-white, nearly unspotted. Length, 9.50 to 10.00; extent, 23.50 to 25.00; wing, 6.75 to 7.75; tail, 3.50 to 4.00. Iris yellow; bill horn-color; toes gray.

Hab. North America west of the Rocky Mountains; South America (Peru, Chili, Buenos Ayres, etc.).

Probably one of the most common birds in California, and known to almost everybody, as they are visible at all times of the day, and not timid. Wherever the large ground-squirrel (*Spermophilus Beecheyi*) is found, — and that is in almost every valley west of the Sierra, — this owl is its constant companion, living in the deserted burrows of the squirrels, and apparently in perfect harmony with them, although the young squirrels doubtless occasionally furnish the owl with a meal, for which the squirrel perhaps takes its pay by sucking the owl's eggs, after the manner of its relative, the rat.

The owl, however, undoubtedly burrows for itself in regions where there are no squirrels large enough to make burrows for it. I found one living near the Colorado, in a burrow which it had apparently just made, as there

NOTE. — Specimens obtained by Mr. Hepburn, near Mariposa, and one from Vancouver Island, closely resemble *A. hypugæa.*

was no animal in that region that made such burrows. The soil was gravelly, and yet the holes were so deep that I could not reach the bottom with a ramrod. The difficulty of burrowing, and comparative scarcity of food, make this owl scarce in that region. The specimen I preserved was like those found near the coast, in colors and size.

At San Diego, as early as January 11th, they were preparing to lay, and I saw one chase a large hawk (*F. polyagrus ?*) away from its burrow, towards which the hawk had darted with the intention of catching the owl.

On February 4th the males began to utter a call in the early evening, as they sat at the mouth of their burrow. This note sounded like the word "cŭc-kōō" slowly pronounced, with an accent on the first syllable, and the last very long. Europeans told me that it was very much like the cry of the well-known European cuckoo, and I have since seen a statement in one of the newspapers that the writer had discovered "the cuckoo" in California, knowing it by its note, which he heard near the tulé marshes of the San Joaquin! His cuckoo was doubtless this owl; for though I have not before seen any record of this fact, I heard and saw the birds too frequently to be mistaken. The tone is soft and musical, but has the property of seeming to be much farther off than the bird really is, and therefore may have been often heard when its origin was unknown. The owl continued this call during the month of March, and then sometimes uttered it during the day at intervals.

On the 4th of April I dug out two fresh eggs from a burrow, which I had to follow for a depth of three feet, and then five feet horizontally, where I found an enlarged chamber in which were the eggs, deposited on a few feathers. They were smaller than pigeons' eggs, nearly round, and pure white. The Eastern species lays four eggs, according to Audubon.

About the end of April the young begin to appear at the mouth of their burrow, where they sit in company with the parents during the early morning, easily distinguished by their darker plumage and almost banded breasts. Their actions are very ludicrous, as they allow one to come quite near, and bow repeatedly to the approaching visitor, uttering at the same time a low cackling note. If alarmed they either fly off to another burrow or suddenly dive down into one. Their flight is rapid, by repeated flaps and undulating sailings, like that of a woodpecker, showing, as does their call-note, a relationship to the *Scansores*.

Their food consists wholly of mice, small birds, and insects, and they must be among the most useful birds we have in destroying gophers. They hunt chiefly or altogether at night, and then often make a loud cackling as they fly. They are constant residents in most parts of this State, and extend north to the Columbia River.

In October, 1863, I heard the call-note of these birds almost every evening about Drum Barracks, Wilmington, showing that it is not confined to the spring.

From Monterey north this species becomes very rare, or entirely absent on the west side of the Coast Range.

Athene hypugæa, BONAPARTE.

THE BURROWING OWL.

Strix hypugæa, BONAPARTE, Am. Orn. I. 1825, 72. — *Athene hypugæa*, CASSIN, BAIRD, Birds N. Amer. 1858, 59. — COOPER and SUCKLEY, XII. iii. Zool. of W. T. 151.

Athene socialis, GAMBEL, Pr. Acad. Phil. III. 1846, 47. — SCLATER, Pr. Zool. Soc. 1857, 201. (Xalapa, Mexico.) — DRESSER, Ibis, 65, 330.

FIGURES. — BONAPARTE, Am. Orn. I. pl. 7, fig. 2; AUDUBON, Birds Amer. pl. 432, fig. 1: oct. ed. I. 31 (upper figure).

SP. CHAR. Tarsi long, slender, thinly covered in front only with short feathers, generally with its lower half nearly bare, and frequently almost entirely naked, and with small circular scales laterally and posteriorly; toes with a few hairs.

Adult. Upper parts light ashy-brown, with numerous partially concealed circular, cordate, and ovate spots of dull white, which spots are enclosed with a narrow edge of dark brown. Throat white; a transverse band of dark brown and reddish-white on the neck in front, succeeded by a large patch of white; breast light brown, with large spots of white, like the upper parts; abdomen yellowish, with transverse narrow bands of red-

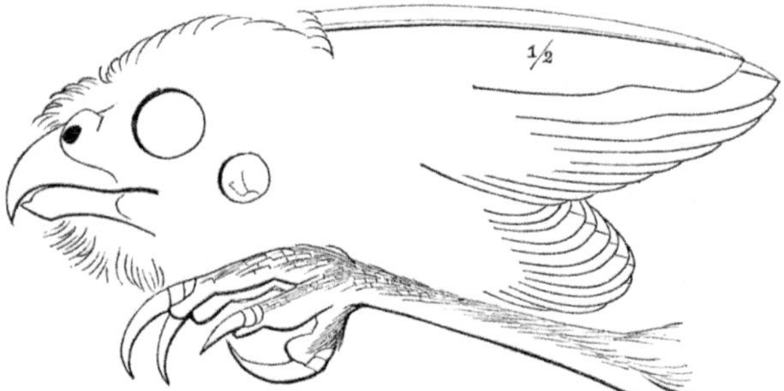

A. hypugæa.

dish-brown; under tail coverts, feathers of the tibia and tarsus, and under wing coverts, yellowish-white. Quills light brown, with semicircular spots of reddish-white on their outer webs, and with oval or irregular spots of the same on their inner webs. Tail light brown, with about five or six irregular transverse bands of yellowish-white. Bill dark bluish at base, yellow at tip and on the ridge of the upper mandible. This is the most mature plumage, and is that represented in Audubon's figure cited above. The most usual plumage is, however, as follows: —

Adult. Upper parts like the preceding, but lighter colored, and much more tinged with dull yellow, having generally a faded or bleached appearance. White spots more numerous and irregular in shape, and frequently giving the predominating color to the head. Rump and tail strongly tinged with reddish. Under parts like the preceding, but lighter, and with the reddish-brown of the abdomen assuming the form of semicircular or hastate spots. This plumage has very nearly the same characters as the preceding, but is lighter and has a faded appearance, and is much the most frequently met with in specimens. This plumage is not figured.

Another plumage is: Upper parts much less tinged with yellow or reddish, being nearly light grayish-brown; white spots very irregular in shape. Abdomen nearly pure white, or tinged with yellowish, with traces only or but few spots of reddish-brown. This plumage is given in Bonaparte's plate, cited above; but it is unusual for the abdomen to be so nearly pure white and unspotted as represented. Total length, female (of skin), about 9.50; wing, 7.00; tail, 3.50. Male, total length, about 9.00; wing, 6.50; tail, 3.00.

Hab. From the Mississippi River to the Rocky Mountains. Mexico.

In this species the feathers on the tarsus are restricted to a narrow longitudinal band or stripe in front, generally quite imperfect in the lower half, leaving that portion nearly bare to the toes, though it is quite unusual to find the tarsus so entirely uncovered, as represented in Audubon's figure cited above. This bird is rather smaller than *cunicularia*, and has the tarsus shorter as well as much less feathered. It inhabits the countries east of the Rocky Mountains, while the *cunicularia* appears to belong exclusively to the west of the same range. (Cassin.)

Whether we really have more than one species of burrowing owl in North America is still a question, the difference, if any, being exceedingly slight. To furnish the means of comparison, however, we reproduce above the remarks of Mr. Cassin on the subject. Mr. James Hepburn, an excellent authority on the birds of the Pacific Coast, thinks that there are two species, and states that both are found within the limits of the State of California. If two species exist, their habits appear to be very similar. (Baird.)

Genus **MICRATHENE**, Coues.

Micrathene, Coues, Pr. A. N. Sc. 1866, 57. (Type, *Athene Whitneyi*, Cooper.)

Gen. Char. Bill small and weak, compressed at base. Facial disk not conspicuously defined; imperfect behind the eyes. Wings very long; from the carpal joint, measuring two thirds the total length of body; exposed portion of first primary only two thirds that of longest; third and fourth quills longest, fifth little shorter, second about equal to sixth. Tarsi moderate; feathered only a short distance below the upper joint, the remaining portion clothed only with sparse bristly hairs, as is the superior surface of the toes. Claws very weak, moderately curved. Middle toe and claw about as long as tarsus. (Coues.)

Of the size and general appearance of *Glaucidium*, this genus has a smaller bill, much longer wings, and much shorter tail, with different proportions of tarsus and toes. From *Athene* it differs in proportion of toes, of quills, etc.; the tarsus being shorter, and the outer primaries much more elongated. The bristly bare legs are the only point common to the two genera. (Coues.)

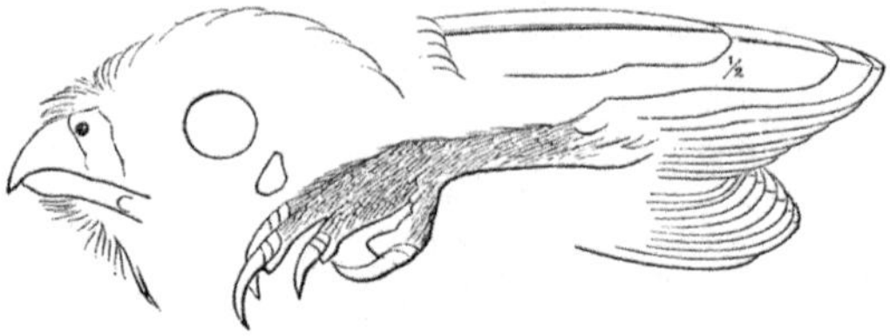

M. Whitneyi.

Micrathene Whitneyi, Cooper.

WHITNEY'S OWL.

Athene Whitneyi, Cooper, Proc. Cal. Acad. N. Sc. II. 1863, 118. — *Micrathene Whitneyi*, Coues, Pr. A. N. Sc. 1866, 51. — Elliot, Ill. B. N. A. XII.; plate.

Sp. Char. Above light brownish-gray, thickly spotted with angular pale brown dots, most densely on head, but those on back largest; back also somewhat barred with waving

A. Whitneyi.

lines of the same color. A concealed white bar crosses the middle of feathers on back of neck, forming a collar, the bases of these feathers being plumbeous like those of the rest. Quills with three to six spots on each web, those on the inner web white, as are those on outer web of the second, third, and fourth quills; the rest of the spots light brown. A row of white spots on edges of lesser coverts, four on the upper, seven on the lower series, with a row of light brown spots between. A few white spots also on outer secondaries. A white stripe on outer row of scapulars, edged by light brown stripes towards the middle of back. Rest of wing feathers dark brown, the secondaries with light ashy dots towards their ends. Tail feathers colored like the quills, the light spots forming five broken bars, and another narrower bar at the tip. Wings and tail ashy-brown beneath, with white bars; edge of wing white; lower wing coverts white, tinged with yellowish, and with a dark brown patch at their ends.

Stiff feathers above eye white, with black spots on the middle of their shafts. Feathers below orbit light brown, faintly barred with darker; bristles around bill black for their outer half. Chin and throat feathers white, their bases black, and the tips of the lower series light brown; the white thus forming a broad crescent in front of neck, extending between outer angles of orbits, somewhat broken at the median line, and edged with brown, darkest laterally. Sides of neck narrowly barred with ashy, alternating with light and dark brown; a large white patch in front of neck, mottled with blackish. Breast imperfectly barred and blotched with the same colors, the brown forming large patches toward abdomen, margined with gray and white. Sides more grayish, tinged yellow, flanks plumbeous. Tibial feathers narrowly barred with light and dark brown. Tarsal bristles white, those on toes yellowish. Length, 6.25; extent, 15.25; wing, 4.50; tail, 2.25; tarsus, 0.90; middle toe, 0.60; with its claw, 0.70; inner lateral claw reaching to base of middle, outer to base of inner; hind toe and claw, 0.50; gape of bill, 0.45; its height, 0.30; with at base, 0.40. Iris bright yellow; bill pale green (drying black with yellow edges); soles yellow; claws horn-brown.

Hab. Colorado Valley, California.

No. 208, State collection, male (?), shot at Fort Mojave, lat. 35°, Colorado Valley, on April 26, 1861, is as yet a unique specimen. It is the smallest owl yet discovered within the United States, being considerably less than the little *Glaucidium gnoma.* In colors it much resembles that species, but differs essentially in the generic characters, these being closely like those of our *Athene,* the burrowing owl. Its habits are probably, however, entirely different, as I found it in a dense thicket, on a very windy morning, where it may perhaps have merely taken a temporary refuge, after being blown down from some of the caverns in the barren mountains surrounding the valley. Its stomach contained insects and some small feathers.

GENUS **GLAUCIDIUM**, BOIE.

Glaucidium, BOIE, Isis, 1826, 970.

GEN. CHAR. Size very small; head moderate, untufted; wings moderate or rather short; tail short; facial disk nearly obsolete; bill short, rather wide, strong; tarsi fully feathered; claws rather long and curved, very sharp.

Glaucidium Californicum, SCLATER.

THE PYGMY OWL.

Strix passerinoides, AUDUBON, Orn. Biog. V. 271, pl. 432, f. 4, 5: oct. ed. I. pl. 30. (Surnia.) NUTTALL, Man. I. 148.
Glaucidium infuscatum, CASSIN, B. of Cal. and Tex. I. 175. — NEWBERRY, P. R. Rep. VI. ii. 77. ("Not of Temminck, which is same as preceding." CASSIN.)
Glaucidium gnoma, WAGLER, Isis, XXV. 1832, 275. — CASSIN, P. R. Rep. Birds, IX. 62. — HEERMANN, X. vi. 34. — COOPER and SUCKLEY, XII. iii. Zool. of W. T. 158. — COUES, Pr. A. N. Sc. 1860, 50. — LORD, Intellectual Observer, 1865, 409. (Habit.)
Glaucidium Californicum, SCLATER, Proc. Zool. Soc. Lond. 1857, 4.

SP. CHAR. Spot before the eye, and extending over it, white. Upper parts entirely brownish-olive, with small circular spots of dull white or pale rufous, numerous on the head, largest on the scapulars. An irregular, partly concealed band of white on the neck behind, succeeded by a black one. Throat white, a band of brownish-olive across breast; other under parts white, striped with dark olive brown; quills dark brown, with small

G. gnoma.

white spots on their outer webs, and large circular or oval ones on their inner; tail dark brown, with about six or seven pairs of spots on each feather; larger on the inner webs. Under wing coverts white, with black spots.

Female, with rather larger and more numerous spots above. Length, 6.50 to 7.50; extent, 13.00 to 14.00; wing, 3.50 to 3.75; tail, 2.75 to 3.00. Iris yellow; bill greenish-yellow; feet pale yellow.

Hab. Oregon, California. No authentic case of its occurrence south of the United States.

Not uncommon in the middle parts of California, but I did not meet with it southward. The one I obtained near the Columbia River, in 1854, was flying about on a cloudy day among some sparrows, not much smaller than itself; and sometimes, according to Townsend, this owl makes a meal of them, though my specimen had nothing but insects in its stomach. Its flight is rapid and easy, and its motions quick for an owl. Drs. Heermann and Newberry mention its frequent occurrence in the Sierra Nevada, but of its habits nothing further has been recorded.

Mr. J. K. Lord found this bird on Vancouver's Island, and found the nest

G. Californicum.

of the pair, or rather two round rough eggs, laid on the decayed wood in a large knot-hole of an oak near a small lake, early in May. By watching in the morning twilight he observed them to fly with a short, quick, jerking flight in search of insects, which they usually captured on the ground, often pouncing upon them from the lower branches. After obtaining enough food the pair would sit close together in the shade during the heat and glare of the sunshine, and about dusk again pursued their prey for a short time, retiring at dark into the knot-hole, thus showing that their habits are not nocturnal.

The Indians seem to have a greater superstitious dread of this owl than of any other animal, and to consider the killing of one as sure to cause horrible misfortunes.

Sub-Family NYCTEININÆ, The Day Owls.

Char. General form compact and robust. Head moderate, without ear-tufts; wings and tail rather long; tarsi strong, which, with the toes, are more densely covered than in any other division of this family.

This division embraces two species only, which inhabit the Arctic regions of both continents, migrating southward in the winter.

GENUS **NYCTEA**, STEPHENS.

Nyctea, STEPHENS, Cont. of Shaw's Zool. XIII. 1826, 62. (Type, *Strix nyctea*, L.)

GEN. CHAR. Large; head rather large, without ear-tufts; no facial disk; legs rather short, and with the toes covered densely with long hair-like feathers, nearly concealing

N. nivea.

the claws. Bill short, nearly concealed by projecting feathers, very strong; wings long; tail moderate, or rather long, wide; claws strong, fully curved.

But one species of this genus is at present known to naturalists, occurring in the north of both hemispheres.

Nyctea nivea, Daudin.

THE SNOWY OWL; THE WHITE OWL.

Strix nivea, Daudin, Traité d'Orn. II. 1800, 190. — *Nyctea nivea*, Brehm, Isis, 1834, 108. — Cassin, Baird, Birds N. Amer. 1858, 63.
Strix nyctea, Linnæus, Syst. Nat. I. 1766, 132.
Strix candida, Latham, Ind. Orn. Supp. 1801, 14. — Dresser, Ibis, 1865, 330. (San Antonio, Texas.)
Strix erminea, Shaw, Gen. Zool. VII. 1809, 251.
Strix arctica, Bartram, Travels, 289. (1791, but not of Sparrmann, 1789.)
Strix scandiaca, "Linnæus," Malmgren, Cab. Jour. 1865, 396.

Figures. — Wilson, Am. Orn. IV. pl. 32, fig. 1; Audubon, Birds Amer. pl. 121: oct. ed. I. pl. 28; Nat. Hist. New York, Birds, pl. 9, fig. 20; Gould, B. of Eur. I. pl. 43.

Sp. Char. Bill nearly concealed by projecting plumes; eyes large. Entire plumage white, frequently with a few spots, or imperfect bands, only on the upper parts, dark brown, and on the under parts, with a few irregular and imperfect bars of the same; quills and tail with a few spots or traces of bands of the same dark brown. The prevalence of the dark brown color varies much in different specimens; frequently both upper and under parts are very distinctly banded transversely, and sometimes this color predominates on the back. Plumage of the legs and toes pure snowy-white; bill and claws dark horn-color; irides yellow. Total length, 24.00 to 27.00; wing, 16.00 to 17.00; tail, 10.00.

Hab. Northern regions of both continents; migrating southward in the winter almost to the Gulf of Mexico.

Although there is as yet no intimation of the occurrence in California of this large owl, its abundance throughout the Arctic regions generally, and the extreme southward migrations along the Atlantic and Mississippi States to the Gulf of Mexico, render it very probable that it will yet be detected south of British Columbia in the winter season, especially along the mountains. Like the hawk owl, this species is abroad at all hours of the day, and in the high north makes its prey of ptarmigan, or white grouse, rabbits, mice, and other animals so abundant in the same region. The nest is made on the ground, and the eggs are usually five or six in number, white as in all owls, and rather small for the apparent size of the bird.

Genus SURNIA, Dumeril.

Surnia, Dumeril, Zoologie Analytique, 1806, 34. (Type. *Strix ulula*, Linnæus.)

Gen. Char. General form rather long, but robust; size median. Head moderate, without ear-tufts; facial disk obsolete; bill moderate, curved from the base, covered with projecting plumes; wings long; tail long, wide, graduated; legs rather short, and with the toes densely feathered; contains one species only, which inhabits the Arctic regions of both continents.

Surnia ulula, Linnæus.

THE HAWK OWL; THE DAY OWL.

Strix ulula, Linnæus, Syst. Nat. I. 1766, 133. — *Surnia ulula*, Bonaparte, Cassin, Baird, Birds N. Amer. 1858, 64. — Lord, Pr. R. Art. Inst. Woolwich, IV. 111.
Strix Hudsonia, Gmelin, Syst. Nat. I. 1788, 295.
Strix doliata, Pallas, Zoog. Ross. As. I. 1811, 316.
"*Strix funerea*, Linnæus, Fauna Suecica, 1761, 22." — Audubon, Birds N. Amer. pl. 378.

Sp. Char. Wings rather long; first three quills incised on their inner webs; tail long, with its central feathers about two inches longer than the outer; tarsi and toes densely feathered. Upper parts fuliginous-brown, with numerous partially concealed circular spots of white on the neck behind scapulars and wing coverts. Face grayish-white; throat white, with longitudinal stripes of dark brown; a large brown spot on each side of

the breast; other under parts with transverse lines or stripes of pale ashy-brown; quills and tail brown, with transverse bands of white; bill pale yellowish; irides yellow. Color of upper parts darker on the head, and the white markings more or less numerous in different specimens. Total length, female, 16.00 to 17.00; wing, 9.00; tail, 7.00. Male rather smaller.

Hab. Northern regions of both continents.

This species, a day hunter like the snowy owl, has not yet been found within the limits of California, but will doubtless be met with sooner or later, as it is so abundant in the regions to the north of it. It nests in trees, and lays from three to six eggs. It is resident in Alaska and the Mackenzie River region, and is one of the best known species of its order.

FAMILY FALCONIDÆ, THE FALCONS.

CHAR. Bill curving uniformly from the base, or only straight for a short distance, provided with a thick and colored cere, perforated by the nostrils. Eyes sunken under a projecting brow. Head feathered. Claws strong and hooked.

Found in all parts of the world.

The exact limits of this family are hard to define. The sub-family, *Polyborinæ*, is about as closely related to the vultures as to the eagles, while others approach to the diurnal owls.

SUB-FAMILY AQUILINÆ, THE EAGLES.

CHAR. Size large, form very robust and strong. Bill large, compressed, straight at base, curved, and acute at tip; wings long, pointed; tail ample, generally rounded; tarsi moderate, very strong; claws much curved, sharp, and strong.

GENUS **AQUILA**, MŒHRING.

Aquila, MŒHRING, Gen. Av. 1752, 49, et auctorum.

GEN. CHAR. Wings long and pointed; tarsi rather short, very strong, feathered to the toes; middle and outer toes connected by a membrane.

Aquila Canadensis, LINNÆUS.

THE AMERICAN GOLDEN EAGLE.

Falco Canadensis, LINNÆUS, Syst. Nat. I. 1766, 125. — *Aquila Canadensis*, CASSIN, P. R. Rep. Birds, IX. 41. — HEERMANN, X. vi. 30. — COUES, Pr. A. N. Sc. 1866, 49.
Aquila chrysætos, RICH. and SWAINSON, Fauna Bor. Amer. II. 12. — AUDUBON, Birds Amer. pl. 182: oct ed. 50, pl. 12. (Name of European species.) — NUTTALL, Man. I. 56. — MAXIMILIAN, Cab. Jour. 1858, 9. — BLASIUS, Bericht Verein Deutsch. Orn. 1862, 83. (These say it is absolutely identical with European.)
Falco fulvus, WILSON, Am. Orn. VII. 13, pl. 55, f. 1. (Not of Linnæus ?)

SP. CHAR. Head and neck behind light brownish-fulvous, varying in shade, frequently light yellowish, generally darker. Tail at base white, often for the greater part of its length, the terminal portion glossy black. Other parts rich purplish-brown, frequently very dark, nearly clear black on under parts of body. Primaries shining black;

secondaries purplish-brown; tibiæ and tarsi brownish-fulvous, generally mixed with dark ashy.

Younger. Entire plumage lighter, and mixed with dull fulvous; under parts of body nearly uniform with upper.

*A. Chrysætos.**

Length, 30.00 to 40.00; wing, 20.00 to 25.00; tail, 12.00 to 15.00. Iris brown; bill horn-color; cere and feet yellow.

Hab. All of North America.

This noble bird is quite common in almost all parts of the State during the colder months, but much less so than the white-headed eagle. They are, however, more of a mountain bird, and their descents into the plains or to the sea-coast are not frequent. Nuttall observed them about the lofty cliffs of the Upper Missouri, and says that they built on a rocky bluff, and laid two, or rarely three eggs, white blotched with reddish. These measure 3.50 × 2.50 inches, according to Audubon. They prey on young fawns, lambs, hares, etc. Recent authors declare that there is not the least difference between European and American specimens.

* The European *Aquila chrysætos* is not in a cut distinguishable from the American bird.

GENUS **HALIÆTUS**, SAVIGNY.

Haliætus, SAVIGNY, Hist. Nat. d'Egypte, I. 1809, 85.

GEN. CHAR. Size large; tarsi short, naked, or feathered for only a short distance below the joint, and like the toes covered with scales; toes rather long; claws very strong, curved and sharp. Bill large, very strong, compressed; margin of upper mandible slightly lobed; wings long, pointed; tail moderate.

Haliætus leucocephalus, LINNÆUS.

THE WHITE-HEADED EAGLE.

Falco leucocephalus, LINNÆUS, Syst. Nat. I. 1766, 124. — WILSON, Am. Orn. IV. 89; pl. 36. — AUDUBON, Birds Amer. pl. 31, 126. — *Haliætus leucocephalus* (SAVIGNY), CUVIER, Règne Anim. 2d ed. 326. — AUDUBON, oct. ed. I. pl. 14. — NUTTALL, Man. I. 74. — NEWBERRY, P. R. Rep. VI. ii. 75. — CASSIN, P. R. Rep. Birds, IX. 43. — HEERMANN, X. vi. 30. — COOPER and SUCKLEY, XII. iii. Zool. of W. T. 151. — DRESSER, Ibis, 1865. (Texan birds.) — COUES, Pr. A. N. Sc. 1866, 49.
Falco ossifragus, WILSON, Am. Orn. VII. 16; pl. 55. (Young.)

SP. CHAR. *Adult*. Head, tail, and its upper and under coverts, white; rest of plumage brownish-black, generally with the edges of the feathers paler; bill, feet, and iris, yellow.

H. leucocephalus.

Younger. Entire plumage dark brown; paler on the throat, edges of the feathers paler or fulvous, especially below; tail more or less mottled with white, which as age advances extends over a large portion of the tail, especially on the inner webs. Bill brownish-black; iris brown.

Length, 30.00 to 43.00; extent, 78.00 to 88.00; wing, 20.00 to 25.00; tail, 13.00 to 15.00.

Hab. All of temperate North America. Greenland; Iceland; accidental in Europe.

This species, adopted as the bird of America, is much less noble in its disposition than the golden eagle. It lives in great part on squirrels, rabbits, wounded birds, and fish stolen from the industrious fish-hawk, rarely venturing to attack large birds or quadrupeds, and having less strength of wing or swiftness than is required for it to catch many large birds. It is a very abundant species where not exterminated by the foolish ambition to "kill an eagle," which inspires most gunners. It is, in fact, so tame around many of the Spanish ranches that it is far easier to kill than the quail or other more legitimate game. The Spanish inhabitants rather encourage its presence, on account of the great number of squirrels it kills; and I have been told of instances where young ones raised from the nest have been kept for several years in a domestic state, going out daily to kill squirrels, and returning to the house at night.

Dr. Gambel states that they were held sacred by the Indians, which will in a measure account for their abundance and protection by the natives.

About the ranches this eagle also depends in great measure on the weakly lambs, calves, or larger animals that often die in great numbers during seasons of drought; but I have never known of their eating carrion.

On the sea-shore, and around lakes or rivers, it depends more on fish, especially salmon, which it obtains in abundance at most seasons, as immense numbers of them die after spawning. I have seen an eagle pick up a dead fish from the surface of the water, and have also seen one catch a flying-fish during its short flight above the surface, but have never known of their diving for fish.

The manner in which they rob the fish-hawk after its successful plunge and laborious ascent into the air is well known, and may be observed in many places where both birds are found, especially along the sea-coast. I have not, however, found this habit universal, as both species seem to live in harmony about the islands of the southern coast. I have seen more than thirty of these eagles in young plumage, soaring about the north end of Catalina Island on the 9th of July, and their nests were numerous among the inaccessible cliffs of that island. They seem, however, to prefer to build in trees, where there are any large ones, and continue to build one nest above another year after year, until the pile becomes large enough to fill a wagon. One such I have seen in Santa Clara Valley, near a farm-house, and I am not certain that the magpie does not sometimes take advantage of these piles to construct its own nest in their recesses, even while the eagle occupies the upper story. I have known the little red-headed linnet (*Carpodacus frontalis*) to build in a similar but smaller nest of a *Buteo*, which I think was occupied at the time by the hawks themselves.

The eggs are said to be laid in February, and to be of a dull white color. Only two are found at a time, and one often hatches much later than the other, though, according to Lawson, they raise several young in a year, laying eggs while the first brood is but half grown, and can assist in hatching by its warmth. This seems somewhat probable from the fact that they are seen about their nests or on them at various seasons, and seem to make them more of a home than do most other birds.

Like all birds of prey this eagle can sustain long periods of fasting; and as it depends so much on chance for a supply of food, its fasts must be quite frequent. According to some authors, they have been known at times to feed on carrion, and even to pursue the vultures, obliging them to disgorge their nauseous food, which the eagle catches before it can reach the ground. Occasionally, made bolder by hunger, they attack full-grown sheep and deer, and Wilson says that one attempted to carry off a child from a garden in New Jersey, while its mother was near by. Nuttall knew of one carrying off an infant to its nest, and though the bird was immediately pursued the child was found dead. Such instances are, however, very rare. The white head and tail are not obtained by this eagle until its third or fourth year.

There is considerable difference in the size of specimens, as is the case with all birds resident over large areas of this continent, the smaller ones being raised in the south. Mr. Cassin thinks it probable that Audubon's *H. Washingtonii* may have been the young of the larger form. It resembled a very large specimen of this in young plumage, but had the bill shorter and more abruptly curved, and the tarsus with broad plates in front instead of scales. No such specimen has since been seen.

GENUS **PANDION**, SAVIGNY.

Pandion, SAVIGNY, Hist. Nat. d'Egypte, I. 1809, 95.

GEN. CHAR. Wings very long; general form heavy; bill short, curved from the base, compressed; tarsi very thick and strong; covered by small circular scales; claws large, curved, very sharp; soles of feet very rough; tail moderate, or rather short. Cere hispid; nostrils obliquely curved; outer toe versatile.

Pandion Carolinensis, Gmelin.

THE FISH-HAWK.

Falco Carolinensis, Gmelin, Syst. Nat. I. 1788, 263.—*Pandion Carolinensis,* Bonaparte, Comp. Eur. and Amer. Birds, 3.—Newberry, P. R. Rep. VI. iv. 75.—Cassin, P. R. Rep. Birds, IX. 44.—Heermann, X. vi. 31.—Cooper and Suckley, XII. iii. Zool. of W. T. 153.—Coues, Pr. A. N. Sc. 1866, 49.

Falco haliætus, Wilson, Am. Orn. V. 14; pl. 37.—Audubon, Birds Amer. pl. 81. Ib. Orn. Biog. I. 415: oct. ed. I. pl. 15. (Not of Linnæus, which is a European species.)

"*Pandion haliætus,* Bonaparte," Nuttall, Man. I. 80.

Sp. Char. *Adult.* Head and entire under parts white; stripe through the eyes, top of the head, and upper parts of the body, wings, and tail, deep umber-brown, the tail with about eight bands of blackish-brown; breast with numerous cordate and circular spots of pale yellowish-brown; bill and claws bluish black; tarsi and toes greenish-yellow.

P haliætus, L European.*

Young. Similar, but with the upper plumage edged and tipped with pale brownish, nearly white. Spots on the breast more numerous and darker colored.

Length, 23.00 to 25.00; extent, 64.00 to 68.00; wing, 19.00 to 21.00; tail, 9.00 to 10.50. Iris yellow; bill black; feet greenish-yellow.

Hab. Throughout the United States; south to Panama.

* From the Society for the Diffusion of Christian Knowledge. Not distinguishable from the American bird.

The fish-hawk is found wherever there is clear water containing fish, and, unlike the white-headed eagle, works for its living, occasionally also providing for that dishonest bird. It may be seen about its resorts, especially along the sea-coast, sailing round in circles high in the air, watching for hours until it sees a fish in a suitable position, when plunging perpendicularly beneath the water, it seizes its prey in its strong, rough-soled talons, and flies off to some convenient perch to feed. It thus obtains abundance of food, and can rest for hours, or amuse itself in carrying materials to its nest, which it does at all seasons, and without apparent reason. I have seen a pair attempt to build on the main-top platform of an old hulk lying in San Diego Bay, and at the time inhabited. Though frequently driven away, and their materials blown off by the wind almost as fast as carried up, they persevered until the keeper of the vessel shot one of them. They build either in high trees, or on cliffs and islands, the nest itself being rather small and compact, but surrounded by a great quantity of sticks, roots, bones, and even wire, of which I found a piece several yards long carried up by them from an old fence. I climbed to this nest on the 20th of June, supposing, from the apparent solicitude of the old birds, that there might be a second brood of eggs in it, but did not find any. Still, they may sometimes raise more than one brood in a year. At that time the young birds about Catalina Island appeared to be fully fledged. Of the eggs I must quote Nuttall's account, who says that they are from two to four, a little larger than hen's eggs, and vary from reddish or yellowish cream-color to nearly white, marked with large blotches and points of reddish-brown. Many nests are often built near together, and small birds sometimes build in their interstices. The fish-hawk is timid, and will allow even the raven to rob it. Many interesting anecdotes and details of habits are given by the authors above quoted, to whom I must refer the reader for further information.

This is one of the many species of Raptores which, at one time supposed distinct from their European congeners, are now considered to be the same.

SUB-FAMILY FALCONINÆ, THE NOBLE FALCONS.

CHAR. Form very compact and robust; tail and legs short; wings long; bill and feet very strong; tooth of upper mandible very acute. Size small or moderate.

These are the swiftest, boldest, and most typical of the birds of prey, often killing other birds and quadrupeds larger than themselves.

The species here included in the single genus *Falco* have been distributed in three genera by some late naturalists, *F. columbarius* being a *Hypotriorchis*, and *F. sparverius* a *Tinnunculus*. The distinctions are, however, slight, and, with so few species, inconvenient.

Genus FALCO, Linnæus.

Falco, Linnæus, Syst. Nat. I. 1766, 124.

Gen. Char. General form robust and compact. Bill short, curved strongly from base to point, which is very sharp, and with a distinct, generally prominent tooth near the tip; nostrils circular, with a central tubercle. Wings long, pointed; tail rather long and wide; legs short, robust, covered with circular or hexagonal scales; middle toe long; claws large, strong, curved, and very sharp. (Cassin.)

This genus includes the most typical species of the sub-family, fifteen or twenty in number, and found in most parts of the world.

Falco nigriceps, Cassin.

THE WESTERN DUCK-HAWK.

Falco nigriceps, Cassin, Birds of Cal. and Tex. I. 1853, 87. In. U. S. Astron. Exp. to Chili, II. pl. xiv. 1855. In. P. R. Rep. Birds, IX. 1858, 8. — Cooper and Suckley, XII. iii. Zool. of W. T. 142; pl. xi. 1859.

Sp. Char. Above dark slaty, with narrow bands of black; quills brownish-black; tail light ashy at base, banded like the back. Beneath reddish-white, with circular spots and transverse bands of black. A white frontal band; top of head, neck, and cheeks nearly black.

Younger. Above dark brown, tail barred with rufous on inner webs. Beneath dull reddish-yellow, paler on the throat, and with broad longitudinal stripes of black; flanks and under wing coverts with transverse bars and circular spots of reddish-white. Iris brown; bill whitish-blue; feet yellow, or pale green.

Male. Length, 17.25; extent, 39.50; wing, 11.00; tail, 6.00.

Female. Length 20.50; extent, 43.00; wing, 13.75.

Hab. Western North and South America.

I have found this species along the whole southern coast of California, where it resides constantly, while north of the Columbia it is migratory. They build in cavities of the lofty, inaccessible cliffs overhanging the water, both along the main-land and on the islands. On Santa Barbara Island, in May, 1863, a pair which probably were still feeding their young swept boldly around my head, when I must have been fully half a mile from the nest, and I shot the female, a very fine specimen. I have seen one pursue a swallow, and turning feet upwards seize it flying, with perfect ease. I have also seen them pursue quails near the coast; but their chief prey consists of ducks and other water-birds, which they seize on the wing or on the water, frequently carrying off birds heavier than themselves.

We give no figure of this species, as it is scarcely to be distinguished in a cut from *anatum* or *peregrinus*.

Falco anatum, BONAPARTE.

THE DUCK-HAWK.

Falco anatum, BONAPARTE, Comp. List, 1838, 4. — CASSIN, BAIRD, Birds N. Amer. 1858, 7. — BLAKISTON, Ibis, 1861, 315. — MARCH, Pr. Phil. Ac. 1863, 304. (Jamaica.) "*Falco peregrinus*," WILSON, AUDUBON, and other authors.

FIGURES. WILSON, Am. Orn. IX. pl. 76; AUDUBON, Birds Amer. pl. 16: oct. ed. I. pl. 20; Lembeye B. of Cuba, pl. 1, fig. 2; DE KAY, Nat. Hist. New York, Birds, pl. 3, fig. 8.

Adult. Frontal band white. Entire upper parts bluish cinereous, with transverse bands of brownish-black, lighter on the rump. Under parts yellowish-white, with cordate and circular spots of black on the breast and abdomen, and *transverse* bands of black on the sides, under tail coverts, and tibiæ; quills and tail brownish-black, the latter with

F. peregrinus. European.*

transverse bars of pale cinereous. Cheeks with a patch of black; bill light blue; legs and toes yellow. Sexes alike.

Younger. Entire upper parts brownish-black, frontal spot obscure, large space on the

* Not distinguishable in a cut from the American bird.

cheeks black. Under parts dull yellowish-white, darker than in adult, and with *longitudinal* stripes of brownish-black; tarsi and toes bluish lead-color.

Total length, 18.00 to 20.00; wing, 14.00 to 15.00; tail, 7.00 to 8.00.

Hab. North America, east of the Rocky Mountains. Jamaica, Cuba, and South America.

It was at one time supposed that there were many species of hawks allied to the peregrine falcon of Europe, America, Japan, Australia, etc., all having their presumed representatives. The tendency, however, of modern research is to call all these one; and the different forms referred to above seem to be considered merely as varieties, if even entitled to that rank. It is very difficult to define with precision the races referred to, and individuals occur within the region of each one that have the characteristics of nearly all the others.

Without attempting here to solve this much-vexed question, I may remark that North America has a large race of the Peregrine falcon of Europe, known as the duck-hawk in the Atlantic States, and very abundant in the whole region north of the boundary of the United States. This is the *Falco anatum*, described at the head of this article. It appears to be replaced on the Pacific Coast by the smaller variety, *F. nigriceps*, just referred to. Occasionally, however, a bird is killed on the Farallones, and elsewhere, which is more nearly related to the *anatum*, differing from *nigriceps* in considerably larger size, and particularly robust bill.

The *F. anatum* has much the habits described for *nigriceps*, and, like it, is a terror to all land animals weaker than itself. It breeds abundantly in the far north, chiefly on cliffs or rocks, and lays three or four chocolate-colored eggs. (Baird.)

Falco polyagrus, CASSIN.

THE PRAIRIE HAWK.

Falco polyagrus, CASSIN, Birds of Cal. and Tex. I. 1853, 88; pl. 16. — IB. P. R. Rep. Birds, IX. 12. — SUCKLEY, XII, iii, Zool. of W. T. 143. — KENNERLY, X. iv. 19. — HEERMANN, X. vi. 31. — DRESSER, Ibis, 1865, 323. (Texas.) — COUES, Pr. A. N. Sc. 1866, 42.
Falco Mexicanus, SCHLEGEL.

SP. CHAR. A narrow white frontal band; line over the eye and entire under parts white; breast and abdomen with longitudinal stripes and spots of brown, a large brown spot on the flanks. Upper parts brown, paler on the rump; tail above grayish-brown, narrowly tipped with white, and with transverse bands of white; quills dark grayish-brown; edge of wing at shoulder white. Beneath spotted with brown, the brown of the back extending somewhat on to the breast at the wing.

Young, with the white parts much obscured by brown; above paler with rufous streaks, and below with a dark brown stripe on nearly every feather.

Length, 18.00 to 20.00; extent, 39.00 to 45.00; wing, 13.00 to 14.00; tail, 7.50 to 8.00. Bill bluish-white; iris brown; cere and feet yellow (lead-color in young). (Cassin.)

Hab. Western North America.

This hawk rarely visits the coast border, although Dr. Heermann records having shot a straggler on the Farallones. At San Diego I saw two or three times what I supposed to be this bird, but could not get near enough to be certain of it. At Martinez, in December, 1863, I succeeded in shooting one as it flew away from its perch on the approach of a wagon in which I was riding. It seems to be the shyest of hawks, and is also one of the swiftest, flying with rapid flappings of the wings, like other falcons. It prefers the borders of prairies, where it catches hares, quails, and other larger game, which these nobler falcons prefer to the smaller and more taste-

F. polyagrus.

less or disagreeable vermin on which other hawks prey. It extends its migrations in summer to the Upper Columbia, but avoids the densely forest-clad regions. I have seen but few of the species in the State, though my explorations of the interior have not been sufficient to determine its true range of migration and habits. Its nest and eggs are still unknown to naturalists. Dr. Heermann saw a young unfledged one at San Francisco, so that it doubtless breeds within the State. It extends its range eastward to the plains east of the Rocky Mountains.

It is still a question whether the name *polyagrus* or *Mexicanus* should have priority.

Falco columbarius, LINNÆUS.

THE PIGEON-HAWK.

Falco columbarius, LINNÆUS, Syst. Nat. I. 1766, 128. — WILSON, Amer. Orn. II. 107; pl. 15, f. 3. — AUDUBON, Birds Amer. I. 88; pl. 21 and 92. — CASSIN, P. R. Rep. Birds, IX. 1858, 9. — COOPER and SUCKLEY, XII. iii. Zool. of W. T. 142. — *Hypotriorchis columbarius*, (KAUP?) HEERMANN, P. R. Rep. X. vi. 31. — NEWBERRY, VI. ii. 74. — COUES, Pr. A. N. Sc. 1866, 42.
Falco temerarius, AUDUBON, Orn. Biog. I. 381.

SP. CHAR. Small, and of stout form. Above bluish slate-color, with a black central line on each feather; forehead and throat white; rest of under parts pale yellowish or reddish-white, each feather with a central brownish-black line, the tibial feathers light ferruginous, with lines of black. Quills black, tipped with ashy-white; tail light bluish-ashy barred with black and tipped with white; inner webs nearly white.

Male. *Female.*

Younger, dusky or light brown, sometimes with ferruginous markings; beneath dull white with light brown stripes. Tail pale brown, with about six white bands.

Young, brownish-black, the stripes wider, the white parts dusky, sides with black bands and white spots; quills and tail nearly black; tail unspotted, or with about four white bands. Female like the young in color.

Length of female, 12.00 to 14.00; extent, 27.00 to 28.00; wing, 8.00 to 9.00; tail, 5.00 to 6 00. Male, 10.00 to 12.00; wing, 7.00 to 8.00. Iris brown; bill bluish-black; cere and feet yellow or greenish in the young.

Hab. Most of North and South America.

This well-known little falcon ranges over the whole State in the colder months, but I have not observed it, even in the high Coast Range, in summer, though some probably remain in the cooler regions in the breeding-season. I shot a fine specimen in winter at Fort Mojave, which differed from the usual form only in being of a very pale, almost ashy color; but this was doubtless the effect of the dry hot climate of the interior, which affects the plumage of many birds in the same way.

Though small, the pigeon-hawk has all the fierceness and courage of a true falcon, and captures birds fully as large as itself. It, however, chiefly follows the flocks of gregarious birds, such as black-birds, doves, etc., and preys much on mice, gophers, and squirrels. I have not heard of its attacking domestic poultry, and those farmers who shoot every "chicken-hawk" that comes around the house would do well to observe them more closely, and will discover that these small species are not the young of the larger ones, and should rather be encouraged than destroyed.

Audubon found a nest in Labrador on June 1st, built in a low fir-tree, ten or twelve feet from the ground, composed of sticks, slightly lined with moss and feathers. The eggs were five, 1.75 by 1.25 inches, elongated. Their ground color was dull yellowish-brown, with irregular thickly clouded blotches of dull dark reddish-brown. (Oct. ed. I. 89.)

Falco femoralis, TEMMINCK.

THE ARIZONA HAWK.

Falco femoralis, TEMMINCK, Pl. Col. I. liv. 21. — CASSIN, BAIRD, Birds N. Amer. 1858, 11; pl. i. — DRESSER, Ibis, 1865, 333. — COUES, Pr. A. N. Sc. 1866, 42.
Falco thoracicus, (ILL.) LICHT. Verz. 1823, 62.

SP. CHAR. Head above, and entire upper parts light cinereous; darker, and with transverse bars of white on the upper tail coverts; front and line over the eye to the back of the neck white, tinged with orange on the latter; a wide band under and behind the eye, and another short band running downwards from the base of the under mandible, dark cinereous. Throat and breast very pale yellowish-white; a wide band across the body beneath, black, with narrow transverse stripes of white; abdomen, tibiæ, and under tail coverts light rufous. Under wing coverts pale yellowish-white, spotted with black;

primaries ashy-black, with numerous transverse bands of white on their inner webs; secondaries light cinereous, tipped with white; two middle feathers of the tail light cinereous, with transverse bands of ashy-white; other feathers of the tail brownish-black, tipped

H. femoralis.

with white, and having about eight transverse bands of white. Total length about 15.00; wing, 10.50; tail, 7.50. Bill yellow at base, tipped with light bluish horn-color; legs yellow.

Hab. New Mexico, Mexico, and South America.

This species is extremely rare in the United States, though occasionally seen and captured in Arizona and New Mexico. Nothing special is known of its habits, except that they resemble those of the other true falcons.

Falco sparverius, Linnæus.

THE SPARROW-HAWK.

Falco sparverius, Linnæus, Syst. Nat. I. 1766, 128. — Wilson, Am. Orn. II. 117, pl. 16, f. 1; IV. 57; pl. 32, f. 2. — Audubon, Birds Amer., oct. I. 90; pl. 22; fol. pl. 42. — Cassin, P. R. Rep. IX. Birds, 13. — Cooper and Suckley, XII. ii. Zool. of W. T. 143. — *Tinnunculus sparverius*, Vieillot, Ois. Amer. Sept. 40; pl. 12 and 13. — Newberry, P. R. Rep. VI. ii. 74. — Kennerly, X. iv. 19. — Heermann, X. vi. 1859, 31. — Coues, Pr. A. N. Sc. 1866, 42.

Sp. Char. *Adult.* Frontal band and space, including eyes, and throat, white; a spot on the neck behind, two others on each side of neck, and a line running down from before the eye, black. Spot on top of the head, and upper parts behind neck, light rufous or cinnamon. Under parts generally, paler rufous, frequently nearly white, with numerous circular oblong black spots. Quills brownish-black, with white bars on their inner webs.

Tail tipped with white, and with a broad subterminal black band. Back generally with transverse bars of black, but they are frequently very few or wanting. Rufous spot on the head variable in size, sometimes wanting.

Younger male, similar, but with the wing coverts and tail ferruginous-red, with numerous transverse bands of brownish-black. Beneath with numerous longitudinal stripes, and

Male. *Female.*

on the sides with transverse bands of brownish-black, external tail feathers palest; the broad subterminal black band obscure or wanting.

Young, with all the rufous parts of the plumage more widely banded with brownish-black; wing coverts dark bluish-cinereous, with large circular spots of black; beneath with longitudinal stripes and large circular black spots.

Length, 10.00 to 12 00; extent, 19.00 to 22.00; wing, 7.00 to 8.00; tail, 5.00 to 5.50. Iris yellow; bill blue, black at tip; feet orange.

Hab. Entire continent of America.

This little hawk resides constantly in California, frequenting chiefly the plains, and feeding on grasshoppers, mice, gophers, etc. It must be considered one of the farmer's best friends, and is seldom killed by observing persons. Unlike most hawks, it builds its nest in deserted holes of the woodpecker, in hollow trees, and is said to lay four or five eggs, light brownish-yellow spotted with brown. Its graceful form, varied and rapid flight, and inoffensive habits, combined with its beautiful plumage, make it one of the most interesting inhabitants of our rural districts.

Sub-Family ACCIPITRINÆ.

Char. Form rather long and slender; tail and legs long; wings rather short; bill short, hooked; upper mandible with a rounded lobe instead of a tooth. (Cassin.)

Forty or fifty species are known, distributed throughout the globe.

Genus **ACCIPITER**, Brisson.

Accipiter, Brisson, Orn. I. 1760, 310.

Gen. Char. Size rather small; wings short, and tail rather long; tarsi long and slender, with the scales in front frequently nearly obsolete.

This genus contains about twenty species, all much alike in color, frequenting woods, and preying chiefly on smaller birds.

Accipiter Cooperii, Bonaparte.

COOPER'S HAWK.

Falco Cooperii, Bonaparte, Am. Orn. II. 1; pl. 10, f. 1. (Young.)—*Astur Cooperii*, (Bonaparte) Nuttall, Man. I. 89.—Audubon, Birds Amer. pl. 36.—Newberry, P. R. Rep. VI. ii. 74.—*Accipiter Cooperii*, Gray, List Brit. Mus. 38—Cassin, P. R. Rep. Birds, IX. 16.—Cooper and Suckley, XII. ii. Zool. of W. T. 145.—Coues, Pr. A. N. Sc. 1866, 43.

Sp. Char. Above ashy-brown, darker on head, mixed with white and rufous on neck; beneath white, throat with narrow streaks, other parts with light rufous bars. Quills ashy-brown with darker bands, and white irregular marks on inner webs; tail dark cinereous, tipped with white, and with four wide bands of brownish-black.

Young, light umber-brown above, with more white and rufous; upper tail coverts tipped with white; beneath white, with narrow light brown stripes.

Length, 16.00 to 20.00; extent, 26.00 to 32.00; wing, 9.50 to 10.00; tail, 8.00 to 9.75. Bill bluish horn-color; cere and feet yellow; iris orange or yellow.

Hab. All of temperate North America.

This is the largest of the genus found in North America, and approaches the falcons in strength of wing, swiftness, and audacity. It frequently comes about the farm-house, and seizes the fowls from the very door. It is a common species during the winter months in all wooded portions of the State, but retires to the north or to the high mountains in the spring to breed, the fledged young appearing in the lower country in August. Its nest and eggs have not yet been described.

A. Cooperii.

Its flight is chiefly by quick flappings, and its home is by preference near or among trees.

Accipiter Mexicanus, Swainson.

THE MEXICAN HAWK.

Accipiter Mexicanus, Swainson, Fauna Bor. Amer.; Birds, 1831, 45. — Cassin, P. R. Rep. Birds, IX. 17. — Cooper and Suckley, XII. ii. Zool. of W. T. 146. — Coues, Pr. A. N. Sc. 1866, 43.

Sp. Char. Very similar to *A. Cooperii*, but smaller; head bluish-black above, back more brownish; throat and under tail coverts white; other under parts fine light rufous, darkest on the tibiæ, spotted and barred transversely with white, chiefly on abdomen; dark central streaks on breast feathers only; otherwise as in *A. Cooperii*.

Young, much tinged with yellowish-red above; beneath yellowish, each feather centrally streaked, and often transversely barred with brown near its base.

Length, 14.00 to 18.00; wing, 9.00 to 10.00; tail, 8.00 to 9.00.

Hab. Western North America and Mexico.

This smaller species has been found at Auburn, Placer County, in spring, by Mr. F. Gruber; and at Bodega, in February, by Mr. T. A. Szabo. It therefore, probably, has much the same habits of migration as *A. Cooperii*, and doubtless resembles it closely in general habits. I have not myself obtained a specimen in this State. It may prove, however, that the bird cannot be separated specifically from *Cooperii*. As it could not be distinguished in a cut, we do not give a figure of it.

Accipiter fuscus, Gmelin.

THE SHARP-SHINNED HAWK.

Falco fuscus and *Falco dubius*, Gmelin, Syst. Nat. I. 1788, 280, 281.

Falco velox and *Falco Pennsylvanicus* (in part), Wilson, Am. Orn. V. 116; pl. 45, f. 1 : VI. 13; pl. 46, f. 1.

Astur fuscus, Audubon, Birds Amer. pl. 374: oct. ed. pl. 25. — Nuttall, Man. I. 90. — *Accipiter fuscus*, Bonaparte, Comp. List, 5. — Cassin, P. R. Rep. Birds, IX. 18. — Cooper and Suckley, XII. ii. Zool of W. T. 146. — Coues, Pr. A. N. Sc. 1866, 43.

Sp. Char. Very similar to *A. Mexicanus*, but smaller; above colored like *A. Cooperii*; tail more brownish; under parts similar; secondary and tertiary quills with large partially concealed spots of white.

Young, brown, darker than that of *A. Cooperii*; beneath with longitudinal, ovate, and

A. fuscus.

circular spots of reddish-brown, becoming transverse bands on flanks and tibiæ; under tail coverts white; otherwise as in preceding.

Length, 10.00 to 14.00; extent, 20.00 to 25 00; wing, 6.00 to 6.50; tail, 5.00 to 5.50.

Hab. Throughout North America and south to Panama.

This little hawk much resembles *A. Cooperii* in habits, allowing for the great difference in its size. The eggs are said by Audubon to be four or five, grayish-white, blotched with reddish-brown; the nest is built in a tree. He found another in a hole in a rock, composed of a few sticks, and lined with grass, while in one case the eggs were deposited on rotten wood in a hollow branch.

They probably breed more generally towards the southern and lower parts of the State than that species, as I have seen a few of them (or perhaps *A. Mexicanus*) in the warmer months, and they are known to build in the Middle Atlantic States.

They are generally found in the woods, flying short distances, and pouncing on their prey from some perch or a low branch. On the Sierra Nevada I have seen one pursuing a striped squirrel, coming down with a zigzag course as if to prevent it from escaping, by appearing on all sides of it at once. They frequently take young chickens, and have been seen by Nuttall to drive away a red-tail hawk from the vicinity of their nest with all the courage of the king-bird. They occasionally soar upward to a great height, apparently for pleasure, like other hawks, and, though their wings are short, seem to ascend with very little effort; while near the earth their flight is by rapid flappings and short sailings, as if laborious.

GENUS **ASTUR**, LACEPÈDE.

Astur, LACÈPEDE, Mém. Inst. III. 1800, 506.

GEN. CHAR. Size rather large but slender; wings rather short; tail long; tarsi long, covered in front with rather wide transverse scales; toes and claws moderate, the latter much curved and sharp. Bill short, nostrils large, ovate.

About twelve species are known, inhabiting the forests of all countries. (Cassin.)

Astur atricapillus, WILSON.

THE AMERICAN GOSHAWK.

Falco atricapillus, WILSON, Am. Orn. VI. 1812, 80; pl. 52, f. 3. — *Astur atricapillus*, BONAPARTE, Obs. Cuvier Règ. An. 33. — NUTTALL, Man. I. 87. — NEWBERRY, P. R. Rep. VI. iv. 74. — CASSIN, P. R. Rep. Birds, IX. 15. — LORD, Pr. R. Art. Inst. Woolwich, IV. 1860, 110. — COOPER and SUCKLEY, XII. ii. Zool. of W. T. 144.

Falco (Astur?) palumbarius, AUDUBON, Birds Amer. pl. 141: oct. ed. I. pl. 23. (Not of Linnæus, European species.)

SP. CHAR. Top of head, hind neck, and a line behind eye, black, generally tinged with ashy. Other upper parts dark slate-color; shafts of feathers black, and frequently their edges also. Stripe over the eye, and a partly concealed patch on occiput, white. Beneath mottled with white and ashy-brown, in numerous irregular and narrow bars on each feather, and a central stripe on the shaft crossing them. Quills brown, with deeper brown bands, and ashy-white on their inner webs. Tail of same color as other upper parts; beneath very pale, nearly white, with about four obscure bands of deeper ashy-brown, and narrowly tipped with white; under tail coverts white.

Young, dark brown above, the feathers edged and spotted with light reddish, nearly

A. atricapillus.

white. Tail light ashy, with about five wide bands of brown, narrowly tipped with ashy-white; quills brown, with wide darker brown bars, and wide bands of reddish-white on their inner webs. Beneath white, generally tinged yellowish or reddish, every feather with a longitudinal stripe, ending in an ovate spot of brown. Sides and tibiæ, with circular and long spots, and irregular bands of brown; under tail coverts white, with a few large spots.

Length, 20.00 to 24.00; extent, 38.00 to 44.00; wing, 12.50 to 14.00; tail, 9.50 to 11.50. Bill bluish-black and white; iris, cere, and feet, yellow.

Hab. Northern parts of North America.

This is a northern representative of the family, frequenting dense forests, and only coming into our valleys in winter. I have seen a dead one on the summit of the Sierra Nevada in September, and one shot at Michigan Bluffs by Gruber.

The European goshawk is readily distinguishable from the American species by the much sharper definition of the transverse bars across the breast.

SUB-FAMILY BUTEONINÆ, THE BUZZARDS.

CHAR. General form heavy; size moderate or large; wings rather short and broad; tail moderate. Flight vigorous, but not swift.

About twenty-five species are known in all countries; they feed on small animals.

GENUS **BUTEO**, CUVIER.

Buteo, CUVIER, Règne Anim. I. 1817, 323.

GEN. CHAR. Bill short, wide at base, not very acute; edges of upper mandible with slight rounded lobes; nostrils large, ovate; wings with fourth and fifth quills usually longest, giving them a rounded form; tail moderate, wide; tarsi moderate, robust, with transverse scales before and behind, small circular and hexagonal scales on sides; toes moderate or short, claws strong.

The number of species is still undetermined, partly on account of the great variations in their plumage, which seem analogous to those of the wolves and foxes.

§ 1. Very stout and heavy. Length over twenty inches. First quill shorter than eighth. (Tails always red in the adults?) "*B. borealis et vars.*" (Bryant.)

Buteo montanus, NUTTALL.

THE RED-TAILED HAWK.

Falco buteo, RICHARDSON and SWAINSON, Fauna Bor. Amer. Birds (in part); not the European *Falco buteo*, LINNÆUS.

Buteo vulgaris, AUDUBON, Orn. Biog. IV. 508; pl. 372: oct. ed. I.

Buteo montanus, NUTTALL, Man. I. 1840, 112. — CASSIN, P. R. Rep. Birds, IX. 26. — NEWBERRY, VI. iv. 75. — BAIRD, X. iii. 12. — KENNERLY, X. iv. 19. — HEERMANN, X. vi. 32. — COOPER and SUCKLEY, XII. iii. Zool. of W. T. 147.

Buteo Swainsonii, CASSIN, Birds of Cal. and Texas, I. 98. (Not of Bonaparte, Comp. List, 1838, 3.)

SP. CHAR. Above dark umber brown, with partially concealed ashy-white and pale fulvous spots and transverse bands, especially on the scapulars and shorter quills. Upper tail coverts reddish-white, with transverse bands of dark brown. Tail above bright rufous, narrowly tipped with white, with a subterminal band of black, and a few short, narrow bars across shafts of feathers, sometimes absent. Throat and neck in front dark brown, mixed with white; abdomen, tibiæ, and under tail coverts deeply tinged with rufous; the tibiæ transversely barred with rufous of a darker shade. Sometimes the breast and abdomen are nearly white, with some central lines of dark brown. Under wing coverts pale yellowish-white, with brown spots; sides with numerous narrow lanceolate and oblong spots of dark brown and rufous; abdomen with a broad transverse band of similar spots varying in numbers, sometimes nearly all wanting.

B. montanus.

Young, darker brown above, the feathers edged and spotted with white, tinged with rufous. Beneath white, with large ovate and sagittate spots of dark brown, more numerous on the abdomen. Tibial and under tail coverts white, with transverse bars and spots of dark brown. Tail above ashy-brown, with about twelve to fourteen bars of brownish-black.

Length, 19.00 to 24.50; extent, 48.00 to 54.50; wing, 15.00 to 17.00; tail, 8.50 to 10.00. Iris brown; bill bluish; cere and feet yellow. *Young*, with bill horn-black; feet greenish.

Hab. Western North America.*

This, which is the most common of the genus, may be taken as a representative of the habits of all the other large species. They are common in nearly all parts of the State and neighboring regions, where not destitute of trees, and reside permanently in most places, merely scattering in pairs

* *Buteo borealis*, GMELIN, Syst. Nat. I. 1778, 266. — WILSON, Am. Orn. VI. 75; pl. 52, f. 1: and 78; pl. 52. — AUDUBON, Birds Amer. pl. 51: oct. ed. 1, pl. 7. — CASSIN, P. R. Rep. Birds, IX. 25.

during the nesting season. Their nests are numerous in the valleys and on the lower mountains. They are generally built in the forks of a sycamore or other large tree, and formed of twigs, pretty firmly constructed, and with a distinct cavity, unlike the vultures. The eggs I found near San Diego, laid about March 20th, were three in number, measuring 2.28 × 1.76 inches, and dull yellowish-white, with faint brown spots. On my climbing to the nest the old birds darted towards me from a neighboring bluff, but when within a few feet of my head they turned away, and did not attempt to make an assault upon me.

They prey on rabbits and the smaller animals, including lizards, and occasionally seize a fowl from the farm-yard. They usually become very fat, especially where squirrels are numerous, and spend hours in sitting stupidly on some tree, dozing while they digest their food. They also soar high into the air during the middle of the day, sometimes in large numbers during the cold months, occasionally rising to so great a height as to disappear entirely.

The Eastern red-tailed hawk is so similar, that sometimes the differences are scarcely appreciable, but, according to Cassin, it may be distinguished by smaller size (19.00 to 24.00 inches long; wing, 14.00 to 16.00; tail, 7.50 to 8.50), also by whiter under parts, fewer spots below, and fewer bars on tail in the young. Dr. Bryant considers them the same, but the majority of Western and Eastern specimens differ as above described. A specimen from Fort Mojave, Colorado Valley, may perhaps belong to this.

Buteo calurus, Cassin.

THE RED-TAILED BLACK HAWK.

Buteo calurus, Cassin, Pr. A. N. Sc. Phil. VII. 1855, 281. Ib. P. R. Rep. Birds, IX. 22. — Baird, X. iii. 11; pl. xiv. — Coues, Pr. A. N. Sc. 1866, 43.

Sp. Char. Entire plumage of body brownish-black, deeper and clearer on back and abdomen, paler on throat and breast. Above with concealed transverse bands of white at the base of feathers, beneath also, with circular white concealed spots; quills brownish-black, with a large portion of their inner webs white, banded and mottled with pale ashy-brown; under tail coverts transversely barred with brownish-black and pale rufous. Tail bright rufous above, white at base, with eight or ten irregular and imperfect narrow bands, and one subterminal wider band of brownish-black, narrowly tipped with reddish-white; beneath silky reddish-white.

Young, duller brown above and below; tail brown with ten or twelve narrow transverse bands; ashy below. Primaries white at base, black at end.

Length, 21.00 to 22.00; extent, 48.00 to 53.00; wing, 16.50 to 17.50; tail, 9.00 to 10.50. Iris brown; bill horn-color; cere and feet yellow.

Hab. New Mexico and California.

Although this hawk much resembles the common Red-tail, I am inclined to think it a distinct species, as it is much heavier and stouter about the chest than that bird, besides the differences of plumage. I have never seen them associated during the breeding-season, and think that this is a more southern species in its range, as I have heard of none northward of Petaluma, where they were found breeding by Mr. Samuels, who obtained the eggs. I have myself not met with its nest, nor seen the bird during the warm season. In the cold months they are common in the southern counties, and in habits much resemble the Red-tail at that time.

The young plumage, not described by Mr. Cassin, I have found as above mentioned, and have little doubt that the bird was of this species.

In the "Proc. Bost. Soc. Nat. Hist. March, 1861," Dr. Henry Bryant gives a synopsis of the "varieties of *B. borealis*," including *montanus, calurus, Cooperi?* and *Harlani* (of Cassin). Very dark specimens of *montanus* from the north connect with a ferruginous *calurus* from Fort Tejon, and light ones from Cape St. Lucas approach the Eastern *B. borealis*. The possibility of occasional hybrids should be considered.

Buteo Cooperi, Cassin.

THE CALIFORNIA HAWK.

Buteo Cooperi, Cassin, Pr. A. N. Sc. Phil. VIII. 1856, 253. Ib. P. R. Rep. Birds, IX. 1858, 31. — Cooper, XII. iii. Zool. of W. T. 1859, 148; pl. xvi.

Sp. Char. Immature bird. Head above, and hind-neck with the feathers white at base, and with a central stripe and tip of brown; back and rump brownish-black, tinged with cinereous; upper tail coverts white, transversely barred with dark brown, and tinged with rufous; wing coverts and quills brownish-cinereous, the latter lighter; inner webs of quills white, mottled, and irregularly banded with light ashy; ends of quills nearly white. Tail white at base; external feathers with their outer webs cinereous, the inner white, mottled with cinereous, other tail feathers mottled and striped with white, rufous, brown, and cinereous, darker on outer web. A subterminal wide brown band, above this chiefly rufous, and basal part of feathers white, tips whitish. Under parts white, with narrow dark brown stripes on throat, neck, and flanks. A large spot of brownish-black on under wing coverts, near upper edge of wing. Tibiæ tinged with reddish-yellow. Iris dark brown; bill bluish; cere and feet yellow.

Female? Length, 20.50; extent, 51.00; wing, 15.00; tail, 9.00.

This specimen (the original) still remains unique in collections, and during my late explorations I have never seen any like it, except what I mistook for the same in 1855, and mentioned as such in my Report on the Zoology of Washington Territory, etc. I am now satisfied that those were merely the *Archibuteo ferrugineus*, which closely resembles this in the color of the tail, suggesting the possibility of its being a hybrid. The type specimen I shot near Mountain View, Santa Clara Valley, California, November,

B. Cooperi.

1855, and have been about there at various seasons since, without seeing others. It is possible that this may be a species common in Mexico or South America.

Buteo Harlani, Audubon.

HARLAN'S HAWK.

Falco (Buteo) Harlani, Audubon, Orn. Biog. I. 441; pl. 86: oct. ed. I. 1830; pl. 8. — Nuttall, Man. I. 104. — *Buteo Harlani* (Bonaparte), Lawrence, Ann. N. Y. Lyc. of N. H., V. 220. — Cassin, P. R. Rep. Birds, 24.

Sp. Char. Brownish or chocolate, nearly black, with a purplish lustre above. Forehead white, and feathers of head and occiput white at base. Quills with transverse lighter bands, and a portion of their inner webs ashy-white, this color prevailing on their under surface. Tail mottled with brownish-black, ashy, and white, a wide subterminal band nearly black, with tinges of bright rufous; tip paler, under surface like that of wings.

Young. Above light brown, dull fulvous and white; tips of feathers generally with a large ovoid brown spot, bordered with fulvous, and white at base; rump with transverse bands of brown and reddish-fulvous; quills dark brown, their under surface white; under wing coverts white, spotted with brown and rufous; tail light ashy, tinged with reddish, and with six or seven imperfect narrow brownish-black bands, all edged with dark reddish-fulvous; the tip white. Beneath white, with ovate and sagittate large spots on the front of neck, sides, and abdomen; tibiæ and under tail coverts with nearly regular bands of brown and pale reddish; under surface of tail ashy.

Length, about 21.00; wing, 16.00; tail, 9.50; tarsus, 3.00. Bill dark bluish; legs greenish-yellow. (Cassin.)

Hab. Mexico north to Louisiana (Audubon); New Mexico (Dr. Henry); California (Lawrence).

This is another form referred to *B. borealis* by Dr. Bryant, as a variety, but it is scarcely well enough known for us to decide on this point. The best marked specimens, above described, are very different. I shot a young specimen at Fort Mojave, agreeing very well with that above described, which, however, Dr. Bryant considers *not* the young of *B. Harlani*, of Audubon.

§ 2. Length generally under twenty inches, less robust; tarsi and bill weaker; first quill longer than eighth; tail never red. ("*B. Swainsonii et vars.*" Bryant.)

Buteo insignatus, Cassin.

THE BROWN HAWK.

Buteo insignatus, Cassin, Birds of Cal. and Tex. 1854, 102, 198; pl. 31. Ib. P. R. Rep. Birds, IX. 23. — Heermann, X. vi. 32.

Sp. Char. Entire plumage above and below brown, each feather with a darker central line. Quills above brown, with a purplish lustre, their under surface pale ashy, with white shafts and irregular white bars. Tail above dark brown, with an ashy tinge, and about ten darker bands; beneath nearly white, with brown bands, the widest subterminal; tip paler or nearly white. Under wing and tail coverts white, striped with reddish-brown; edges of wings at shoulder nearly pure white; tibiæ rufous, irregularly barred with brown; throat, and a few feathers of forehead, white, each feather having a dark brown line.

Younger, with upper plumage darker, and the under parts dark rufous chestnut; darker on breast, quite uniform on flanks and abdomen; throat, forehead, under wing and tail coverts white.

Young, beneath reddish-white; each feather with a large terminal oblong brown spot; abdomen and tibiæ with numerous brown bars. Above with nuchal feathers white at base, edged with reddish; scapulars and greater wing coverts with large, partly concealed, rufous spots. Under wing coverts pale reddish-white, with large brown spots; under tail coverts pale reddish, with a few brown stripes.

Length, 17.00 to 19.75; extent, 50.00; wing, 14.50 to 15.75; tail, 7.50 to 9.00. Iris brown; bill horn-color and bluish; feet yellow.

Hab. Western North America. Canada.

I found this species pretty common near San Diego, in March, 1862, when they were apparently migrating northward. Two which I shot agreed closely with Cassin's description, and were very nearly alike. In habits they resembled very much the larger species.

I afterwards saw a large number of hawks flying over in a scattered flock towards the north. Among them were variously colored specimens which I saw only at a distance, as they did not come near enough to be shot.

Bryant considers this merely a variety of *B. Swainsonii*, next described; but as none of that style of coloration have been found on the Pacific slope, I think it as well to consider this for the present as distinct. *Buteo Bairdii*, of Hoy, is, however, the young bird of *B. Swainsonii*.

The whole subject of the American buzzard hawks, however, requires a careful revision, and it is very probable that naturalists will ultimately conclude to accept a much smaller number of species than they do at the present time.

B. Swainsonii. Adult.

Buteo Swainsonii, Bonaparte.

SWAINSON'S HAWK.

Buteo vulgaris, Rich. and Swainson, Fauna Bor. Amer. Birds, 47. (Not *B. vulgaris* of Europe.)

Buteo Swainsonii, Bonaparte, Comp. List, 1838, 3. (Not of Bonaparte, Consp. Av. 19, nor of Cassin, Birds of Cal. and Texas, I. 98, which is *B. montanus*, Nuttall, according to Cassin.) — Cassin, P. R. Rep. Birds, IX. 19. — Baird, P. R. Rep. X. iii. 11; pl. xii. and xiii. — Dresser, Ibis, 1865, 324. — Coues, Pr. Phil. Ac. 1866, 43.

Buteo Bairdii, Hoy, Pr. A. N. Sc. Phil. — Cassin, Birds N. Amer. 1858, 21. (Young.)

Sp. Char. Above rufous or dark brown, the feathers edged with rufous; tail brown, tinged with ashy and barred with ten or twelve narrow bands of blackish, the subterminal widest; tip white, sometimes more ashy and with few bars. Throat white with longitudinal brown lines, sometimes yellowish, the lines faint; neck in front and breast ashy-brown, nearly as in tail; some of the feathers edged with reddish; sometimes nearly

B. Swainsonii. Young.

black; other under parts white with transverse irregular bars of rufous on tibiæ and flanks, and darker bars on abdomen; sometimes sagittate spots, or nearly pure white, unspotted; under wing coverts white, or with a few spots of brown. Quills brownish-black, with wide cinereous bands on their inner webs, becoming nearly pure white towards bases; the shafts of quills and tail, white beneath. Length, 19.50 to 22.50; extent, 46.50 to 51.50;

wing, 15.50 to 16.50; tail, 8.00 to 8.50; tarsus, 2.50. Bill dark slate-color; tarsi, toes, and cere, yellow. (Cassin.)

Hab. Rocky Mountains and British America to Lower Missouri. California?

Average size equal to those of the previously described division.

Should this prove the same as *B. insignatus*, the name *Swainsonii* will have precedence. Dr. Bryant thinks that the adult *B. Harlani* of Audubon may be also the same, but Professor Baird informs me by letter, that Mr. Kennicott has obtained what he considers a black form of *B. Swainsonii* in British America, from which I suppose that he does not consider *B. Harlani* (Audubon) the same. There are also two or three Mexican black hawks described, which may be the adult plumage of some of those described by Cassin; as, for instance, Professor Baird thinks *B. fuliginosus*, Sclater, the adult of *B. oxypterus*, Cassin, neither of which has yet been found in California.* Many specimens and notes on these difficult birds will be required to determine whether the black ones are distinct or merely the perfect plumage of others which may themselves be merely different stages of plumage, of a few true species whose limits are undetermined. At any rate, the various stages can as well be described under six names as under two.

Buteo elegans, Cassin.

THE ELEGANT HAWK.

Buteo elegans, Cassin, Pr. A. N. Sc. Phil. VII. 1855, 281. Ib. P. R. Rep. Birds, IX. 28. — Newberry, VI. iv. 75. — Kennerly, X. iv. 19. — Heermann, X. vi. 32; pl. ii. iii. — Dresser, Ibis, 1865, 325. — Coues, Pr. Phil. Ac. 1866, 45. — Cooper, XII. iii. Zool. of W. T. 147.

Sp. Char. Above dark brown, feathers of head and back edged with rufous; upper tail coverts narrowly tipped with white. Shoulders dark rufous, each feather with a narrow central line of dark brown; upper wing coverts dark brown, their inner webs edged with rufous, with transverse stripes and partly concealed circular spots and tips of white. Primary and secondary quills brownish-black, with numerous irregular bands of white, running obliquely on their inner webs, all the quills tipped with white. Tail brownish-black, white at base, with four white bands and a white tip. Throat brownish-black, with a few white feathers; breast dark rufous, unspotted, and other under parts of the same color, with numerous, nearly regular, transverse bars of reddish-white. Under wing coverts dark rufous, transversely barred with reddish-white. Dark lines on shafts of breast-feathers.

Young. Above dull brown, many feathers edged with reddish-white or rufous, especially on the back and wing coverts. Quills brownish-black, their inner webs barred with white. Tail brown, tinged ashy, with ten or twelve narrow darker bars, and tipped with white.

* Figured by Baird, Birds of N. Amer. 1860; pl. xv.

Throat dark brown, with narrow white stripes. Under parts yellowish-white, each feather with wide irregular confluent bands of dark brown, and at the end a brown sagittate spot (more elongated in young males), less numerous on abdomen and under tail coverts.

B. elegans.

Tibial feathers and under wing coverts pale rufous (dark in males), under tail coverts tinged with the same.

Length, 18.50 to 19.50; extent, 40 00 to 41.00; wings, 12.00 to 13.00; tail, 8.00 to 9.00. Iris brown; bill horn-blue; cere and feet yellow.

Hab. Western North America. (Not in Oregon ?)

This is the Pacific Coast representative of the red-shouldered hawk (*B. lineatus*) of the Atlantic States, and resembles that species closely in habits as well as in plumage. The principal difference appears to be in the young. I have found it common in the southern part of the State, especially near San Diego, but did not see any in the Colorado Valley, though Dr. Kennerly obtained one in New Mexico. On my approach they would always fly off from their usual perch, circling up high into the air, and uttering short, shrill screams in rapid succession, like the Eastern bird. From finding the same pair constantly at one place during my visit to Judge Witherby's ranch in March, I supposed that they were about building there; but I did not succeed in finding their nest.

Buteo zonocercus, Sclater.

THE BAND-TAILED BLACK HAWK.

Buteo zonocercus, Sclater, Proc. Zool. Soc. Lond. 1859, 263; pl. 59. — Coues, Pr. A. N. Sc. 1866, 46. — Elliot, Illus. B. N. A. II. pl. 33.

Sp. Char. *Male*. Of the size of *B. insignatus*, Cassin, moderately robust; bill small; lobe of upper mandible slight; wings and tail long; feet rather weak; tarsus feathered in front more than one third of its length, and with eleven transverse scales; naked behind; claws moderate, curved, the hind claw forming nearly a semicircle.

Color. Almost entirely black, slightly tinged with brown, the head and neck somewhat

B. zonocercus.

ashy; feathers not lustrous. A narrow white band across base of upper mandible; feathers of head and neck white at base, and mixed with white down over rest of the body. Quills brownish, inner side mottled with ashy, but not banded; under wing coverts black. Tail black above, tipped with ashy-white, and with two ashy bands, half an inch wide, at two inches from the tip, the other a quarter of an inch wide at four and a half inches from tip. Also a rudimentary band of light spots on the two outer feathers two inches higher. All these are pure white on the inner webs, and beneath. Length, 20.25; extent, 56.50; wing, 16.25; tail, 10.00; tarsus, 2.50. Iris reddish-brown; bill black, whitish at base; cere and feet yellow.

Hab. Southern California and Mexico.

Differs from *B. Harlani* of Cassin in smaller size, generally weaker form, darker colors, longer wings and tail, and shorter tarsus, with fewer scales. Still, it may not differ so much from the male, Cassin's specimens being probably all females. It belongs to the smaller type, of which some have been taken for *B. Harlani*, Audubon, though his plates represent a larger and browner bird, with various other differences of color, which may indicate immaturity, the tail having several narrow black bars and no white ones. His type specimen is also said to belong to the large forms, though he gives the length as only twenty-one inches.

I shot the specimen above described, the first found within the United States, on the 23d of February, 1862, thirty miles north of San Diego, and five from the coast. It was associating with many of *B. insignatus* and other hawks wintering there, and was rather sluggish and tame as long as I observed it. I saw no more black hawks in that region.

Buteo oxypterus, Cassin.

THE SHARP-WINGED HAWK.

Buteo oxypterus, Cassin, Pr. Phil. Ac. VII. 1855, 282. Ib. Baird, Birds N. Amer. 1858, pl. xv. f. 1.

Buteo fuliginosus, Sclater, Proc. Zool. Soc. Lond. 1858, 356. Ib. Trans. IV. 1869, 267; pl. lxii.

Sp. Char. About the size of *Buteo Pennsylvanicus*, Wilson. Bill rather long and compressed, edge of upper mandible slightly waved in its outline, but scarcely lobed; wing long, pointed, third quill longest; tail moderate or rather short; legs rather long; tarsus feathered in front for about one third of its length, naked behind, naked portion in front having about fourteen narrow, transverse scales; claws large, strong, fully curved.

Young bird. Sex unknown. Entire plumage above dark brown, nearly black on the back. Feathers of the head white at base, and edged laterally with the same; upper plumage with partially concealed spots and transverse bands of white. Quills nearly black, with the inner webs dark cinereous, barred with brown; tail above ashy-brown, white at base, and having about ten transverse bands of dark-brown, outer feathers ashy-white on their inner webs; tail beneath silky, ashy-white, with a bronzed yellowish olive lustre.

Behind and under the eye a stripe of rufous brown. Under parts pale yellowish-white; throat with lines and narrow stripes of brownish-black, and on other under parts every feather with a large lanceolate, cordate, or circular spot of dark brown, some feathers on the flanks and sides having also some irregular bands of the same color. Nearly all the feathers on the under parts with lines of dark brown on their shafts. Quills, with their inner webs on the under surface, grayish or dark ashy, and near the shafts with a bronzed olive lustre; shafts white (on the under surface). Inferior coverts of the wing white, with

B. oxypterus.

sagittate spots of dark brown. Tibial feathers yellowish-white, tinged with rufous, and having irregular transverse bars of dark brown.

Total length (sex unknown) about 16.00; wing, 13.50; tail, 7.00.

Hab. Fort Fillmore, New Mexico.

A single specimen of this species, considered by Mr. Cassin to be different from all described North American raptores, was obtained at Fort Fillmore New Mexico. Subsequently Dr. Sclater, under the name of *Buteo fuliginosus,* described a bird of the same species from Mexico which was entirely brownish-black, and which Mr. Cassin considers to be the adult. Additional specimens and observations will be necessary before the question of relationship can be positively decided. (Baird.)

Genus **ARCHIBUTEO**, Brehm.

Archibuteo, Brehm, Isis, 1828, 1269.

Gen. Char. Tarsi densely feathered to the toes, but more or less naked and scaly behind. Wings long and wide; toes short, claws moderate; tail rather short and wide. Form robust and heavy, generally larger than in *Buteo.*

Archibuteo ferrugineus, Lichtenstein.

THE RUSTY SQUIRREL-HAWK.

Buteo ferrugineus, Lichtenstein, Trans. Acad. Berlin, 428. — *Lagopus ferrugineus*, Frazer, Pr. Zool. Soc. 1844, 36. — *Archibuteo ferrugineus*, Gray, Gen. Birds, 2d ed. 3; pl. 6. (Named *A. regalis* in plate.) — Cassin, P. R. Rep. Birds, IX. 34. Ib. Birds of Cal. and Texas, I. 159; pl. 26. — Heermann, P. R. Rep. X. vi. 32. — Cooper and Suckley, XII. iii. Zool. of W. T. 149. — Coues, Pr. A. N. Sc. 1866, 46.
Buteo Californicus, Grayson, Hutchings's California Magazine, March, 1857.

Sp. Char. Above light rufous and dark brown, palest on rump and wing coverts; quills ashy-brown, the greater part of their inner webs white; tail above reddish-white, mottled with ashy-brown; beneath pale yellowish-white. Under parts of body white, with narrow longitudinal lines and lanceolate spots of reddish-brown on the breast, and

A. ferrugineus.

with narrow irregular transverse lines of brown and black on the abdomen; flanks and axillary feathers bright ferruginous; tibial and tarsal feathers the same, with narrow transverse black stripes.

Young. Umber-brown above, slightly mixed with fulvous; upper tail coverts white, spotted with brown; under parts pure white, with a few streaks of brown on the breast; and sagittate brown spots on sides and abdomen, larger and more numerous on the flanks; tibiæ white; tarsi dark brown, mixed with white; under wing coverts and edges of wings white.

Length, 21.00 to 25.00; extent, 52.00 to 56.00; wing, 17.00 to 17.75; tail, 8.00 to 9.00. Iris brown; bill black and horn; cere and feet yellow.

Hab. Western North America, Mexico.

This large and powerful bird abounds in spring and fall in the south-western parts of California, migrating in summer through the interior plains at least as far north as the Dalles of the Columbia and the Platte Rivers. I found it common in December at Martinez, and few probably migrate beyond this State. It is usually seen slowly sailing over the plains, sometimes in circles, occasionally pouncing down obliquely on its prey, which consists chiefly of the large ground-squirrels. It usually alights on the ground, but often on trees also. I have not discovered its nest and eggs, but Dr. Heermann thus describes it as found by him on the Cosumnes River, in 1851: "While climbing a tree to examine some magpies' nests, the hawk in flying from her own betrayed her retreat. It was placed in the centre of a bunch of mistletoe springing from the forks of the oak, and was composed of coarse twigs, lined with grasses and moss; the eggs, two in number, being white, marked with faint brown dashes."

This hawk appears to attack poultry but rarely, limiting its prey to the wild animals, and should therefore be considered rather as a friend than an enemy to the farmer. This may be said, indeed, of all the birds of this sub-family, which may be seen in large numbers in the thinly settled districts watching from the tree-tops, but rarely approaching the farm-yard.

Archibuteo lagopus, Gmelin.

THE ROUGH-LEGGED HAWK.

Falco lagopus, Gmelin, Syst. Nat. I. 1788, 260. — Wilson, Am. Orn. IV. 59; pl. 33 (young). — *Buteo lagopus*, Audubon, Birds Amer. pl. 166 and 422. — Nuttall, Man. I. 99. — *Archibuteo lagopus*, Gray, Gen. Birds, 2d ed. 3. — Cassin, P. R. Rep. Birds, IX. 32. — Kennerly, X. iv. 19. — Cooper and Suckley, XII. iii. Zool. of W. T. 148. — Coues, Pr. A. N. Sc. 1866, 48.

Sp. Char. Head above yellowish-white, with reddish-brown stripes; back, scapulars, and shorter quills, pale ashy, with partly concealed transverse bands of white and dark brown, the latter frequently prevailing on back; rump dark umber-brown; primaries edged with ashy, a large space on their inner webs at base white. Under parts white, throat with dark brown stripes, breast with larger spots and concealed reddish-brown stripes; abdomen with numerous narrow bars of brownish-black, most conspicuous on flanks, and tinged with ashy; tibiæ and tarsi barred transversely with white and dark brown, tinged with reddish; under tail coverts white; upper tail coverts white at base, tipped with brownish-black; tail white at base, with a wide subterminal band of black, and about two other black bands alternating with light cinereous. Under wing coverts white, with brownish-black spots, and a long patch of ashy-brown on the longer ones.

Young. Light umber-brown above, much edged with yellowish and reddish-white. A wide band on abdomen brownish-black, other under parts yellowish-white, with a few

A lagopus.

lines and spots of brownish-black. Quills ashy-brown, a large portion of their inner webs white at base, with a subterminal band of light umber-brown, tip white; tibiæ and tarsi pale reddish-yellow, with stripes and spots of dark-brown.

Length, 19.00 to 23.00; extent, 50.00 to 53.00; wing, 15.00 to 17.00; tail, 8.00 to 9.00 Iris pale brown; bill slate-color; cere and feet yellow.

Hab. Northern America and Europe.

This species is only a winter visitor in California, as far as known, and I have not seen them south of Santa Clara Valley, though I think some may breed in the high mountains, as they are seen at the Columbia River in July. In habits they are in winter somewhat gregarious and rather stupid, sitting like owls on the tops of low trees, and allowing a near approach. East of the mountains they go south to Mexico. They breed in great numbers in the northern part of the American continent, especially north of Great Slave Lake. The nest is built on a tree, of sticks and twigs, lined with leaves and grass, and usually contains three, sometimes four eggs, thickly blotched with brownish, though sometimes nearly white.

Archibuteo Sancti-Johannis, Gmelin.

ST. JOHN'S BLACK HAWK.

Falco Sancti-Johannis, Gmelin, Syst. Nat. I. 1788, 273. — *Archibuteo Sancti-Johannis*, Gray, Gen. Birds, Vol. I. — Cassin, P. R. Rep. Birds, IX. 33. — Blakiston, Ibis, III. 318 (eggs).

Falco niger, Wilson, Am. Orn. VI. 1812, 82; pl. 53, f. 1, 2.

Buteo lagopus, Audubon, Birds Amer. pl. 422, and 166 (young ?). — Nuttall, Man. I. 99 (in part), and of some other authors, who consider it the adult plumage of that species.

Sp. Char. Entire plumage glossy black, often with a brown tinge; forehead, throat, and large partly concealed spot on occiput, white. Tail with one transverse band of white, and irregular markings of the same towards the base. Quills with their inner webs

A. Sancti-Johannis.

white, conspicuous from below. Head sometimes more or less striped with yellowish-white and reddish-yellow, and tail sometimes with several transverse bands of white more or less irregular.

Young, like that of *A. lagopus*, but larger, and with more numerous dark spots on the under parts.

Length, 20.00 to 24.00; wing, 16.00 to 17.50; tail, 8.00 to 9.00. Bill blackish; iris, cere, and feet, yellow.

Hab. Northern America; south in winter.

There is good reason to believe that this is only a more perfect plumage, or variety of *A. lagopus*, but it is rare, and I have seen only one, obtained by Mr. Lorquin at San Francisco.

GENUS **ASTURINA**, VIEILLOT.

Asturina, VIEILLOT, Analyse, 1816, 24. (Type, *Falco nitidus*, LATH.)

GEN. CHAR. Smaller than in the preceding two genera. General form compact, and adapted to greater activity of habits and swifter flight. Bill rather thick, strong; cere large, extending somewhat into the feathers of the forehead; wings moderate, third and fourth quills longest; tail rather long; legs rather long; claws strong, fully curved. This genus contains a few species, all of which are South American.

Asturina nitida, LATHAM.

THE GRAY HAWK.

Falco nitidus, LATHAM, Ind. Orn. I. 41. — *Asturina nitida*, BONAPARTE, Consp. Av. 1850, 30. — CASSIN, BAIRD, Birds N. Amer. 1858, 35; pl. 64.
Asturina cinerea, VIEILLOT, Analyse, 1816, 68. — OWEN, Ibis, 1861, 69 (egg).
Astur striolatus, CUVIER, Règ. An. I, 332.
Asturina plagiata, SCHLEGEL, Rev. Mus. Pays-Bas, 1862.
FIGURES. Temm. Pl. Col. 87, 294; Vieillot, Gal. I. pl. 20.

SP. CHAR. *Adult.* Upper parts light cinereous; darker, and sometimes nearly black on the rump; upper tail coverts white; quills ashy-brown, with obscure dark bands,

A. nitida

and widely edged with white on their inner webs; tail brownish-black, with about three transverse bands of white. Under parts with numerous narrow transverse bands of cinereous and white, the former predominating and darker on the breast; under tail coverts white; cere and legs yellow.

Young. Entire upper parts umber-brown; darker on the rump, and much mixed with white on the head; upper tail coverts white; tail light brown, with about eight bands of brownish-black. Under parts white, with longitudinal stripes of umber-brown; under wing and tail coverts white; cere and legs yellow.

Total length, female, about 18.00; wing, 10.00; tail, 7.50. Male smaller.

Hab. Northern Mexico and South America.

This handsome hawk was found in the state of New Leon, one of the most northern provinces of Mexico, by Lieutenant D. N. Couch, United States Army, and very probably extends its range northward into the territory of the United States. In fact, a specimen is said to have been taken in Arizona by Lieutenant Whipple's party. Not much is known of its habits, except that the nest is made in the top of a lofty tree, and that the eggs are greenish-white, like those of *Astur.*

If Schlegel be correct in distinguishing two allied species, the bird here referred to will be known as *A. plagiata.* (Baird.)

SUB-FAMILY MILVINÆ, THE KITES.

CHAR. Size usually moderate or small. General form rather slender and not strong; wings and tail long; bill short, weak, hooked and acute; tarsi and toes slender and weak.

E. leucurus.

Of this genus but few species are known, mainly confined to the warmer regions of the Old World and the New, including Australia. All have the same general appearance as to color and size, and might indeed not improperly be considered local or geographical varieties of one species.

GENUS **ELANUS**, SAVIGNY.

Elanus, SAVIGNY, Hist. Nat. d'Eygpte, I. 1809, 97.

GEN. CHAR. Wings long, pointed; tail moderate, emarginated; tarsi short. Bill moderately strong, compressed, and rounded above; mouth cleft behind the eyes; lobe of upper mandible obtuse; cere villous; nostrils oval. Tarsi short, thick, reticulated, feathered half-way down in front; toes distinct; claws large and acute, the outer very small.

Elanus leucurus, VIEILLOT.

THE BLACK-SHOULDERED HAWK.

Milvus leucurus, VIEILLOT, Nouv. Dict. XX. 1818, 563. — *Elanus leucurus*, BONAPARTE, Comp. List, 4. — CASSIN, P. R. Rep. Birds, IX. 37. — HEERMANN, X. vi. 33. — COOPER and SUCKLEY, XII. iii. Zool. of W. T. 149.

Falco dispar, TEMM. Planches col. I. liv. 1824, 54. — AUDUBON, Orn. Biog. IV. 367; pl. 352: oct. ed. pl. 16. — BONAPARTE, Am. Orn. II. 18; pl. 11, f. 1. — *Elanus dispar*, NUTTALL, Man. I. 95. — AUDUBON, oct. ed. 70; pl. 16.

SP. CHAR. Head, tail, and entire under parts, white. Upper parts fine bluish-cinereous; lesser wing coverts glossy black, forming a large oblong patch on the shoulder; lower wing coverts white, with a smaller black patch. Middle tail feathers light ashy like the back. Length, 15.00 to 17.00; extent, 39.00 to 41.00; wing, 11.00 to 12.00; tail, 7.00 to 7.50. Iris orange red; bill black; feet orange.

Hab. Southern Atlantic States; California; South America.

This beautiful and harmless species is quite abundant in the middle districts of California, remaining in large numbers during winter among the extensive *tulè* marshes of the Sacramento and other valleys. (Heermann.) I did not see one during winter at Fort Mojave, nor do they seem to have been collected by any one in the dry interior, or the southern part of California. I have seen them as far north as Baulines Bay, and near Monterey, but always about streams or marshes. Their food consists entirely of mice, gophers, small birds, snakes, etc., and they rarely, if ever, attack poultry. Of their nest and eggs no observations have yet been recorded in this State; but, according to Audubon, they build on the Santee River, South Carolina, in March.

GENUS **CIRCUS**, LACEPÈDE.

Circus, LACEPÈDE, Mem. de l'Inst. Paris, III. cxi. 1803, 506.

GEN. CHAR. Face partly encircled by a ruff of short projecting feathers, as in the owls. Head rather large; bill short, compressed, curved, from the base; nostrils large;

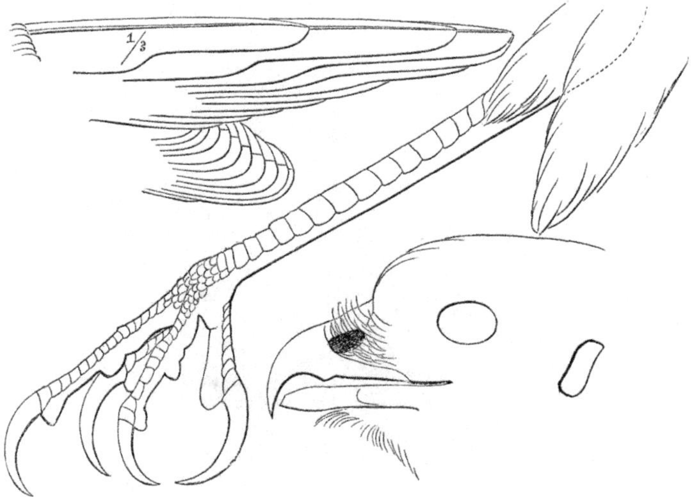

C. Hudsonius.

wings long, pointed; tail long; tarsi long and slender; toes moderate; claws rather weak.

Circus Hudsonius, LINNÆUS.

THE MARSH HAWK.

Falco Hudsonius, LINNÆUS, Syst. Nat. I. 1766, 128. — *Circus Hudsonius*, VIEILLOT, Ois. Am. Sept. pl. 9. — CASSIN, P. R. Rep. Birds, IX. 38. — NEWBERRY, P. R. Rep. VI. iv. 74. — BAIRD, X. iii. 12. — KENNERLY, X. iv. 19. — HEERMANN, X. vi. 33. — COOPER and SUCKLEY, XII. iii. Zool. of W. T. 150. — COUES, Pr. A. N. Sc. 1866, 491. (Arizona.)

Falco uliginosus, GMELIN, Syst. Nat. I. 1788, 278. — WILSON, Am. Orn. VI. 67, pl. 51, f. 2. — BONAPARTE, Am. Orn. II. pl. 11, f. 1. (Adult male.)

Falco cyaneus, AUDUBON, Orn. Biog. IV. 396; pl. 356: oct. ed. I. pl. 26. — Nuttall, Man. I. 113 (but not the same as *C. cyaneus*, of Europe).

SP. CHAR. *Adult.* Entire upper parts, head and breast, pale bluish-cinereous, on the back of head mixed with dark fulvous; upper tail coverts white. Beneath white, with small cordate or hastate spots of light ferruginous; quills brownish-black, their outer webs tinged with ashy, and a large portion of their inner webs white; tail light cinereous, nearly white on the inner webs of the feathers, and with obscure transverse bands of brown; its under surface silky-white; under wing coverts white.

Female. Male.

Younger. Entire upper parts dull umber-brown, many feathers edged with dull rufous, especially on the neck; beneath dull reddish-white, with longitudinal brown stripes, most numerous on front of throat and neck; tibiæ tinged with reddish; upper tail coverts white.

Young. Entire upper parts dull umber-brown, except white tail coverts. Beneath rufous, with stripes of brown on breast and sides; tail reddish-brown, with about three wide bands of dark fulvous, paler on the inner webs.

Length, 16.00 to 21.00; extent, 37.00 to 45.00; wing, 13.50 to 15.50; tail, 9.00 to 10.00. Bill slate-blue; iris, cere, and feet, yellow.

Hab. All of North America. South to Costa Rica.

One of the most abundant hawks throughout the unwooded country, and about every marsh, even in the dense forest. The brown plumaged specimens are much more numerous everywhere, and it appears as if only very old specimens attained the blue plumage, as these become very scarce in the older settled districts, where few are allowed to live many years. I have obtained both male and female in this plumage, but have also seen a pair, which had a nest near San Diego, one of which was of each color, showing that they breed in the brown as well as the blue plumage. The nest was built on the ground, on the slope of a grassy hill, neatly constructed, chiefly of grass, and contained four white eggs, laid about April 10th. The old birds flew around, making a sort of cackling noise, but did not dare to attack me. Sometimes, however, these birds show considerable boldness, seizing wounded birds almost from the sportsman's grasp, and occasionally catching young poultry. Their chief prey consists, however, of small birds, mice, snakes, and even grasshoppers, and they are often pursued and driven away by birds as small as the blackbirds, when too near their nest.

Sometimes they are seen in large numbers in fall, as if migrating from the north, but I have not observed them more abundant in the southern part of California in winter. They probably do not remain in the hot interior valleys in summer as they are then found about prairies at the summits of the mountains.

Sub-Family POLYBORINÆ, The Vulture Eagles.

Char. Size large; form combining the character of the vultures and eagles. Bill strong, rather short, and much hooked; legs rather elongated, and toes and claws rather weak. Head covered with feathers, or naked near the eyes.

Genus **POLYBORUS**, Vieillot.

Polyborus, Vieillot, Analyse, 1816, 22.

Gen. Char. About the size of *Pandion*; bill rather long, compressed; cere large; wings long, pointed; tail moderate or rather long; tarsi long, rather slender; claws long, rather weak, and but slightly curved; space in front of and below the eye naked.

But two other species are known, inhabiting South America.

Polyborus Audubonii, Cassin.

THE CARACARA EAGLE.

"*Polyborus Brasiliensis*, Swainson," Audubon, Birds, oct. ed. I. 21 ; pl. 4. — Nuttall, Man. I. 52.
Polyborus vulgaris, Vieillot, Nouv. Dict. V. 1816, 357. — Audubon, Birds Amer. pl. 161.
Polyborus tharus, Cassin, P. R. Rep. Birds, IX. 45. — Heermann, X. vi. 30.
Polyborus Audubonii, Cassin, Pr. A. N. Sc. 1865, 2. — Coues, Pr. A. N. Sc. 1866, 49.

Sp. Char. *Adult.* Head and body above, and a wide belt on the abdomen and tibiæ, brownish-black ; neck, breast, upper and under tail coverts, yellowish-white ; the breast with narrow transverse bands of black. Tail white at base, with numerous black bands, and widely tipped with black ; primaries banded with white.

Younger. Head and body above dull brown, darker on the head, and many feathers having paler edgings ; under parts dark brown, with dull yellowish white stripes ; throat

P. Audubonii.

dull white ; tail for the greater part, and its coverts above and below, white, with numerous ashy-brown bars, tipped with brownish-black.

Bill pale bluish, edged with yellow ; space before eye, cheeks, and cere, bright red ; feet yellow ; iris dark brown.

Length, 23.00 to 25.00 ; extent, 48.00 ; wing, 15.50 to 17.00 ; tail, 9.00 or 10.00.

Hab. Central America, north to Florida, Texas, New Mexico, Colorado Valley, California.

According to Dr. Heermann, as quoted above, this bird visits Fort Yuma in company with the vultures, which it resembles in habits, being scarcely worthy of the name of eagle, though resembling it in form. He also says that he found its nest in an oak-tree on the Medina River, Texas, and that it is abundant there. As he does not mention the *Craxirex*, which much resembles this, and obtained no specimen, it is possible that some of his observations may properly belong to that species. It is, however, quite probable that they sometimes reach the Colorado Valley, though I saw none there myself. Mr. Cassin has lately separated the North and Central American race of this species from the South American, as *P. Audubonii*.

Audubon found their nests in Florida on tall trees, and also on bushes, composed of sticks, flat, lined with roots, grass, and moss, the eggs two (color not given).

He also states that they catch ducks when wounded, and, unlike the vultures, carry their prey in their talons.

Genus **CRAXIREX**, Gould.

Craxirex, Gould, Voy. Beagle, Birds, 1841, 22.

Gen. Char. About the size of the larger buteos; legs long; tarsi and toes strong. Bill rather long, abruptly curved at the tip; edges of upper mandible festooned; wings and tail long. Tarsi with wide transverse scales in front; claws moderate.

One or two species only are known.

Craxirex Harrisii, Audubon.

HARRIS'S BUZZARD.

Buteo Harrisii, Audubon, Orn. Biog. V. 30; pl. 392: oct. ed. I. pl. 5. — Nuttall, Man. I. 111. — *Craxirex Harrisii*, Cassin, Pr. A. N. Sc. 1865, 2. — Coues, Pr. A. N. Sc. 1866, 49. (Arizona.)

Craxirex unicinctus, Cassin, P. Rep. Birds, IX. 46. — Kennerly, X. iv. 20.

Sp. Char. *Adult.* Shoulders, wing coverts and tibiæ, reddish-chestnut; upper and under tail coverts white; tail white at base and tip, between which is a very wide band of brownish-black, with a violet tinge. Body above and below dark brown, sometimes nearly black on the under parts.

Younger. Upper parts dull umber-brown, much mixed with fulvous; shoulders chestnut-red, spotted with dark brown; entire under parts yellowish-white, with large oblong and circular spots of brown; upper and under tail coverts white; tail brown, with many bands of a deeper shade; the inner webs yellowish and reddish-white; base and tip of tail yellowish-white.

C. Harrisii.

Length, 20.00 to 24.00; wing, 13.00 to 15.00; tail, 9.00 to 10.00. Bill light blue, darker at the tip; cere and feet yellow.

Hab. Tropical America to Southern United States, New Mexico, Colorado Valley, California.

Dr. Kennerly, while with Lieutenant Whipple's Pacific Railroad Expedition, obtained one of these birds from the Indians, in Colorado Valley, near Fort Mojave, February 27, 1854. It may have been kept by the Indians as a pet, caught in summer, or raised from the nest, as I did not see any of the species during my five months' residence at the Fort in 1860–61. It is said to be found in large numbers in Texas, and described by Colonel McCall as "habitually frequenting the ground in the vicinity of water. It is slow and heavy in flight, and a dull, sluggish bird in all its habits, partaking in these respects of the general characters of the vultures."

Mr. Cassin considers the North and Central American specimens of this bird as distinct from the South American, and retains Audubon's name of *Harrisii* for the former.

Family VULTURIDÆ, The Vultures.

Char. Bill contracted or indented on the anterior border of the cere, the culmen ascending again anteriorly; somewhat bow-shaped. Eyes projecting even with sides of head. Head sparsely covered with downy feathers only, or partially naked. Claws weak, rather slender, and but moderately curved ; tarsi and bases of the toes reticulated.

These birds feed wholly on carrion, and are cowardly and gregarious. They are numerous in hot climates, both in species and individuals. Some kinds kill small animals.

GENUS **CATHARTES**, ILLIGER.

Cathartes, ILLIGER, Pr. Syst. Mamm. et Av. 1811.

GEN. CHAR. Bill long and straight to the curved tip ; cere extending beyond middle of bill ; nostrils narrow, elongated, and pervious, situated near middle of bill ; tongue channelled and serrated. Head elongated, flattened, and wrinkled ; neck with patches of naked skin ; tail of twelve feathers. Tarsus rather slender ; toes long, the lateral ones equal, united by a web at base. Third quill longest.

This genus is the American representative of the true vultures (*Vultur*), of the old continent. There are five species known besides the two found within this State.

Of the vultures there are two principal divisions, one with the nostrils on opposite sides separated by a bony partition, the other lacking such a wall. The first embraces the Old World species, the other the New.

C. Californianus. Young.

Cathartes Californianus, Shaw.

THE CALIFORNIAN VULTURE.

Vultur Californianus, Shaw, Nat. Misc. IX. 1779, 1; pl. 301.—*Cathartes Californianus*, Cuvier, Règne Anim. II. 316.—Nuttall, Man. I. 39.—Audubon, Birds Amer. pl. 411: oct. ed. I. 12; pl. 1.—Baird and Cassin, P. R. Rep. Birds, IX. 5.—Newberry, P. R. Rep. VI. iv. 73.—Heermann, X. vi. 59.—Cooper and Suckley, XII. iii. Zool. of W. T. 141.—Coues, Pr. A. N. Sc. 1866, 42. (Fort Yuma.)—Taylor, Hutchings's Magazine, IV. 1859, 537. (Figure of egg and young.)—Gurney, Catal. Raptoral Birds, 1864, 39.—Sclater, Pr. Zool. Soc. 1866, 366 (figure from life); 1868, 183 (young in down, from life).

Sp. Char. Plumage black, lustrous above, duller below; secondary quills with a grayish tinge; greater wing coverts tipped with white, forming a transverse band on the wing.

C. Californianus.—Female, not quite mature.

Bill yellowish; head and neck orange-yellow and red (drying brown). A semicircular spot of short black feathers at base of upper mandible, and a few scattered hair-like feathers on other parts of the head. Ruff of long lanceolate feathers extending on to breast.

Length, 45.00 to 56.00; extent, about 100.00; wing, 30.00 to 35.00; tail, 15.00 to 18.00. Iris carmine (hazel in young female); feet and tarsi bluish-black (dirty white in male, Taylor.)

Hab. Western North America, to lat. 49°. (Douglas.)

This large bird, second in size only to the condor of South America, among the Raptores, appears to be limited to the western part of the United States, not having been yet obtained in Mexico, and rarely north of the Columbia River.

It is most abundant in the hot interior valleys of California, where the large herds of cattle furnish abundance of food; but I saw none along the Colorado, or east of the San Bernardino Mountains, the scarcity of large animals there being a barrier to their migration, although from their lofty flight and extensive vision they probably sometimes see a dead or sickly antelope and follow it to the more desert regions of the State, in which they may find also some mountain sheep. The cattle killed at Fort Mojave attracted but two turkey-buzzards there during five months, and *no* vultures.

I have not seen many of these birds along the sea-coast where most of my later collections were made, and none on the islands or in the highest Sierra Nevada. They are said, however, when other food is scarce, to feed on dead seals and whale meat, though I have not seen them do so.

At Monterey I saw in Dr. Canfield's possession a full-grown living specimen, which he had raised from the nest. Being fed on fresh meat, it had no offensive smell, and its plumage was clean and shining. It was gentle and familiar, but stupid, spending most of its time dozing on the fence.

The following are Douglas's remarks on this bird in the "Zoölogical Journal," Vol. IV., as quoted by Audubon, omitting some incorrect parts: —

"Food, carrion, dead fish, or other dead animal matter. In no instance will they attack any living animal, unless wounded and unable to walk. Their senses of smelling and seeing are very acute. In searching for prey they soar to a very great altitude, and when they discover a wounded deer or other animal they follow its track, and when it sinks, precipitately descend on their object. Although only one is seen at first occupying the carcass, few minutes elapse before the prey is surrounded by great numbers, and it is then devoured to a skeleton within an hour, even though it be one of the larger animals, as the elk or horse. Their voracity is almost insatiable, and they are extremely ungenerous, suffering no other animal to approach them while feeding. After eating they become so sluggish and indolent as to remain in the same place until urged by hunger to go in quest of another repast. At such times they perch on decayed trees, with their head so much retracted as to be with difficulty observed through the long loose feathers of the collar; the wings at the same time hang down over the feet. This position they invariably observe in dewy mornings, or after the rains. Except after eating, or while protecting their nest, they are so excessively wary, that the hunter can scarcely approach sufficiently near even for buckshot to take effect upon them, the fulness of the plumage affording them a double chance of escaping uninjured. Their flight is slow, steady, and particularly graceful; gliding along with scarcely any apparent motion of the wings, the tips of which are curved upwards in flying. Preceding hurricanes or thunder-storms, they appear most numerous and soar the highest. The quills are used by hunters as tubes for tobacco-pipes."

Dr. J. K. Townsend informed Audubon that "the California vulture inhabits the region of the Columbia River, to the distance of five hundred miles from its mouth, and is most abundant in spring, at which season it feeds on the dead salmon that are thrown upon the shores in great numbers."

I never saw them north of the Columbia, nor near its mouth, and that river may be considered as usually their northern limit. "It is also met with near the Indian villages, being attracted by the offal of the fish thrown around their habitations. It associates with the *Cathartes aura*, but is easily distinguished from that species in flight, both by its greater size and the more abrupt curvature of the wing." To this I may add, by the large white patch. "Indians, whose observations may generally be depended upon [!], say that it ascertains the presence of food solely by its power of vision."

"In walking they resemble a turkey, strutting over the ground with great dignity; but this dignity is occasionally lost sight of, especially when two are striving to reach a dead fish, which has just been cast upon the shore; the stately walk then degenerates into a clumsy sort of hopping canter, which is anything but graceful. When about to rise, they always hop or run for several yards, in order to give an impetus to their heavy body; in this respect resembling the condor of South America, whose well-known habit enables it to be easily taken in a pen by the Spaniard."

Hutchings's California Magazine for June, July, and August, 1859 (Vol IV. No. 36 – 38) contains a series of articles on this vulture by Alexander S. Taylor, then of Monterey, and well known for his many interesting contributions to the zoölogy of this State, as well as its ethnology.

Mr. Taylor calls it the "California condor," putting it in the genus *Sarcoramphus*, on account of its great size, nearly equalling that of the South American condor (*S. gryphus*). Size, however, is insufficient as a generic distinction without other characters, in which our vulture agrees closely with the *Cathartes*, while it differs very much from *Sarcoramphus*, in wanting the fleshy comb on the head.

The illustrations given with the articles, representing the male flying, the female standing, the newly hatched young, and the eggs, are the best I have ever seen of this bird, having been taken from life by W. M. Ord. The male, however, is represented as carrying off a hare in its *claws*, — a doubtful circumstance, as these vultures are not addicted to carrying dead animals, for which their straight, weak claws and toes are poorly adapted. The young are fed by food disgorged from the crop of the old bird, not carried in its claws.

Mr. Taylor writes: "In January, 1858, a large condor was killed by Mr. S. B. Wright, near St. Helena, in Napa County, while flying off with a nine-pound hare it had killed." [?] "The bird measured fourteen feet from tip to tip of wings." He also mentions others *said* to measure eleven and twelve feet; but as those he measured himself only measured eight to nine feet in extent, and knowing the tendency of newspaper contributors to exaggerate, we may set this down as their usual stretch of wings. Douglas's largest was nine feet three inches in extent.

The following are extracts from Mr. Taylor's article: —

"One of the rancheros of the Carmelo, in hunting among the highest peaks of the Santa Lucia Range, during the last week of April, disturbed two condors from their nests, and at great risk of breaking his neck, etc., brought away a young bird of six or seven days old, and also an egg, — the egg from one tree, the chick from another. There was, properly speaking, no nest, but the egg was laid in the hollow of a tall old robles-oak, in a steep barranca, near the summit of one of the highest peaks in the vicinity of the Tularcitos, near a place called 'Conejos.' The birds are said by some hunters not to make nests, but simply to lay their eggs on the ground, at the foot of old trees, or on the bare rocks of solitary peaks; others say they lay in old eagles' or buzzards' nests, while some affirm they make nests of sticks and moss; but the truth seems to be, they make no nests." (Only a slight one "of a few loose sticks thrown negligently together," according to Heermann, who saw several in the Sierra Nevada.) "The egg weighed ten and a half ounces, and the contents eight and three quarter ounces. The color of the egg-shell is what painters call 'dead, dull white,' the surface not glossy, but slightly roughened. Its form is very nearly a perfect ellipse. It measured four and a half inches in length by two and three eighth inches in breadth, and was eight and three quarters inches in circumference. The egg-shell, after the contents were emptied (which were as clear, fine, bright, and inodorous as those of a hen's egg, with a bright yellow yolk), held as much as nine fluid ounces of water. Some of the old hunters say the egg is excellent eating." The weights used were avoirdupois, except the fluid ounce. "The young mentioned above is from five to seven days old, and weighed ten ounces. The whole skin is of an ochreous yellow, and covered with a dull white fine down ; the beak was horn-colored, — the skin of the head and neck entirely bare of down, and of an ochreous yellow, — the color of the legs a deeper shade of that of the body ; it had the musky smell of the old birds ; the size and appearance similar to that of a two-month old gosling ; it had only been dead a couple of hours." In the following points Mr. Taylor's description of the bird differs from that of others : "*Male*, with bright lemon-yellow head and upper neck. *Female*, with dark coppery olive head and neck, covered with feathery down on head and most of the neck ; both birds about same weight." This pair, killed in July and November, were probably young birds of the year, as they had horn-colored bills, while, according to Douglas, the bill becomes yellow, and the head deep orange, the neck brownish yellow with blue tints. (Douglas described also from fresh specimens, as did Townsend, whose accounts were quoted by Nuttall and Audubon.) The female, however, retains a darker hue of head and neck through life, and also, as observed by Taylor, a row of black spots on the white portion of wing-feathers. From its size, Douglas's specimen must have been a female, and he states that the sexes are alike in colors, probably from observation. "Dr. Canfield tells me that he has seen as many as one hundred and fifty at one time and place, in the vicinity of antelopes he had killed ; he invariably observed that they sighted their prey." "It is often killed by feeding on animals, such as bears, when poisoned with strychnine by the rancheros ; the poisoned meat kills them readily. The rancheros have very little fear of their depredations on young cattle, though it has been within my knowledge for five or six to attack a young calf, separate it from its mother and kill it ; the Californians also say they are often known to kill lambs, hares, and rabbits."

"A large grizzly bear being killed, the vaquero left it on the plains near the sea-shore, to return to the house, about three miles distant, for assistance in skinning the animal. Before his return, which was in about two hours, a flock of vultures had cleaned the entire carcass of its flesh and viscera, leaving nothing but the skin and skeleton."

"The 'condors' and turkey-buzzards often feed together over the same carcass, and generally in such cases do some fighting and biting ; they may sometimes be seen soaring and circling together in the air.

"Many of them make their nests in the high mountains east and south of the Carmelo Valley, and also near Santa Cruz, and in the Santa Lucia Range, where they may be seen at all seasons of the year, but in greater numbers from July to November. These huge creatures may often be seen fighting each other over a carcass on the beach ; generally striking with their outstretched wings, and running along the ground like the common turkey-buzzard.

"A few days ago we got within about seventy yards of a number of the male and female vultures. They were feeding on the carcass of a whale on the sea-shore, and must have been gorged, as we could make out every feature of both sexes with distinctness, except that the color of the head and neck of the male appeared of an orange-color instead of a bright lemon-yellow." Probably this was an old female, and the supposed females of Taylor were young of the year. "We got within thirty yards of the male, but he kept his position on a pine-tree hard by, without moving more than his head, in great anxiety; he appeared incapable of flight.

"Sometimes they make a smothered and squeaking noise or hiss, but they are generally mute."

Dr. Newberry's observations in the northern part of California were as follows: —

"A portion of every day's experience in our march through the Sacramento Valley was a pleasure in watching the graceful evolutions of this splendid bird. Its colors are pleasing; the head orange, body black, with wings brown and white and black; while its flight is easy and effortless, almost beyond that of any other bird.

"This vulture, though common in California, is much more shy and difficult to shoot than its associate the turkey-buzzard (*C. aura*), and it is never seen in such numbers, nor exhibiting such familiarity as that species. We had, however, on our first entrance into the field, many opportunities of shooting this bird, but were unwilling to burden ourselves with it. After we left the Sacramento Valley, we saw very few in the Klamath basin, and *none* within the limits of Oregon. It is sometimes found there, but much more rarely than in California."

This confirms the observations of Dr. Suckley and myself, as we saw none during a long residence and travels near the Columbia, except one which I *supposed* to be this, seen at Fort Vancouver, in *January*. Like several other birds seen there by Townsend and Nuttall, they seem to have retired more to the south since 1834.

Dr. Heermann gives the following, on its habits towards the south: —

"Whilst unsuccessfully hunting in the Tejon Valley, we have often passed several hours without a single one of this species being in sight, but on bringing down any large game, ere the body had grown cold, these birds might be seen rising above the horizon, and slowly sweeping towards us, intent upon their share of the prey. Nor, in the absence of the hunter, will his game be exempt from their ravenous appetite, though it be carefully hidden, and covered by shrubbery and heavy branches; as I have known these maurauders to drag forth from its concealment and devour a deer within an hour. Any article of clothing thrown over a carcass will shield it from the vulture, though not from the grizzly bear, who little respects such flimsy protection. The California vulture joins to his rapacity an immense muscular power; as a sample of which it will suffice to state that I have known four of them, jointly, to drag off, over a space of two hundred yards, the body of a young grizzly bear, weighing upwards of a hundred pounds.

"A nest of this bird, with young, was discovered on the Tuolumne River, by some Indians who were sent there in search of a horse-thief. It was about eight feet back from the entrance of a crevice in the rocks, completely surrounded and masked by thick underbrush and trees, and composed of a few loose sticks thrown negligently together. The effluvium arising from the vicinity was overpowering. We found two other nests of a like construction and similarly situated; one at the head of the Merced River, and the other in the mountains near Warner's Ranch. From the latter the Indians annually rob the young, and, having duly prepared them by long feeding, kill them at one of their great festivals."

At Santa Cruz I saw three or four pairs of vultures constantly, from February to October. At almost all times they could be seen sailing far overhead; but I did not, after much watching, trace them to their nests. They are doubtless constant residents.

The figure of the adult vulture we give is taken from a cast of the cut in the Proceedings of the Zoölogical Society quoted above. The bird was presented by Dr. Canfield, of Monterey, to the Zoölogical Society of London, and the figure taken from life. The figure of the young bird in the down is from the same specimen in infancy, taken from a photograph furnished by Dr. Canfield, and a duplicate of the cut sent to the Zoölogical Society of London, in exchange for that of the adult.

Cathartes aura, Linnæus.

THE TURKEY-BUZZARD, OR VULTURE.

Vultur aura, Linnæus, Syst. Nat. I. 1766, 122. — Wilson, Am. Orn. IX. 95; pl. lxxv. f. 1. — *Cathartes aura*, Illiger, Prod. Syst. 1811, 236. — Audubon, Birds Amer. 106; pl. cli.: oct. ed. I. pl. 2. — Nuttall, Man. I. 44. — Baird and Cassin, P. R. Rep. IX. Birds, 6. — Newberry, VI. iv. 73. — Cooper and Suckley, XII. iii. Zool. of W. T. 140. — Heermann, X. vi. 59. — Dresser, Ibis, 1865, 322. (Texas.) — Coues, Pr. A. N. Sc. 1866, 42.

Sp. Char. Brownish-black, darkest above, with a purplish lustre, many feathers with a pale border. Bill yellowish; head and neck bright red. A ruff of projecting feathers at base of neck. Tail rather long, rounded. Length of female, about 30.00; wing, 23.00; tail, 12.00. Male smaller, length about 27.00.

Hab. All of the United States and Mexico, and throughout South America.

This vulture is exceedingly numerous throughout the Pacific States, wherever it can find food; and though none may be seen for weeks, in places where animals are scarce, they very soon appear when a large carcass has been for a short time exposed. At Fort Mojave I saw none until March, though they remain in considerable numbers in the middle parts of the State during winter. Their lofty soaring and sharp sight enable them to perceive dead animals at a great distance, and, with the preceding species, they soon

C. aura.

gather around one from their airy watch-towers, in which they seem to the eye no larger than swallows, while objects on earth must appear to them plainly perceptible.

I saw numerous nests of this species on the cliffs near San Diego, but did not succeed in finding eggs or young in them, although I examined them from January to May. The eggs are said to be yellowish-white, irregularly blotched with brown spots, and larger than those of a turkey.

The black vulture (*C. atratus*, Bart.) has been several times reported as a species found on the Pacific Coast, but no late collectors have obtained or seen it. The young turkey-buzzard, which has a black head, may have been mistaken for it.

ORDER RASORES.

CHAR. Bill not longer than the head; the terminal portion more or less vaulted, hard, with or without a soft skin intervening between it and the head. Nostrils with an overlapping fleshy or leathery scale or valve extending over their upper edge (except in some tropical families).

Though there are many points of resemblance between the pigeons and gallinaceous birds, the differences are still so great that they should probably constitute different orders, representing each other in the two sub-classes of birds formed by Bonaparte, viz. the *Altrices*, or those which feed their young in the nest, and the *Præcoces*, or those whose young run about as soon as hatched.

Still, as many tropical pigeons are closely similar to the Gallinæ (though probably none are really *Rasores*, or scratchers), and as the divisions of *Altrices* and *Præcoces* present still greater difficulties among the herons, cranes, and some aquatic families, I retain this as a convenient though unnatural grouping. The *Columbæ* have some affinity to the parrots and cuckoos, and should perhaps be put among *Insessores*.

SUB-ORDER COLUMBÆ, THE PIGEONS.

CHAR. The basal portion of the bill covered by a soft skin, in which are situated the nostrils, overhung by an incumbent fleshy valve, the apical portion hard and convex. The hind toe on the same level with the rest; the anterior toe without membrane at the base. Tarsi more or less naked; covered laterally and behind with hexagonal scales.

FAMILY COLUMBIDÆ.

CHAR. Bill horny at tip. Tail feathers twelve, occasionally fourteen. Head uncrested.

SUB-FAMILY COLUMBINÆ.

CHAR. Tarsi, stout, short, with transverse scutellæ anteriorly; feathered for the basal third above, but not at all behind. Toes lengthened, the lateral decidedly longer than the tarsus. Wings lengthened and pointed. Size large. Tail feathers twelve. (Baird.)

GENUS **COLUMBA**, LINNÆUS.

Columba, LINNÆUS, Syst. Nat. 1735. (Type, *Columba livia*, L., the domestic pigeon.)

GEN. CHAR. Bill stout and rather short; culmen from the base of the feathers about two fifths the head. Lateral toes and claws about equal, reaching nearly to the base of

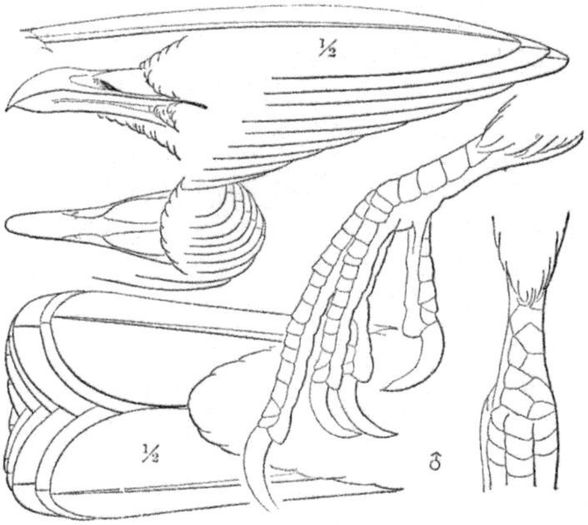

C. fasciata.

the middle claw; the claws rather long and not much curved. Tail rather short, rounded, or nearly even; as long as from the carpal joint to the end of secondaries in the closed wing. Second and third quills longest. (Baird.)

This genus, and several closely allied to it, have numerous representatives scattered in nearly all parts of the world, especially in the tropics. Our species is the only one, however, found in temperate North America, except two which straggle to Texas and Florida. The Eastern wild pigeon belongs to a different genus (*Ectopistes*), with a very small head, and long, pointed tail. That species has been obtained near the Pacific Coast, at latitude 49°, and may yet be found to wander to California.

Columba fasciata, Say.

THE BAND-TAIL PIGEON.

Columba fasciata, Say, Long's Exped. R. Mts. II. 1823, 10. — Bonaparte, Amer. Orn. I. 1825, 77; pl. viii. — Nuttall, Man. I. 1832, 624. — Audubon, Orn. Biog. IV. 1838, 479; pl. 367. Ib. Syn. 1839, 191. Ib. Birds Amer. IV. 1842, 312; pl. 279. — Newberry, Zool. Cal. and Or. Route; P. R. Rep. VI. iv. 92. — Baird, P. R. Rep. Birds, IX. 597. — Cooper and Suckley, XII. iii. Zool. of W. T. 217. — Coues, Pr. A. N. Sc. 1866, 93.

Columba monilis, Vigors, Zool. Beechey's Voy. 1839, 26; pl. x.

Sp. Char. Above, olivaceous tinged with ash, changing on the wing coverts to bluish-ash, of which color are the hinder part of the back, rump, and basal portion of the tail.

C. fasciata.

The terminal third of the tail is whitish-brown, with a tinge of ash, succeeding a narrow bar of dusky. Head all round, sides of neck, and under parts, including the tibia, purplish-violet, the middle of the abdomen, anal region, and crissum, whitish. Tibia and throat tinged with blue. Quills brown, narrowly margined with white. A conspicuous narrow half-collar of white on the nape; the feathers below this to the upper part of the back metallic golden green.

Female similar, with less purple; the nuchal collar of white obsolete or wanting.

Length, about 15.50; extent, 25.00, wing, 9.00; tail, 6.20. Iris carmine, edged with gold; bill yellow, black at tip; feet yellow.

Hab. From Rocky Mountains to Pacific Coast; south to Orizaba.

This large and beautiful bird is common in the mountains of most parts of California, wherever there are trees, and descends to the valleys in quest of grain during autumn and winter. North of San Francisco I have seen them in flocks in the grain-fields as early as July, and at the Columbia River they spend the summer in the valleys as well as throughout the mountains. They are there migratory, leaving in October, but in California their wanderings are guided chiefly by want of food. I have found them building in the Coast Range as far south as Santa Cruz, though I did not succeed in finding any nests. I was told that they built in companies on low bushes in unfrequented parts of the mountains, but Townsend found their eggs on the ground near the banks of streams in Oregon, numbers congregating together. I have myself found eggs which I supposed to

be of this bird in a similar situation. They are white and about the size of those of the tame pigeon. Like all their tribe, they lay but one or two at a time. Their cooing is very much like that of the tame pigeon, and easily distinguishable from that of the dove. From their similarity of habits, there seems no reason why they should not be easily domesticated.

They feed on acorns, which they swallow whole, even when very large; also on berries, especially those of the Madrona (*Arbutus*), grain, and seeds of various kinds. Being large, and delicate food, they furnish much sport for the fowler in certain districts, but soon become so watchful and shy that they are shot with difficulty, except when young or where they can be watched for in ambush.

In Oregon they collect in flocks of thousands in the autumn, but I have never seen more than a hundred together in this State.

Columba flavirostris, WAGLER.

THE RED DOVE.

Columba flavirostris, WAGLER, Isis, 1831, 519. — LAWRENCE, Ann. N. Y. Lyc. V. May, 1851, 116. — DRESSER, Ibis, 1866, 23. — BAIRD, Birds N. Amer. 1858, 598; pl. 61. IB. U. S. Mex. Bound. V. 21; pl. 23. — *Chlorœnas flavirostris*, BONAPARTE, Consp. Av. II. 1854, 52.

? *Columba solitaria*, McCALL, Pr. A. N. Sc. Phil. III. July, 1847, 233. Rio Grande, Texas. Description referring probably to the female of this species.

SP. CHAR. Second and third quills equal, and decidedly longer than the first and fourth, also nearly equal. Tail truncate, slightly rounded. Head and neck all round, breast, and a large patch on the middle and lesser wing coverts, light chocolate-red, the

latter deeper and more opaque-red; the middle of the back, scapulars, and tertials, olive; the rest of body, wings, and tail very dark slaty-blue; the inferior and concealed surfaces of the latter black. Bill and legs yellow in the dried skin, said to be purple in life; eyes purple. Length, 14.00; wing, 8.00; tail, 5.70.

Hab. Lower Rio Grande and Northern Sonora. Cape St. Lucas; south to Costa Rica.

There is no trace of any metallic scale-like feathers on the neck of this species. The wing feathers, including the greater coverts, are whitish on their external border. There is a tinge of the red on the inside of the wing.

This beautiful pigeon is about the size of the common passenger pigeon, although without the long graduated tail of the latter species. It is common in the Rio Grande of Texas, and extends across to the Gulf of California, and has been taken so near to the southern boundary of Arizona, as to render it quite certain that it must cross it at times. Little is known of its habits, save that it is quite solitary, keeping in pairs or alone, usually near the water.

GENUS **ECTOPISTES**, SWAINSON.

Ectopistes, SWAINSON, Zool. Jour. III. 1827, 362. (Type, *Columba migratoria*, L.)

GEN. CHAR. Head very small. Bill short, black; culmen one third the rest of head. Tarsi very short, half covered anteriorly by feathers. Inner lateral claw much larger than outer, reaching to the base of the middle one. Tail very long and excessively cuneate; about as long as the wings. First primary nearly equal to the second or longest.

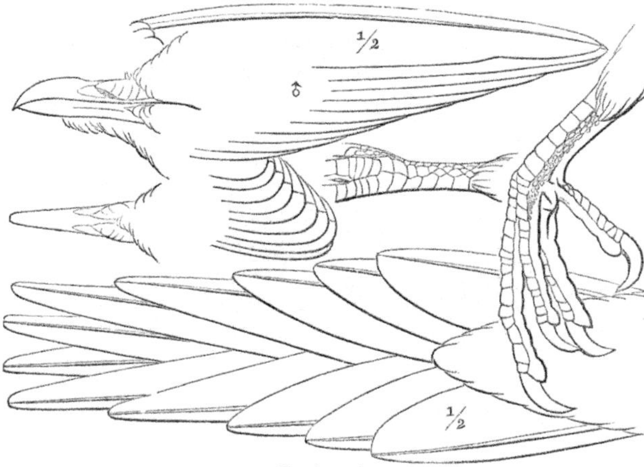

E. migratoria.

Ectopistes migratoria, SWAINSON.

THE WILD PIGEON; THE PASSENGER PIGEON.

Columba migratoria, LINNÆUS, Syst. Nat. I. 1766, 285. — GMELIN, Syst. Nat. I. 389. — FORSTER, Phil. Trans. LXII. 1772, 398. — WILSON, Am. Orn. I. 1808, 102; pl. xliv. — BONAPARTE, Obs. Wils. 1825, No. 179. — WAGLER, Syst. Av. 1827, No. 91. — AUDUBON, Orn. Biog. I. 1831, 319; V. 561; pl. 62. — *Ectopistes migratoria*, SWAINSON, Zool. Jour. III. 1827, 355. IB. Fauna Bor. Amer. II. 1831, 363. — BONAPARTE, List, 1838. IB. Consp. Av. II. 1854, 59. — AUDUBON, Syn. 1839, 194. IB. Birds Amer. V. 1842, 25; pl. 285. — "REICH. Icones Av. tab. 249, figs. 1377, 1379." — BAIRD, Birds N. Amer. 1858, 600. — COOPER and SUCKLEY, XII. iii. Zool. of W. T. 218.

Columba Canadensis, LINNÆUS, Syst. Nat. I. 1766, 284. — GMELIN, Syst. Nat. I. 1788, 785. Female or young. (Prior name ?)

Columba Americana, "KALM, It. II. 527."

Passenger Pigeon, PENNANT, II. 322. — LATH. Syn. II. ii. 661.

SP. CHAR. Tail with twelve feathers. Upper parts generally, including sides of body, head, and neck, and the chin, blue. Beneath purplish brownish-red, fading behind with a violet tint. Anal region and under tail coverts, bluish-white. Scapulars, inner tertials, and middle of back, with an olive-brown tinge; the wing coverts, scapulars, and inner tertials, with large oval spots of blue-black on the outer webs, mostly concealed, except on the latter. Primaries blackish, with a border of pale bluish, tinged internally with red.

E. migratoria.

Middle tail feathers brown; the rest pale blue on the outer web, white internally; each with a patch of reddish-brown at the base of the inner web, followed by another of black. Sides and back of neck richly glossed with metallic golden violet. Tibia bluish-violet. Bill black; feet yellow.

The female is smaller; much duller in color; more olivaceous above; beneath, pale

blue instead of red, except a tinge on the neck; the jugulum tinged with olive, the throat whitish.

Length of male, 17.00; wing, 8.50; tail, 8.40.

Hab. North America to High Central Plains. Straggles westward to West Humboldt Mountains, and the vicinity of Puget's Sound.

The account of the birds of the Western United States would not be complete without the introduction of the passenger pigeon, which, although never known in such flocks as have rendered it so conspicuous in the East, is yet met with in small numbers along the northern boundary of the United States to Puget's Sound. With its habits every one is familiar, — the vast hosts in which it formerly appeared, its roosts in large forests, where huge trees are broken down by the weight of the clustering masses, the devastation it sometimes causes, and the varied modes in which its capture is accomplished. It is strictly confined to North America, the only record of its occurrence elsewhere being in Cuba.

Sub-Family ZENAIDINÆ.

Char. Tarsi stout, lengthened; always longer than the lateral toes, and entirely without feathers; the tibial joint usually denuded. Tarsus sometimes with hexagonal scales anteriorly. Tail feathers sometimes fourteen.

Genus **ZENAIDURA**, Bonaparte.

Zenaidura, Bonaparte, Consp. Av. II. 1854, 84. (Type, *Columba Carolinensis*, L.)
Perissura, Cabanis, Jour. für Orn. IV. 1856, 111.

Gen. Char. Bill weak, black; culmen from frontal feathers, about one third the head above. Tarsus not quite as long as middle toe and claw, but considerably longer than the

Z. Carolinensis.

lateral ones; covered anteriorly by a single series of scutellæ. Inner lateral claw considerably longer than outer, and reaching to the base of middle. Wings pointed; second

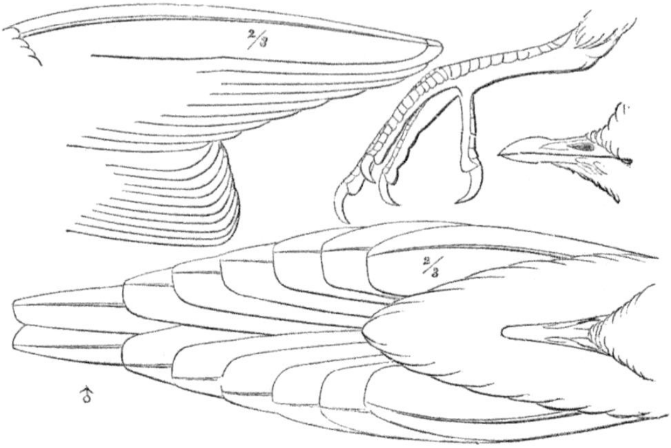

Z. Carolinensis.

quill longest; first and third nearly equal. Tail very long, equal to the wings; excessively graduated and cuneate, of fourteen feathers.

Zenaidura Carolinensis, Linnæus.

THE CAROLINA OR COMMON DOVE.

Columba Carolinensis, Linnæus, Syst. Nat. I. 1766, 286, No. 37. — Wilson, Am. Orn. V. 1812, 91; pl. xliii. — Audubon, Orn. Biog. I. 1831, 91; V. 1839, 555; pl. 17. — Nuttall, Man. I. 1832, 626. — *Ectopistes Carolinensis*, Rich. List, 1837. — Audubon, Syn. 1839, 195. Ib. Birds Amer. V. 1842, 36; pl. 286. — Newberry, P. R. Rep. VI. iv. 92. — Heermann, X. vi. 60. — *Zenaidura Carolinensis*, Bonaparte, Consp. Av. II. 1854, 84. — Baird, P. R. Rep. IX. Birds, 604. — Kennerly, X. iv. 33. — Cooper and Suckley, XII. iii. Zool. of W. T. 218. — Coues, Pr. A. N. Sc. 1866, 93.

Ectopistes marginellus, Woodhouse, Pr. A. N. Sc. VI. June, 1852, 104. Ib. Exp. Zuñi and Color. 1853, 93; Birds, pl. v. Canadian River, Ark. Immature bird.

Sp. Char. Above bluish, although this is overlaid with light brownish-olive, leaving the blue pure only on top of the head, the exterior of the wings, and the upper surface of the tail, which is slightly tinged with this color. The entire head, except the vertex, the sides of the neck, and the under parts generally, light brownish-red, strongly tinged with purple on the breast, becoming lighter behind, and passing into brownish-yellow on the anal region, tibia, and under tail coverts. Sides of the neck with a patch of metallic purplish-red. Sides of body, and inside of wings, clear light blue. Wing coverts and scapulars spotted with black, mostly concealed, and an oblong patch of the same below the ear. Tail feathers seen from below blackish, the outer web of outermost white, the others tipped with the same, the color becoming more and more bluish to the innermost, which is brown. Seen from above, there is the same gradation from white to light blue in the tips; the rest of the feather, however, is blue, with a bar of black anterior

to the light tip, which runs a little forward, along the margin and shaft of the feather. In the sixth feather the color is uniform bluish, with this bar; the seventh is without bar.

Female smaller and with less red beneath.

Length of male, 12.75; extent, 18.00; wing, 5.75; tail, 6.70. Iris brown; bill black; feet red.

Hab. Throughout the United States from the Atlantic to the Pacific. Cuba, Gundlach; south to Costa Rica.

This dove is abundant throughout California, and north at least to lat. 49° in summer, while a few winter in California as far north as San Francisco, lat. 38°, though I think most of them leave the State.

They arrive from the south in large flocks in March and April, and spread over the whole country, even those barren desert mountains towards the Colorado, where scarcely any birds are to be found. I there noticed them in May coming from all directions about sunset to drink at the springs which are scattered at long intervals in that region. From early in April to June their nests and eggs may be found in various situations, on the ground, on fences, stumps, large branches, and among the foliage of trees and bushes. The nest consists of a few twigs carelessly laid together, is about 4.50 inches wide, with scarcely any depression, and so open that the two white eggs may be seen through the bottom. These measure about 1.12 × 0.90 inches.

Being delicate food, and easily killed, they are much shot, and being unprotected by law, are barbarously killed even during the breeding season, when pot-hunters take advantage of their affectionate disposition and shoot them in pairs even on the nest. Such barbarism should be universally condemned by all true sportsmen.

GENUS **MELOPELIA**, BONAPARTE.

Melopelia, BONAPARTE, Consp. II. Dec. 1854, 81. (Type, *Columba leucoptera*, L.)

GEN. CHAR. Similar to *Zenaida*; the orbital region naked; the bill longer; the mid-

M. leucoptera.

dle toe longer; the hinder shorter. Tarsal scutellæ in a single series anteriorly. First quill nearly as long as the second and third.

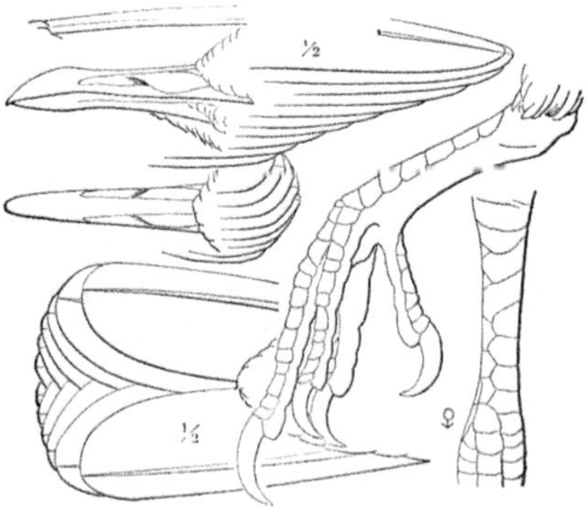

M. leucoptera.

Melopelia leucoptera, Bonaparte.

THE WHITE-WINGED DOVE.

Columba leucoptera, Linnæus, Syst. Nat. I. 1766, 281. (Not the locality,—"Asia.") — Gmelin, Syst. Nat. I. 1788, 773. — Wagler, Syst. Av. 1827, *Columba*, No. 71. — M'Call, Pr. A. N. Sc. IV. 1848, 64. — *Zenaida leucoptera*, Gray, Gen. — *Turtur leucopterus*, Gosse, Birds Jam. 1847, 304. — *Melopelia leucoptera*, Bonaparte, Consp. Av. II. 1854, 81. — Baird, Birds N. Amer. 1858, 603. — Coues, Pr. A. N. Sc. 1866, 93. — March, Pr. A. N. Sc. 1863, 302. (Jamaica.)
? *Columba hoilotl*, Gmelin, Syst. Nat. I. 1788, 777.
Columba trudeaui, Audubon, Birds Amer. VII. 1843, 352; pl. 496.

Sp. Char. Tail moderately graduated on the sides. Second and third quills longest; first a little shorter; fourth considerably shorter. In the female the upper parts generally are light olive brown; the head and neck above purplish, with a black spot below the ear; the lower part of the neck with scale-feathers of metallic golden green. Forehead and under parts light bluish-gray; more blue on the sides. Tail feathers, except the middle, bluish above, black beneath, broadly terminated with white; the upper surface with a bar of black in the end of the blue. Quills (except inner tertials) black, margined or tipped with white; a broad white patch along the exterior of the greater wing coverts and alular feathers. Bill black, base pinkish-purple; iris purple. Length (female), 11.00; wing, 6.00; tail, 4.75.

Hab. Valley of Rio Grande, westward through New Mexico and Southern Arizona, to Cape St. Lucas. South to Costa Rica. Cuba and Jamaica.

Not much is known of the habits of this dove, although a common species in the region above mentioned. According to Mr. March, the species is gregarious, sometimes collecting in very large flocks. The eggs are white, like those of all the pigeons; size 1.30 × 0.90. Its geographical distribution is somewhat peculiar in extending from Cape St. Lucas across the southern border of the United States to the West Indies, in this range accompanied by *Chamæpelia passerina*.

C. passerina.

GENUS **CHAMÆPELIA**, SWAINSON.

Chamæpelia, SWAINSON, Zool. Jour. III. 1827, 361. (Type, *Columba passerina*, L.)

GEN. CHAR. Size very small. Bill slender, elongated. Culmen more than half the head measured from frontal feathers. Legs stout. Tarsi longer than lateral toes; equal to the middle without its claw; covered anteriorly by a single series of scutellæ. Wings

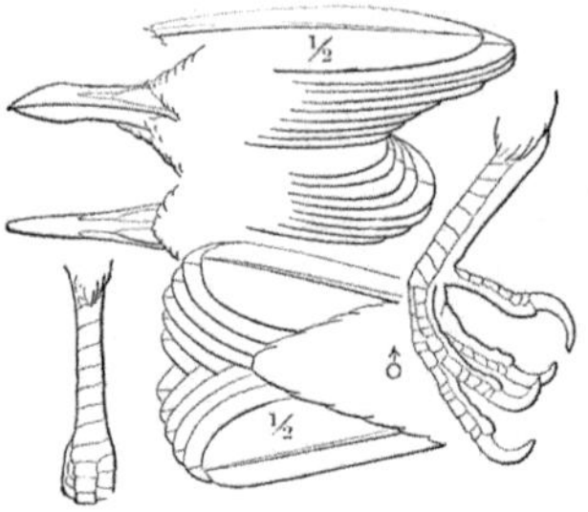

C. passerina.

broad; the tertials excessively lengthened; nearly as long as the primaries; quite equal to the first primary. Tail nearly as long as the wings; rounded laterally.

Chamæpelia passerina, LINNÆUS.

THE GROUND DOVE.

Columba passerina, LINNÆUS, Syst. Nat. I. 1766, 285. — WILSON, Am. Orn. IV. 1811, 15; pl. xlvi. — AUDUBON, Orn. Biog. II. 1834, 471; V. 1839, 558; pl. 182. IB. Birds Amer. V. 1842, 19; pl. 283. — NUTTALL, Man. I. 2d ed. 767. — *Chamæpelia passerina*, SWAINSON, Zool. Jour. iii. 1827, 358. — BAIRD, P. R. Rep. IX. Birds, 606. — COUES, Pr. A. N. Sc. 1866, 93.

SP. CHAR. Back, rump, exposed surface of tertials, and tail above, uniform grayish-olive; neck above and occiput tinged with bluish; forehead, sides of head and neck, under parts generally, and lesser upper wing coverts, light purplish-red, tinged with dusky towards the tail. Feathers of the head, neck, and fore-breast, margined with a darker shade of the ground color, the forehead and chin, only, nearly uniform. Feathers of the breast dusky brown in the centre, — this most conspicuous on the jugulum. Under wing coverts, axillars, and quills brownish orange; the latter margined externally, and tipped with dusky-brown, — the tertials almost entirely of this color. Middle tail feathers like the back; the others mostly black, the outer one edged towards the tip with white. The exposed surface of the wing variously marked with blotches exhibiting black, steel blue, and violet. Bill and feet yellow, the former tipped with brown.

Female with little or none of the purplish-red.

Length, 6 50; extent, 10.50; wing, 3.50; tail, 2.80.

Hab. South Atlantic and Gulf Coasts. Fort Yuma, California, to Fort Mojave? lat. 35°. Also, West Indies, Mexico, and Central America.

This pretty little dove, which is common in the South Atlantic States, Mexico, and the West Indies, undoubtedly visits Fort Yuma, and I have heard of what I supposed to be the same as far north in the Colorado Valley as Fort Mojave, lat. 35°, though I did not myself see them there.

In Florida they associate in small flocks, and come familiarly around the door to feed. I found their nests there, both on the ground and in trees, built like those of the common dove, and with two white eggs. They are in miniature very similar, both in appearance and habits, to the common pigeon, and are often kept in cages, — mostly to fatten for food.

Chamæpelia passerina, var. **pallescens,** BAIRD.

THE CAPE GROUND DOVE.

Chamæpelia passerina? var. *pallescens*, BAIRD, Phil. Acad. 1859, 305.

SP. CHAR. Shade of color considerably lighter than in *C. passerina*. Chin and anal region nearly white; the latter considerably lighter than the belly, instead of nearly the same color. Bill darker. Legs stouter.

Hab. Cape St. Lucas.

This bird, hardly a species, but a decided variety, has hitherto been only found at Cape St. Lucas, forming one of the many forms peculiar to that part of Lower California. The eggs and nest resemble those of *C. passerina*, and the habits are probably identical. (Baird.)

GENUS **SCARDAFELLA**, BONAPARTE.

Scardafella, BONAPARTE, Consp. Av. II. 1854, 85. (Type, *Columba squamosa*, TEMM.)

GEN. CHAR. Bill lengthened; culmen more than half the length of head measured from the frontal feathers. Feet as in *Chamæpelia*. Wings with the tertials nearly as

S. inca.

long as the primaries; shorter, however, than the first primary. Tail considerably longer than the wings; much graduated; the feathers narrow, linear, or tapering towards the end.

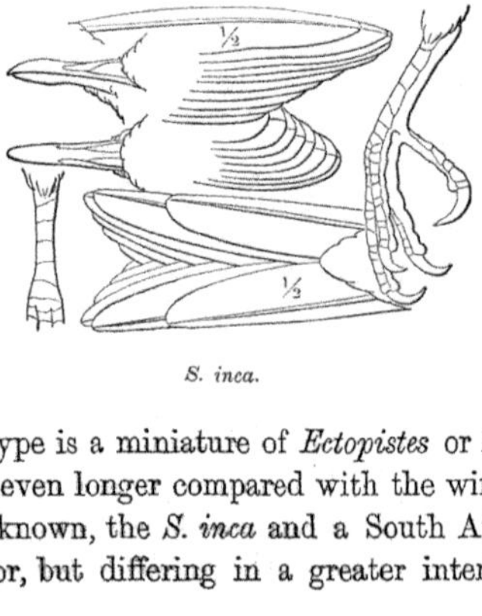

S. inca.

This remarkable type is a miniature of *Ectopistes* or *Zenaidura* in respect to the tail, which is even longer compared with the wings.

Two species are known, the *S. inca* and a South American ally, *S. squamosa*, similar in color, but differing in a greater intensity of the dark borders to the feathers.

Scardafella inca, Bonaparte.

THE INCA DOVE.

Scardafella squamosa, Baird, Birds N. Amer. 1858, 605. (Not of Temminck.)
Scardafella inca, Bonaparte, Consp. 1850. — Elliot, B. N. A. II. pl. 37. — Butcher, Pr. A. N. Sc. 1868.

Sp. Char. *Female.* Above ashy-olive, changing to purer ashy on the wings. Beneath ashy-white, changing on the breast and throat to pale violaceous. All the feathers on the head and body abruptly margined with dark brown, except on the forehead and

chin. All the quills, except the innermost tertials, orange-brown; the outer margins and tips dusky brown; the under coverts orange-brown; the axillars strongly tinged with sooty. Tail feathers blackish, tinged with gray above; all (except the innermost) broadly tipped with white; the exterior with the white extending backwards on the outer web. *Female.* Length, 8.00; wing, 3.75; tail, 410. Iris purple; bill black; feet flesh-color.

Hab. Valley of Rio Grande; westward to the Gulf of California, and south to Honduras and Guatemala.

This species, now known to be an inhabitant of Southern Texas, is not rare there, and is met with throughout Mexico; we have no special account of its habits, which are probably similar to those of *Chamæpelia*, though it is perhaps not so terrestrial. It has not yet been detected in Arizona. (Baird.)

Order GALLINÆ.

Char. Bill usually rather short and stout, and less than the head. Basal portion hard, generally covered with feathers, and not by a soft naked skin. Legs lengthened; the hind toe generally elevated above the level of the rest, and short; when lower down it is longer. Toes connected at the base by a membrane. The feathers of forehead not extending on the culmen in a point, but more restricted, and parted by the backward extension of the culmen.

This division, of which the common barn-yard fowl (*Gallus*) is the type, has numerous representatives in the tropics, most of which belong to families different from those found here.

The *Penelopidæ*, peculiar to Middle and South America, have the hind toe even with the others, as in most pigeons, and are mostly large birds known as Curassows, Guans, etc.

The *Phasianidæ*, including the common fowl, are mostly Asiatic, but have one genus in America, namely, *Meleagris*, including the turkeys, of which the *M. Mexicana*, found as far northwest as Tucson, is now believed to be the original stock of the domestic turkey.

Family PHASIANIDÆ, The Pheasants.

Char. Bill moderate, the legs, toes, and nasal fossæ, bare; the tarsus usually with one or more spurs, in the male. The hind toe elevated above the level of the others. Tail feathers more than twelve. Face generally more or less naked.

Of the entire family of *Phasianidæ*, as above described, but a single genus, *Meleagris*, belongs to America, the others being found entirely in the Old World. It includes the different pheasants, Jungle fowl, the domestic chicken, the turkeys, the peacocks, and other well-known birds, among them by far the most important and interesting species domesticated by man.

SUB-FAMILY MELEAGRINÆ.

CHAR. Tail moderate, truncate. Head and neck nearly naked, and more or less carunculated or with fleshy lobes.

The preceding diagnosis is quite sufficient to distinguish the *Meleagrinæ* of Gray from his other sub-families, the *Pavoninæ* having the tail and its coverts much developed and depressed, but broad and rounded; the *Phasianinæ* have the tail greatly lengthened and attenuated, cuneate, compressed; the *Gallinæ* have the tail moderate, arched, and compressed, the sides of the head only naked; and the *Lophophorinæ* have the head feathered, except immediately around the eye; the tail moderate, broad, and rather depressed.

M. gallopavo.

But one of these groups is represented by indigenous species in the New World, although species of all the others are familiar inhabitants of the barn-yard or the aviary. Three species of the genus *Meleagris* are known, — two closely allied, the third, *M. ocellata* of Yucatan, being quite different. This is one of the most beautiful of the gallinacea, in the brilliancy of its plumage almost rivalling the peacock, and especially in the eye-shaped spots on its feathers.

66

GENUS **MELEAGRIS**, LINNÆUS.

Meleagris, LINNÆUS, Syst. Nat. 1735. (Type, *Meleagris gallopavo*, LINN.)

SP. CHAR. Legs with transverse scutellæ before and behind; reticulated laterally. Tarsi with spurs. Tail rounded, rather long, usually of eighteen feathers. Forehead with

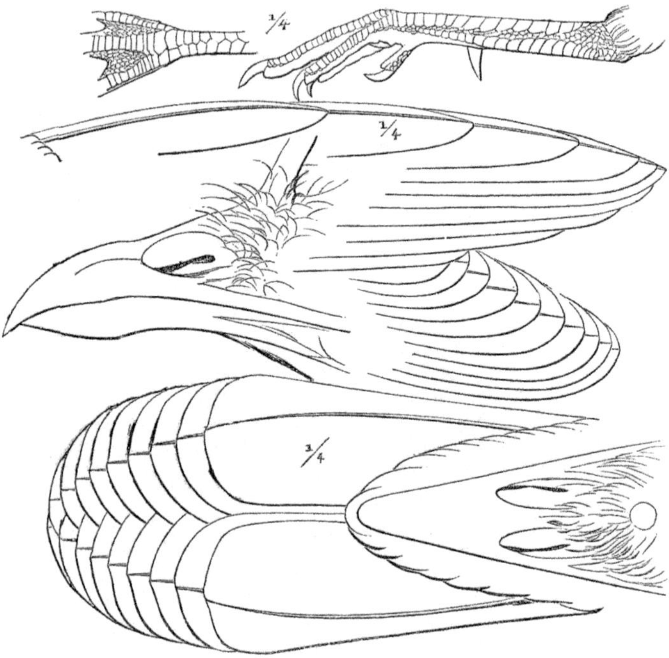

M. gallopavo.

a depending fleshy cone. Head and the upper half of the neck without feathers. Breast of male in most species with a long tuft of bristles.

The above diagnosis will be sufficient to distinguish the true turkeys from their allies, the nearest being *Numida*, according to most authors. Three species are known to naturalists; the wild turkey of the Eastern United States (*M. gallopavo*), the *M. Mexicana*, and the *M. ocellata*, or peacock turkey of Honduras. The latter species is the smaller, and by far the most varied in coloration, resembling the peacock in the pattern of some of its markings.

Meleagris Mexicana, Gould.

THE MEXICAN TURKEY.

Meleagris Mexicana, Gould, Pr. Zool. Soc. 1856, 61.—Baird, Birds N. Amer. 1858, 618. Ib. Rep. U. S. Ag. Dept. for 1866, 288.—Elliot, Ill. B. N. A. II. pl. 38.—Coues, Pr. A. N. Sc. 1866, 93.
Meleagris gallopavo, Gray, Catal. Br. Mus. Gallinæ, 1867, 42.

Sp. Char. Naked skin of head and neck livid blue; legs red; general color copper bronze, with copper and green reflections, each feather with a narrow black border. All the quills brown, closely barred with white. Tail feathers chestnut, narrowly barred with black, with a very broad subterminal black bar, and the tips light brownish-yellow or cream-color. Upper tail coverts also with pale, almost white tip.

Female smaller, and less brilliant.

Length, 48.50; wing, 19.00; tail, 14.60; tarsus, 6.50; bill from nostril, 1.04.

Hab. Southern Rocky Mountains, and from Western Texas to Arizona, south to Orizaba.

Persons familiar with the wild turkey of the United States, east of the Missouri Plains and of Eastern Texas, have been struck on going westward, after crossing the plains of the Pecos River, in Texas, and reaching New Mexico and Arizona, to find a wild turkey which, while presenting the general features of the Eastern bird, has yet various peculiarities, the most striking consisting in the creamy-white or fulvous tips to the tail feathers and the upper tail coverts, these being chestnut brown in the other, and in the white instead of dark flesh. Other differences, as the lighter shade of the chestnut, and the tendency to a greenish rather than purplish gloss, seem of less importance.

A careful consideration of the subject brings us to the conclusion that these two series form, if not distinct species, at least strongly marked and permanent races, and furthermore, that the Rocky Mountain bird, extending as it does southward through Mexico, is really the original of our domestic turkey. It is well known that at the period of the Spanish discovery the native turkey was widely domesticated in Mexico, and was introduced thence first into Europe, and then into North America. Furthermore, the native bird of Eastern North America does not occur in Mexico at all. The markings of the domestic turkey are sometimes exactly like those of wild birds of *M. Mexicana*, while they never assume the plumage of the wild *M. gallopavo* of the North.

According to Dr. Sartorius, the wild Mexican bird breeds in March or April, the female laying three to twelve spotted eggs, which are hatched out in thirty days. Dr. Coues and Dr. Palmer state that the species is not rare in Arizona, although difficult to procure. Some specimens were of enormous size, fully equalling the largest of the Eastern race.

FAMILY TETRAONIDÆ, THE GROUSE.

CHAR. The *Tetraonidæ* are characterized among gallinaceous birds by their densely feathered tarsi, and by the feathers of the nasal fossa or groove, which fill it completely, and conceal the nostrils. The toes are usually naked (feathered to the claws in the genus *Lagopus*), and with pectinations of scales along the edges. The number of tail feathers varies from sixteen to eighteen and even twenty; the tail is rounded, acute, or forked. The orbital region is generally somewhat bare, with a naked stripe above the upper eyelid, beset by short fringe-like processes.

There are two genera of this family found in North America, which do not occur in California. These are the *Cupidonia cupido*, or pinnated grouse, east of the Mississippi, and the species of *Lagopus* or ptarmigan, inhabiting the Arctic regions and snowy summits of mountains. The *L. leucurus*, Swainson, has been found about the snowy peaks near the Columbia River, and in the Rocky Mountains down to lat. 39°, but not in California. It may, however, yet be detected in the Sierra Nevada.

T. obscurus.

GENUS **TETRAO**, LINNÆUS.

Tetrao, LINNÆUS, Syst. Nat. 1744. (Type, *T. urogallus*, L., an European species.)

GEN. CHAR. Tail lengthened, slightly narrowed to the square or somewhat rounded tip; about two thirds the wing; the feathers with stiffened shafts. Tarsus feathered to

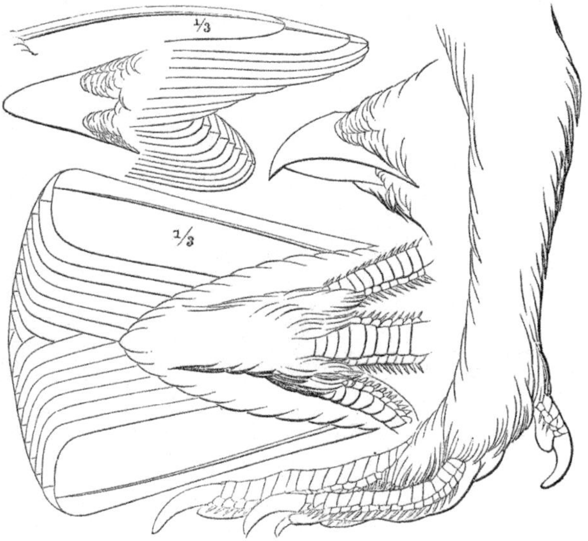

T. obscurus.

and between the bases of the toes. No unusual feathers on the side of throat. Culmen between the nasal fossæ nearly half the total length. Color mostly black.

Inhabit wooded regions.

Mr. D. G. Elliott, in Proc. Acad. Nat. Sc. of Phil. XVI. 23, Feb. 1864, proposes to separate the *T. obscurus* and *T. Richardsonii* from the genus *Tetrao* (which is then represented only in Europe), on account of their greater number of tail feathers (twenty), and other differences. The name proposed for the new genus is *Dendragapus,* from the Greek for tree-lover, *δένδρον, ἄγαπω.*

The other American species with sixteen tail feathers, etc. (*Canadensis* and *Franklinii*) had before been separated under the name of *Canace,* by Reichenbach.

Tetrao obscurus, Say.

THE DUSKY GROUSE.

Tetrao obscurus, Say, Long's Exped. R. Mts. II. 1823, 14. — Bonaparte, Am. Orn. III. 1830; pl. xviii. — Swainson, Fauna Bor. Amer. II. 1831, 344; pl. lix., lx. — Nuttall, Man. I. 1832, 666. — Audubon, Orn. Biog. IV. 1838, 446; pl. 361. Ib. Birds Amer. I. 1842, 89; pl. 295. — Newberry, P. R. Rep. VI. iv. 1857, 93. — Baird, P. R. Rep. IX. Birds, 620. — Heermann, X. vi. 61. — Cooper and Suckley, XII. iii. Zool. of W. T. 219. — Lord, Pr. R. Art. Inst. Woolwich, I. 1863, 122. (British Columbia.) — *Canace obscura*, Bonaparte, Comptes Rendus, XLV. 1857, 428. — *Dendragapus obscurus*, Elliot, Pr. A. N. Sc. 1864, 23. Ib. Mon. Tetraonidæ, plate.

Sp. Char. Sexes dissimilar. Above bluish-black; plumbeous or black beneath. Tail uniform black, finely and obscurely mottled above, and broadly tipped with light slate. Beneath uniform plumbeous. A dusky half-collar on the throat. The chin and throat above white, varied with black. Tail about two thirds the length of the wings, broad,

Male.

rounded, composed of twenty broad, even, and truncated feathers. Tarsi feathered to the toes, the feathers extending along the sides of the basal half of the first joints of the toes. Pectinations on the sides of the toes very short. Length, 19.00; extent, 28.00; wing, 9.50; tail, 7.50. Iris brown; bill horn-color; feet grayish.

Hab. Oregon and Washington; through the mountains of Northern California, and eastward to the Black Hills and southward to New Mexico.

This fine game-bird is common in Oregon and Northern California, extending in the Coast Range nearly to San Francisco Bay, and in the Sierra Nevada to about lat. 38°. They are brought to market in winter from the mountains near Napa, and are said to come down at times into the valleys, but have never been met with in California south of San Francisco.

In the Sierra Nevada, lat. 39°, I found them rather scarce, and in September only above an elevation of six thousand feet, but was informed that they went down much lower in winter, probably about as far as the snows fall, or to about two thousand feet, in that latitude.

I think their range is more dependent on the prevalence of spruce and other dense coniferous forests than on the climate, as it is much milder near Napa than at that elevation in the Sierra Nevada, and towards the north they frequent valleys. I have seen them near the Columbia River at all seasons, usually inhabiting the dense forests, and therefore hard to shoot.

Female.

They, however, come out on the borders of prairies and openings when not molested, especially in the early morning, and then afford a fine shot for the sportsman, as they fly straightforward on being flushed, only requiring a quick eye and heavy shot to bring them down. If in or near the woods, they often alight in a tree, and conceal themselves in the dark foliage so perfectly as to require a long search to find them. They stand so perfectly motionless at such times as to be mistaken often for a bunch of dry leaves, and I have seen them fly off from a tree where I had been looking for them in vain for some time. In flying downward they usually sail silently with spread wings, but in starting from the ground make a loud whirring like other grouse.

In spring the males may be heard " tooting " in the trees, — making a low hollow sound, something like the cooing of a pigeon, and which seems to come from some place quite distant from where the bird really is.

Their food consists of various berries, nuts, and seeds, besides grain around farms. Their eggs are numerous, laid in a slightly constructed nest on the ground, and of an ashy-brown color, blunt at both ends. (Nuttall.) The young at Puget's Sound are full grown about August, and probably earlier in California. They are then easily killed, and very white and tender, being fit for shooting about September 1st. In the Sierra Nevada they frequented such difficult ground, covered with dense thickets, and on steep slopes of the mountains, that I did not succeed in shooting any, but had a good view of one, quite sufficient to determine its species, as it flew down from the top of a lofty spruce, and sailed half a mile away, while I was looking down on it from the top of a mountain ridge.

In winter they live much on the buds of trees, even those of the coniferæ, and in Oregon are seen so seldom on the ground at that season that they have been supposed by some to emigrate. Several may sometimes be shot from one tree by taking care to kill the lowest ones first. They weigh three or four pounds.

Tetrao Richardsonii, Douglas.

RICHARDSON'S GROUSE.

Tetrao Richardsonii, Douglas, Linn. Trans. XVI. 141. — Wilson, Zool. Illust. pl. 30, 31. — Gray, Cat. Gallinæ, Br. Mus. 1867, 86. — Lord, Pr. R. Art. Inst. I. 1863, 122. — *Dendragapus Richardsonii*, Elliot, Mon. Tetraonidæ, plate.

Tetrao obscurus, Swainson and Rich. Fauna Bor. Amer. II. 1831, 344; pl. 59, 60. (Also of Audubon in part.)

Sp. Char. Similar to *T. obscurus;* differs in having the tail more square at end, and entirely black, or without the gray tip.

Hab. Central Rocky Mountains, from South Pass, and northward to Fort Liard. H. B. T.

This bird appears, in general habits and characteristics, to be undistinguishable from the common dusky or blue grouse, except as stated above. Its distribution is quite peculiar, being encircled as it were by the *T. obscurus*, to the east, west, and south, and reaching northward at least as far as Fort Liard, west of Fort Simpson, perhaps as far as the forest extends. It is extremely abundant at Fort Liard, a post of the Hudson's Bay Company, situated on the Liard River, and not very far from Fort Simpson, likewise one of the company's posts on the Mackenzie River. At certain seasons of the year its meat is excellent, and much sought after by the Indians and whites. With the exception of the tail, there appears but little to separate it from its ally, *Tetrao obscurus*, and it was described as the last-mentioned species in the "Fauna Boreali Americana" of Swainson and Richardson, and the two again confused by Audubon. (Baird.)

Tetrao Franklinii, Douglas.

FRANKLIN'S GROUSE.

Tetrao Franklinii, Douglas, Trans. Linn. Soc. XVI. 1829, 139. — Rich. F. Bor. Amer. II. 1831, 348; pl. lxi. — Gray, Cat. Gallinæ, 86. — Baird, Birds N. Amer. 1858, 623. — Cooper and Suckley, N. H. Wash. Terr. 220. — *Canace Franklinii*, Elliot, Pr. A. N. Sc. 1864, 23. Ib. Mon. Tetraonidæ, plate.

Tetrao Canadensis, var. Bonaparte, Am. Orn. III. 1830, 47; pl. xx.

? *Tetrao fusca*, Ord, Guthrie's Geog. 2d Am. ed. II. 1815, 317. Based on "Small brown pheasant" of Lewis and Clark, II. 182, which very probably is this species.

Sp. Char. Prevailing color in the male black; each feather of the head, neck, and upper parts generally, having its surface waved with plumbeous gray. This is in the form of two or three well-defined concentric bars parallel to each other, one along the exterior edge of the feather, the others behind it. The sides of the body, the scapulars, and outer surface of the wings are mottled like the back, but more irregularly, and with a browner shade of gray, the feathers with a central white streak expanding towards the tip (on the wing these streaks seen only on some of the greater coverts). There is no white above, except as described. The under parts are mostly uniform black, the feathers of the sides

Male.

of the belly and breast broadly tipped with white, which sometimes forms a pectoral band. There is a white bar across the feathers at the base of the upper mandible, usually interrupted above; a white spot on the lower eyelid, and a white line beginning on the cheeks and running into a series of white spots in the feathers of the throat, the lower feathers of which are banded terminally with whitish. The feathers at the base of the bill, and the head below the eyes and beneath, are pure black. The quills are dark brown, without any spots or bands, the outer edges only mottled with grayish. The tail feathers are similar, but darker, and the tail is entirely black to the tip. Upper tail covert broadly tipped with white; feathers of the legs mottled brown and whitish; dirty white behind the tarsi. The bill is black.

The female is smaller but somewhat similar, the black bars above broader, the inner gray bars of each feather, including the tail, replaced by broader ones of brownish-orange. The under parts have the feathers black, barred with the brownish-orange, which, on the

Female.

tips of the belly feathers, is pure white. The clear continuous black of the head and breast are wanting. The scapulars, greater coverts, and sides are streaked as in the male. The tail is tipped with whitish, as are the upper coverts.

Length, about 15 00; wing, 7.35; tail, 5.62.

Hab. Northern Rocky Mountains of the United States, and westward through Washington Territory to the Cascade Mountains.

This interesting variety of the Canada grouse, if it be not a distinct species, replaces the other form in the Rocky Mountains of the United States, but does not seem to extend much northward, as all specimens received at the Smithsonian Institution, from the Mackenzie River region, and Alaska, including Sitka, are true *T. Canadensis.* The differences consist mainly in a rather longer, broader, and more truncated tail, which is pure black to the ends, instead of being tipped with brownish-orange. The upper tail coverts also are sharply ended with white. The female has the tail tipped with whitish, instead of orange-brown.

The species is abundant in the Rocky and Bitter Root Mountains, and extends to the Cascade Range. It lives among the spruce and pine vegetation, principally in swampy tracts, feeding on leaves and buds of the coniferæ. The eggs are light chocolate-brown, spotted with darker, much like those of the Canada grouse. (Baird.)

GENUS **PEDIŒCETES**, BAIRD.

Pediœcetes, BAIRD, P. R. Rep. IX. Birds, 1858, 625. (Type, *Tetrao phasianellus.*)

GEN. CHAR. Tail short, graduated, the shorter part of it half the length of the full rounded wing. Tarsi densely feathered to the toes and between their bases. Neck without peculiar feathers. Culmen between nasal fossæ not half the length of bill. (Elliot.)

P phasianellus

Two species of the genus are recognized by later authors.

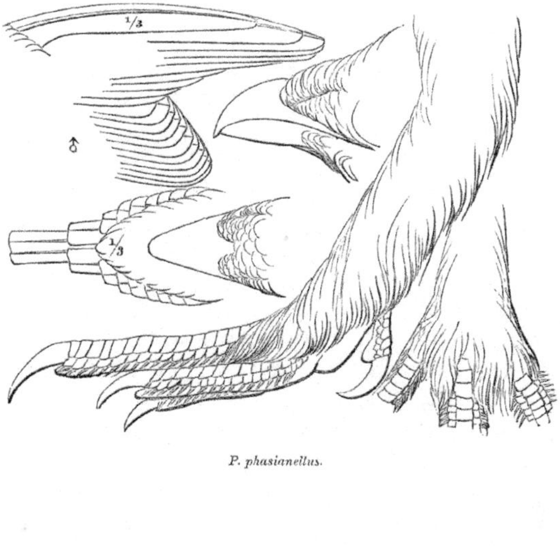

P. phasianellus.

Pediœcetes Columbianus, Ord.

THE SHARP-TAILED GROUSE, OR PRAIRIE CHICKEN.

Phasianus Columbianus, Ord, Guth. Geog. 2d Am. ed. II. 1815, 317. — *Pediœcetes Columbianus*, (Ord) Elliot, Pr. A. N. Sc. Phil. XIV. 1862, 403. Ib. Mon. Tetraonidæ, plate.

Tetrao phasianellus, Ord, same place, and of all other authors since 1800, except Bonaparte, Amer. Orn. III. (text only), which is true *phasianellus*, Linnæus (Elliot). Figures (all of *Columbianus*), Bonaparte, Amer. Orn. III. pl. xix. (supposed female). — Audubon, Orn. Biog. IV. pl. 382, p. 569. Ib. Birds Amer. V. 110; pl. 298. Ib. Synopsis, 205. — Nuttall, Man. I. 669. — Newberry, P. R. Rep. VI. iv. 1857, 94. — Swainson and Rich. F. Bor. Am. II. 361. — *Pediœcetes phasianellus*, Baird, P. R. Rep. IX. Birds, 1858, 626. — Cooper and Suckley, XII. iii. Zool. of W. T. 223.

Sp. Char. General color white and brownish-yellow with irregular black markings. Beneath pure white, the feathers on the breast and flanks with brown V-shaped markings. Throat buff. (Elliot.) Length, 18.00; extent, 28.00; wing, 8.50; tail, 5.24.

Hab. Northern prairies and plains from Wisconsin, west to Cascade Mountains and Sierra Nevada.

Though well known as a game-bird, and a delicate luxury for the table, in the Upper Mississippi Valley (where it is confounded by sportsmen generally with the pinnated grouse, *Cupidonia cupido*), this prairie-chicken has not yet been enrolled among the game-birds of California, on account of the remote and thinly inhabited part of the State which it inhabits. It is worthy, however, of an attempt at naturalization in the lower and western districts. From what I have seen of their resorts and habits along the Upper Columbia, I see no reason why they cannot flourish in many of the valleys west of the Sierra Nevada, unless there is some unknown agent fatal to them either when young or full grown, such as the peculiarities of the climate or some poisonous plant or seed which they eat. Their eggs might, at any rate, be easily obtained and hatched under a hen for trial.

According to Dr. Newberry, the prairie-chicken is first met with in California, near Canoe Creek, about fifty miles northeast of Fort Reading, on the east side of the Sierra Nevada. Thence they are abundant north and eastward, and I think extend south as far as lat. 39°, though this is somewhat uncertain. I have not heard of them in the southeastern parts of California, though there are tracts apparently suited for them, such as Owen's Valley. Towards the Columbia they are numerous in all the grassy prairies about the upper part of the river east of the Cascade Range. It has been lately ascertained that their range to the north ends about lat. 49° ; Mr. Kennicott having obtained specimens from British America which prove different, and turn out to be the true *T. phasianellus* of Linnæus, though described also as *P. Kennicottii* by Suckley. It is a resident species throughout all its range, or only descends from the high lands to the warmer valleys in the cold months.

I have seen the nest of this species near the Upper Missouri River in May. It is a mere cavity in the grass under a log or bush, lined with a few feathers ; the eggs are about twelve, white, obscurely spotted with brownish. In the spring the male is said by Nuttall to utter a shrill but rather feeble crowing ; and at all times when started they have a loud cry of *kuk-kuk-kuk*, which assists the whirring of their wings and their unexpected appearance in throwing the inexperienced sportsman off his guard.

In the autumn they often congregate about favorite spots in great numbers, and where not much hunted furnish excellent sport for a cool and quick shot. They usually scatter in the morning among the long grass in the neighborhood of thickets, and lie so close as to be almost trodden under foot before they fly, when they rise with a few whirring flaps of the wing to a height of about six feet, and then sail off direct for a hundred yards or more, unless they are much hunted, when they make for the nearest thicket. Sometimes, when unused to the hunter's devices, they merely fly into a low tree, and, as if perfectly safe there, look down coolly while several are successively shot around them, the only precaution being to shoot the lowest first, so that their fall may not alarm the others.

In frosty and snowy weather they often sun themselves in the morning on the trees, and live much in winter on the buds of the alder, poplar, etc.

We give above a figure of the head of the *P. phasianellus* to illustrate the supposed differences between this race and *P. Columbianus*.

GENUS **CENTROCERCUS**, SWAINSON.

Centrocercus, SWAINSON, F. Bor. Amer. II. 1831, 496. (Type, *Tetrao urophasianus*, BONAPARTE.)

GEN. CHAR. Tail excessively lengthened, cuneate, longer than the wings, the feathers all lanceolate. Tarsi feathered to the joint and between the bases of the toes. Lower

C. urophasianus.

throat and its sides with stiffened spinous feathers. Wings with the feathers rather sharp; the primaries longer than the secondaries. Nasal fossæ extending very far for-

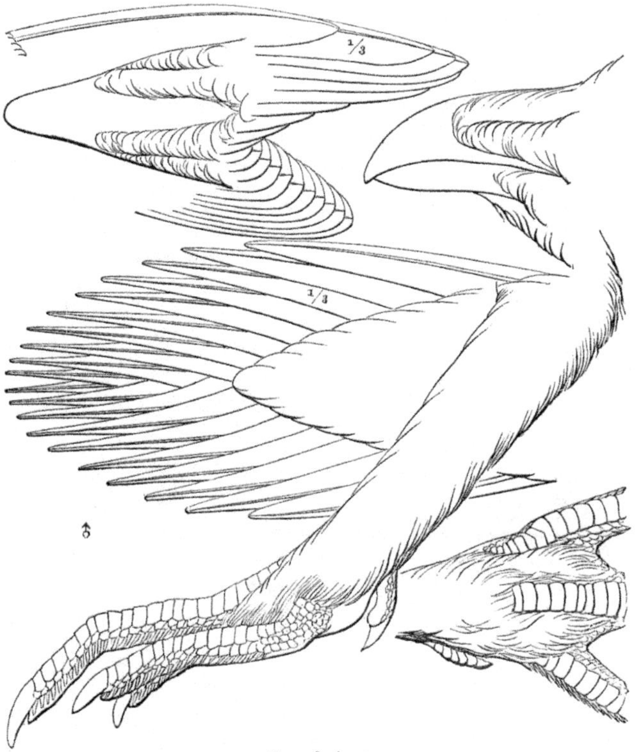

C. urophasianus.

ward; the length of culmen between them two thirds the total length. Color mottled yellowish above, with large black patches beneath.

But one species is yet known, though the "sage fowls," said to inhabit Lower California near lat. 32°, may very possibly prove distinct, few of this family having such a long range north and south as from 32° to 49°.

The genus *Centrocercus* presents several peculiarities among its allies, as shown by the above cut, that readily distinguish it, and embraces much the largest species of true grouse found in America. This form of grouse is less known to sportsmen than any of the others belonging to the fauna of the New World, as it only makes its first appearance at a considerable distance to the west of the Mississippi.

Centrocercus urophasianus, Bonaparte.

THE SAGE-COCK; COCK OF THE PLAINS.

Tetrao urophasianus, Bonaparte, Zool. Jour. III. Jan. 1828, 214. Ib. Am. Orn. III. 1830; pl. xxi. f. 1. — Wilson, Illus. 1831; pl. 26, 27. — Nuttall, Man. I. 1832, 666. — Audubon, Orn. Biog. IV. 1838, 503; pl. 371. Ib. Birds Amer. V. 1842, 106; pl. 297. — Newberry, Zool. Cal. and Or. Route, Rep. P. R. R. Surv. VI. iv. 1857, 95. — Maximilian, Cab. Jour. 1858, 431. — *Tetrao (Centrocercus) urophasianus*, Swainson, F. Bor. Amer. II. 1831, 358; pl. lviii. — *Centrocercus urophasianus*, Baird, P. R. Rep. IX. Birds, 624. — Cooper and Suckley, XII. iii. Zool. of W. T. 222.

Sp. Char. Tail feathers twenty. Above varied with black and brownish-yellow; coverts having all the feathers streaked with the latter. Beneath black; the breast white, the upper feathers with spiny shafts, the lower streaked with black; tail coverts

Male.

with white tips; the sides also with much white. Length, 29.00; extent, 42.00; wing, 12.00; tail, 11.50. Iris of adult yellow, of young brown; bill and toes blackish.

Hab. Sage plains of the northwest, to Eastern California. South to Lower California?

Although the largest and most beautiful of American grouse, this species has not been much noticed as a game-bird, on account of the barren, uninhabited regions it frequents, and also from the bitterness of its flesh, caused by feeding on the "wild sage" (*Artemisia*). It is said, however, that the latter objection may be overcome by stripping off the skin as soon as the bird is killed, and emptying the contents of the crop and stomach. Probably if trapped and fed for a few days this trouble would be entirely obviated.

I have not myself seen this bird in California, but there is a fine specimen in the possession of Mr. Dillon, of Oakland, which he says was shot near the Mojave River. I saw what I thought must be the tracks of this bird among the scrub-oaks near the head-waters of that river, and have heard of birds resembling them having been seen near the boundary of Lower California, — a country similar to that inhabited by them east of the Sierra. I have heard that they are common near Virginia City, Nevada.

Female.

Dr. Newberry found them common in the northeast corner of California, along Pit River, near "Round Valley," and thence abundant towards the north, and east of the Cascade Mountains. In the plains north of the Upper Columbia I found some of them in 1853, but they were not abundant, and in 1860 I saw none there, while Mr. J. Hepburn, who visited that district especially in search of them, was equally unsuccessful. Nuttall, however, speaks of seeing flocks of hundreds of them south of Snake River, and their chief range is probably in Eastern Oregon, Utah, and Nevada. They are only found in those barren, sandy, and almost waterless districts where the chief vegetation is *Artemisia*, and other plants which cattle will not touch. They feed on the leaves of this bitter shrub, as well as others, and probably on various seeds of grasses, etc.

68

They do not fly readily, but run; and if they think they are not seen, squat close to the ground, allowing travellers to pass very near without flying, and often without being detected, although scarcely concealed by vegetation. When they fly it is sometimes with a loud whir, sometimes without much noise, and they generally continue their flight for a long distance before they alight.

When they fly they make a cackling noise, very much like that of the common fowl, when flushed, but rather hoarser. Their flight is rather slow and laborious, compared to that of other grouse.

According to Nuttall, they "pair" in March and April, but are also polygamous, like most of the order. At that time they assemble on eminences near the banks of streams, the male lowering his wings and strutting about with a humming sound, the wings dragging on the ground, tail spread out like a fan, and the bare skin of the breast inflated like an orange, this being the apparatus that produces the hollow sound, like blowing into a hollow cane. The blue grouse has this blowing organ also connected with the windpipe.

The nest is on the ground, made of dry grass and slender twigs, under low bushes or tufts of high grass. The eggs, from thirteen to seventeen, about the size of hens'-eggs, are wood-brown, with irregular chocolate blotches at the thick end. They are hatched in twenty-one or twenty-two days, and the young at once run about. (Nuttall.) According to Newberry, the male weighs five or six pounds; the female is a third smaller, as usual in some birds of this family.

B. umbellus.

GENUS **BONASA**, STEPHENS.

Bonasa, STEPHENS, Shaw's Gen. Zool. XI. 1819. (Type, *Tetrao bonasia*, L., of Europe.)
Tetrastes, KEYSERLING and BLASIUS, Wirbelthiere Europas, 1840, lxiv.

GEN. CHAR. Tail widening to the end, its feathers very broad, as long as the wings; the feathers soft, and eighteen in number. Tarsi naked in the lower half; covered with two rows of hexagonal scales anteriorly, as in the *Ortyginæ*. Sides of toes strongly pectinated. Naked space on the side of throat covered by a tuft of broad soft feathers. Portion of culmen between the nasal fossæ about one third the total length. Top of head with a soft crest.

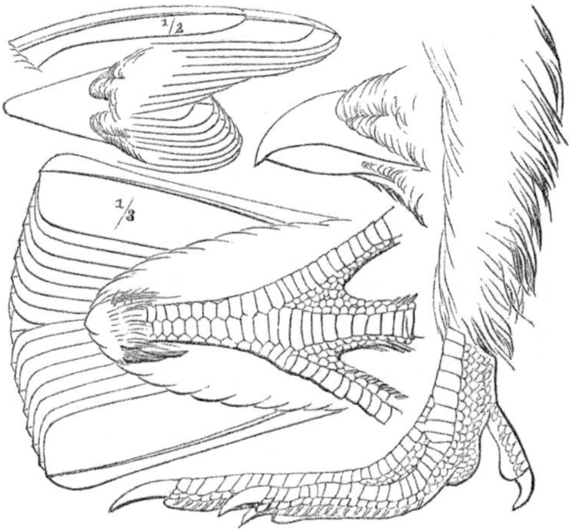

B. umbellus.

The ruffed grouse — "partridge," or "pheasant," of the Atlantic States — is different, though very much like ours. That of the Northern Rocky Mountains is probably also different, being very pale gray, with but little brown in its plumage. (*B. umbelloides*, Douglas.) There is also one species at least in Europe. The genus, in its partly naked tarsi, with two rows of scutellæ anteriorly, indicates a close degree of affinity to the American partridges or quails, as will be seen by a comparison of the generic illustrations of the two forms. Both have the flesh white and tender, and are equally sought after as a delicacy.

Bonasa Sabinii, BAIRD.

THE OREGON GROUSE.

Tetrao Sabinii, DOUGLAS, Trans. Linn. Soc. XVI. 1829, 137. — RICH. F. Bor. Amer. II. 1831, 343. — BAIRD, P. R. Rep. Birds, IX. 631. — COOPER and SUCKLEY, XII. iii. Zool. of W. T. 224. — LORD, Pr. R. Art. Inst. 1864, 123. — ELLIOT, Mon. Tetraonidæ, plate.

Tetrao umbellus, RICH, F. Bor. Amer. II. 1831, 342. — NEWBERRY, Zool. Cal. and Or. Route, Rep. P. R. R. Surv. VI. iv. 1857, 94. And of other authors on the West Coast species (not of Linnæus).

SP. CHAR. Dark orange-chestnut above; the back with cordate spots of lighter. Beneath reddish-yellow, transversely barred with dull brown. Tail tipped with gray, and

with a subterminal bar of black. Broad feathers of the ruff black. Length, 18.00; extent, 23.00; wing, 7.30; tail, 6.70; middle toe, 2.30. Iris brown; bill and toes horn-color.

Hab. Rocky Mountains to Pacific Coast of Oregon and Washington. Northern California?

Although this bird has not yet been obtained in California, as far as I can learn, it is so abundant at the Columbia River and in Oregon, that I think it must inhabit the northern part of this State, where there are districts well suited for its habits. It is an inhabitant of the forests, especially those of deciduous trees along streams, and about the borders of prairies, but never ventures far from the woods. At times they feed about grain-fields, and early in the morning are fond of dusting and sunning themselves on roads. From the dense covert they usually inhabit they are not easy to shoot, but often alight in trees, and if quickly shot at, give time for killing them before flying.

In January and February the males may be heard "drumming" often during the day, and sometimes at night. This they do by standing on a log and beating it with their short concave wings, beginning slowly, and gradually striking more rapidly, until the sound becomes a hollow whir. They may sometimes be seen doing this, but are very difficult to approach in the woods. They make no other sound, except a cackling, sometimes when they fly, and the low notes of the female when in charge of her young. The Eastern species (*B. umbellus*), very similar to ours, lays ten to fifteen dull yellowish eggs, in a depression among the dead leaves near a log or bush. The young fly in about a week, and at Puget's Sound are hatched in May.

According to Lord, in his "Naturalist in Vancouver's Island," etc., this species forms a nest in May, on the ground under a fallen log or bush, of dry leaves, lined with grass, bits of moss, and a few feathers, laying ten to fourteen eggs, of a dirty whitish, unspotted. He found about ten nests in one swamp near the Spokan Prairie, Washington Territory.

It is somewhat remarkable that the ruffed grouse of the Lower Yukon, in Alaska, should belong to the variety *B. umbelloides*, rather than to one so characteristic of the coast region of Washington and Oregon.

Lagopus albus.

GENUS **LAGOPUS**, VIEILLOT.

Lagopus, VIEILLOT, Analyse, 1816. (Type, *Tetrao lagopus*, L.)

GEN. CHAR. Nasal groove densely clothed with feathers. Tail of sixteen or eighteen feathers. Legs feathered densely to the claws. All the American species becoming white in winter.

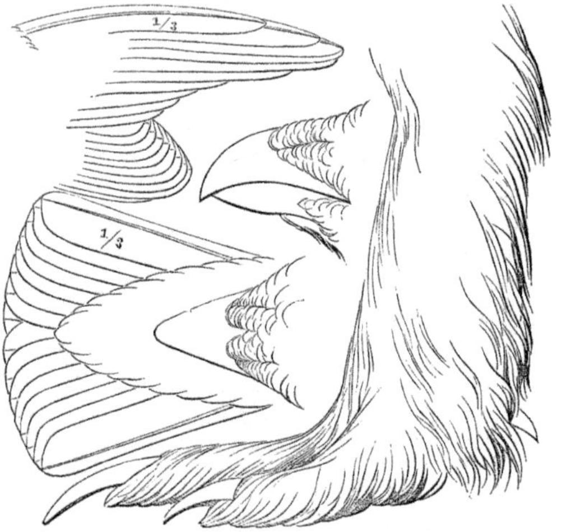

L. albus.

Of this genus three, if not four, well-marked species occur in North America, belonging principally to the high north, and especially abundant in and near the Arctic circle.

Lagopus leucurus, SWAINSON.

THE WHITE-TAILED PTARMIGAN.

Tetrao (*Lagopus*) *leucurus*, SWAINSON, Fauna Bor. Amer. II. 1831, 356 ; pl. lxiii. — NUTTALL, Man. II. 1834, 612. IB. I. 2d ed. 1840, 820. — BAIRD, Birds N. Amer. 1858, 636. — GRAY, Cat. Gallinæ, Br. Mus. 1867, 93. — *Tetrao leucurus*, AUDUBON, Orn. Biog. V. 1839, 200 ; pl. 418. — *Lagopus leucurus*, AUDUBON, Syn. 1839. IB. Birds. Amer. V. 1842, 125 ; pl. 302. — ELLIOT, Mon. Tetraonidæ, plate.

SP. CHAR. Bill slender. Plumage in summer barred with brownish-yellow. In winter pure white, including the tail feathers. Length, 13 00 ; wing, 7.00 ; tail, 4.25.

Hab. High alpine regions of the Rocky Mountains, from New Mexico to the Arctic Ocean. Also highest peaks of Washington Territory, and British Columbia.

The fact that the tail feathers, which on the other species of *Lagopus* are always black throughout the year, are here entirely white, will serve to distinguish the species at once from all its congeners. Formerly but little known in the United States, and its first introduction based on specimens collected in the Cochetope Pass of the Rocky Mountains (lat. 39°), by Captain Marcy, in 1858, this species has now come to be a familiar member of the alpine regions of the Rocky Mountains, as depicted by artists and authors. (Baird.)

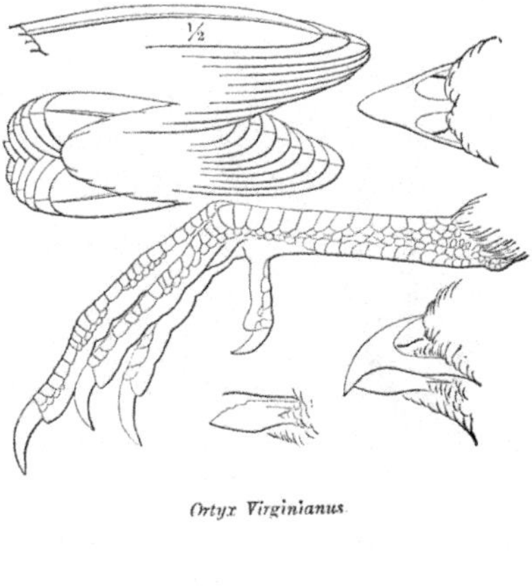

Ortyx Virginianus.

FAMILY PERDICIDÆ, THE PARTRIDGES.

CHAR. Nostrils protected by a naked scale. The tarsi bare and scutellate.

The partridges differ from the grouse in the bare legs and naked nasal grooves. They have a very extended distribution, being found, in one genus or another, over almost the whole globe. The Old and the New World forms are quite different, the latter forming the sub-family *Ortygina*.

SUB-FAMILY ORTYGINÆ.

CHAR. Bill stout; the lower mandible more or less bidentate on each side near the end.

The toothed or serrated character of the edge of the lower jaw, although an apparently trifling feature, yet marks a constant distinction from the smooth-edged bills of the Old World sub-families. There are several genera recognized by naturalists, many of them represented by species of the United States, and the total number of species is nearly fifty. They have a close resemblance to each other, and the differences of form pointed out

Ortyx Virginianus.

appear, in many cases, to be specific rather than generic. The crested quails are principally Californian.

The type of this sub-family is the well-known quail of the Atlantic States, of which a figure is given on the preceding page.

GENUS **OREORTYX**, BAIRD.

Oreortyx, BAIRD, Birds N. Amer. 1859, 642. (Type, *Ortyx picta*.)

GEN. CHAR. Body stout, broad; bill large; crest as in *Lophortyx;* tail short, broad, scarcely more than half the wing, rounded, the longest feathers not much exceeding the

O. pictus.

coverts. Legs developed, the claws extending beyond the tip of the tail; the lateral toes short, the outer claw falling considerably short of the base of the middle. Very similar to *Ortyx*, except in the crest.

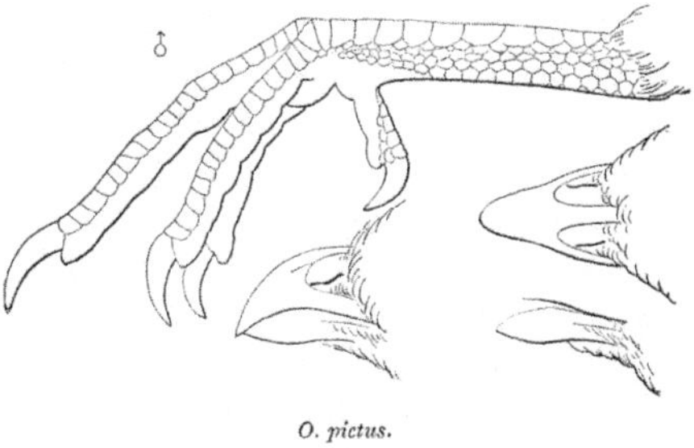

O. pictus.

The Eastern quails belong to the genus *Ortyx*, Stephens, and are not crested.

Oreortyx pictus, Douglas.

THE PLUMED PARTRIDGE; THE MOUNTAIN QUAIL.

Ortyx picta, Douglas, Trans. Linn. Soc. XVI. 1829, 143. — *Callipepla picta*, Gould, Mon. Odont. pl. xv. — Newberry, P. R. Rep. VI. iv. 1857, 93. — Heermann, X. vi. 61. — *Oreortyx pictus*, Baird, P. R. Rep. IX. Birds, 642. — Cooper and Suckley, XII. iii. Zool. of W. T. 225.

Ortyx plumifera, Gould, Pr. Zool. Soc. V. 1837, 42. — Audubon, Birds Amer. V. 1842, 69; pl. 291. — *Perdix plumifera*, Audubon, Orn. Biog. V. 1839, 220; pl. 422. — *Lophortyx plumifera*, Nuttall, Man. I. 2d ed. 1840, 791.

Sp. Char. Head with a crest of two straight feathers, much longer than the bill and head. Anterior half of the body grayish-plumbeous; the upper parts generally olivaceous brown, with a slight shade of rufous, this extending narrowly along the nape to the crest. Head beneath the eyes and throat orange-chestnut, bordered along the orbits, and a short distance behind by black, bounded anteriorly and superiorly by white, of which color is a short line behind the eye. Posterior half of the body beneath white, a large central patch

anteriorly (bifurcating behind), with the flanks and tibial feathers orange-chestnut-brown, the sides of body showing black and white bands, the former color tinged with chestnut. Upper tail coverts black, streaked with orange-chestnut. Upper tertials margined internally with whitish. Length, 11.50; extent, 18.00; wing, 5.75; tail, 3.25. Iris brown; bill bluish horn-color; feet brownish-white.

Hab. Mountain ranges of California and Oregon. Not in Coast Range south of San Francisco.

This bird, one of the most beautiful of its family, is common in the higher mountain ranges of California and Oregon, and I think a few are found north of the Columbia. South of San Francisco they are unknown near the coast, unless some birds seen by members of the survey in the Mount Diablo Range at an elevation of over three thousand feet, were of this species. In the Sierra Nevada they have been obtained at Fort Tejon, about four thousand feet elevation, and seen at Cajon Pass in winter, about the same elevation, and in lat. 34°. They probably extend farther south in the mountains. At lat. 39° they descend in winter to about three thousand feet, and are found lower towards the north, until in Oregon they frequent the borders of the Willamette Valley, but little above the sea-level. They are not common anywhere within the range of the gunners who supply the San Francisco market; all those I have seen there having been brought alive from the Sierra Nevada. They abound in summer up to seven thousand feet in that range, where I found them in September, the young not quite full grown, and the old birds moulting. In habits and flight they have considerable resemblance to our other quails, but their cries are quite different. Their note of alarm is a rather faint chirp, scarcely warning the sportsman of their presence before they fly. They scatter in all directions when flushed, and then call each other together by a whistle, very much like that of a man calling his dog. According to Newberry, the hen has a *cluck*, much like that of the common hen, when calling together her young brood about the first of August. The chickens also uttered a piping note, scattering and concealing themselves in the grass. Their notes and habits in spring have not yet been described.

They do not seem anywhere to associate in flocks of more than fifteen or twenty, and from the rugged, shrubby character of the country they inhabit they are not easily shot, except in the early morning, when they come out into the roads and openings to feed. They live on seeds, berries, and insects, and are very good for the table. As with the other species, more are taken in traps than by the gun. When hunted in the thick brush they generally run some distance before flying, and then rise singly, scattering so that only one can be killed at a shot. I have never seen them perch in trees, like the other species.

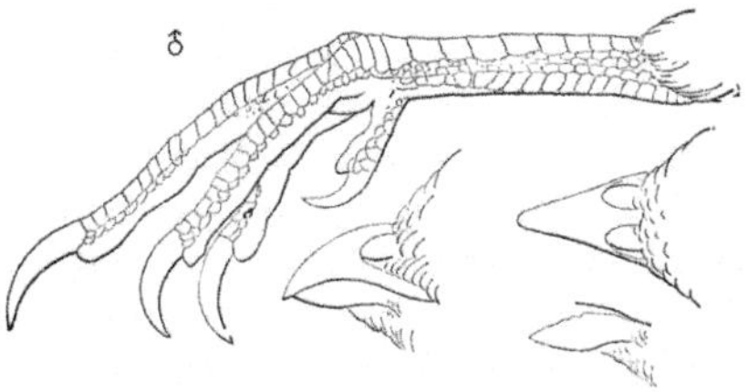

Lophortyx Californicus.

Genus LOPHORTYX, Bonaparte.

Lophortyx, Bonaparte, Geog. and Comp. List, 1838. (Type, *Tetrao Californicus*, Shaw.)

Gen. Char. Head with a crest of lengthened feathers springing from the vertex, the shafts in the same vertical plane, the webs roof-shaped, and overlapping each other; the number varies from two to six or more; they widen to the tip, where they are slightly recurved. Tail lengthened and graduated; nearly as long as the wing, composed of twelve stiff feathers. Wings with the tertials not as long as the primaries; the coverts without any unusual development; claws rather short; the lateral reaching to, but scarcely beyond the base of the middle; the outstretched toe not reaching the tip of the tail.

L. Californicus.

Several other species of this genus are found in Mexico and Lower California. Of these *Lophortyx neoxenus*, Vigors (Nuttall); *cristatus*, Linnæus; *L. elegans*, Nuttall; *L. Douglasii*, Bonaparte; and "*Ortyx*" *fasciatus*, Natterer, have been attributed to Upper California, but not found by any recent collectors.

The two species of this genus belonging to the Western United States are very similar in pattern of color, and differ rather in tints. The males may be best distinguished by the following diagnosis.

L. Californicus. Vertex and occiput light smoky olive-brown; forehead whitish; spot in middle of belly orange-chestnut; sides of body olivaceous-plumbeous.

L. Gambelii. Vertex and occiput clear chestnut-brown; forehead blackish; spot in middle of belly black; sides of body orange-chestnut.

Lophortyx Californicus, Shaw.

THE CALIFORNIA QUAIL.

Tetrao Californicus, Shaw, Nat. Misc. pl. 345 (prior to 1801). — *Perdix Californica*, Latham, Suppl. Ind. Orn. II. App. 1801; pl. lxii. — Audubon, Orn. Biog. V. 1839, 152; pl. 413. — Hutchings, Cal. Mag. II. 1857, 241. (Woodcut of bird and egg.) — *Ortyx Californica*, Stephens, in Shaw's Zool. XI. 1819, 384. — Audubon, Syn. 1839, 199. Ib. Birds Amer. V. 1842, 67; pl. 290. — *Lophortyx Californica*, Bonaparte, List, 1838. — Nuttall, Man. I. 2d ed. 1840, 789. — Baird, P. R. Rep. IX. Birds, 644. — Cooper and Suckley, XII. iii. Zool. of W. T. 225. — *Callipepla Californica*, Gould, Mon. Odont. pl. xvi. — Newberry, P. R. Rep. VI. iv. 1857, 92. — Heermann, X. vi. 60.

Sp. Char. Crest black. Anterior half of body and upper parts plumbeous; the wings and back glossed with olive brown. Anterior half of head above brownish-yellow, the shafts of the stiff feathers black; behind this is a white transverse band, which passes back along the side of the crown; within this white, anteriorly and laterally, is a black suffusion. The vertex and occiput are light brown. Chin and throat black, margined laterally

Male.

and behind by a white band, beginning behind the eye. Belly pale buff anteriorly (an orange-brown rounded patch in the middle), and white laterally, the feathers all margined abruptly with black. The feathers on the sides of body, like the back, streaked centrally with white. Feathers of top and sides of neck with the margins and shafts black. Under tail coverts buff, broadly streaked centrally with brown.

Female similar, without the white and black of the head; the feathers of the throat brownish-yellow, streaked with brown. The buff and orange-brown of the belly wanting. The crest short.

Female.

Length, 10.00; extent, 14.00; wing, 4.50; tail, 4.20. Iris brown; bill black; feet olive-brown.

Hab. Plains and lowlands of California and Oregon towards the coast, and as far as the Columbia River. Mojave River.

The California quail, valley quail, or common crested quail, is so abundant a species in all the lower parts of the country, that its appearance is perfectly familiar to every one, and it has had numerous illustrations in books of natural history, etc., during the last seventy years. It abounds from the Columbia River to Lower California in the valleys, and on the lower parts of the mountain-slopes up to about three thousand feet in lat. 39°, above which it is replaced by the mountain quail. It is also numerous on Catalina Island, but was probably carried there originally, as a flight of eighteen miles at once would probably be too far for a bird with so short wings. It is not found on any of the other more southern islands.

At most seasons these quails can be found in flocks, sometimes of several hundreds, and they do not scatter in pairs even in the nesting season, except for a short period, the males soon collecting again in small flocks. It has been stated that in the warm valleys of the interior they commence to lay as early as March, but I have never found eggs near the coast before April. At San Diego I found the first on the 22d of that month, about thirty days later than Gambel's quail begins to lay at Fort Mojave. At Santa Barbara I found the first on May 1st, and at Santa Cruz observed the young just hatched, May 25th, so that the usual time for laying seems to be throughout April and May, perhaps extending two weeks earlier and later in the extremes of its range. The number of eggs in a nest I have found to vary from eleven to twenty, and several females probably lay in one nest at times, as in the case with nearly all gallinaceous birds. Their color is white or yellowish, with numerous blotches and spots of various sizes and shades of brown, scattered mostly towards the larger end. The shade and distribution of spots varies in such a manner as to lead to the opinion that they are laid by different birds. As an evidence of their habit of laying at times in strange nests, I have found two of their eggs in a nest containing four others of a *Pipilo*, on the summit of the Santa Cruz Mountains, in which case the nest was deserted by both species. The size of the egg is about 1.20 or 1.24 × 0.90 or 0.96.

A writer in Hutchings's California Magazine for December, 1857, mentions that a female kept in San Francisco laid during one summer the astonishing number of seventy-nine eggs, being of course deprived of them and not allowed to sit, while supplied with abundance of food. Probably in many instances the wild birds raise a late brood after losing their first nest, and perhaps at times two broods annually. In the instance just mentioned the male bird showed at times a very tyrannical disposition, driving the female for refuge into corners for several days together. I have not, however, observed any pugnacity among them in a wild state, or much even when confined in cages, unlike many of their foreign allies which have become celebrated for their fighting propensities.

The period of incubation is probably about four weeks, as with its nearest allies, but I have had no opportunity to verify this. They are very easily tamed, either when taken wild or raised from the eggs. In San Francisco they are constantly to be found exposed for sale alive, and many escape from the cages, scattering about the city, flying from roof to roof, and occasionally descending into quiet gardens, especially if there are caged birds to call them. A few rods from the suburbs, flocks of quails are frequent among the dense undergrowth which covers the sand-hills.

They are carried East in almost every steamer, but no account of their successful naturalization in our more severe Atlantic climate has yet been published. They are, however, said to be very numerous in France, where they have found a suitable climate, and are carefully protected. Experiments made in England have not been so successful.

In the spring the males sit on a bush and utter at short intervals a single loud, almost screaming note, which seems to be a call peculiar to that season. At other times they have a triple call-note rather soft and pleasant, capable of being interpreted in various words, according to the fancy of the listener. Perhaps the nearest approach to it is in the syllables *túck-kè-téu*, quickly whistled. This is their usual call to assemble when scattered; if suddenly startled they fly off with several loud chirps, *pip-pip-pip-pip*, exactly like the alarm-note of the robin. The young when alarmed chirp similarly, but more faintly.

When hunted they usually run a good deal before flying, and scatter, rising in small numbers at a time or singly; but it is probable that hunting them with dogs may change this habit, especially in open grounds, and induce them to lie closer, like the species of older countries. When there are trees near, some of them almost always fly into them, and if hard pressed conceal themselves so closely in the foliage that they are discovered with great difficulty.

Their habit of scattering makes it difficult to shoot more than one or two at a time, even on the ground, and market-hunters rely chiefly on traps, by which the quails are very soon exterminated from a district. Sportsmen are obliged to destroy these traps wherever found, to prevent this destruction in more populous districts.

The whir they make when started up is sufficient to put an inexperienced shot quite off his guard; and I think the birds where much hunted discover this fact, and lie closer, starting up in larger numbers and nearer to their pursuer, in order to destroy the effect of his shot. The Eastern quail shows similar differences in habits in different districts.

From their frequenting the vicinity of bushes and other cover, there seems to be no necessity for the female to use artifice in protecting her young brood, and I have always seen them scatter at once, the old bird flying off as if confident that the young could take care of themselves. The Eastern species (*Ortyx Virginianus*) has often been raised by the domestic hen, and there seems no reason why this should not be also.

The only point east of the Sierra Nevada, where I have seen this bird, is along the upper part of the Mojave River, and it is doubtful whether they occur elsewhere, unless it is in Owen's Valley. Towards the south, Dr. Heermann states that the Colorado desert forms an impassable barrier to their extension eastward. Near the sink of Mojave River they meet with the nearly allied species, *L. Gambelii*, and it would be highly interesting to discover whether the two species hybridize in a wild state.

Attempts have been made to naturalize this species at Puget's Sound, but they probably could not flourish there for many years. They have been established, however, in the Sandwich Islands, and it is quite probable that persistent and systematic effort will result in making them familiar inhabitants of many regions to which they do not belong naturally.

The flesh of the California quail is white and delicate, but rather deficient in flavor, though capable of improvement by fattening in confinement.

Lophortyx Gambelii, NUTTALL.

GAMBEL'S QUAIL.

Lophortyx Gambelii, "NUTTALL," GAMBEL, Pr. A. N. Sc. Phil. I. 1843, 260. — McCALL, Pr. A. N. Sc. V. June, 1851, 221. — BAIRD, P. R. Rep. IX. Birds, 645. — KENNERLY, X. iv. 33. — COUES, Ibis, 1866, 46 (habit). IB. Pr. A. N. Sc. Phil. 1866, 94. — DRESSER, Ibis, 1866, 28. — *Callipepla Gambelii*, GOULD, Mon. Odont. pl. xvii. — CASSIN, Illust. I. ii. 1853, 45; pl. ix. — HEERMANN, P. R. Rep. X. v. 19; vi. 60.

SP. CHAR. Head with a crest of five or six purplish-black feathers, about as long as the bill and head together, or a little longer. Upper parts, with the neck all round, and the breast, plumbeous gray; the shafts of the feathers brown; those on the neck above and on the sides edged with the same. Anterior half of head all round, with the chin and upper part of throat, and a large spot on the belly, black; the forehead streaked with hoary gray. Top of the head chestnut, bordered anteriorly and laterally by black, imme-

Male.

diately succeeded by an abruptly defined white stripe. A second stripe starts from the posterior corner of the eye, and borders the black on the sides of the head and on the throat all round. Belly pale brownish-yellow; the sides of the body dark orange-brown, broadly streaked centrally with white. Inner edges of tertials light brownish-yellow. Tail light plumbeous.

Female without the black and white of the head and the black of the belly, and only a slight trace of the chestnut crown; the crest shorter and of fewer feathers.

Length, 11.00 ; extent, 14.50 ; wing, 4.75 ; tail, 4.25. Iris brown ; bill black, pale below ; feet lead-gray.

Female.

Hab. Upper Rio Grande and Gila to the Colorado of California. A few as far west as the sink of Mojave River.

This species is almost exactly like the better known *L. Californicus* in habits as well as in pattern of plumage, but differs in both quite sufficiently to constitute a good species, having also an entirely different range of distribution. They seem more able to endure the droughts and heat of the interior deserts, as I have seen small flocks on the barren mountains west of the Colorado, and they occur in a generally more inhospitable country. Their range toward the north is not known to be above lat. 36°, though possibly farther. Though more southern in their habitat, their average size is larger than that of the coast bird, as seen by the measurements above given, from many specimens, while birds of the same species are always found smaller toward their southern limits. They doubtless extend far into Mexico, but not into Lower California.

At Fort Mojave they are numerous, and have all the calls of the coast species, except the alarm chirp like a robin's, which I never heard them utter. There is, however, a slight difference in their notes, which is recognizable by strangers.

They are also equally sociable, soon becoming tame, and after a few days' feeding in a cage may be released to associate with the fowls, coming to be fed, and making the vicinity of the house their home. One thus tamed went off in the spring, hatched a brood of young, and brought them to the house for protection. Another laid its eggs in the hen-house, which were soon hatched by a hen, which, however, immediately killed the young, perhaps considering them helpless dwarfs. These eggs required as much as twenty-four days to hatch, which is, however, less than those of the Eastern quail, said to need four weeks.

I obtained the first eggs March 26th, and as late as May 25th saw a nest with fourteen eggs on which the hen was sitting. It was formed like that of the California quail, but more carefully, being well lined with soft grass bent around in a circle about eighteen inches wide, and concealed under a tuft of tall grass.

The young when hatched are prettily marked with several brown stripes on a yellow ground like many chickens, but have also little crests of downy plumes.

The eggs present all the varieties shown by those of the coast bird, and in fact are scarcely distinguishable from them. They measure 1.18 to 1.22 × 0.90 to 0.98 inch.

As a game-bird this species is also similar to the California quail, though perhaps less inclined to fly, and rarely if ever taking to trees. Its flesh is similar, though, from the more barren districts it inhabits, generally drier and more insipid, probably also tougher.

Genus **CALLIPEPLA**, Wagler.

Callipepla, Wagler, Isis, 1832. (Type, *Ortyx squamata*, Vigors.)

Gen. Char. Head with a short depressed tufted crest of soft thick feathers springing from the vertex. Other characters as in *Lophortyx*.

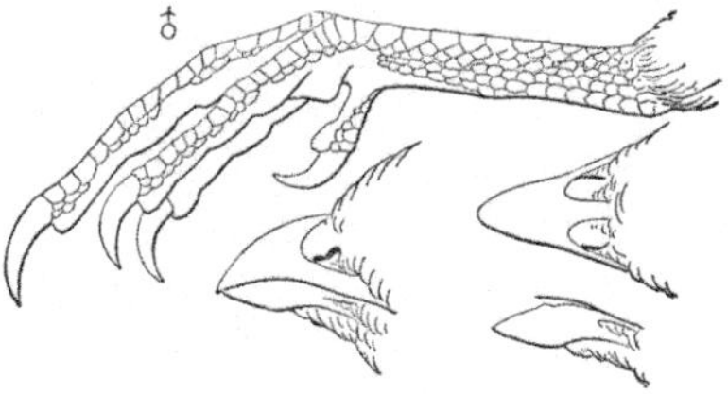

C. squamata.

The single United States species is of a bluish tint, without any marked contrast of color. The feathers of the neck, breast, and belly have a narrow edging of black.

Callipepla squamata, Gray.

THE SCALED OR BLUE PARTRIDGE.

Ortyx squamatus, Vigors, Zool. Jour. V. 1830, 275. — Abert, Pr. A. N. Sc. III. 1847, 221. —*Callipepla squamata*, Gray, Gen. III. 1846, 514. — McCall, Pr. A. N. Sc. V. 1851, 222. — Cassin, Illust. I. v. 1854, 129; pl. xix. — Gould, Mon. Odont. pl. xix. — Baird, Birds N. Amer. 1858, 476.
Callipepla strenua, Wagler, Isis, XXV. 1832, 278.
Tetrao cristata, De la Llave, Registro trimestre, I. 1832, 144.

Sp. Char. Head with a full broad flattened crest of soft elongated feathers. Prevailing color plumbeous-gray, whitish on the belly, the central portion tinged with brownish;

C. squamata.

the exposed surface of the wings tinged with light yellowish-brown, and very finely and almost imperceptibly mottled. Head and throat without markings, light grayish-plumbe-

ous, throat tinged with yellowish-brown. Feathers of neck, upper part of back and under parts generally, except on the sides and behind, with a narrow but well-defined margin of blackish, producing the effect of imbricated scales. Feathers on the sides streaked centrally with white. Inner edge of inner tertials, and tips of long feathers of the crest, whitish. Crissum rusty white, streaked with rusty. *Female* nearly similar. Length, 9.50; wing, 4.80; tail, 4.10.

Hab. Valley of Rio Grande of Texas. Not yet detected farther west. Most abundant on the high broken table-lands and mezquite plains. Southward into Mexico.

This species has not yet, I believe, been actually taken west of the Rocky Mountains, but there is no reason why it may not occur in Arizona, and it is introduced here for the purpose of completing the history of the family. It is found high up in the mountains of New Mexico, and is there exposed to considerable intensity of cold, while in the low lands it resists successfully an equal extreme of heat. They afford considerable sport to the hunter, but rarely lie to a dog, running away with great swiftness when approached. The flesh, like that of other species of the family, is said to be excellent.

GENUS **CYRTONYX**, GOULD.

Cyrtonyx, GOULD, Mon. Odont.? 1845. (Type, *Ortyx massena*, LESSON.)

GEN. CHAR. Bill very stout and robust. Head with a broad soft occipital crest of short decumbent feathers. Tail very short, half the length of the wings, composed of soft feathers, the longest scarcely longer than the coverts; much graduated. Wings long and

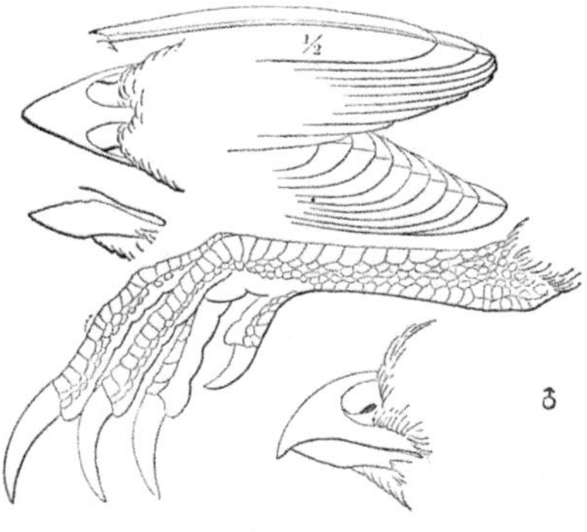

C. massena.

broad, the coverts and tertials so much enlarged as to conceal the quills. Feet robust, extending considerably beyond the tip of the tail. Claws very large, the outer lateral reaching nearly to the middle of the central anterior. The toes without the claws, however, are very short.

C. massena.

This genus differs very much from its North American allies in the great development of the feathers composing the wing coverts, the very short and soft tail, and the very short toes and long claws. It is almost worthy of forming the type of a distinct sub-family, so many and great are its peculiarities. The single North American representative is the only one of our species with round white spots on the lower surface and black ones above. A second species, *C. ocellatus*, is found in Mexico.

Cyrtonyx massena, Gould.

THE MASSENA PARTRIDGE.

Ortyx massena, Lesson, Cent. Zool. 1830, 189. — *Cyrtonyx massena*, Gould, Mon. Odont. 1850, 14; tab. vii. — McCall, P. A. N. Sc. V. 1851, 221. — Cassin, Illust. I. i. 1853, 21; pl. xxi. — Reichenbach, Syst. Av. 1850, pl. xxvii. — Baird, Birds N. Amer. 1858, 647. Ib. Mex. Bound. Rep. II. 23. — Coues, Pr. A. N. Sc. 1866, 95. — Dresser, Ibis, 1866, 29. — Gray, Catal. Gallinæ Br. Mus. 1867, 74.
Ortyx montezumæ, Vigors, Zool. Jour. V. 1830, 275.
Odontophorus meleagris, Wagler, Isis, XXV. 1832, 279.
Tetrao guttata, De la Llave, Registro trimestre, I. 1832, 145.

Sp. Char. Head striped with white, black, and lead-color; chin black. Feathers above streaked centrally with whitish, those on the outer surface of the wings, with two series of rounded black spots. Central line of breast and belly dark chestnut; the abdomen, thighs, and crissum black; the sides of breast and body lead-color, with round white spots. Legs blue. Length, 8.75; wing, 7 00; tail, 2.50.

Hab. Chiefly on the Upper Rio Grande from the high plains of the Pecos. Westward to Fort Whipple, Arizona, and the Gulf of California.

This, the most strikingly marked of all the American partridges, is abundant along the southern border of the United States, and is met with in small flocks or coveys, principally in sterile or rocky districts. It is said to be very gentle and unsuspicious, scarcely disturbed by the presence of man,

Male.

and capable perhaps, as suggested by Mr. Cassin, of being readily domesticated. Nothing definite is known as to their mode of breeding. (Baird.)

Female.

APPENDIX.

By S. F. BAIRD.

I.

ADDITIONAL SPECIES.

THE following species were omitted in their proper place, and are given here to render the work more complete: —

Dendroica Graciæ, COUES.

GRACIE'S WARBLER.

Dendroica Graciæ, COUES, MS. BAIRD, Rev. Amer. Birds, 1865, 210. — COUES, Pr. A. N. Sc. Phil. 1866, 67.

SP. CHAR. (No. 36,988, ♂ ?) Bill shorter than the head; gonys slightly convex. Color of upper parts, with sides of neck, ash-gray; the middle of back, and less conspicuously, the upper tail coverts, streaked with black. A line from nostrils to above the eye (passing into white for a short distance behind it), eyelids, a crescentic patch beneath the eye, the chin, throat, and centre of jugulum, bright yellow; the rest of under parts, including inside of wing, axillars, and tibiæ, white; the border of the yellow, and the sides of body streaked with black. A line from bill, through the eye, the cheeks (enclosing the

yellow crescent), the sides of the vertex, the forehead, and the centres of feathers on top of head, blackish. Wings and tail blackish, the outer edges of the larger feathers pale bluish-gray; two white bands across the wing coverts. Lateral tail feather white, except the inner web at extreme base, the shaft, and a narrow streak at the end of the outer web; the next feather similar, but the basal blackish extending farther along third feather with edge of outer web, and a wedge-shaped patch in end of inner web, only, white.

Autumnal specimens similar; the black markings less distinct; the back tinged with olivaceous.

Very young birds do not differ materially from the adult, showing nothing of the spotting and mottling of the *Turdidæ*.

Total length, 4.60; wing, 2.75; tail, 2.30.

Hab. Fort Whipple, near Prescott, Arizona, and Belize, Honduras.

This species is almost exactly like *D. nigrescens* in the color and markings of the back (with its blackish interscapular streaks), wings, and tail, as well as of the under parts, except that the chin and throat are yellow, margined with black, instead of black margined with white. The heads are, however, very differently marked.

This interesting addition to the fauna of Arizona was made by Dr. Coues, during his residence in that Territory. It is an inhabitant of the pines about Prescott, where it breeds, though its nest was not obtained. It has much the habits of other warblers, without any very striking or conspicuous note.

Junco annectens, Baird.

THE PINK-SIDED SNOW-BIRD.

Sp. Char. Whole upper parts, with head and neck all round, and upper part of breast, nearly uniform ash-color; this on the under surface with a convex outline behind, and not extending under the wings. Whole interscapular region, outer webs of greater wing coverts, and sides of body from bend of wing, light chestnut rufous; paler beneath, and of a pinkish tinge, and contrasting quite abruptly with the white of middle of breast, belly, and anal region. Lores abruptly darker; outer two tail feathers white, the next with longitudinal white stripe Bill yellow, with dusky tip; legs yellow. Length, 5 50; wing, 3.10; tail, 2.80.

Hab. Rocky Mountains to Arizona. (Fort Bridger, Fort Burgwyn, and Fort Whipple.)

This species or race is so decidedly intermediate in character between *J. Oregonus* and *caniceps*, that I have always considered it (with concurrence of Dr. Coues, who met with it in life) as a hybrid between the two. The characters, however, appear very constant, the relative share of the two species being always preserved, and I think, therefore, that it is entitled to a provisional appellation, even if a hybrid. In the case of the Upper Missouri hybrid *Colaptes* (between *Mexicanus* and *auratus*), no two specimens are alike in the proportional share of features of the parent species, while in the present instance there is a decided constancy.

The general appearance is that of *caniceps*, in the light, uniform ash-color of the upper and anterior portion of body, and the blackish lores in appreciable contrast with the rest of the face. The posterior outline of the ash on the breast, however, is not concave, extending along the sides as in *caniceps* and *hyemalis*, but convex, transverse, and on the sides washed with pinkish-rufous, as in *Oregonus*. This latter color, however, is usually deeper, more extended, and better defined against the white of the belly. The chestnut or rufous of the back is not confined to the interscapular region, as in *caniceps*, but extends on the wing coverts as in *Oregonus* and *cinereus*. Its color is, however, that of *caniceps*, not of *Oregonus*.

II.

EXPLANATION OF TERMS USED IN DESCRIBING THE EXTERNAL FORM OF BIRDS.

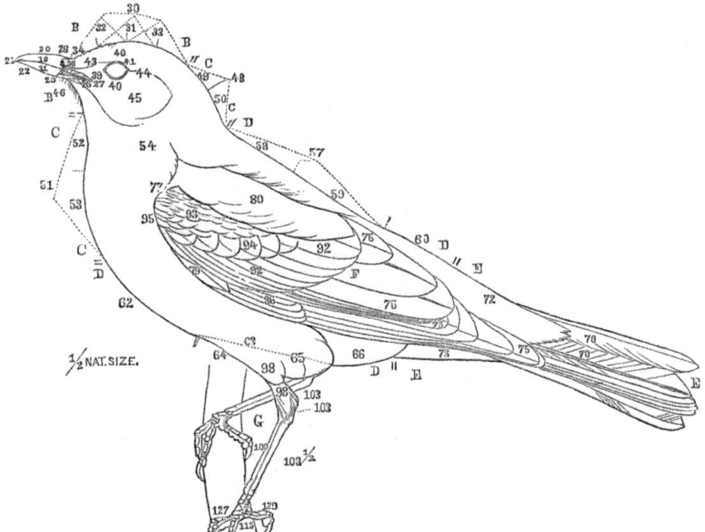

Turdus migratorius, L.

REFERENCES TO THE FIGURE.

N. B. In the figure the adjacent regions are separated by a double bar, with the letters belonging to each affixed.

A.	The body in general.	E.	The region of the tail.
B.	The region of the head.	F.	" " " wings.
C	" " " neck.	G.	" " " legs.
D.	" " " trunk.	H.	The feathers.

NOTE.—I am under obligations to Professor Sundevall of Stockholm and Dr. Sclater of London for assistance in correcting and improving the present article.—S. F. BAIRD.

B. Head.

9. Bill in general.
10. Maxilla.
11. Mandible.
20. Ridge.
21. Tip of maxilla.
22. Keel.
23. Angle of chin.
27. Angle of mouth.
28. Commissure.
28½. Nostrils.
30. Cap (pileus), includes 32, 33.
31. Crown (vertex).
32. Front head (sinciput).
33. Hind head (occiput).
34. Forehead.
36. Frontal points.
39. Lores.
40. Ophthalmic region.
41. Orbits.
42. Cheeks.
43. Eyebrows.
44. Temples.
45. Parotics.
46. Chin.

C. Neck.

48. Hind neck (includes 49, 50).
49. Nape.
50. Scruff.
51. Fore neck (includes 52, 53).
52. Throat.
53. Jugulum.
54. Side neck.

D. Trunk or Body.

57. Back (includes 58, 59).
58. Upper back.
59. Lower back.
60. Rump.
61. Mantle (back and wings together)
62. Breast.
63. Abdomen (includes 64, 65)
64. Epigastrium.
65. Belly.
66. Crissum.

E. Tail.

70. Tail feathers (or rectrices).
72. Upper tail coverts.
73. Lower tail coverts.

F. Wings.

75. Primary quills.
76. Secondary quills.
77. Bend of wing.
79. False wing (alula).
80. Scapulars.
86. Primary coverts.
89. Secondary coverts (include 92, 93, 94).
92. Greater wing coverts.
93. Lesser wing coverts.
94. Middle wing coverts.
95. Edge of wing.

G. Legs.

97. Thigh (concealed under skin).
98. Shin (tibia).
103. Heel joint.
103½. Tarsus.
112. Foot.
116. Toes.
126. Outer toe.
127. Inner toe.
128. Middle toe.
129. Hind toe.

For the purpose of defining the form, markings, coloration, and other peculiarities of birds, the different regions of the body have received names by which intelligible reference can be made to any portion. It is, perhaps, hardly necessary to say that all living birds have a head supported on a neck, with jaws extended into a bill covered with a horny sheath, or with skin, the two jaws situated one above the other, and always destitute of teeth. The anterior pair of limbs is developed into wings which, however, are not always capable of use in flight; the posterior serve as legs for the support of the body in an oblique or nearly erect position. The body is covered with feathers of variable structure and character, both in the young bird and the old. (The wings are apparently wanting in some fossil species.)

The following terms, English and Latin, are those most generally employed in describing the external form of birds, and are principally as defined by Illiger. In cases where there is no suitable English word in use, the Latin equivalent only is given. The figure selected for illustration, drawn by Mr. R. Ridgway, is that of the common American robin (*Turdus migratorius*, L.), and will be familiar to most students of ornithology.

A. Body in General (*Corpus*).

1. **Feathers** (*Plumæ*). A dry elastic object, with a central stem at one end forming a hollow horny tube implanted in the skin at its tip, the other feathered on opposite sides.
2. **Quills** (*Pennæ*). The large stiff feathers implanted in the posterior edge of the wing and in the tail.
3. **Plumage** (*Ptilosis*). The general feathery covering of the body.
4. **Unfeathered** (*Implumis*). A portion of skin in which no feathers are inserted.
5. **Upper parts** (*Notæum*). The entire upper surface of the animal. (Sometimes restricted to the trunk.)
6. **Lower parts** (*Gastræum*). The entire lower surface of the animal. (Sometimes restricted to the trunk.)
7. **Anterior portion** (*Stethiæum*). The forward part of the body (about half), both upper and under surfaces, including the chest.
8. **Posterior portion** (*Uræum*). The hinder portion of the body (about half), including the abdominal cavity.

B. The Head (*Caput*).

9. **Bill** (*Rostrum*). The projecting jaws, one above the other, united by a hinge joint behind, and covered by a horny sheath, or a skin, and enclosing the mouth.

10. **Maxilla**, or **upper jaw** (*Maxilla*).

11. **Mandible**, or **lower jaw** (*Mandibula*).

12. **Ramphotheca.** The horny covering, or sheath of the jaws.

13. **Rhinotheca.** The covering of the upper jaw.

14. **Gnathotheca.** The covering of the lower jaw.

15. **Cere** (*Cera*, or *Ceroma*). A skin at the base of the maxilla, in certain birds. (In birds without a horny sheath to the bill, the cere may be considered as extending to its very tip.)

16. **Edges of bill** (*Tomia*). The margins of upper and lower jaws where they come in contact. We have thus a

17. **Maxillary tomium**, and a

18. **Mandibular tomium.**

19. **Gape** or **Commissure** (*Commissura*). The junction of the tomia, or of the two bills.

Head (*Continued*).

20. **Ridge** (*Culmen*). The upper outline of the bill when viewed laterally; extending from base of bill to the

21. **Tip** (*Dertrum*).

22. **Keel** (*Gonys*). The lower outline of the bill viewed laterally; extending from the angle of the chin to the tip.

23. **Angle of the chin** (*Angulus mentalis*). The point where the two branches, or rami, of the lower jaw

24. (**Gnathidia**, *Rami*) unite, thence to be continued to its tip as the

25. **Myxa** (*Symphysis*).

26. **Malar region** (*Regio malaris*). The outside of the base of lower jaw; usually covered with feathers.

27. **Angle of the mouth** (*Angulus oris*). The angle formed by the mandible and maxilla; the posterior boundary of the gape or commissure, the tip of bill being the anterior.

28. **Nostrils** (*Nares*).

29. **Head,** as restricted (*Caput*). The head, exclusive of the bill.

30. **Cap** (*Pileus*). The whole top of head from the base of bill to nape.

31. **Crown** (*Vertex*). The highest central portion of the top of head (between the ears).

32. **Sinciput** (*Sinciput*). The anterior half of cap (from bill to middle of crown).

33. **Occiput** (*Occiput*). The posterior half of cap, (from middle of crown to the nape).

34. **Forehead** (*Frons*). From base of bill to crown (usually anterior to line of eye).

35. **Nape** (*Nucha*). See 49.

36. **Frontal points** (*Antiæ*). The two projecting feathered angles of the forehead embracing the base of the culmen, or included between the frontal angle of the maxilla and the angle of the mouth (not always present).

37. **Mastax** (*Mastax*). The side of the fore part of the head, adjacent to the base of the maxilla, and distinguished by its feathers or its color.

38. **Capister** (*Capistrum*). The anterior portion of the head all round the base of the bill.

39. **Lore** (*Lorum*). Narrow space between the bill and the eye, on each side.

40. **Ophthalmic region** (*Regio ophthalmica*). Space round the eye.

41. **Orbits** (*Orbita*). The innermost portion of the ophthalmic region immediately adjacent to the eye.

42. **Cheeks** (*Genæ*). See Malar region.

43. **Eyebrow** (*Supercilia*). A longitudinal stripe immediately above the eye.

44. **Temples** (*Tempora*). Whole side of the head behind the eye or between the eye, top of head, and the ear.

45. **Parotic region** (*Regio parotica*). Space around the ears.

46. **Chin** (*Mentum*). Space embraced between the branches of lower jaw.

C. The Neck (*Collum*).

47. **Neck** (*Collum*). The part connecting the head and trunk, enclosing the neck vertebræ.

48. **Hind-neck** (*Cervix*). The upper or posterior portion of the neck, from occiput to back.

49. **Nape** (*Nucha*). The portion of hind neck nearest the head.

Neck (*Continued*).

50. **Auchenium** (*Auchenium*). The portion of hind neck nearest the back, the "scruff" of the neck.

51. **Fore-neck** (*Guttur*). The inferior or anterior portion of neck, from the chin to the breast.

52. **Throat** (*Gula*). The upper part of fore neck, or that nearest the chin.

53. **Jugulum** (*Jugulum*). The lower part of fore neck, between the throat and the breast. (Divided into upper, middle, and lower.)

54. **Side neck** (*Parauchenium*). The sides of the neck, between the front and the hind neck.

55. **Collar** (*Torques*). A ring of any kind encircling the neck.

D. The Trunk (*Truncus*).

56. **Trunk** (*Truncus*). That portion of the body enclosing the viscera and intestines, and carrying the neck and head at one end, the tail at the other, as also the four limbs.

57. **Back** (*Dorsum*). The portion of the upper surface of the trunk, from the neck to the rump, and corresponding to the dorsal and sacral vertebræ.

58. **Upper back** (*Interscapulium*). The upper portion of the back, or along the dorsal vertebræ; between the shoulder-blades, and opposite the breast, sometimes called *dorsum anticum*.

59. **Lower back** (*Tergum*). The lower portion of the back along the sacral region, from the upper back to the rump, and opposite the belly, sometimes called *dorsum posticum*.

60. **Rump** (*Uropygium*). The portion of the upper side of the trunk corresponding to the caudal vertebræ.

61. **Mantle** (*Stragulum; Pallium*). The back and the outside of the folded wings taken together.

61½. **Ventral region** (*Regio ventralis*). Under side of body, including breast and abdomen.

62. **Breast** (*Pectus*). The most anterior portion of the lower surface of trunk, representing the region of the sternum or breast bone (between the jugulum and the abdomen).

63. **Abdomen** (*Abdomen*). The under side of body, between the breast and the anal region.

64. **Epigaster** (*Epigastrium*). The anterior portion of abdomen, next to the breast.

65. **Belly** (*Venter*). The hinder portion of the abdomen, next to the anal region or crissum.

66. **Anal region** (*Crissum*). The region around the anus, below the tail, and opposite to the rump. Frequently includes under tail coverts.

67. **Flanks** (*Hypochondria*). The sides of the soft parts of the body.

68. **Humeral region** (*Regio humeralis*). The anterior portion of the sides; that in which the wing is implanted.

E. The Tail (*Cauda*).

69. **Tail** (*Cauda*). The feathers forming the posterior extremity of the body, implanted on the os coccygis, or rump bone.

70. **Tail feathers** (*Rectrices*). The long individual feathers belonging to the tail.

NOTE. — *Tegmina* (73½) are coverts in general, whether of wing or tail. *Calypteria* are tail coverts. *Tectrices* (83), wing coverts.

Tail (*Continued*).

71. **Tail coverts** (*Calypteria*). The feathers overlying and covering the base of the tail feathers; the

72. **Upper** (*superiores*) being those above, and

73. **Lower** (*inferiores*) those below; sometimes concealing or projecting beyond the tail.

F. The Wings (*Alæ*).

74. **Wings** (*Alæ*). The anterior pair of limbs of the bird, used in flight.

75. **Primary quills**, or **quills of the first series**; **Hand-quills** (*Remiges primariæ*). The (usually) ten stiff feathers inserted on the hand or first joint (metacarpus and digit), or from the bend of the wing to the tip.

76. **Secondary quills** or **quills of the second series**; **Arm-quills** (*Remiges secundarii; Pennæ cubili*). The inner quills, or those inserted along the posterior edge of the forearm or *cubitus*. The innermost of these quills, sometimes longer or different from the rest, are frequently called tertials. (*Pennæ tertiariæ.*)

77. **Bend of the wing** (*Flexura, Plica*). The angle of junction of the hand-joint and that of the forearm.

78. **Armpit** (*Axilla*). The under side of the insertion of the wing into the body.

79. **False** or **Bastard wing** (*Alula*). A series of several stiffened feathers on the edge of the wing, overlying and exterior to the primary quills, and inserted on the thumb joint of the hand.

80. **Scapulars** (*Pennæ scapulares*). Stiffened feathers inserted on the shoulder-blade or the insertion of arm (*humerus*), and filling up the interval between the secondary quills and the body.

81. **Axillars** (*Pennæ axillares*). Similar feathers connecting the under surface of the wing and the body, and concealed in the closed wing.

82. **Speculum**, or **Mirror** (*Speculum alæ*). A brilliantly colored portion of the wing especially in the ducks, over the extremities of the secondary quills, and framed in on one side (in the closed wing) by the primary quills, or the other by the scapulars.

83. **Wing coverts** (*Tectrices*). The smaller feathers of the wing. The

84. **Upper** (*superiores*), side above or outer. The

85. **Lower** (*inferiores*), below, or inside, and overlying the bases of the quills. These and the quills form the surfaces of the wings.

86. **Primary coverts** (*Tectrices primariæ*). The feathers, which either

87. **Upper** or

88. **Under** overlie the bases of the primary quills. These are on the upper or under surface of the wing; not often distinguished in descriptions.

89. **Secondary coverts** (*Tectrices secundariæ*). The feathers which, as

90. **Upper** and

91. **Under**, cover the bases of the secondary quills, on the upper or under surface of the wings, being generally those referred to as "coverts."

92. **Greater coverts** (*Tectrices majores*). The longest coverts projecting beyond the rest, and resting directly upon the bases of the secondary quills.

93. **Lesser coverts** (*Tectrices minores*). The succession of many series of small feathers beginning at and covering the anterior edge of the wing, very small at first and increasing in size behind.

94. **Middle coverts** (*Tectrices mediæ*). One or more rows of coverts, intermediate in size as well as position, between the lesser and greater coverts.

Wings (*Continued*).

95. **Edge of the wing** (*Campterium; Margo carpi*). The small feathers covering the anterior edge of the wing, both along the forearm and the hand or first joint, including the bend of the wing.

G. The Legs (*Pedes*).

96. **Legs** (*Pedes*). The posterior pair of limbs inserted in the pelvis, and used in walking or running.

97. **Thigh** (*Femur*). The basal joint of the leg, its head articulating with the pelvis. This is generally imbedded in the flesh, and covered by the skin so as not to be appreciable, especially in the prepared specimen.

98. **Shin** (*Tibia*). The second or middle joint of the leg, articulated above to the thigh, below to the tarsus. The upper part, sometimes the whole, is enveloped in flesh, and covered by skin and feathers; sometimes the lower extremity is covered by horny plates, the

99. (**Cnemidium**).

100. **Foot joint** (*Podarthrum*). The junction of the tarsus below with the foot.

101. **Podotheca** (*Podotheca*). The horny or skinny covering of lower tibia, tarsus, and feet.

102. **Knee** (*Genu*). The junction of the thigh with the leg, usually concealed by the skin.

103. **Heel joint** (*Suffrago*). The junction of the leg with the tarsus.

103½. **Tarsus** (*Tarsus*). The third joint of the leg and next to the tibia; covered generally with horn, sometimes with naked skin or feathers, never with flesh; the toes are articulated to its lower extremity. This joint corresponds to the ankle joint of the human body.

104. **Instep**, or **Front of tarsus** (*Acrotarsium*). The anterior face of the tarsus, usually covered by small plates, which in the higher groups are united into one; sometimes covered by skin.

105. **Side of tarsus** (*Paratarsium*).

106. **Back of tarsus** (*Planta tarsi*). Homologically the **Sole** (*Planta*).

107. **Heel** (*Calcaneus; Talus*). The upper posterior extremity of the tarsus.

108. **Spur** (*Calcar*). Any bony sharp process or spine implanted on the tarsus, as in the rooster.

109. **Scutellæ** (*Scutella*). The succession of small, usually rectangular plates, applied against the anterior face of the tarsus, and the upper surface of toes. These sometimes encircle the tarsus completely, meeting on the inner side; sometimes reach half round with similar half-rings on the back side of the tarsus; are sometimes divided into polygonal plates; are sometimes fused into a continuous plate, either anteriorly or laterally. Modifications of structure in this respect indicate differences in rank and systematic position of the highest value.

110. **Scutellate tarsus.** When the tarsus is covered with transverse or polygonal scales, as described above.

111. **Booted tarsus.** Where the anterior face is covered with a continuous horny plate not divided into scutellæ.

112. **Foot** (*Pes*). The toes and tarsus taken together.

113. **Top of foot** (*Acropodium*). The entire upper surface of the foot.

114. **The track** (*Pelma*). The entire lower surface of the foot.

115. **Heel pad** (*Pterna; Tuber*). The posterior portion of the *pelma*, immediately under the joint of the foot, and frequently prominent.

Legs (*Continued*).

116 **Toes** (*Digiti*). The, usually four, sometimes three, very rarely two, articulated portions of the leg hinged on the lower extremity of the tarsus. When all four are present, one is usually behind, sometimes two before, and two behind.

117. **Top of toes** (*Acrodactylum*). The upper surface of the toes individually.

118. **Soles of toes** (*Hypodactylum*). The lower or plantar surface of the toes individually.

119. **Side of toes** (*Paradactylum*). The sides, in any way distinguished from the soles.

120. **Phalanges.** The several bones composing a toe.

121. **Claw** (*Ungues*). The horny tips sheathing the last joint of the toes.

122. **Claw joint** (*Rhizonychium*). The terminal bone of the toe, carrying or armed with the claws.

123. **Pads** (*Tylari*). The swellings or bulbs on the under side of the phalanges.

Those **Toes** are

124. **Anterior** which are directed forwards;

125. **Posterior**, directed backwards;

126. **Exterior**, on the outer side of the foot;

127. **Interior**, on the inner side of the foot; the

128. **Middle toe** is the central of three toes directed forwards.

129. **Hind toe** (*Hallux*). The single toe directed backwards. This is homologically the first or great toe directed backward. It is

130. **Insistent** (*insistens*), when the tip at least touches the ground, but the base raised above the level of the rest;

131. **Incumbent** (*incumbens*), when its whole under surface touches the ground; and

132. **Elevated** (*elevatus; amotus*), when raised so high that the tip does not touch the ground at all.

133. **Unarmed toe** (*Digitus muticus*). Toe without a claw. The tarsus is unarmed when without a spur.

134. **Fringed toe** (*Digitus lomatinus*). A lateral membranous margin to the toes. This

135. **Fringe** (*Loma*) may be

136. **Continuous** (*continuum*), or

137. **Lobed** or **Scolloped** (*lobatum*).

138. **Membrane** (*Palama*). A skin either soft or covered with scales or feathers connecting two adjacent toes together at the base, and sometimes extending to or beyond their tips. The foot so constructed is called

139. **Palmate** (*palmatus*) when the anterior toes only are so connected and

141. **Oared** (*Steganopus*), where all the toes, including the hinder, are so connected in the cormorants, etc. The feet may be half, or semipalmate; entirely or totipalmate.

NOTE. In the usual arrangement of the toes, of three before and one behind, the hinder corresponds to the great toe of man, or the first; the inner anterior is the second; the middle is the third; and the outer is the fourth. When the toes are in pairs or two before and two behind, it is the outer or fourth toe that is turned backwards, as is the woodpeckers. In the Trogons, however, the inner toe is revered. With scarcely an exception in birds, the hinder or first toe has two joints; the inner (2d) has three; the middle (3d) has four; and the outer (4th) has five, or a formula of 2.3.4.5. In the typical *Caprimulgidæ* the outer toe has only four phalanges the formula being 2.3.4.4. Finally, in some *Cypselidæ* (*Cypselus* and *Panyptila*), we have the middle and outer toes with three joints only each, the formula being 2.3.3.3. When there are but three toes, the hinder or first is wanting; the ostrich (*Struthio*) has but two toes, lacking the first and second.

H. The Outer Covering (*Indumentum*).

142. **Outer covering** (*Indumentum; Ptilosis*). The exterior of the bird in detail.

143. **Feathers** (*Plumæ*). Composed of the *stem* and the *webs*.

144. **Stem** (*Scapus*). The entire central axis of the feather.

145. **Quill** (*Calamus*). The hollow horny basal portion of the feather.

146. **Shaft** (*Rhachis*). The solid terminal portion of the stem in which the fibres are implanted.

147. **Webs** (*Pogonia*). The series of fibres implanted on each side the shaft, generally stiff, and having little

148. **Hooks** or **barbules** along the edges, by which adjacent ones interlock; sometimes soft, with the barbules not interlocking, the barbules sometimes wanting. The

149. **Inner web** (*Pogonium internum*) is situated on the inner side of the shaft; the

150. **Outer** (*externum*), on the outer side.

151. **Vane** (*Verillum*). The shaft and webs taken together, or the portion of the feather left when the barrel or quill is cut away.

III.

GLOSSARY OF TECHNICAL TERMS.

N. B. The figures refer to the paragraphs of the preceding systematic explanation of terms used in describing the external form of birds.

A.

Abdomen, 63.
Abnormal. Different from the usual character.
Acrodactylum, 117.
Acropodium, 113.
Acrotarsium, 104.
Acuminate. With a narrowed or tapering point, the sides of which are usually more or less concave.
Adult. A bird which has assumed its perfect or final plumage and shape.
Alæ, 73.
Alar. Relating to the wings.
Alula, 79.
Amotus, 132.
Anal. About the anus.
Anal region, 66.
Angle of chin, 23.
Angle of mouth, 27.
Annotine. A bird one year old, or which has renewed its feathers once.
Anterior. The forward portion; in front of.
Anterior portion, 7.
Anterior toes, 124.
Anticæ, 36.
Anus.
Armpit, 78.
Ash or *Ashy* (color). The color of wood-ashes.
Attenuated. Lengthened out, and gradually becoming narrower.
Auchenium, 50.
Auricular. Pertaining to the region about the ears.
Autumnal. The dress or plumage of a bird in the autumn.
Axilla, 78.
Axillaries, 81.
Axillars, 81.

B.

Back, 57.
Back of tarsus, 106.
Band. A mark perpendicular or transverse to the axis of the body.
Barbules (of feathers), 148.
Basal. At the base; in a feather, the end next the body.
Bearded. With elongated feathers or hairs hanging from the chin or throat.
Belly, 65.
Belt. A band completely encircling the body.
Bend of wing, 77.
Bevelled. Where two surfaces meet obliquely or other than at a right angle.
Boat-shaped (tail). The planes of the two sides of tail forming an angle with each other, the concavity above.
Boot; Booted. The horny covering or envelope of the tarsus when continuous and not divided transversely into scales or scutellæ.
Booted tarsus, 111.
Breast, 62.
Buff. The color of yellow buckskin.

C.

Calamus, 145.
Calcaneus, 107.
Calcar, 108.
Calypteria, 71.
Campterium, 99.
Canthus. The corners of the eye where the two eyelids meet.
Cap, 30.
Capistrum, 38.
Capitate (feather). Where the end of a linear feather is slightly expanded.
Caput, 9, 29.

Caruncles. Naked fleshy portion of the head usually wrinkled, or warty, and highly colored.
Cauda, 69.
Cera, 15.
Cere, 15.
Ceroma, 15.
Cervix, 48.
Cheeks, 26.
Chin, 46.
Ciliated. Fringed with hairs or bristles, like the eyelashes.
Cinereous. Ashy; the color of wood-ashes.
Circular. Shaped like a circle; round.
Claw joint, 122.
Claws, 121.
Collar, 55.
Collum, 47.
Coloration. The tint or color of an object.
Comb. Naked fleshy crest on the upper part of the bill or hood.
Commissure, 19.
Compressed. Brought together so as to be higher than wide.
Compressed (tail). The planes of the two sides of the tail forming an angle with each other concave below.
Concave. The inner side of a curved line or inner face of a curved surface; the opposite of convex.
Confluent. Running together; as when spots run into each other.
Connate. Growing together, or united.
Continuous. Without any interruption or break.
Convex. The outer side of a curved line, or outer face of a curved surface; the opposite of concave.
Cordate. Heart-shaped; ovate, with a notch in the broader end producing two rounded lobes, the other end sometimes acutely pointed, sometimes rounded.
Crest. Lengthened feathers in the cap or nape.
Crissum, 66. The under tail coverts; generally used for the region behind the anus.
Crown, 31.
Culmen, 20.
Cuneate. Wedge-shaped, or of angular outline.
Cuneate (tail). The central tail feather longest, the others diminishing rapidly, the outline forming an angle with the sides straight.
Cuspidate. With a sharp and rigid point, not formed entirely by a projection of the shaft.

D.

Decumbent. Bent, or hanging downwards.
Decurved. Bent, or curved downwards.
Dentate; Dentated. Notched, leaving nearly regular tooth-like points, with the two sides equal, or forming nearly an isosceles triangle.
Denuded. Laid bare (of the skin, when freed from feathers).
Deplumate. Bare of feathers.
Deplumatus, 4.
Depressed. Flattened down; broader than deep.
Dertrum, 21.
Digit. The finger or toe portion of the hand or foot.
Digiti, 116.
Disk. (In owls.) A space around the eye, with the feathers different in texture and character from those of the rest of the head.
Divaricated (tail). The tail feathers curved outwards on each side.
Dorsum, 57.
Down. Short, soft, concealed feathers inserted between the bases of the stiffened feathers of the body in some birds (especially the *Anatidæ*). The basal portion of the webs of stiff feathers, sometimes developed into a similar substance.
Dusky. Of a dark color.

E.

Edge of wing, 95.
Elevated toe, 132.
Elliptical. Considerably longer than wide, and rounded equally at the ends.
Elongated. Lengthened, or longer than common.
Emarginated. With a notch at the end, as of a leaf or feather.
Emarginated (tail). When the outline of the end of the tail forms a slightly concave curve; the outermost feathers the longer, and the others diminishing gradually and slightly.
Epigastrium, 64.
Erectile. Capable of being stiffened up, or raised to a more or less upright position, — as the tail of a peacock or turkey, etc.
Exterior. The outer portion or surface, as distinguished from the interior or inner.
Exterior toes, 126.
Even (tail). The ends of all the feathers in the same straight line.
Eyebrow, 42.

F.

Falcate. Shaped like a scythe or sickle.
False wing, 79.
Fastigiate. Gathered up into a conical bunch or head.
Feathers, 1, 143.
Femur, 97.
Fibrillæ. Little fibres.
Filiform. Thread-like.
Flanks, 67.
Flexura, 77.
Flexure. The bend (of the wing).
Foot, 112.
Foot-joint, 100.
Forehead, 33.
Fore-neck, 51.
Forked (tail). The exterior feathers considerably the longer; the intermediate ones of nearly the same length. When the central feathers of a forked tail are longer than those next to them, the tail becomes *doubly forked.*
Fossa. A place hollowed out, as in that portion of the bill in which the nostril is situated.
Fringe, 135.
Fringed toe, 134.
Frons, 33.
Frontal points, 36.
Front of tarsus, 104.
Fulvous. A brownish-yellow.

G.

Gape. The opening of the mouth, 28.
Gastræum, 6.
Genæ, 43.
Genu, 102.
Gnathidea, 24.
Gnathotheca, 14.
Gonys, 22.
Gorget. A crescent-shaped patch on the neck.
Graduated. Diminishing or increasing regularly, and rather rapidly in length.
Graduated (tail). The central feathers longest, the others diminishing regularly and considerably to the outermost one; the outline of end of tail forming a curve longer than broad.
Greater coverts, 92.
Gula, 52.
Guttur, 51.

H.

Hallux, 129.
Humerus, 68.
Hand-quills, 75.
Head, 9, 29.
Heel, 107.
Heel-joint, 103.
Heel-pad, 115.
Hexagonal. Six-sided and six-angled.
Hind-neck, 48.
Hind-toe, 129.
Hoary. A light silvery gray.
Hooks of feathers, 148.
Hornotine. A bird of the brood of the same year.
Hypochondria, 67.
Hypodactylum, 118.

I.

Incubation. The covering an egg by the parent bird, in hatching it.
Incumbent. Bent downwards so that the ends touch, or rest on something else.
Incumbent toe, 131.
Indumentum, 142.
Inner toes, 127.
Inner web, 148.
Insistent toe, 130.
Instep, 104.
Interior. Inner.
Interior toes, 127.
Interrupted. Not continuous, but broken up.
Interscapular. Between the shoulder-blades.
Interscapulum, 58.
Immaculate. Without spots or marks; of an uniform color.
Immature. Said of a bird which has not assumed its final color or shape.
Isthmus. A neck, or narrow stripe, connecting two larger portions of the same region or color.

J.

Jugulum, 53.

K.

Keel, 22.
Knee, 102.

L.

Lamella. A little plate, as in the inside of the edge of a duck's bill.
Lamellate. Provided with lamellæ.
Lamina. A thin plate.
Lanceolate. Tapering rather narrowly towards either end, like a lance-head.

Lanceolate. Lance-shaped; rather narrow and tapering gradually to one end; more rapidly to the other.
Larynx. The organ of voice in the windpipe.
Lateral. Towards one side or the other; not in the middle.
Legs, 96.
Lesser coverts, 93.
Linear. With straight outlines parallel to each other, and close together.
Lobe. A projecting division, rounded or blunt at the end.
Lobed, 137.
Loma, 135.
Lomatinus, 134.
Longitudinal. In the direction of the greatest dimension, or of the axis; the direction from head to tail.
Lore, 39.
Lorum, 39.
Lower back, 59.
Lower jaw, 11.
Lower parts, 6.
Lower wing coverts, 85.
Lunulate. Crescent-shaped.

M.

Mala, 26.
Mandible, 11.
Mandibula, 11.
Mantle, 61.
Mastax, 37.
Mature. See *Adult.*
Maxilla, 10.
Median. In the middle, as distinguished from lateral.
Membrane, 138.
Mentum, 46.
Middle coverts, 94.
Middle toe, 128.
Minor, 82.
Mottled. When one or more colors are laid on in rather well-defined spots or patches, more or less irregularly.
Mucronate. In a feather when the shaft, or mid-rib, projects slightly beyond the rest of the feather.
Muticus, 133.
Myxa, 25.

N.

Nape, 35, 49.
Nares, 28½.
Neck, 47.
Normal. Of the ordinary or usual character.
Nostrils, 28½.
Notæum, 5.
Nucha, 35, 49.
Nuchal. Pertaining to the nape.

O.

Oared-foot, 140.
Ob (in composition) signifies that the usual condition is inverted, or that an attribute, generally posterior, is now anterior, or *vice versa.*
Obsolete. With only a faint trace or indication (as of a spot).
Obtuse. With the point rounded off.
Occipital. Relating to the occiput.
Occiput, 33.
Olivaceous. A greenish-brown color, similar to that of an olive.
Ophthalmic region, 40.
Orbita, 41.
Orbits, 41.
Oscine. A special group of birds, pre-eminent for the power of singing.
Outer covering, 142.
Outer web, 150.
Oval. But little longer than wide, and rounded equally at both ends; shaped like the outline of an egg.
Ovate. A more pointed oval.

P.

Pads, 123.
Palama, 138.
Palmate (feet). A membrane connecting the middle toe to the two lateral, and extending as far as the claws or beyond.
Palmate foot, 139.
Paradactylum, 119.
Paratarsium, 105.
Parauchenium, 54.
Parotic region, 45.
Parotics, 45.
Pectus, 62.
Pectinated; Pectination. Shaped like the teeth of a comb.
Pedes, 96.
Pelma, 114.
Penna, 2.
Perforate. Pierced through.
Phalanges, 120.
Pileus, 30.
Planta, 106.
Plica, 77.

Pluma, 1, 143.
Plumage. The coating of feathers.
Plumbeous. The color of tarnished lead.
Podarthrum, 100.
Podium, 112.
Podotheca, 101.
Pogonium, 147.
Polygon. A continuous figure of many sides and angles.
Posterior. The hinder, or most backward.
Posterior portion, 8.
Posterior toes, 125.
Prehensile. Capable of taking hold of an object or of grasping it.
Primary coverts, 86.
Primary quills, 75.
Protractile. Capable of being moved, or extended forward, as the tongue of a bird.
Pterna, 115.
Pteromata, 89.
Ptila, 86.
Ptilosis, 3.

Q.

Quadrangular. Four-sided.
Quadrate. Squared, with four equal sides and angles.
Quills, 2, 75, 76, 145.
Quills. The large, stiff feathers of the wing, inserted in its posterior edge.

R.

Rhampotheca, 12.
Rectrices, 70.
Recurved. Bent gradually upwards.
Reflection. A play of color on a surface, generally differing with the direction of vision.
Remiges, 75, 76.
Reticulated. Lines uniting into meshes, as in a net.
Retractile. Capable of being drawn back and driven forward, as the claw of a hawk.
Rhachis, 146.
Rhinotheca, 13.
Rhizonychium, 122.
Rictus. The gape of the mouth.
Ridge, 20.
Rounded (tail). The central feathers the longer, the other diminishing regularly and gradually, but slightly to the exterior. A tail is sometimes emarginated in the centre, and rounded at the sides.
Rudimentary. Reduced from the usual condition to a very small size, as the first quill of many birds, etc.
Rump, 60.

S.

Sagittate. An elongated cordate figure, with the lobes acutely pointed, not rounded or shaped like the head or barb of an arrow.
Scapula. The shoulder-blade.
Scapular. Pertaining to the region of the shoulder-blade.
Scapulars, 80.
Scapus, 144.
Scissor-shaped (tail). The outer tail feather considerably the longer, the others diminishing regularly to the central.
Scolloped, 137.
Scutella, 109.
Scutellæ. The small plates, or scales, on the leg.
Scutellate. Provided with scutellæ.
Scutellate tarsus, 110.
Secondary coverts, 89.
Secondary quills, 76.
Semipalmate. When a membrane connects the middle toe to the lateral, for the basal half only.
Setæ. Bristles.
Shaft, 146.
Shin, 98.
Shoulders, 68.
Side-neck, 54.
Side of tarsus, 105.
Sides of toes, 119.
Serrated. Marked or shaped like the teeth of a saw.
Sinciput, 32.
Sole, 106.
Soles of toes, 118.
Spatulate (feather). When the shaft is naked, or narrowed for a certain distance, and is then webbed or widened at the end.
Speculum, 82.
Spur, 108.
Spurious (quill). A quill reduced to a very small size.
Stegani, 141.
Steganopus, 141.
Stethiæum, 7.
Stem, 144.
Stragulum, 61.
Streak. A short longitudinal mark, or one nearly in the direction of the line of the body.

Stripe. A continuous longitudinal mark, with the sides nearly parallel.
Sub (in composition). Somewhat or slightly.
Suffrago, 103.
Suffusion. A "running" of color, or its gradual fading around the outside of a spot or mark.
Sulcus. A groove or furrow.
Superciliary. Pertaining to the region of the eyebrow (as a stripe over the eye).
Supercilium, 42.
Syndactyle. The toes more or less united together directly and without intermediate membrane.

T.

Tail, 69.
Tail coverts, 71.
Tail feathers, 70.
Tarsus, 103½.
Tarsus, back of, 106.
Tarsus, booted, 111.
Tectrices, 72, 73, 83.
Tegmina, 72, 73.
Temples, 44.
Tergum, 59.
Terminal. At the end.
Tempora, 44.
Tertials, 80.
Thigh, 97.
Throat, 52.
Tibia, 98.
Tibial. Pertaining to the tibia.
Toes, 116.
Top of toes, 117.
Tomia. The edges of the two jaws that shut together.
Torques, 55.
Track, 114.
Transverse. In a direction perpendicular to or across that of the line of the body.
Truncate. As if abruptly cut off, either perpendicularly to the general line, or oblique to it.
Truncus, 56.
Trunk, 56.
Tubercle. A little lump or elevation
Tylari, 123.

U.

Unarmed toe, 133.
Ungues, 121.
Upper back, 58.
Upper jaw, 10.
Upper parts, 5.
Upper wing coverts, 84.
Uræum, 8.
Uropygium, 60.

V.

Vane, 151.
Venter, 65.
Vernal. Pertaining to the spring.
Versatile. Capable of being changed or turned from one position to another, as a lateral toe of some birds.
Vertex, 31.
Vexillum, 151.

W.

Washed. As if overlaid, or coated with a thin layer of another color.
Wattle. A naked fleshy, and usually wrinkled and highly colored skin hanging from the chin or throat.
Web, 147.
Wedge-shaped. See *Cuneate*.
Whiskered. With elongated feathers, or bristles, on the cheeks.
Wings, 73.
Wing, bend of, 77.
Wing coverts, 83.

IV.

SPANISH NAMES OF CALIFORNIAN BIRDS.

The following are names in use among the Spanish population of Upper and Lower California. The list has been furnished by John Xantus, Esq.

Cathartes aura.	Aura morena.
Polyborus Auduboni.	Queleli.
Accipiter Cooperi.	Aguililla pinta.
Buteo Swainsoni.	Gavilan blanco.
Archibuteo ferrugineus.	Gavilan cerrano.
Circus Hudsonius.	Halcon bravo.
Craxirex Harrisii.	Halcon negro.
Falco polyagrus.	Halcon acuátil.
Tinnunculus sparverius.	Hembrilla.
Pandion Carolinensis.	Aguila pescadora.
Bubo Virginianus.	Tecolote grande.
Strix pratincola.	Lechuza blanca.
Scops Mc Callii.	Lechuza cerrana.
Pyrocephalus Mexicanus.	Muscicapa colorada.
Sayornis nigricans.	Muscicapa oscura.
Phænopepla nitens.	Cazador de moscas.
Tyrannus verticalis.	Cazador amarillo.
Myiarchus cinerascens.	Cazador común.
Vireo, Sylvicolidæ, Troglodytes, } without distinction.	Reyezuelo.
Polioptila plumbea.	Burioncito.
Auriparus flaviceps.	Paro amarillo.
Helminthophaga ruficapilla.	Paro colorado.
Anthus ludovicianus.	Paro acuátil.
Agelaius phœniceus.	Mirlo colorado.
Xanthocephalus icterocephalus.	Mirlo negro.
Scolecophagus cyanocephalus.	Mirlo pequeño.
Chondestes grammaca.	Gorrion de huertas.
Calamospiza bicolor.	Gorrion pinto.
Poospiza bilineata.	Gorrion prieto.
Collurio excubitoroides.	Carnicero.
Campylorhynchus affinis.	Vitacoche.
Pipilo albigula.	Gorrion blanco.
Pipilo chlorura.	Gorrion verde.
Selasphorus rufus (?)	Chuparosa colorado.
Heliopædica xantusii ♀.	Chuparosa pinto.
" " ♂.	Chuparosa esplendente.
Melanerpes formicivorus.	Carpintero pinto.

Picus lucasanus.	Carpintero negro.
Centurus uropygialis.	Carpintero fresno.
Colaptes chrysoides.	Picamadero.
Cyanocitta Californica.	Gallo azul, *or* Pajaro azul.
Corvus ossifragus (?).	Cuervo pescador.
Hirundo thalassina.	Golondrina verde.
Hirundo bicolor.	Golondrina blanca.
Progne purpurea.	Tragazon.
Chordeiles texensis.	Tapa camino.
Melopelia leucoptera.	Paloma común.
Zenaidura Carolinensis.	Paloma zorrita.
Chamæpelia passerina *var.* pallescens.	Torcacita.
Lophortyx Californicus.	Codorniz.
Geococcyx Californianus.	Churea.
Ceryle alcyon.	Pescador del Rey.
Cardinalis igneus.	Cardenal.
Pyrrhuloxia sinuata.	Rosepecho.
Carpodacus frontalis.	Burion colorado.
Carpodacus (?) (a green species).	Burion verde.
Carpodacus Californicus.	Burion cerrano.
Cyanospiza versicolor.	Trinador.
Harporhynchus cinereus.	Vitacochon.
Guiraca melanocephala.	Gorrion grande.
Guiraca cœrulea.	Gorrion azul.
Mimus polyglottus.	Sinsonte.
Oreoscoptes montanus.	Cercion.
Icterus cucullatus.	Calandria amarilla.
Icterus parisorum.	Calandria negra.
Icteria longicauda.	Calandria acuátil.
Pyranga ludoviciana.	Acanta orilla.
Zonotrichia leucophrys.	Acanta pinta.
Melospiza ——.	Acanta prieta.
Turdus mustelinus ?	Alondra del monte.
Sturnella neglecta.	Alondra cerrana.
Fulica Americana.	Gallineta negra.
Rallus —— ?	Gallineta morena.
Gallinago Wilsonii.	Zopenco prieto.
Macrorhamphus griseus.	Zopenco gris.
Ægialitis vociferus.	Filtir, *or* Filetir.
Gambetta, Calidris, } without distinction.	Agachadiza.
Himantopus nigricollis.	Sarapico negro.
Ibis falcinellus.	Sarapico verde.
Tantalus loculator.	Sarapico grande.
Limosa.	Bobo.
Numenius.	Chorlito.
Tringa.	Engaño.
Anser.	Ganzo.
Anas boschas.	Anade grande.
Dafila acuta.	Puntiacola.
Spatula clypeata.	Palapico.
Querquedula cyanoptera.	Cerceta.
Querquedula Carolinensis.	Zarceta.
Bucephala ——.	Pato buzo.

Grus Americanus?	Grulla cerrana.
Botaurus lentiginosus.	Garza morena.
Nyctherodius violaceus.	Garza llamadora.
Ardea herodias.	Garza grandisima.
Florida cœrulea.	Garza azul.
Garzetta candidissima.	Garza pequeña.
Herodias egretta.	Garza blanca.
Pelecanus fuscus.	Alcatráz negro.
Pelecanus erythrorhynchus.	Alcatráz blanco.
Carbo —— ?	Cormorán.
Lophodytes cucullatus.	Merganzar.
Tachypetes aquila.	Pájaro de fragata.
Larus (no distinction).	Gaviota.
Sterna (no distinction).	Golondrina del mar.
Podiceps (no distinction).	Buzo.
Brachyrhamphus.	Paloma del mar.

INDEX OF SCIENTIFIC NAMES.

N. B. Specific names begin with a small letter; all others have capital initials.

INDEX OF ENGLISH AND SPANISH NAMES.

END OF VOL. I.

Cambridge: Electrotyped and Printed by Welch, Bigelow, & Co.

Printed in Dunstable, United Kingdom

91222338R00342